FESTKÖRPERPROBLEME XXI

ADVANCES IN SOLID STATE PHYSICS

W0255453

FESTKÖRPER PROBLEME XXI

ADVANCES IN SOLID STATE PHYSICS

Plenary Lectures of the Divisions
"Semiconductor Physics"
"Metal Physics"
"Low Temperature Physics"
"Thermodynamics and Statistical Physics"
"Thin Films"
"Magnetism"
"Quantum Optics"
of the German Physical Society
Münster, March 9–14, 1981

Edited by
J. Treusch, Dortmund

With 221 figures

Springer Fachmedien
Wiesbaden GmbH

CIP-Kurztitelaufnahme der Deutschen Bibliothek

Festkörper-Probleme: plenary lectures of the Divisions Semiconductor Physics ... of the German Physic. Soc. = Advances in solid state physics. – Braunschweig: Vieweg
ISSN 0430-3393
Bis Bd. 15 auch verlegt von Pergamon Press, Oxford, Elmsford, Toronto, Sydney

21. Münster, March 9–14, 1981.

NE: Deutsche Physikalische Gesellschaft / Fachausschuß Halbleiterphysik; PT

ISSN 0430-3393

All rights reserved
© Springer Fachmedien Wiesbaden 1981
Originally published by Friedr. Vieweg & Sohn Verlagsgesellschaft mbH, Braunschweig 1981.

No part of this publication may be reproduced, stored in a retrieval system or transmitted, mechanical, photocopying or otherwise, without prior permission of the copyright holder.

Set by Vieweg, Braunschweig
Printed by E. Hunold, Braunschweig
Bookbinder: W. Langelüddecke, Braunschweig
Cover design: Barbara Seebohm, Braunschweig

ISBN 978-3-662-16088-6 ISBN 978-3-540-75368-1 (eBook)
DOI 10.1007/978-3-540-75368-1

Foreword

Volume XXI of "Festkörperprobleme/Advances in Solid State Physics" comprises a selection of the invited papers delivered at the Spring meeting of the Solid State Division of the German Physical Society, held in Münster, 1981.

The first contribution is by K. v. Klitzing. He was awarded the Walter-Schottky prize for his determination of the Fine Structure Constant through measurements of the quantized Hall-resistance in a quasi-twodimensional electron gas. The latter was realized through a MOSFET, thus yielding a particularly nice example for the close relationship between fundamental physics and technical applications. A theoretical and an experimental contribution to surface excitations and their spectroscopy demonstrate the vital interest to understand the physics of surfaces. Four reviews deal with tetrahedral semiconductors, namely with ternary compounds, with the electronic structure of amorphous hydrogenated Silicon, with deep impurity levels and with self-interstitials and vacancies.

Talks on impurities in transition metals, on intercalated graphite, on the dynamics of ion conductors and on thermoelectric effects in superconductors yield a glimpse on the wide range of physical phenomena dealt with at this conference. Finally, aspects of applied physics are touched in talks on nonlinear spectroscopy and on magnetic components for microwaves and optics. The last paper concerning Josephson devices closes the circle, demonstrating once more the close connexion between fundamental physical phenomena and powerful technical applications evolving from them.

I take the opportunity to express my gratitude for six years of successful and speedy cooperation to the Vieweg Verlag and all authors, this volume being the last before my retirement as editor. Editing this series was, as my predecessor, Hans Joachim Queisser, predicted, fun and excitement. This will go to Peter Grosse, the elected successor as chairman of the Semiconductor Division.

Joachim Treusch

Dortmund, May 1981

Contents

The Fine-Structure Constant α
A Contribution of Semiconductor Physics to the Determination of α

Klaus von Klitzing

Physik-Department, Technische Universität München, Garching, Germany

Summary: The knowledge of an accurate value for the fine-structure constant α is necessary in order to check the quantum electrodynamic theory and to reduce the uncertainty of other fundamental constants. The various established methods for the determination of α are briefly reviewed and an introduction is given to a new solid state quantum effect (the quantized Hall resistance of a two-dimensional electron gas), which allows α to be calculated directly. After a summary of the properties of a two-dimensional electron gas in a strong magnetic field, measurements of the quantized Hall resistance are presented. Correction terms, due to carrier localization, scattering processes, or the finite geometry of the sample, are analyzed.

1 Introduction

The fine-structure constant α is one of the most important fundamental physical constants, since this dimensionless constant, which characterizes the interaction of charged particles with electromagnetic fields, can be expressed by different combinations of other fundamental constants. Any change in the value of the fine-structure constant will influence the value of other fundamental constants like the Compton wavelength of the electron λ_c, the Faraday constant F, the gyromagnetic ratio of the proton γ_p, the elementary charge e, the electron restmass m_e, and the Planck constant h.

The small value of $\alpha \approx 1/137$ allows to use α as an expansion parameter in quantum electrodynamics and the corresponding theoretical equations are developed to a high accuracy [1].

The purpose of this paper is to demonstrate that a new solid state quantum effect [2], the quantized Hall resistance of a two-dimensional system, may be used to determine directly the value of the fine-structure constant. This additional path for the determination of α by a new type of experiment increases the number of independent results for the fine-structure constant which should give informations about the over-all consistency in our physical description of nature and may lead to a new set of fundamental constants with higher accuracy.

2 Established Methods for the Determination of the Fine-Structure Constant

Originally, the fine-structure constant α was introduced in 1916 in Sommerfeld's theory of the fine-structure of the optical spectrum of atomic hydrogen [3], where α was defined as the ratio of the electron velocity in the lowest Bohr orbit to the velocity of light c. In SI units, this ratio is given by the expression (μ_0 = permeability of vacuum)

$$\alpha = \frac{e^2}{h}\frac{\mu_0 c}{2} \qquad \begin{matrix}(\mu_0 = 4\pi \times 10^{-7}\ \mathrm{H/m}) \\ (c = 2.99792458 \times 10^8\ \mathrm{m/s})\end{matrix} \tag{1}$$

This fundamental constant plays an important role in quantum electrodynamics (QED) in the calculation of the following quantities known from highly accurate measurements:

a) hyper-fine splitting (hfs) in the ground state of a hydrogen-like system (hfs (H) for hydrogen, hfs (M) for Muonium, and hfs (P) for positronium);
b) fine-structure splitting in hydrogen ΔE_H or $(\Delta E - \delta)_H$ (if Lamb shift is included) and fine-structure splitting ν_{01} in atomic helium;
c) anomalous magnetic moment of the electron (a_e), positron (a_p), and muon (a_μ).

Under the assumption that the QED theory is correct, the fine-structure constant α can be determined from a comparison between theory and experiment [4]. Fig. 1

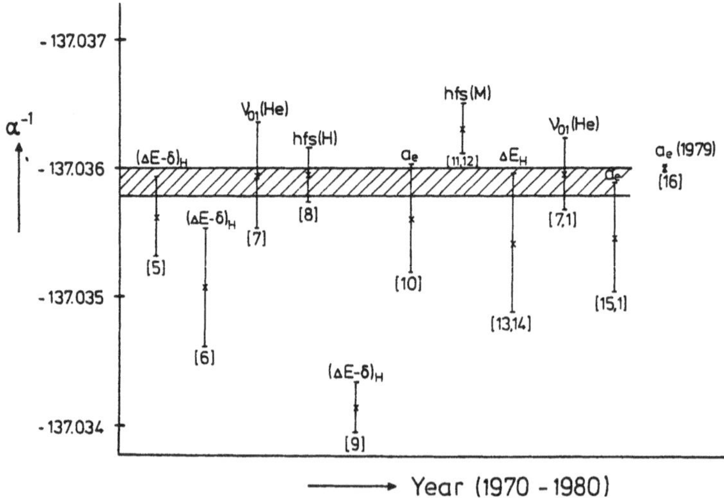

Fig. 1 Summary of fine-structure measurements using quantum electrodynamic theory (QED-data). The mean value of α^{-1} (based on data published up to 1976 [1]) is characterized by the shaded region.

shows a summary of α-values derived from different experiments [4–16] using quantum electrodynamic theory (QED-data). The shaded region characterizes the mean value of α (± one standard deviation) for QED-data published up to 1976 [1]. Very recently, a highly accurate experimental value for the anomalous magnetic moment has been reported [17] which leads to a fine-structure constant of [16]

$$\alpha^{-1}(a_e) = 137.036007\,(0.11\ \text{ppm}).$$

Most of the uncertainty originates from the estimated uncertainty in the QED theory which predicts [18]:

$$a_e = \frac{\alpha}{2\pi} + C_4\left(\frac{\alpha}{\pi}\right)^2 + C_6\left(\frac{\alpha}{\pi}\right)^3 + C_8\left(\frac{\alpha}{\pi}\right)^4 + \ldots$$

with $C_4 = -0.3284784458$ and $C_6 = 1.184 \pm 0.007$. C_8, which consists of 891 Feynman diagrams, has not been calculated up to now.

For comparison, Fig. 2 shows a summary for the value of α determined from different measurements [19–29] which do not require quantum electrodynamic theory for its analysis (WQED-data). In this case α is determined from the equation

$$\alpha^{-1} = \frac{h}{e^2}\,\frac{2}{\mu_0 c}\,.$$

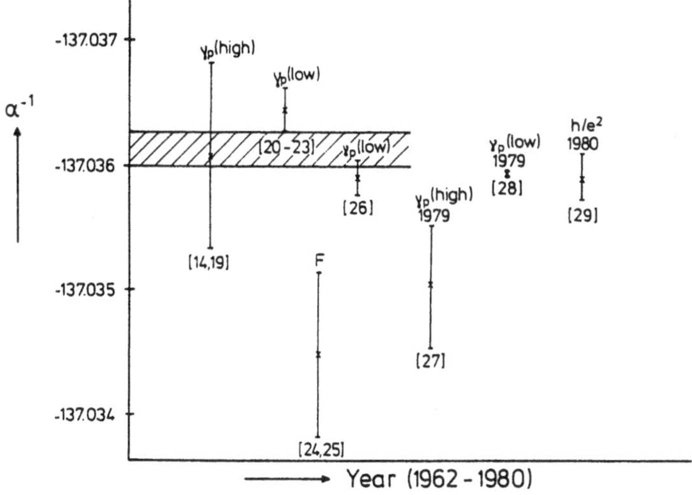

Fig. 2 Summary of fine-structure measurements determined without using quantum electrodynamic theory (WQED-data). The shaded region characterizes the α-value of the last least square adjustment (1973).

This expression is identical with the equations [14] $\alpha^{-1} = C_1 \sqrt{1/\gamma_p}$, γ_p = gyromagnetic ratio of protons and $\alpha^{-1} = C_2 \sqrt{1/F}$, F = Faraday constant, where C_1 and C_2 are combinations of fundamental constants which are known with uncertainties much smaller than the uncertainties of γ_p and F. γ_p (low) in Fig. 2 means that the (low) magnetic field in this experiment is produced by a current through a solenoid with known dimensions, whereas the (high) magnetic field in the γ_p (high)-method is calibrated by measuring the force on a current carrying conductor placed in it.

Fig. 2 shows that the result deduced from measurements of the Faraday constant is incompatible with the remaining data and is therefore not included in the official least square adjustment of the fundamental constants [4]. The resulting best value for α^{-1} (1973) is characterized by the dashed region. In addition to the γ_p-data used for the official adjustment of the fundamental constants, new γ_p (high) [27], γ_p (low) [28], and h/e^2-data [29] are plotted too, where h/e^2 means the determination of α using the quantized Hall resistance. This new method, which will be discussed in detail in this publication, yields a value for the fine-structure constant with an uncertainty comparable with other methods. A further reduction in the uncertainty of α^{-1} (h/e^2) is expected in the near future which should lead presumably to the incorporation of this new method in the next least square adjustment of fundamental constants and to a reliable value for α with an uncertainty of the order 10^{-7} without using quantum electrodynamic theory.

3 The Quantized Hall Resistance

In this chapter we will present a new quantum effect which allows h/e^2 to be determined directly and thereby the fine-structure constant α (eq. 1). Since h/e^2 has the dimension of a resistance ($h/e^2 \approx 25.813$ kΩ), the new method is equivalent to the realization of a device with fixed resistance values which depend (based on theoretical grounds) only on the ratio h/e^2. This quantity appears in different theories, for example in the minimum metallic conductivity [30, 31] for a metal-insulator transition or in the peak value for the conductivity in magneto-quantum-oscillations [32] of a two-dimensional electron gas (2DEG). But only one effect, the quantized Hall resistance of a 2DEG [2], seems to depend exclusively on fundamental constants and not on other quantities like the geometry of the sample, the temperature, the scattering mechanism or other parameters. Such a result is not expected because the Hall resistance R_H of a three-dimensional degenerate electron gas depends on inaccurate parameters like the thickness d of the sample, the strength of the magnetic field B and the carrier concentration n_c

$$R_H = \frac{B}{e n_c d} \,. \tag{2}$$

However, for a 2DEG the parameter d is incorporated in the two-dimensional carrier density n_{inv} and it will be shown that the ratio n_{inv}/B becomes quantized in a strong magnetic field in integer multiples of e/h. This leads to a quantization of the Hall resistance with resistance values of

$$R_H = \frac{h}{e^2 i} \qquad i = 1, 2, \ldots . \tag{3}$$

The accuracy of the new method for the determination of h/e^2 (and therefore α) is basically limited by the uncertainty in the realization of the SI unit Ohm. Fortunately, the SI unit Ohm is the best-known electrical unit with an uncertainty of less than 0.1 ppm, since this SI unit can be deduced from the calculable cross capacitor by Thompson and Lampard [33] and is not linked to the inaccurate values for the SI units Volt and Ampère. If eq. (3) is correct, h/e^2 can be measured in SI units with a relatively high accuracy by comparing the Hall resistance with the SI unit Ohm.

On the other hand, if the fine-structure constant α and therefore h/e^2 is known from other experiments, our new device with a resistance value of $R_H = h/ie^2$ ($i = 1, 2, \ldots$) represents a resistance standard. Even if its value is not known exactly in SI units, the new 'quantum resistor' may be used as a resistance reference similar to the realization of a reproducible, time-independent voltage by the ac Josephson effect.

Such applications of the quantized Hall resistance are only possible, if all corrections are known. Up to now, the experimental verification of a parameter-independent value for the quantized Hall resistance justifies the assumption that eq. (3) is correct. Recent theoretical publications [34, 35] support this assumption. In order to derive eq. (3) and to discuss possible corrections, experimental data and theories related to the quantized Hall resistance are summarized in the following paragraphs. For a general review concerning the properties of a quasi two-dimensional electronic system, we refer to other publications [36].

3.1 The Quasi Two-Dimensional Electron Gas

A quasi two-dimensional electron gas is formed, if electrons can move freely within a plane (x-y-plane) but are trapped in z-direction within a narrow potential well. The potential well must be so narrow that the resulting energy eigenvalues E_i for the motion in z-direction (electric subbands) are so well separated that the kinetic energy of the electron is smaller than the subband separation.

Such a quasi two-dimensional electron gas can be formed at the helium surface [37], at interfaces between two semiconductors [38, 39] or at the insulator-semiconductor interface of a Metal-Oxide-Semiconductor (MOS) structure [40]. Most of the experiments concerned with the quantized Hall resistance were done on (100) Silicon MOS field-effect transistors at helium temperatures, and the following discussion is mainly related to this system.

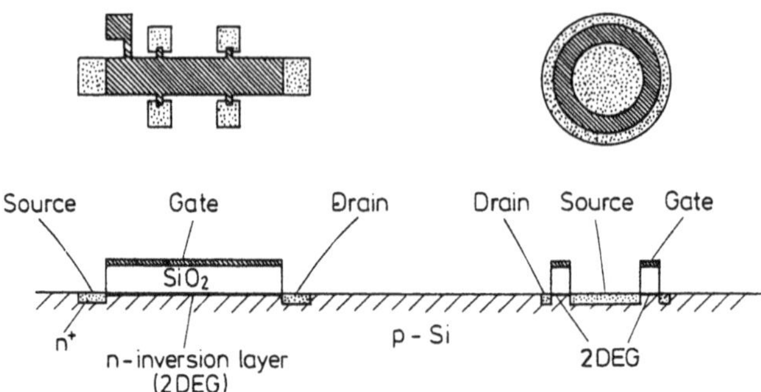

Fig. 3 Top view and cross-section of typical MOS-devices used in our experiments. Left side: long device. Right side: circular device.

Fig. 3 shows a cross-section and a top view of typical MOS devices used in our experiments. On a single crystal of p-Si (substrate material with an acceptor concentration N_A) a SiO_2 layer is thermally grown with a thickness of typically 0.1 μm– 1 μm. A metal (usually Al) on the top of the oxide forms the gate. With a positive gate voltage relativ to the semiconductor a quasi two-dimensional electron gas (inversion layer) close to the interface between semiconductor and insulator can be influenced.

Highly doped n^+ regions at the semiconductor surface are used as ohmic contacts to the inversion layer. The depletion layer between the n^+-contacts or the n-type inversion layer and the p-substrate separates electrically the surface channel from the bulk. Therefore the current from source to drain flows (at least at low temperatures) exclusively through the inversion layer and not through the substrate. The geometry of the gate and the arrangements of the n^+ contacts shown in Fig. 3 allow measuring in the presence of a magnetic field either the conductivity component σ_{xx} (circular device) or the resistivity components ρ_{xx} and ρ_{xy} (long device) [41].

An important parameter in an analysis of Hall effect data is the surface charge density. Contrary to bulk semiconductors the carrier concentration in the surface layer can be controlled by the gate voltage V_g. The relation between V_g and the electron charge density in the inversion layer $-Q_{inv}$ is approximately given by the equation

$$-Q_{inv} = C_{ox}(V_g - V_{thr}) \tag{4}$$

where C_{ox} is the oxide capacitance per unit area. The threshold voltage V_{thr} is determined (if we neglect the Aluminium-Silicon work function difference and the interface charge density) by the charge Q_d in the depletion region and the voltage drop φ_d across the depletion layer at threshold [42]

$$V_{thr} = \varphi_d - \frac{Q_d}{C_{ox}} \tag{5}$$

$-Q_d$ can be varied through an external substrate bias voltage V_{SB} across the depletion layer according to the relation $Q_d = -\sqrt{2\epsilon_0 \epsilon_{Si} e N_A (V_i - V_{SB})}$.The built-in voltage V_i at low temperatures is about 1.2 V.

Since the relation between gate voltage V_g and the surface charge density $\dot{Q}_{inv}$ is important for an analysis of Hall effect data $R_H(V_g)$, the limitation of the usually accepted proportionality between Q_{inv} and V_g has to be regarded. Generally, the connection between Q_{inv} and V_g can be deduced from the condition for charge neutrality $Q = -Q_{inv} - Q_d$ (Q = total charge density at the gate electrode) and the condition that the applied gate voltage is the sum of the voltage drop across the oxide and the voltage drop across the depletion and the inversion layer [43].

$$V_g = \frac{Q}{C_{ox}} + \varphi - \frac{Q_{inv}}{\epsilon_{Si}\epsilon_o} \langle z \rangle \tag{6}$$

Eq.(4) is then obtained from epps. (5) and (6) under the assumptions that the mean distance $\langle z \rangle$ of the 2DEG from the interface is negligibly small compared with the thickness of the oxide depletion layer, and that the depletion layer thickness remains constant at gate voltages above the threshold voltage. Only under these approximations and the assumption that oxide charges do not contribute to a change in the voltage drop across the oxide the usually adopted linear relation between a change in the gate voltage and a change in the inversion carrier concentration is correct.

Since the density of states for a 2DEG is independent of energy [36], an increase of the gate voltage corresponds approximately to a linear increase of the Fermi energy (relative to lowest electric subband E_0) with a slope of about 6 meV/10^{12} electrons cm^{-2} for an n-type silicon (100) inversion layer.

The energy separation between the two lowest electric subbands is so large that at surface carrier concentrations below $6 \times 10^{12}\ cm^{-2}$ only the lowest electric subband E_0 is occupied (electric quantum limit). This may be seen from Fig. 4 where the energy separation $E_1 - E_0$ for two different substrate bias voltages is determined from optical experiments [44]. This result agrees with theoretical calculations. The question whether the higher electric subband E_1 or a new subband series E_0' (belonging to valleys with a lower effective mass for the motion perpendicular to the surface) becomes occupied at extremely high surface carrier concentrations has not been solved unambiguously [45, 46]. At least at inversion carrier concentrations below $6 \times 10^{12}\ cm^{-2}$ or for measurements with a negative substrate bias voltage ($V_{SB} < -2$ V) the condition for the electric quantum limit is fulfilled. This means that the energy eigenvalues can be written

$$E = E_0 + \frac{\hbar^2 k_x^2}{2m_x} + \frac{\hbar^2 k_y^2}{2m_y} \tag{7}$$

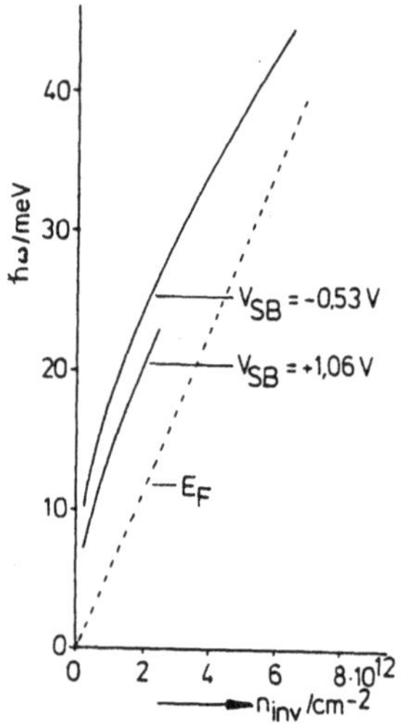

Fig. 4

The resonat positions of intersubband transitions ($E_1 - E_0$) as a function of the inversion carrier concentration for two different substrate bias voltages V_{SB}. The dotted line corresponds to the Fermi energy E_F relative to the lowest subband E_0.

with the two-dimensional wavevector (k_x, k_y) and the effective masses m_x and m_y. The density of states per unit area for this two-dimensional electron gas is [36]

$$D(E) = \frac{\sqrt{m_x m_y}}{2\pi\hbar^2} \tag{8}$$

If the spin and the valley degeneracy for a (100) silicon inversion layer is included, the density of states increases by a factor of 4.

Eq. (7) und eq. (8) are not only applicable to a MOS device, but also to the 2DEG at the interface between two semiconductors (for example GaAs-$Ga_{1-x}Al_xAs$ interface [47]), which is shown in Fig. 5. Contrary to the MOS system, the carrier concentration of this 2DEG is determined by the doping of the $Ga_{1-x}Al_xAs$ and not by a gate voltage. Apart from this, the following discussion is valid for all two-dimensional electronic systems and not only for the (100) silicon MOS inversion layer.

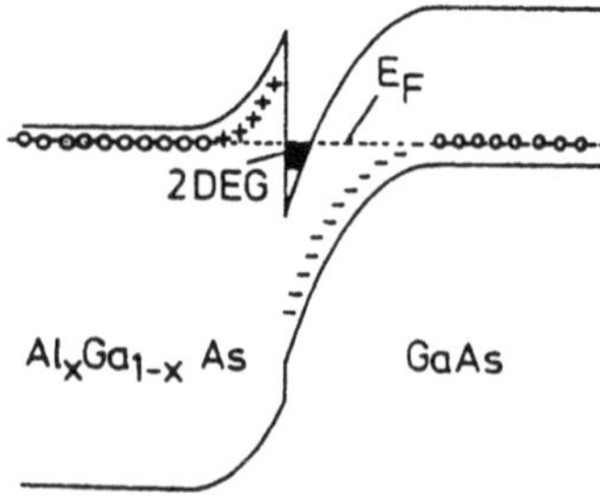

Fig. 5

Schematic energy level diagram of a GaAs-$Al_xGa_{1-x}As$ heterostructure. The p-type GaAs is weakly doped compared to the doping level of the n-type $Al_xGa_{1-x}As$.

Recently, Tsui [73] has demonstrated that the quantized Hall resistance can be measured in GaAs-$Ga_{1-x}Al_xAs$ heterostructures, too.

3.2 Quantum Transport of a 2DEG in Strong Magnetic Fields

3.2.1 Density of States

A strong magnetic field B with a component B_z normal to the surface causes the inversion layer electrons to move in cyclotron orbits parallel to the interface.

As a consequence of the orbital quantization the energy levels of a 2DEG in a strong magnetic field can be written schematically in the form (silicon (100)-inversion layer):

$$E_{n,s,v} = E_0 + \left(n + \frac{1}{2}\right)\hbar\omega_c + sg\mu_B B + vE_v \tag{9}$$

where $s = \pm 1$ (spin quantum number), g is the Landé g-factor [64], $v = \pm 1/2$ (valley quantum number), E_v is the valley splitting [65, 66] and $\hbar\omega_c = \hbar(eB_z/\sqrt{m_x m_y})$ is the energy difference between Landau levels with quantum numbers n and n + 1. The Landau splitting $\hbar\omega_c$ depends on B_z, the magnetic field component perpendicular to the surface, whereas the spin splitting depends on the total magnetic field B. This allows one to change their energies independently by changing the orientation of the magnetic field relative to the surface normal.

The wavefunction of a 2DEG in a strong magnetic field may be written in a form where one of the center coordinates of the cyclotron motion is a good quantum number [48]. The degeneracy of each level $E_{n,s,v}$ is given by the total number of center coordinates of the system. Since only one center exists in a region of area $2\pi l_0^2$ [49] ($l_0 = \sqrt{\hbar/eB}$ is the radius of the ground Landau orbit) the degeneracy factor per unit area is given by

$$N = \frac{eB}{h} \tag{10}$$

This degeneracy factor for each energy level $E_{n,s,v}$ can also be calculated from the number of states (for B = 0) in the energy range $\hbar\omega_c$, since the condition of total state conservation leads to a condensation of the states within this energy range on one Landau level

$$N = D(E)\,\hbar\omega_c = \frac{eB}{h}$$

This degeneracy factor (number of states per unit area for each energy level) plays an important role in the analysis of the quantized Hall resistance because the surface carrier concentration for fully occupied Landau levels becomes quantized leading to a Hall resistance given in eq. (3).

3.2.2 Magneto-Conductivity σ_{xx}

The modification of the density of states in a strong magnetic field leads to a drastic change in the conductivity. Ando has calculated the level broadening and the transverse conductivity σ_{xx} of a 2DEG in strong magnetic fields [32, 48] and finds that the density of states of each Landau level $E_{n,s,v}$ is not a delta function, but has an elliptic form in the self-consistent Born approximation (SCBA) with energy gaps between the levels. Additional tails in the density of states at the edges of the levels appear in a higher approximation [50]. The conductivity σ_{xx} at low temperatures is mainly determined by the square of the density of states at the Fermi energy E_F and has peak values (if one level is half filled) of $\sigma_{xx}^{max} = (e^2/\pi^2 \hbar)(n + \frac{1}{2})$, if the scatterers are of short-range. In the case that the scatterers are of finite range, this peak value is reduced [48].

The characteristic features of the conductivity can be understood by a simple diffusion of the center coordinate of the cyclotron orbit, where the conductivity is proportional to the product of the diffusion constant and the density of states at the Fermi energy. Without scattering, the electrons cannot diffuse in the direction of the electric field (they are drifting perpendicular to the electric and magnetic field). Therefore an observed vanishing conductivity σ_{xx} in strong magnetic fields corresponds to a vanishingly small scattering rate.

Fig. 6 shows a typical result for the conductivity σ_{xx} (measured with a circular device) as a function of the gate voltage at two different magnetic field values B. The oscillations are related to the variation of the density of states at the Fermi energy. The valley splitting is resolved only for the Landau levels up to n = 1, whereas the spin splitting is visible for Landau levels up to n = 3. The period in the gate voltage is constant within 1 % and increases linearly with increasing magnetic field within the experimental accuracy of 0.5 %.

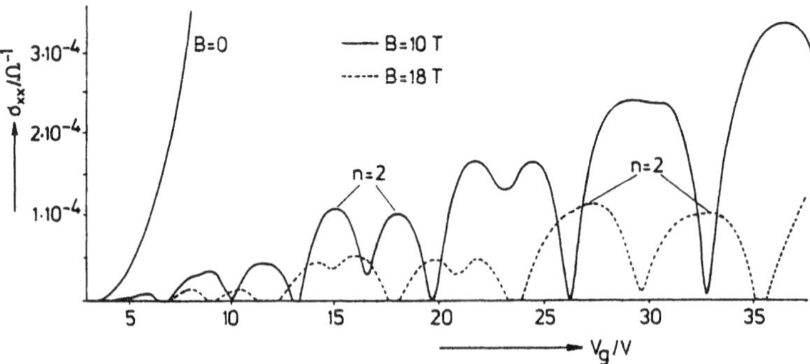

Fig. 6 The conductivity σ_{xx} as a function of gate voltage V_g at different magnetic field values B (T = 1.5 K).

This result is compatible with eqs. (4) and (10) which means that the changes of inversion carrier concentration are proportional to the changes in gate voltage and the degeneracy of each level is proportional to the magnetic field. At gate voltages where σ_{xx} drops to zero, an energy level $E_{n,s,v}$ is fully occupied. This means that exactly at these gate voltages the inversion carrier concentration is a multiple of the degeneracy N of a Landau level

$$n_{inv} = -\frac{Q_{inv}}{e} = iN = i\frac{eB}{h} \qquad (i = 1, 2, 3, \ldots) \tag{11}$$

At these gate voltages a Hall resistance $R_H = h/e^2 i$ (i = 1, 2, 3, ...) is expected. For high precision measurements of this quantized resistance an accurate adjustment of the gate voltage V_g corresponding to fully occupied Landau levels is necessary. Such an adjustment of V_g with an uncertainty of less than 10^{-6} seems to be impossible due to instabilities in both the gate voltage and the threshold voltage. In addition, the σ_{xx} curve close to σ_{xx}^{min} is strongly asymmetric (Fig. 7, see also [58]) and the position of the minimum is not well definied. Moreover, at very high magnetic fields (Fig. 6, B = 18 T) σ_{xx} becomes immeasurably small ($\sigma_{xx} < 10^{-9}\ \Omega^{-1}$) within a certain region ΔV_g, usually referred to as the localized region. Therefore, the gate voltage position of the minimum in σ_{xx} cannot be determined. Kawaji and Wakabayashi investigated this localization in detail and they found that for Landau quantum numbers $n \geqslant 1$ approximately a fraction 1/2n + 1 of a Landau level is localized [52]. Different theories (localization due to lattice imperfections [53–55] or electron-electron interaction [56, 57] are able to explain the observed 'localization' of electrons.

Up to now, it is not clear at which gate voltage V_g within the localized region all states of the energetically lower lying Landau levels are occupied. The results in

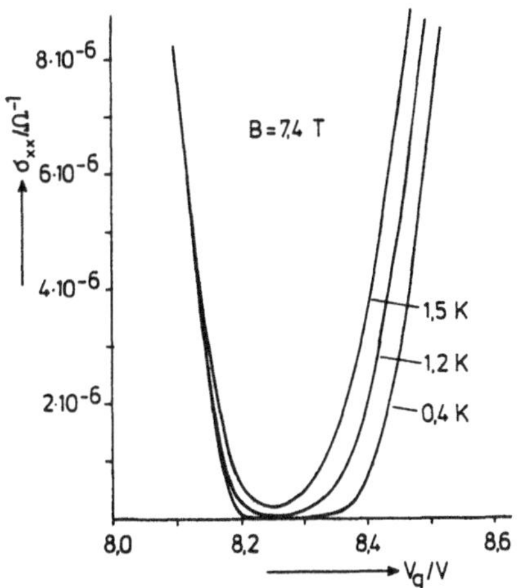

Fig. 7
Temperature dependence of σ_{xx} (measured at B = 7.4 T) close to the gate voltage of a fully occupied Landau level n = 1.

Fig. 7 (T = 0.4 K) seem to indicate that in the localized region states of the higher energetic Landau level become occupied (in agreement with calculations for a two-dimensional disordered electron system in strong magnetic fields [55]), but measurements by other authors [59] or on other samples (Fig. 12) do not confirm this result.

The uncertainty in the location of the gate voltage for which eqs. (11) and (4) are fulfilled, seems to prevent a highly accurate adjustment of $n_{inv} = i\,eB/h$ and therefore an accurate determination of the quantized Hall resistance. Fortunately, Hall effect measurements show that the Hall resistance remains extremely constant in the gate voltage region where localization is observed and has a value corresponding to a complete occupation of an integer number of energy levels [29].

3.2.3 Hall Conductivity σ_{xy}

Experimentally, the Hall conductivity σ_{xy} cannot be determined from measurements with a circular device, because the azimuthal component of the current $j_\varphi = \sigma_{xy} E_{SD}$ cannot be measured (E_{SD} = source-drain field). However, for a long device both the current along the long axis (source-drain current) and the electric field perpendicular to the current (Hall field) can be measured. If the current has only one component ($j = j_x$), the potential drop along the current direction is proportional to the component ρ_{xx} of the resistivity tensor, and the Hall field is directly proportional to ρ_{xy} [41]. It should be noted that in two-dimensional systems ρ_{xy} is given directly by the ratio of the measured quantities U_H (Hall voltage) and I_x (source-drain current) and is (apart from small corrections) independent of the dimension of the sample. Therefore, the experimentally determined quantities $R_H = U_H/I_x$ (Hall resistance) and R_x (resistivity of the sample measured between potential probes) are identical with ρ_{xy} and ρ_{xx}, if corrections due to the finite length of the sample are negligibly small (see Figs. 13–15).

The corresponding components of the conductivity tensor σ_{xx} and σ_{xy} can be calculated from the equations ($\rho_{xx} = \rho_{yy}, \rho_{xy} = -\rho_{yx}$):

$$\sigma_{xx} = \rho_{xx}/(\rho_{xx}^2 + \rho_{xy}^2) \tag{12}$$

$$\sigma_{xy} = -\rho_{xy}/(\rho_{xx}^2 + \rho_{xy}^2) \tag{13}$$

A typical result for $R_x \approx \rho_{xx}$ and $R_H \approx \rho_{xy}$ as a function of the gate voltage at a constant magnetic field is shown in Fig. 8. The oscillations in $R_x(V_g)$ are mainly determined by the oscillations in $\sigma_{xx}(V_g)$, because eq. (12) reduces to $\sigma_{xx} = \rho_{xx}/\rho_{xy}^2$ in strong magnetic fields ($\rho_{xy} > \rho_{xx}$). Therefore zeros in σ_{xx} correspond to zeros in ρ_{xx}.

The Hall resistance measurement shown in Fig. 8 demonstrates that $R_H(V_g)$ is a step-like function (quantized Hall resistance). In gate voltage regions ΔV_g where the resistivity R_x is smaller than 1 Ω, the Hall resistance R_H remains extremely constant ($\Delta R_H/R_H < 10^{-5}$), even if the gate voltage is changed by $\Delta V_g/V_g = 3\,\%$.

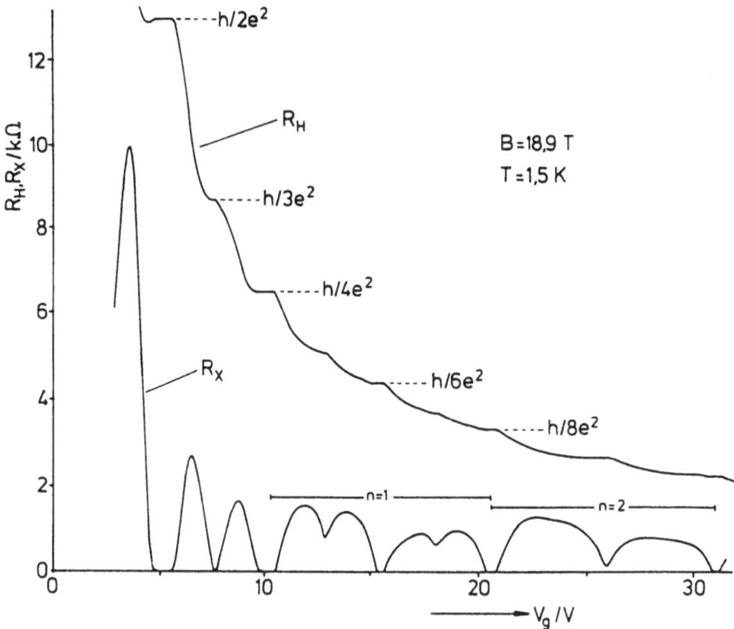

Fig. 8 Gate voltage dependence of the measured Hall resistance R_H ($\approx \rho_{xy}$) and resistivity R_x ($\approx \rho_{xx}$) at B = 18.9 T for a long device (length to width ratio L/W = 8, distance between potential probes Δ = L/3).

Measurements on a large number of different devices, obtained from different laboratories with different geometries and different Si-SiO_2 interface properties, show that within our uncertainty of less than 10^{-5} the resistance values of the plateaus in R_H are independent of sample parameter and independent of the experimental parameter (current, magnetic field, substrate bias voltage, temperature). These experimental results form the basis for the assumption that the quantized Hall resistance depends only on the fundamental constant h/e^2 (as expected for an idealized two-dimensional electron gas), since no correction terms are known which are independent of the parameter varied in our experiments. However, it is clear that at a level of higher accuracy corrections to $R_H = h/e^2 i$ are expected. For example, a finite scattering time $\tau(\sigma_{xx} \neq 0)$ influences the Hall conductivity σ_{xy} [60, 61] and leads to a correction term which vanishes only, if $\tau = \infty (\sigma_{xx} = \rho_{xx} = 0)$. Since σ_{xx} (or ρ_{xx}) is never exactly zero in our experiments, such correction terms must be known for a high accuracy analysis of ρ_{xy}. The first part of our discussion of Hall effect data deals with this question, in the second part the origin of the plateaus in $R_H(V_g)$ (in gate voltage regions where $\rho_{xx} \approx 0$) will be discussed, and finally the influence of the sample geometry on the experimental results will be analyzed.

3.2.3.1 Correction to the Hall Conductivity due to Scattering

The influence of a finite scattering time on the Hall conductivity has been calculated in the self-consistent Born approximation (SCBA). According to the classical expression $\sigma_{xy} = - n_{inv} e/B + \sigma_{xx}/\omega_c \tau$ (τ = relaxation time) the results are written in the form [62]

$$\sigma_{xy} = - \frac{n_{inv} e}{B} + \Delta \sigma_{xy} \tag{14}$$

where n_{inv} is the total number of states below the Fermi energy per unit area. The deviation of σ_{xy} from the fundamental term $-n_{inv} e/B$ comes from the electron scattering, if localization of carriers is not included. In the limit of short ranged potential scatterers, $\Delta \sigma_{xy}$ at zero temperature is written as [48, 62]:

$$\Delta \sigma_{xy} = \frac{\Gamma}{\hbar \omega_c} \cdot \frac{e^2}{\pi^2 \hbar} (n+1) \left[1 - \left(\frac{E_F - E_{n,s,v}}{\Gamma}\right)^2\right]^{3/2} \tag{15}$$

where the with Γ of the Landau level is related to the electron mobility μ in the absence of magnetic field as (B measured in T, μ in m^2/Vs):

$$\Gamma = 0.46 \sqrt{\frac{B}{\mu}} \text{ meV} \tag{16}$$

The maximum value of $\Delta \sigma_{xy}^{max}$ is proportional to σ_{xx}^{max} and can be written

$$\Delta \sigma_{xy}^{max} = \frac{\Gamma}{\hbar \omega_c} \sigma_{xx}^{max} \tag{17}$$

with $\Gamma/\hbar\omega_c < 1$ in strong magnetic fields.

The experimentally observed $\Delta \sigma_{xy}^{max}$-values are at least a factor of two larger than the expected value on the basis of the SCBA. Results are shown in Fig. 9 for two different substrate bias voltage V_{SB} [61].

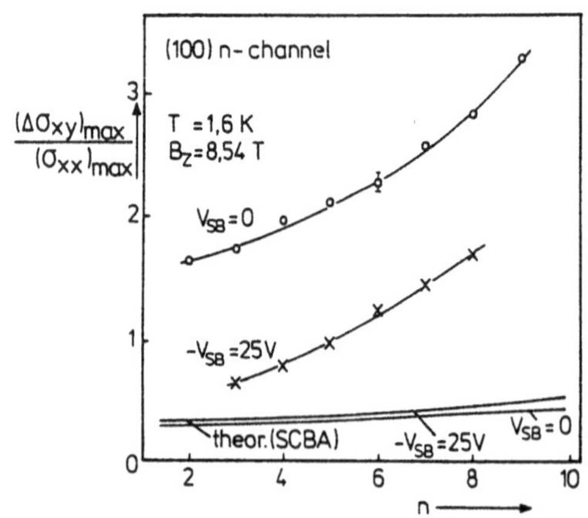

Fig. 9
$\Delta\sigma_{xy}^{max}/\sigma_{xx}^{max}$-measurement for different Landau quantum numbers n and two different substrate bias voltages V_{SB}.

Similar data were obtained by Kawaji et al. [63], and all results show that the ratio $\Delta\sigma_{xy}^{max}/\sigma_{xx}^{max}$ is about 1. This ratio decreases (within the SCBA) with decreasing σ_{xx}, because $\Delta\sigma_{xy}/\sigma_{xx}$ is proportional to the density of states at the Fermi energy. Therefore, a correction term $\Delta\rho_{xy} = \rho_{xx}^{min}$ (equivalent with $\Delta\sigma_{xy} = \sigma_{xx}$) at gate voltage, where ρ_{xx} has a minimum, represents an upper estimate. In our high precision measurements of the quantized Hall resistance, ρ_{min} is smaller than 0.02 Ω. Moreover, ρ_{xx}^{min} (and therefore the correction $\Delta\rho_{xy}$) can be reduced by increasing the magnetic field or decreasing the temperature, which leads practically to a vanishingly small correction $\Delta\rho_{xy}/\rho_{xy}$. Therefore we believe that the correction $\Delta\rho_{xy}$ due to the scattering is unimportant as long as a magnetic field and temperature independent value for the quantized Hall resistance is measured.

3.2.3.2 Hall Conductivity in the Localized Region

The most surprising result of our Hall resistance measurements in Fig. 8 is the existence of plateaus in $R_H(V_g)$. This seems to be in contradiction with the idea that the Hall resistance decreases with the carrier concentration and therefore with the gate voltage (eq. (4)). Up to now the following three mechanisms are able to explain a gate voltage independent value for the Hall resistance:

a) Contrary to eq. (4) the surface carrier concentration $n_{inv} = iN = i\,eB/h$ (eq. (11)) remains constant in the gate voltage region where $\sigma_{xx} = 0$. A change in the gate voltage corresponds mainly to an increase in the band bending φ (connected with a small change in the depletion layer thickness), but not in an increase of Q_{inv} (see eq. (6)). The experimentally observed breakdown in the capacitance [67, 68] and the corresponding calculations [69] demonstrate that a nonlinearity between gate voltage and inversion carrier concentration is present in strong magnetic fields. However, quantitatively this effect is too small to explain the observed plateaus in ρ_{xy}.

b) The plateaus in ρ_{xy} may arise from an occupation of chargeable interface states located energetically within the gaps between Landau levels which do not contribute to the Hall resistance. The resulting concentration of interface states of about $2 \times 10^{13}\,cm^{-2}\,eV^{-1}$ seems to be too high. Such surface states should be measurable as a difference in the carrier concentration deduced on the one hand from the oxide capacitance and on the other hand from the Hall effect. High quality samples don't show this difference [70].

c) The most probable origin for the plateaus in ρ_{xy} are localized states in the tails of the Landau levels. Calculations show that despite the presence of immobile carriers the Hall conductivity is determined by the total number of electrons below the Fermi level E_F, if the Fermi energy lies in the gap between Landau levels [34, 35]. This means that the nonlocalized states carry an extra Hall current which exactly compensates for that not carried by the localized states. Moreover, if the Fermi level lies in the localized region, the contribution $\Delta\sigma_{xy}$ of a localized state $|\alpha\rangle$ is given by

$$\Delta\sigma_{xy}^{\alpha} = f(E_{\alpha})\frac{e}{B} \qquad (19)$$

where f(E) is the Fermi distribution function. Consequently $\Delta\sigma_{xy}^{\alpha}$ cancels exactly the change in the term $-n_{inv}e/B$. Therefore a constant value

$$\sigma_{xy} = -i\frac{e^2}{h} \qquad (20)$$

is expected independent of the number of occupied states, provided that E_F lies in the localized region (or gap) between the (i − 1)th and i-th Landau subband. This conclusion is still valid even if electron-electron interaction is taken into account [35]. Fig. 10 summarizes the theoretical results.

Aoki and Ando [35] pointed out that a mixing of different Landau subbands has not been included in the theory. Such corrections should depend on the ratio $\Gamma/\hbar\omega_c$ which can be changed experimentally by changing the strength of the magnetic field. Since within our uncertainty a magnetic-field dependent value for the quantized Hall resistance has not been observed, this correction is assumed to be small.

Fig. 11 shows a series of measurements of $R_H \approx \rho_{xy}$ and $R_x \approx \rho_{xx}$ for different magnetic fields at gate voltages close to the occupation of the Landau level n = 0 (corresponding to four levels due to spin- and valley degeneracy). The width of the plateaus increases drastically with increasing magnetic field. For high accuracy measurements, the width of the plateaus should be as large as possible, because under this condition inaccuracies in the gate voltage (due to the voltage drop

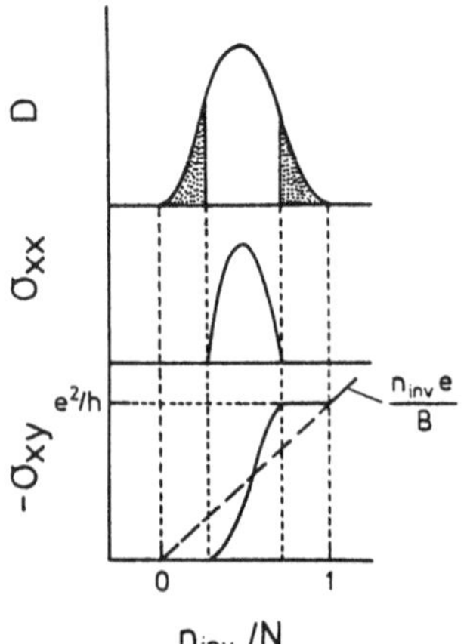

Fig. 10
The density of states (D), dc conductivity (σ_{xx}), and the Hall conductivity (σ_{xy}) are schematically shown as a function of the electron concentration (n_{inv}). Shaded regions in the density of states correspond to the localized regime (Aoki et al. [35]).

across the sample or threshold shifts during the experiment) don't influence the experimental result. Moreover, corrections arising from a finite scattering rate ($\rho_{xx} \neq 0$) decrease with increasing width of the plateaus. Since the activation energy for the localized states is only a few meV [41], the temperature should be as low as possible (typically 1.4 K in our experiments) and the current density should be below 0.05 A/m. Fig. 12 shows that R_H and R_x change strongly as a function of temperature.

In order to get a large Hall voltage needed in high-precision measurements, the current through the sample should be as large as possible. However, at current densities above 0.05 A/m the electrons are heated electrically by more than 0.2 K, and therefore a large width W of the sample is necessary. However, with decreasing length to width ratios L/W, the shortening of the Hall voltage at the source and drain contacts becomes more and more important and leads to a reduction in the measured Hall resistance. This correction seems to be important in measurements on devices with aspect ratios smaller than 1.

3.2.3.3 Corrections due to Sample Geometry

Generally the measured Hall resistance R_H^{exp} is always smaller than the theoretical value $R_H^{theor} = \rho_{xy}$ expected for an infinitely long sample:

$$R_H^{exp} = G \cdot R_H^{theor} \qquad G \leqslant 1$$

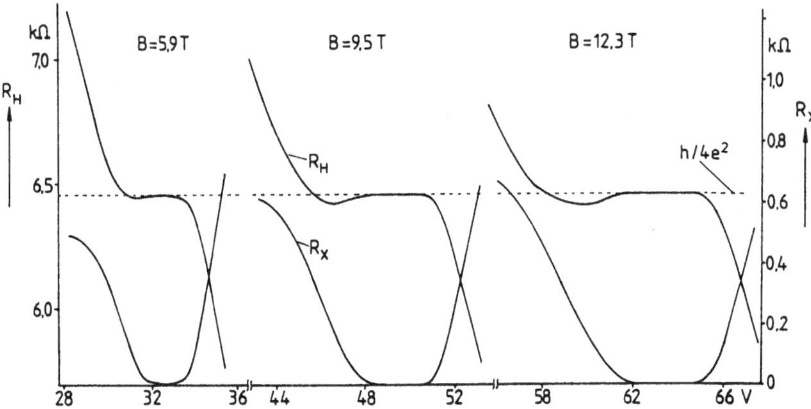

Fig. 11 Measurements of the Hall resistance R_H and the resistivity R_x as a function of the gate voltage V_g at different magnetic field values B. The curves are plotted for gate voltages close to a fully occupied Landau level n = 0 (aspect ratio of the device L/W = 1, distance between potential probes Δ = 0.6 L).

The factor G, which depends on the Hall angel θ ($\mathrm{tg}\,\theta = \sigma_{xy}/\sigma_{xx}$) and the aspect ratio L/W, has been calculated [71] using the conformal mapping published by Lippmann and Kuhrt [72]. Fig. 13 shows results for the correction $1 - G$ for diffe-

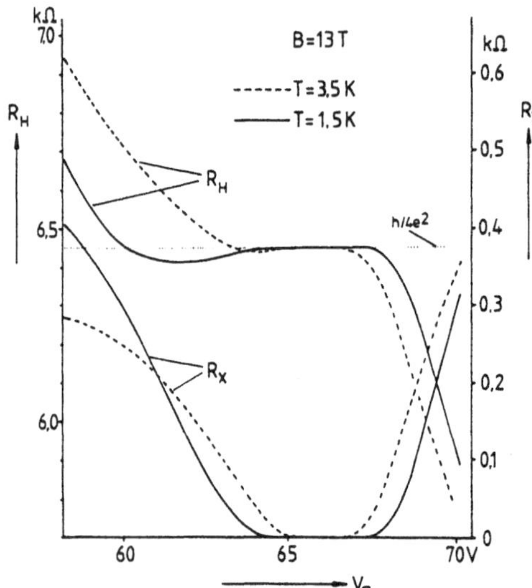

Fig. 12
Temperature dependence of the measured Hall resistance R_H and resistivity R_x.

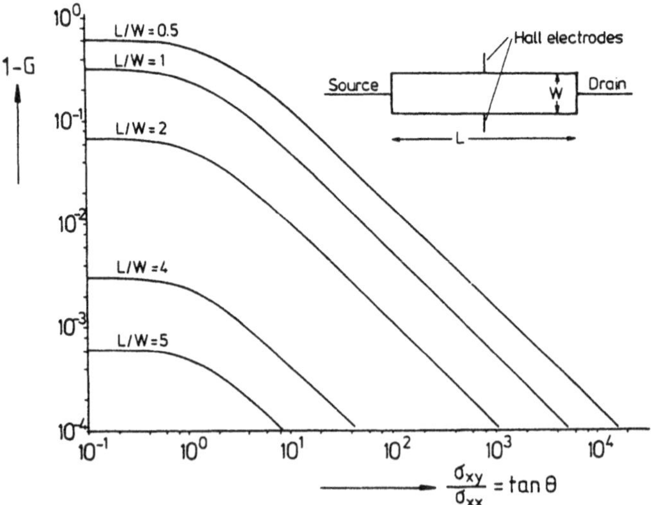

Fig. 13 Calculations of the correction term G in Hall resistance measurements due to the finite length to width ratio L/W of the device (Hall electrodes are assumed to be exactly in the middle between source and drain).

rent values of L/W. In these calculations the position l of the Hall electrodes is exactly in the middle between source and drain ($l/L = \frac{1}{2}$). If the position of the Hall electrodes is asymmetric ($l/L < \frac{1}{2}$), the correction $1 - G$ becomes much larger. Fig. 14 shows results for a sample with an aspect ratio L/W = 5 and l/L values ranging from $l/L = \frac{1}{4}$ to $l/L = \frac{1}{13}$.
For a strongly asymmetric position of the Hall electrodes ($l/L \ll \frac{1}{2}$), the correction $1 - G'$ can be found from Fig. 13 using an aspect ratio $2l/W$:

$$1 - G'\left(\frac{L}{W}, \frac{l}{L}\right) = \frac{1}{2}\left(1 - G\left(\frac{2l}{W}, \frac{l}{L} = \frac{1}{2}\right)\right)$$

These calculations show that for a typical value of $\sigma_{xy}/\sigma_{xx} = 10^6$ an aspect ratio of $L/W > 2$ is necessary, if the correction should be smaller than 10^{-7}. However, at gate voltages outside the plateaus in ρ_{xy} the correction $(1 - G)$ becomes relatively large. For example, the minimum in R_H at about $V_g = 62$ V (T = 1.5 K) in Fig. 12 originates from the finite aspect ratio. For an infinite long sample, the minimum disappears and a Hall resistance corresponding to the curve $R_H(\infty) = \rho_{xy}$ shown in Fig. 15 is expected.
It should be noted that the sign of the correction due to the shortening effect is opposite to the sign of the correction $\Delta\sigma_{xy}$ in eq. (14) and therefore cancel each other partially. We believe that on the basis of the known correction terms to σ_{xy} the experimental ρ_{xy}-values (measured at gate voltages where $\rho_{xx} = \rho_{xx}^{min}$) agree

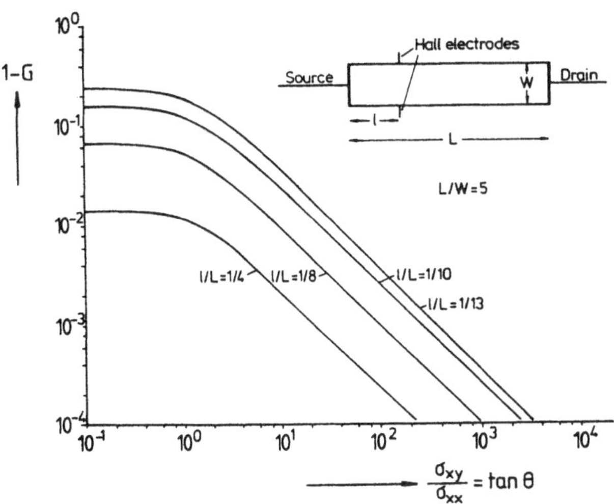

Fig. 14 Calculations of the correction term G in Hall resistance measurements for a device with an aspect ratio L/W = 5 and different positions l/L for the Hall electrodes.

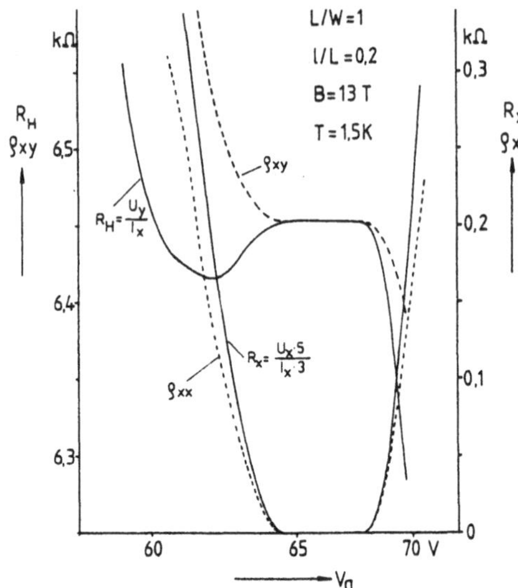

Fig. 15
Comparison between the measured quantities R_H and R_x and the corresponding resistivity components ρ_{xy} and ρ_{xx}, respectively. The device has a length to width ratio $L/W = 1$ and the positions of the Hall electrodes (potential probes) are $l/L = 0.2$ (see Fig. 14).

with $\rho_{xy} = h/e^2 i$ within an uncertainty of 10^{-7}, if $\sigma_{xy} > 10^5 \sigma_{xx}$ and the aspect ratio is larger than three. However, much more high precision measurements of the quantized Hall resistance on different devices and different magnetic field values are necessary in order to eliminate systematic errors not taken into account up to now.

4 Conclusion

Measurements of the Hall resistance R_H of a two-dimensional electron gas, realized with a silicon MOS field effect transistor, demonstrate that at carrier concentrations corresponding to fully occupied Landau levels R_H is (within our uncertainty of less than 10^{-5}) independent of external parameters. We believe that the value of R_H is given by the equation $R_H = h/e^2 i$ (i = number of fully occupied energy levels) as expected for an ideal two-dimensional electron gas, since all known corrections (finite scattering rate, interaction between Landau levels, shortening of the Hall voltage at the contacts) should change with magnetic field, temperature or sample geometry, which is not observed. The experimentally determined value for h/e^2 allows us to calculate the fine-structure constant α with an estimated uncertainty of 1.3 ppm. At this level of uncertainty an inclusion of this new method for the determination of α in the next least square adjustment of fundamental constants is expected, which should lead to a new set of fundamental constants with higher accuracies.

Acknowledgements

The experiments were performed on a large number of MOS devices obtained from different sources. I would like to thank G. Dorda (Siemens Forschungslaboratorien, München), M. Pepper (Cavendish Laboratory, Cambridge and Plessey Company), D. Tsui (Bell Laboratories, Murray Hill), and R. J. Wagner (Naval Research Laboratory, Washington) for providing me with samples. The high precision measurements were not possible without the help of the Physikalisch-Technische Bundesanstalt Braunschweig, and I would like to thank E. Braun, V. Kose, and F. Melchert for the assistance. I am indebted to D. Tsui, R. E. Prange, and T. Ando for communicating results prior to publication, and I am grateful to B. Vinter and J. S. Lass for many helpful discussions. This work was supported by the Deutsche Forschungsgemeinschaft.

References

[1] For a review see: *B. N. Taylor* and *E. R. Cohen*, in Atomic Masses and Fundamental Constants 5, *J. H. Sanders* and *A. H. Wapstra* (ed.), Plenum Press (New York) 1976, p. 663.

[2] *K. von Klitzing, G. Dorda*, and *M. Pepper*, Phys. Rev. Lett. **45**, 494 (1980).

[3] *A. Sommerfeld*, Annalen der Physik (Leipzig) **51**, 1 (1916).

[4] *E. R. Cohen* and *B. N. Taylor*, J. Phys. Chem. Ref. Data **2**, 663 (1973).

[5] *B. L. Cosens* and *T. V. Vorburger*, Phys. Rev. **A2**, 16 (1970).

[6] *T. W. Shyn, T. Rebane, R. T. Robiscoe*, and *W. L. Williams*, Phys. Rev. **A3**, 116 (1971).

[7] *A. Kponou, V. W. Hughes, C. E. Johnson, S. A. Lewis*, and *F. M. J. Pichanick*, Phys. Rev. Lett. **26**, 1613 (1971).

[8] *L. Essen, R. W. Donaldson, M. J. Bangham*, and *E. G. Hope*, Nature **229**, 1106 (1971).

[9] *S. L. Kaufmann, W. E. Lamb, K. R. Lea*, and *M. Leventhal*, Phys. Rev. **A4**, 2128 (1971).

[10] *S. Granger* and *G. W. Ford*, Phys. Rev. Lett. **28**, 1479 (1972).

[11] *P. A. Thomspon, P. Crane, T. Crane, J. J. Amato, V. W. Hughes, G. zu Putlitz*, and *J. E. Rothberg*, Phys. Rev. **A8**, 86 (1973).

[12] *D. Favart, P. M. Mc Intyre, D. Y. Stowell, V. L. Telegdi*, and *R. De Voe*, Phys. Rev. Lett. **27**, 1336 (1971).

[13] *J. C. Baird, J. Brandenberger, K. I. Gondaira*, and *H. Metcalf*, Phys. Rev. **A5**, 564 (1972).

[14] *B. N. Taylor, W. H. Parker*, and *D. N. Langenberg*, Rev. Mod. Phys. **41**, 375 (1969).

[15] *P. Cvitanovic* and *T. Kinoshita*, Phys. Rev. **D10**, 4007 (1974).

[16] *E. R. Cohen*, in Atomic Masses and Fundamental Constants 6, *J. A. Nolen* and *W. Benenson* (ed.), Plenum Press (New York) 1980, p. 525.

[17] *R. S. van Dyck, P. B. Schwinberg*, and *H. G. Dehmelt*, Bull. Am. Phys. Soc. **24** (5), 758 (1979), Abstract D 5.

[18] *T. Kinoshita*, in New Frontiers in High Energy Physics Studies in Natural Sciences, Sanibal Island 1978, *A. Perlmutter* and *L. F. Scott* (ed.), Plenum Press, New York 1978, Vol. 14, p. 127.

[19] *G. K. Yagola, V. I. Zingerman*, and *V. N. Sepetyi*, Izmeritel. Tekhn. **7**, 44 (1966), English Transl.: Meas. Tech. **1967**, 914 (1967); Izmeritel. Tekhn. **5**, 24 (1962), English transl.: Meas. Tech. **1962**, 387 (1962).

[20] *K. Hara, H. Nakamura, T. Sakai*, and *N. Koizumi*, report to Comité Consultatif d'Electricité, Comité International des Poids et Mesures, 12th Session, October 1968.

[21] *P. Vigoureux*, Proc. Roy. Soc. A (London) **270**, 72 (1962).

[22] *T. N. Malyarevskaya, N. V. Studentsov,* and *V. Ya. Shifrin,* in Studies in the Field of Magnetic Measurements, Mendeleev All Union Scientific Research Institute of Metrology, Publications of the Institutes of Metrology, U.S.S.R., No. 120 (180), 1971 (Standards Publications, Moscow-Leningrad 1971), p. 14 [Trudy Metrologicheskikh Institutov, SSSR, No. 120 (180), 1971].

[23] *N. V. Studentsov, T. N. Mayarevskaya,* and *V. Ya. Shifrin,* Izmeritel. Tekhn. **11,** 29 (1968), English transl.: Meas. Tech. **1968,** 1483 (1968).

[24] *D. N. Craig, J. I. Hoffman, C. A. Law,* and *W. J. Hamer,* J. Res. Nat. Bur. Stand. (U.S.), **64A** (Phys. and Chem.), No. 5, 381–402 (Sept.-Oct. 1960).

[25] *G. Marinenko* and *J. K. Taylor,* Analyt. Chem. **40,** 1645 (1968).

[26] *P. T. Olsen* and *R. L. Driscoll,* in Atomic Masses and Fundamental Constants 4, Ed. by *J. H. Sanders* and *A. H. Wapstra* (Plenum Publishing Corp., New York 1972), p. 471.

[27] *B. P. Kibble* and *G. J. Hunt,* Metrologia **15,** 5 (1978).

[28] *E. R. Williams* and *P. T. Olsen,* Phys. Rev. Lett. **42,** 1575 (1979).

[29] *E. Braun, E. Staben,* and *K. von Klitzing,* PTB Mitteilungen **90,** 350 (1980).

[30] *D. C. Licciardello* and *D. J. Thouless,* Phys. Rev. Lett. **35,** 1474 (1975).

[31] *M. Pepper,* Proc. Roy. Soc. **A353,** 225 (1977).

[32] *T. Ando, Y. Matsomuto, Y. Uemura, M. Kobayashi,* and *K. F. Komatsubara,* J. Phys. Soc. Japan **32,** 859 (1972).

[33] *A. M. Thompson* and *D. G. Lampard,* Nature (London) **177,** 888 (1956).

[34] *R. E. Prange,* Phys. Rev. Letters (to be published).

[35] *H. Aoki* and *T. Ando,* Solid State Commun. (to be published).

[36] *F. Stern,* in Critical Reviews in Solid State Sciences **5,** 499 (1974); *G. Landwehr,* in Festkörperprobleme XV, *H. J. Queisser* (ed.), Pergamon Vieweg 1975, p. 49.

[37] *C. C. Grimes,* Surface Sci. **73,** 379 (1978).

[38] *R. Dingle,* in Advances in Solid State Physics **15,** 21 (1975), *J. J. Queisser* (ed.), Vieweg 1975.

[39] *L. Esaki* and *L. L. Chang,* Thin Solid Films **36,** 285 (1976).

[40] *A. B. Fowler, F. F. Fang, W. E. Howard,* and *P. J. Stiles,* J. Phys. Soc. Japan **21,** 331 (1966).

[41] *Th. Englert* and *K. von Klitzing,* Surf. Science **73,** 70 (1978).

[42] *S. M. Sze,* Physics of Semiconductor Devices, Wiley and Sons 1969 (New York).

[43] *K. Ohta,* Jap. Journal of Applied Physics **10,** 850 (1971).

[44] *C. C. Hu, J. Pearse, K. M. Cham,* and *R. G. Wheeler,* Proc. Second Int. Conf. on the Electronic Properties of Two-Dimensional Systems, Berchtesgaden 1977 (unpublished), see also Surc. Science **73,** 207 (1978).

[45] *D. C. Tsui* and *G. Kaminsky,* Phys. Rev. Lett. **35,** 1468 (1975).

[46] *W. E. Howard* and *F. F. Fang,* Phys. Rev. **B13,** 2519 (1976).

[47] *R. Dingle, H. L. Störmer, A. C. Gossard,* and *W. Wiegmann,* Surf. Science **98,** 90 (1980).

[48] *T. Ando* and *Y. Uemura,* J. Phys. Soc. Japan **36,** 959 (1974).

[49] *R. Kubo, S. J. Miyake,* and *N. Hashitsume,* in Solid State Physics **17,** 269 (1965), *F. Seitz* and *D. Turnbull* (ed.), Academic Press (New York) 1965.

[50] *T. Ando,* J. Phys. Soc. Japan **37,** 622 (1974).

[51] *R. J. Nicholas, R. A. Stradling,* and *R. J. Tidey,* Solid State Commun. **23,** 241 (1977).

[52] *S. Kawaji* and *J. Wakabayashi,* Surface Sci. **58,** 238 (1976).

[53] *M. Pepper,* Phil. Mag. **B37,** 83 (1978).

[54] *H. Aoki* and *H. Kamimura,* Solid State Commun. **21,** 45 (1977).

[55] *H. Aoki,* J. Phys. **C11,** 3823 (1978).

[56] *H. Fukuyama,* Solid State Commun. **19,** 551 (1976).

[57] *M. Tsukada,* J. Phys. Soc. Japan **42,** 391 (1977).

[58] *R. J. Nicholas, E. Kress-Rogers, F. Kuchar, M. Pepper, J. C. Portal,* and *R. A. Stradling,* Surface Science **98,** 283 (1980).

[59] *D. C. Tsui,* Solid State Commun. **21,** 675 (1977).

[60] *J. Wakabayashi* and *S. Kawaji,* Surface Science **98,** 299 (1980).

[61] *G. Ebert,* Diplomarbeit Würzburg 1978 (unpublished).

[62] *T. Ando, Y. Matsumoto,* and *Y. Uemura,* J. Phys. Soc. Japan **39,** 279 (1975).

[63] *S. Kawaji, T. Igarashi,* and *J. Wakabayashi,* Progr. of Theor. Physics, Suppl. 57, 176 (1975).

[64] *Th. Englert, K. von Klitzing, R. J. Nicholas, G. Landwehr, G. Dorda,* and *M. Pepper,* Phys. stat. sol. (b) **99,** 237 (1980).

[65] *H. Köhler* and *M. Roos,* Phys. stat. sol. (b) **91,** 233 (1979).

[66] *R. J. Nicholas, K. von Klitzing,* and *Th. Englert,* Solid State Commun. **34,** 51 (1980).

[67] *M. Kaplit* and *J. N. Zemel,* Phys. Rev. Lett. **21,** 212 (1968).

[68] *A. M. Voshchenkov* and *J. N. Zemel,* Phys. Rev. **B9,** 4410 (1974).

[69] *K. Ohta,* Japanese J. of Applied Physics **10,** 850 (1971).

[70] *I. Eisele* and *G. Dorda,* Solid State Commun. **15,** 1391 (1974).

[71] *H. Obloh* (unpublished).

[72] *H. J. Lippmann* and *F. Kuhrt,* Z. Naturschg. **13a,** 462 (1958); Z. Naturforschg. **13a,** 474 (1958).

[73] *D. C. Tsui* and *A. C. Gossard* (to be published).

Festkörperprobleme XXI (1981)

Surface Waves

Alexei A. Maradudin

Max-Planck-Institut für Festkörperforschung Stuttgart, Germany, and Department of Physics, University of California, Irvine CA 92717, U.S.A.

Summary: A survey is given of recent developments in theoretical and experimental studies of vibrational and electromagnetic surface waves. In addition to structurally ideal surfaces, structurally nonideal surfaces and surfaces in which the material properties vary with distance from the surface, are treated, along with effects of spatial dispersion. Nonlinear properties of surface waves are considered, as well as guided surface waves in confined geometries, such as thin films and wedges.

Introduction

Surface waves are waves that propagate along surfaces and whose fields are localized in the vincinity of the surface. They occur in elastic, electromagnetic, hydrodynamic, plasma and other systems in which wave propagation can occur.

In this article it is my intention to present a survey of recent theoretical and experimental studies of two types of surface waves of particular interest in solid state physics, viz. vibrational and electromagnetic surface waves. Although a great deal of interest in the former derives from the utility of surface acoustic waves in device applications, the emphasis in this review will be on the basic physics of surface waves and not on their applications. The selection of topics to be discussed is a somewhat personal one, that reflects to a large extent my own interests and those of my colleagues. That, and limitations of space, have the consequence that what follows cannot be a comprehensive survey of the work done on even the two types of surface waves to which I am limiting myself here. Nevertheless, I hope that the reader will find discussions of and references to the most interesting topics of current interest in the study of vibrational and electromagnetic surface waves.

Part I. Surface Acoustic Waves

1 Introduction

The first type of surface wave to be discovered was a surface acoustic wave on the stress-free, planar surface of a semi-infinite isotropic elastic medium [1]. These waves, now called *Rayleigh waves* after their discoverer, are wavelike in directions parallel to the surface of the solid, but their amplitudes decay exponentially with increasing distance into the solid from the surface. They are acoustic waves, in the sense that their frequencies are linear in the magnitude of the two-dimensional wave vector that characterizes their propagation along the surface, and consequently vanish with vanishing wave vector.

The theory of surface acoustic waves, their experimental study, and their applications have developed subsequently into a sub-field of surface physics that is already rich in accomplishment, but which at the same time is far from being played out. I hope to convey a sense of this in what follows.

2 Surface Acoustic Waves on Various Media

The work of Lord Rayleigh that resulted in the discovery of Rayleigh waves was prompted by the problem of the propagation of seismic shocks in the earth's crust. Because the radius of the earth is much larger than the wavelength of seismic disturbances, the analysis was simplified by neglecting the curvature of the earth, and considering its surface to be the top of a semi-infinite medium.

Although surface acoustic waves have been known and studied for nearly a century now, there has been an enormous increase in the interest shown in these waves in the past decade. This is due largely to the recognition that they can provide a new approach to signal processing, with important applications to radar and communications [2, 3].

This is due to several properties of surface acoustic waves [3]. The most important of these is their extremely low speed, about 10^{-5} that of electromagnetic waves. Because of their low speeds acoustic waves also have a very small wavelength compared with the wavelength of electromagnetic waves of the same frequency. As a result the reduction in device size they offer is also by a factor of the order of 10^{-5}. Surface acoustic waves travel along the surface of a solid and are confined to its vicinity. They are therefore easier to modify or tap into than are bulk acoustic waves. Surface wave devices are compatible with integrated circuit technology, and fabrication by lithographic techniques produces surface acoustic wave devices of comparatively low cost with precise and reproducible characteristics.

Beyond their utility for device applications, surface acoustic waves have so far been comparatively little used for the study of physical properties of materials. However, such examples of this use as do exist will be referred to at appropriate points in the discussion that follows.

We begin by considering surface acoustic waves on the surface of three characteristically different types of materials, and then proceed to a discussion of various properties of these waves.

2.1 Elastic Media

Surface acoustic waves can exist on several different kinds of solids, elastic, piezoelectric, and magnetic. Although such waves on the surface of purely elastic media are perhaps somewhat less interesting from a technological standpoint that are surface acoustic waves on piezoelectric or magnetic media, their study is nonetheless of interest for several reasons. Many materials of physical interest are purely elastic, i.e. are neither piezoelectric nor magnetic. Among these are the nontransition metals and homopolar semiconductors, and the propagation of surface acoustic waves over their surface can be of technological utility, e.g. in nondestructive testing for cracks [4]. At the same time, the theory of surface acoustic waves on purely elastic media is rather simpler than the corresponding theory for piezoelectric and magnetic media, and serves as a good introduction to the latter, as well as to the several types of guided waves that are associated with an overlayer that will be discussed below.

2.1.1 Semi-Infinite Media

We consider an elastic medium occupying the upper half-space $x_3 > 0$. The surface $x_3 = 0$ is assumed to be stress-free. Within the framework of the linear theory of elasticity the equations of motion of the medium are [5]

$$\rho \frac{\partial^2 u_\alpha}{\partial t^2} = \sum_\beta \frac{\partial T_{\alpha\beta}}{\partial x_\beta} \qquad \alpha = 1, 2, 3, \tag{1}$$

where ρ is the mass density of the medium, $u_\alpha(\vec{x}, t)$ is the α Cartesian component of the displacement of the medium at the point $\vec{x}$ and time t, and $T_{\alpha\beta}(\vec{x}, t)$ is the stress tensor. The latter can be expressed as

$$T_{\alpha\beta} = \sum_{\mu\nu} C_{\alpha\beta\mu\nu} \frac{\partial u_\mu}{\partial x_\nu} \qquad \alpha, \beta = 1, 2, 3, \tag{2}$$

where the $C_{\alpha\beta\mu\nu}$ are the elements of the elastic modulus tensor. The latter are symmetric in α und β, in μ and ν, and in the interchange of the pairs $\alpha\beta$ and $\mu\nu$. These equations have to be supplemented by the conditions that the stress components $T_{\alpha 3} (\alpha = 1, 2, 3)$ vanish at the surface $x_3 = 0$.

Since we are interested in waves localized at the surface $x_3 = 0$, we assume a solution of Eqs. (1)–(2) of the form

$$\vec{u}(\vec{x}, t) = (U_1, U_2, -iU_3)\, e^{i\vec{k}_\parallel \cdot \vec{x}_\parallel - \alpha x_3 - i\omega t}, \tag{3}$$

where $\vec{x}_\parallel = \hat{x}_1 x_1 + \hat{x}_2 x_2$ and $\vec{k}_\parallel = \hat{x}_1 k_1 + \hat{x}_2 k_2$ are two-dimensional position and wave vectors in the plane of the surface, and $\hat{x}_1$ and $\hat{x}_2$ are unit vectors in the x_1- and x_2-directions, respectively. When Eq. (3) is substituded into Eqs. (1)–(2), the solvability condition for the resulting set of three, homogeneous, linear equations for the amplitudes $\{U_\alpha\}$ yields a cubic equation for the square of the decay constant $\alpha(\vec{k}_\parallel \omega)$, for each value of $\vec{k}_\parallel$ and ω. There are in general three values of $\alpha(\vec{k}_\parallel \omega)$ with positive real part, denoted by $\alpha_j(\vec{k}_\parallel \omega)$ (j = 1, 2, 3).

To satisfy the boundary conditions the solutions corresponding to the three α_j are superposed,

$$(u_1, u_2, iu_3) = \sum_{j=1}^{3} (\xi_j, \eta_j, \zeta_j)\, K_j e^{i\vec{k}_\parallel \cdot \vec{x}_\parallel - \alpha_j(\vec{k}_\parallel \omega) x_3} e^{-i\omega t}, \tag{4}$$

where the K_j are new amplitudes, and ξ_j, η_j, ζ_j are the cofactors of the elements of the first row of the homogeneous matrix equation for the $\{U_\alpha\}$, evaluated with $\alpha = \alpha_j(\vec{k}_\parallel \omega)$.

When these expressions for the displacement components are substituted into the boundary conditions, one obtains a set of homogeneous equations for the $\{K_j\}$, the solvability condition for which is the dispersion relation for the surface acoustic wave (Rayleigh wave). In general, the solution has the form $\omega = c_R(\hat{k}_\parallel)\, k_\parallel$, where $\hat{k}_\parallel$ is the unit vector in the direction of $\vec{k}_\parallel$. The speed of Rayleigh waves, $c_R(\hat{k}_\parallel)$, is thus independent of the magnitude of the wave vector $\vec{k}_\parallel$ (i.e. these waves are dispersionless) but depends on the direction of propagation.

The possible types of solution are three in number.

1. All $\alpha_j(\vec{k}_\parallel \omega)$ real and positive – *ordinary Rayleigh waves*;
2. Some $\alpha_j(\vec{k}_\parallel \omega)$ complex, but with positive real parts – *generalized Rayleigh waves*;
3. One of the $\alpha_j(\vec{k}_\parallel \omega)$ is pure imaginary, rather than real or complex – *pseudo-surface waves* [6, 7].

In the case of pseudosurface waves there is a piece of each displacement component that does not decay exponentially with increasing distance into the solid from the surface – hence the name pseudosurface wave. The nondecaying part of the displacement field gives rise to a flow of energy away from the surface into the solid. This in turn requires that the components of the wave vector $\vec{k}_\parallel$ have imaginary parts, so that the wave attenuates as it propagates along the surface. However, in many cases the attenuation is so small that the pseudosurface wave is readily observable. Which of the three possible types of surface acoustic wave appears in a particular case depends on the elastic moduli, on the crystallographic plane that constitutes the surface and, to some extent, on the direction of propagation.

The progam outlined above ordinarily has to be carried out numerically. However, in some special cases it can be carried out analytically. Thus, for example, if the elastic medium possesses cubic symmetry, and the cube axes are oriented parallel to the coordinate axes, so that the surface $x_3 = 0$ is a (001) plane, one finds the following cubic equation for the speed of Rayleigh waves propagating in the [100] direction [8]

$$\left[1 - \frac{c_{11}}{c_{44}} R\right] \left[1 - \frac{c_{12}^2}{c_{11}^2} - R\right]^2 = R^2(1 - R), \tag{5a}$$

and the equation

$$\left[1-\frac{c_{11}}{c_{44}}R\right]\left[\frac{c_L}{c_{11}}-\frac{c_{12}^2}{c_{11}^2}-R\right]^2=R^2\left[\frac{c_L}{c_{11}}-R\right], \tag{5b}$$

for propagation in the [110] direction [8]. In Eqs. (5) $R=\rho c_R^2/c_{11}$; c_{11}, c_{12}, c_{44} are the three, independent bulk elastic moduli of a cubic medium in the contracted, Voigt notation [9]; and $c_L=\frac{1}{2}(c_{11}+c_{12}+2c_{44})$.

For isotropic crystals, for which $c_{12}=c_{11}-2c_{44}$, the speed of Rayleigh waves becomes independent of their direction of propagation, and the equation determining their speed, obtained from Eqs. (5) can be written in the form

$$\left(1-\frac{c_R^2}{c_t^2}\right)\left[4\left(1-\frac{c_t^2}{c_l^2}\right)-\frac{c_R^2}{c_t^2}\right]^2=\frac{c_R^4}{c_t^4}\left(1-\frac{c_t^2}{c_l^2}\frac{c_R^2}{c_t^2}\right), \tag{6}$$

where $c_l^2=c_{11}/\rho$ and $c_t^2=c_{44}/\rho$ give the speeds of sound for bulk longitudinal and transverse waves, respectively.

For a hexagonal medium, whose axis of sixfold symmetry is oriented perpendicular to the surface $x_3=0$, the speed of Rayleigh waves is independent of the direction of propagation in the surface, and is a solution of the equation [10]

$$c_{33}\left(c_R^2-\frac{c_{44}}{\rho}\right)\left(c_R^2-\frac{c_{11}}{\rho}+\frac{c_{13}^2}{\rho c_{33}}\right)=c_{44}c_R^4\left(c_R^2-\frac{c_{11}}{\rho}\right). \tag{7}$$

In each of the cases given by Eqs. (5)–(7) the physical solution of the cubic equation is the one with the smallest value of c_R. Indeed, it is a general result that the frequency of a Rayleigh wave of wave vector $\vec{k}_\parallel$ is lower than the lowest bulk wave frequency for the same two-dimensional wave vector $\vec{k}_\parallel$. Rayleigh waves thus propagate in a stop band for the normal modes of the corresponding bulk crystal.

The displacement patterns for the Rayleigh waves whose frequencies are given by Eqs. (5)–(7) are superpositions of only two exponentially decaying terms, and trace out ellipses in the *sagittal plane*, i.e. the plane defined by the direction of propagation of the wave ($\hat{k}_\parallel$) and the normal to the surface ($\hat{x}_3$) (Such modes are sometimes referred to as *shear vertical* polarized modes). However, for other directions of propagation in cubic crystals and in crystals of lower symmetry the displacements are given by superpositions of three exponentially decaying terms, and trace out ellipses whose planes in general are inclined with respect to the sagittal plane.

It is impossible to have a straight-crested surface wave polarized perpendicular to the sagittal plane (*shear horizontal* polarization) on a homogeneous, semi-infinite, isotropic medium bounded by a planar, stress-free surface, since a displacement field of this form cannot simultaneously satisfy the equations of motion and the boundary conditions. Moreover, I knew of no cases of straight-crested surface waves polarized parallel to the surface on any purely elastic (i.e. non-piezoelectric, non-magnetic) medium.

The attenuation lengths of a Rayleigh wave into the elastic medium, α_j^{-1}, are found to be of the order of the wavelength $\lambda = (2\pi/k_{\parallel})$ of the wave along the surface, and this is true for many other types of surface waves as well.

For additional information concerning surface acoustic waves on various surfaces of both cubic and non-cubic elastic materials, the reader may consult the review article by Farnell [11].

The majority of the studies of Rayleigh waves in the laboratory have employed ultrasonic techniques with suitable wedges [12] or interdigital [13] transducers to generate and detect them. A variant of these approaches uses a light beam instead of a transducer for detection [14, 15]. All of these experimental methods possess the common feature of requiring external generation of intense surface waves by transducers. Recently, however, Sandercock [16] has succeeded in measuring thermally excited surface acoustic waves in the 1–20 GHz frequency range using Brillouin scattering techniques. (The theory of such experiments has been worked out by several authors [17–33] and the reader is referred to their papers for the details of this theory.) Sandercock has studied surface acoustic waves on the (100) surface of Si, on crystalline samples of GaAs and Cr, and on polycrystalline samples of Al, Cu, Ta, Ni and Fe (Fig. 1). The speeds of Rayleigh waves measured by Sandercock are consistently 1–5 % below the values calculated from the bulk elastic moduli. It is suggested that this is due to differences of the elastic moduli near the surface from their values in the bulk, although the changes would have to be strong in order to produce a change in a wave that extends several hundred atomic layers into the bulk.

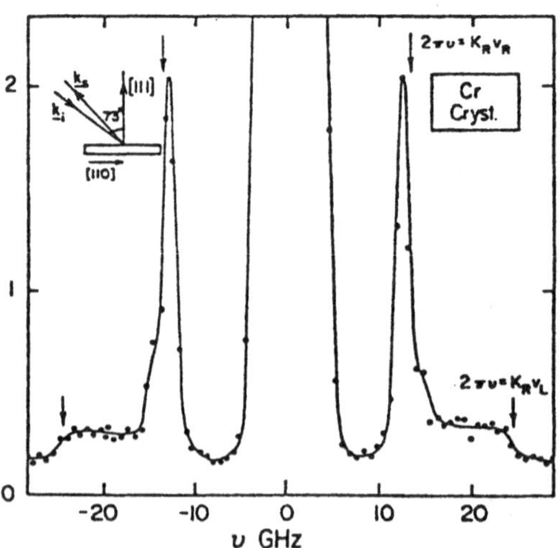

Fig. 1
Brillouin spectrum for scattering from the (111) surface of crystalline chromium. The two prominent peaks labelled $2\pi\nu = K_R v_R$ are due to scattering by thermally excited Rayleigh waves [16].

In addition to being objects of study in themselves, Rayleigh waves have been used recently to study some physical and electronic properties of solids. The temperature dependence of the speed of Rayleigh waves in the (100) direction on the (001) surface of $SrTiO_3$ has been measured at several different frequencies at temperatures just above the temperature of the structural phase transition at 105 K [34] (Fig. 2).

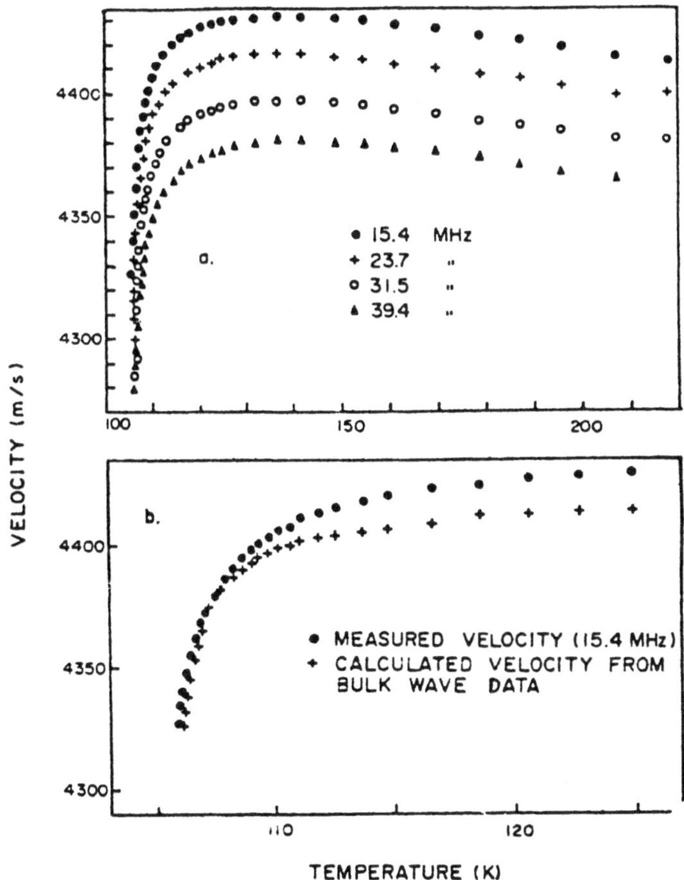

Fig. 2 (a) Measured critical speeds of Rayleigh waves propagating in the (100) direction on a (001) surface of $SrTiO_3$ at different frequencies as functions of temperature in the vicinity of the structural phase transition at 105 K. (b) Measured Rayleigh wave speeds compared with speeds calculated from bulk wave measurements [34].

The aim of the investigation was to determine how the material properties of the crystal differ from their bulk values within a distance from the surface of the order of the correlation length over which the order parameter has a surface-induced spatial dependence. Since this correlation length diverges as the transition temperature is approached from either side, surface waves will be affected more and more by the surface layer as the critical temperature is approached. Such effects were not observed in the experiments of Bjerkan and Fossheim [34], who concluded that the critical properties of $SrTiO_3$ are dominated by the bulk critical properties in the frequency range studied. Nevertheless, the use of Rayleigh waves for the study of surface critical properties remains an attractive one.

The attenuation of Rayleigh waves on Cu/Pb-, Ag/Pb-, and Al/Pb-double layers has been measured as a function of temperature, magnetic field, and frequency [35]. From the results the electron interaction parameter N(0) V for the normal metals Cu and Ag has been determined.

Changes in the speed of Rayleigh waves with as little as a few hundredths of an adsorbed monolayer of gas atoms can be measured [36], which offers the possibility of monitoring atom – surface interaction phenomena in this way.

Additional applications of Rayleigh waves to the study of physical and electronic properties of solids have recently been described by Borisov [37].

2.1.2 A Plate on a Different Half-Space

Surface acoustic waves on semi-infinite media are dispersionless, i.e. their speed is independent of their wavelength. This is a useful property for certain types of applications, and has its origin in the absence of any characteristic length in the structure supporting these waves, If, however, some dispersion in a surface wave is desired, this can be achieved by adding a thin film of one material onto a substrate of a different material. The presence of a characteristic length in such a structure, viz. the thickness of the film, renders surface acoustic waves in it dispersive, in a manner that can be varied by changing the thickness of the film and the material properties of the film and the substrate.

A great deal of work on surface acoustic waves in such a structure has been done in the case that both the plate and the substrate are elastically isotropic. The theory in the case that one or both of the media are elastically anisotropic presents additional computational complexity, but no significantly new features. A good review of surface acoustic waves in the case of an elastic plate on an elastic substrate has been given by Farnell and Adler [38], and I shall confine myself to several brief remarks.

We consider a structure consisting of a plate characterized by a mass density ρ and elastic moduli $C_{\alpha\beta\mu\nu}$ in the region $-d < x_3 < 0$ bonded to a semi-infinite substrate characterized by a mass density ρ' and elastic moduli $C'_{\alpha\beta\mu\nu}$ in the region $x_3 < -d$. The equations of motion are given by Eqs. (1)–(2) for the plate and by the same equations, but with ρ and $C_{\alpha\beta\mu\nu}$ replaced by ρ' and $C'_{\alpha\beta\mu\nu}$, respectively, for the substrate. The boundary conditions are now that the stresses $T_{\alpha 3}$ ($\alpha = 1, 2, 3$) vanish on the surface $x_3 = 0$, and that the displacement components u_α and stress components $T_{\alpha 3}$ ($\alpha = 1, 2, 3$) be continuous across the interface $x_3 = -d$.

Several types of surface waves are possible in this structure. They fall into two general categories: those that are polarized perpendicular to the sagittal plane, and those that are polarized parallel to the sagittal plane. For simplicity, we will discuss only the case that both the plate and the substrate are elastically isotropic. In this case we can assume with no loss of generality that the waves propagate in the x_1-direction with a wave vector $\vec{k}_\parallel = \hat{x}_1 k$.

The simplest type of surface acoustic wave in this structure is the *Love wave* [39]. It is a wave polarized perpendicular to the sagittal plane that has the character of a standing wave in the plate, with an amplitude that decays exponentially with increasing distance into the substrate from the interface. The dispersion relation for Love waves is

$$\frac{\rho' c_t'^2 \alpha'}{\rho c_t^2 \hat{\alpha}} = \tan \hat{\alpha} d, \tag{8}$$

where $\hat{\alpha} = ((\omega^2/c_t^2) - k^2)^{1/2}$, $\alpha' = (k^2 - (\omega^2/c_t'^2))^{1/2}$. There is an infinite number of these waves in the region of the (ω, k) plane defined by $(\omega/c_t') < k < (\omega/c_t)$, so that the existence of these waves requires that $c_t' > c_t$. These modes are dispersive because of the presence of a characteristic length, d, in the structure.

The several types of surface waves polarized in the sagittal plane that are possible when an isotropic plate is bonded to a different isotropic substrate can be referred to collectively as *generalized* Lamb waves [40]. These waves are standing waves in the plate, and their amplitudes decay exponentially with increasing distance into the substrate from the interface. The complications in the analysis in the present case arise from the necessity of satisfying the stress-free coundary condition at the surface $x_3 = 0$. The dispersion relation that is obtained [41] can be solved only numerically. The frequencies of generalized Lamb waves are dispersive because the thickness of the slab enters the theory, through the boundary conditions.

The characteristics of generalized Lamb waves depend strongly on the ratio of the speed of transverse bulk waves in the substrate, c_t', to the speed of bulk transverse waves in the plate, c_t [42]. If $c_t \gg c_t'$ there is only one generalized Lamb wave solution. It reduces to a Rayleigh wave on the substrate in the limit as $kd \to 0$, and exists only over the range of kd for which $(\omega/k) < c_t'$.

In the opposite limit $c_t \ll c_t'$ there is an infinite number of solutions that fall into two families of modes, often called the M_1 series and the M_2 series. These two families of modes reduce to the symmetric and antisymmetric *Lamb waves*, respectively, when the density or elastic moduli of the substrate go to zero, so that we are dealing with the vibrations of a free plate. (Lamb waves [43–46] are waves, polarized in the sagittal plane, in a plate of finite thickness with stress free surfaces. The normal modes of the plate can be classified as symmetric or antisymmetric according to whether the 3-component of the displacement field is odd or even with respect to the midplane of the plate.)

The fundamental modes of the two series of generalized Lamb waves, that are denoted by M_{11} and M_{21}, have particularly intersting properties [42]. For plate thicknesses approaching zero ($kd \to 0$) the M_{11} mode approaches a Rayleigh type of wave on the substrate, while all the higher M_1 modes and all M_2 modes are leaky waves, i.e. they radiate energy into the substrate. As the plate thickness is increased the first additional mode to become trapped is the M_{21} mode, which is called the *Sezawa wave* [47]. For very high frequencies or thick plates ($kd \to \infty$) the M_{11} mode approaches a Rayleigh type of surface wave on the upper (stress-free) surface of the plate. All higher modes ($M_{1i}, M_{2i}, i > 1$) behave like the Lamb waves of a free thick plate as $kd \to \infty$.

The behavior of the mode M_{21} depends critically on the relative material properties of the plate and substrate.

If c_t' and c_t are appreciably different, the mode M_{21} approaches a shear wave polarized in the sagittal plane in the limit as $kd \to \infty$, i.e. it also approaches a Lamb wave of a free, thick plate. However, for certain special combinations of material parameters with $c_t' \approx c_t$, but with c_t still smaller than c_t', it becomes a bound wave at the interface between the plate and the substrate when $kd \to \infty$. This is the *Stoneley wave* [48]. Stoneley waves can also exist when $c_t \approx c_t'$ and $c_t > c_t'$. In this case the Rayleigh wave solution for the unplated substrate ($kd = 0$) becomes a bound state at the interface (with $(\omega/k) < c_t'$) in the limit as $kd \to \infty$.

In a Stoneley wave the amplitude of the wave in the slab has decayed essentially to zero before the surface $x_3 = 0$ is reached, so that the stress-free boundary conditions at the surface $x_3 = 0$ can be ignored. What we have then in effect is two different, semi-infinite elastic media in contact across a planar interface. Numerical methods have to be used to obtain the frequency of Stoneley waves even when both media are isotropic. The frequency of Stoneley waves has the form $\omega = c_S k$, where the speed of these waves, c_s, must lie between the speed of Rayleigh waves and transverse bulk waves in the denser medium, i.e. $c_R' < c_s < c_t'$, where $\rho' > \rho$ [49, 50]. The solutions are very sensitive to the ratio c_l/c_t in the denser of the two media, but are only slightly influenced by this ratio in the less dense medium. Stoneley waves are nondispersive, due to the lack of a characteristic length in the structure supporting them.

The theoretical study of Love waves and of generalized Lamb waves in the limit as $kd \to 0$ has been significantly simplified by Velasco et al. [51] and by Djafari-Rouhani et al. [52]. Using the thinness of the plate d they eliminate the boundary conditions at the plate-vacuum interface ($x_3 = 0$) by modifying in a systematic way the boundary conditions at the plate-substrate interface ($x_3 = -d$) to obtain effective boundary conditions on the vibrations of the substrate at $x_3 = -d$ that incorporate the effects of the plate to first order in kd. In this way these authors obtain the dispersion relation for a Rayleigh wave in the presence of an overlayer and for a Love wave in the form $\omega = c_R k(1 - \lambda kd + o(kd))$ and $\omega = c_t k(1 - A(kd)^2 + o((kd)^2))$, respectively, where the coefficients λ and A are obtained in closed form. This treatment generalizes earlier approaches in the same vein [53–55].

Recently, thermally excited Sezawa waves, as well as Rayleigh waves, have been observed experimentally by Brillouin scattering from GaAs and Si surfaces coated with a thin aluminium film [56], and from Si surfaces coated with thin aluminium and gold films [57] (Fig. 3). This leaves thermally excited Love waves and Stoneley waves yet to be observed by this sensitive experimental technique.

2.2 Piezoelectric Media

A piezoelectric medium is characterized by the fact that deforming it produces an electric field; conversely, applying an electric field to it deforms it. Surface acoustic waves can exist on such media. Piezoelectricity is found to cause only second order effects on the dominantly mechanical nature of surface acoustic waves. However, the elastic deformation of the medium in such waves is accompanied by an electric

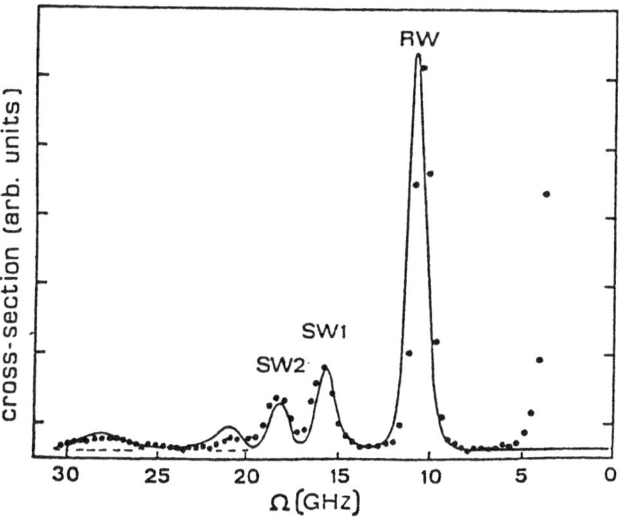

Fig. 3 Experimental (dots) and computed (full line) Brillouin spectrum (Stokes) for scattering from a 2000 Å film of aluminium deposited on a silicon substrate. The peak labelled RW is due to scattering by thermally excited Rayleigh waves; the peaks labelled SW1 and SW2 are due to scattering from the two thermally excited Sezawa waves of lowest frequency [57].

field. It is this electric field that makes surface acoustic waves on piezoelectric media interesting from a technological standpoint, for it represents the way in which a surface wave is coupled to external electrical curcuits, to other surface waves, or to charge carriers in semiconductors.

2.2.1 Piezoelectric Surface waves

The equations of motion of a piezoelectric medium ($x_3 > 0$) are still given by Eqs. (1), except that the stress tensor now possesses an electrical contribution in addition to the usual mechanical one,

$$T_{\alpha\beta} = \sum_{\mu\nu} C_{\alpha\beta\mu\nu} \frac{\partial u_\mu}{\partial x_\nu} - \sum_{\mu} e_{\mu\alpha\beta} E_\mu . \tag{9}$$

In this equation $\vec{E}$ is the macroscopic electric field in the medium, the $\{C_{\alpha\beta\mu\nu}\}$ are the elastic moduli at constant electric field, and $e_{\mu\alpha\beta}$ is the piezoelectric tensor. It is symmetric in the indices α and β. The equations of motion have to be supplemented by the constituitive relation connecting the electric displacement $\vec{D}$ with the displacement gradients and the macroscopic electric field

$$D_\alpha = \sum_{\mu\nu} e_{\alpha\mu\nu} \frac{\partial u_\mu}{\partial x_\nu} + \sum_{\beta} \epsilon_{\alpha\beta} E_\beta , \tag{10}$$

where $\epsilon_{\alpha\beta}$ is the dielectric tensor of the medium (at constant strain). It is symmetric in the indices α and β. Since the disturbances of interest to us propagate with sonic rather than electromagnetic speeds, we can use the quasistatic, or electrostatic, approximation, and write the electric field as the gradient of a scalar potential, $E_\alpha = - \partial\phi/\partial x_\alpha$. The equations of motion now become

$$\rho \frac{\partial^2 u_\alpha}{\partial t^2} = \sum_{\beta\mu\nu} C_{\alpha\beta\mu\nu} \frac{\partial^2 u_\mu}{\partial x_\beta \partial x_\nu} + \sum_{\beta\nu} e_{\nu\alpha\beta} \frac{\partial^2 \phi}{\partial x_\beta \partial x_\nu}, \qquad x_3 > 0. \tag{11}$$

A closed system of a equations is obtained when we include the Maxwell equation $\nabla \cdot \vec{D} = 0$,

$$\sum_{\alpha\mu\nu} e_{\alpha\mu\nu} \frac{\partial^2 u_\mu}{\partial x_\alpha \partial x_\nu} - \sum_{\alpha\beta} \epsilon_{\alpha\beta} \frac{\partial^2 \phi}{\partial x_\alpha \partial x_\beta} = 0, \quad x_3 > 0. \tag{12}$$

In the vacuum below the piezoelectric medium ($x_3 < 0$) there is no displacement field: only the scalar potential is nonzero. It satisfies the Laplace equation $\nabla^2 \phi = 0$ ($x_3 < 0$).

The boundary conditions at the surface $x_3 = 0$ are that the stresses $T_{\alpha 3}$ ($\alpha = 1, 2, 3$) vanish there, and that ϕ and D_3 be continuous across it.

In the present case the most general forms for the displacement field and potential inside the medium are

$$u_\alpha(\vec{x}, t) = \sum_{j=1}^{4} C_j \xi_\alpha^{(j)} e^{ikx_1 - \alpha_j x_3 - i\omega t} \qquad \alpha = 1, 2, 3, \quad x_3 > 0 \tag{13a}$$

$$\phi(\vec{x}, t) = \sum_{j=1}^{4} C_j \xi_4^{(j)} e^{ikx_1 - \alpha_j x_3 - i\omega t} \qquad x_3 > 0, \tag{13b}$$

for propagation in the x_1-direction, while in the vacuum outside the medium we have that

$$\phi(\vec{x}, t) = C_5 e^{ikx_1 + |k| x_3 - i\omega t} \qquad x_3 < 0. \tag{14}$$

In these expressions the $\{\alpha_j(k\omega)\}$ are decay constants that are obtained from the solvability condition that results from substituting solutions proportional to $\exp(ikx_1 - \alpha x_3 - i\omega t)$ into Eqs. (11)–(12), while the $\{\xi_\alpha^{(j)}\}$ are cofactors. There are thus as many unknown $(C_1, C_2, \ldots, C_5)$ as there are boundary conditions. Substitution of Eqs. (13)–(14) into the boundary conditions yields a set of five homogeneous equations in the five unknowns, the solvability condition for which is the dispersion relation for surface acoustic waves on a piezoelectric surface. As is the case for Rayleigh waves, and for the same reason, surface acoustic waves on piezoelectric surfaces are dispersionless. Also, like Rayleigh waves in anisotropic media, the particle motion in a piezoelectric surface wave is not confined to the sagittal plane: we see from Eq. (13a) that although nothing varies with x_2, there is in general a nonvanishing 2-component of the displacement field.

The calculation of the speed of surface acoustic waves on piezoelectric surfaces has to be carried out numerically, and such calculations have been performed by several authors, e.g. by Tseng and White [58] and by Tseng [59] for several piezoelectric surfaces.

In the case that a piezoelectric film is deposited on a semi-infinite piezoelectric substrate one can have the analogues of Love, Stoneley, and Sezawa waves. These will not be discussed here. A comprehensive survey of the properties of such waves has been given by Farnell and Adler [38], to which the interested reader is referred.

2.2.2 Shear Horizontal Surface Acoustic Waves

An important consequence of piezoelectricity for surface acoustic waves is that it gives rise to new types of surface waves that have no counterparts for purely elastic media. We have noted earlier that shear horizontal surface acoustic waves do not exist on the stress-free surface of a semi-infinite substrate, or at the interface between two different elastic media. Such modes can exist, however, in special cases, when the substrate is piezoelectric or when at least one of the two media in contact across the interface is piezoelectric.

The earliest demonstration that surface acoustic waves polarized perpendicular to the sagittal plane can exist on a piezoelectric surface appears to have been given by Kaganov and Sklovskaya [60], whose work, however, seems to have been largely overlooked by subsequent authors. Kaganov and Sklovskaya considered the surface of a cubic piezoelectric. Simpler results are obtained in other cases.

For example, we consider a crystal such as CdS, belonging to the crystal class C_{6v}. We assume that its axis of sixfold symmetry is parallel to the x_3-direction, and that it fills the half-space $x_2 > 0$. We consider a wave that propagates in the x_1-direction, and has a displacement field with only a 3-component, that is a function of only x_1 and x_2. A solution of Eqs. (11)–(12) and the boundary conditions can be found for such a wave that represent a displacement field and electrostatic potential localized at the surface $x_2 = 0$. Such a localized solution also exists if it is assumed that the surface $x_2 = 0$ is coated with a thin metallic film that does not affect the vibrations of the medium but makes the plane $x_2 = 0$ an equipotential surface on which ϕ vanishes. The potential then remains identically zero for $x_2 < 0$. The shear horizontal surface acoustic wave obtained in this geometry for either set of boundary condition is called a *Bleustein-Gulyaev wave* [61, 62]. It was first observed experimentally by Maerfeld *et al.* [63]. The case of a metallized surface is the more interesting one, since the wave on a free surface penetrates so deeply into the substrate as to be of little practical use in signal processing applications [64].

Subsequently, Koerber and Vogel [65] described the criteria for the existence of Bleustein-Gulyaev waves in crystal classes other than C_{6v}, and they have now been studied theoretically in rhombic, and monoclinic [66–68] crystals.

In 1971 Maerfeld and Tournois [69] showed that a surface acoustic wave polarized perpendicular to the sagittal plane can be guided by the interface between two different semi-infinite media when at least one of these media is piezoelectric. This wave is an interface Bleustein-Gulyaev wave; the amplitude of this wave decreases exponentially with increasing distance from the interface into each medium. A Maerfeld-Tournois wave can exist also in the case that the interface between the two media is metallized.

2.3 Magnetic Media

Magnetic media have several formal similarities to piezoelectric media. Straining such a medium, for example, gives rise to a net magnetization in it; conversely, magnetizing the medium gives rise to a strain in it. Surface waves can exist on the surface of a magnetic medium. However, they have properties that are different from those of surface acoustic waves that exist on the surface of a piezoelectric. In particular, surface waves on magnetic media are *nonreciprocal*, i.e. the frequency for a surface wave with wave vector $\vec{k}_{\parallel}$ may be different from that of a surface wave with wave vector-$\vec{k}_{\parallel}$, while those on piezoelectric media are *reciprocal*, $\omega(-\vec{k}_{\parallel}) = \omega(\vec{k}_{\parallel})$. This is due, ultimately, to the difference between the axial vector nature of a magnetic field and the polar vector nature of an electric field. Surface waves have now been studied theoretically and experimentally on ferromagnetic, antiferromagnetic, and paramagnetic surfaces.

2.3.1 Ferromagnetic Surfaces

The dynamical properties of the spins in a ferromagnetic medium are determined by the magnetic dipole coupling between them and by the exchange coupling between them. Both are present simultaneously. However, there are situations in which either one interaction or the other predominates, or in which they play comparable roles. In particular, the dipolar interactions are dominant for long wavelength modes, while the exchange interaction is dominant for short wavelength modes.

When the dipolar coupling is large compared to the exchange coupling we are in the so-called *magnetostatic limit*. In ferromagnets magnetostatic surface waves have been studied both experimentally and theoretically in a wide variety of geometries. They were first studied for ferromagnetic spheres [70], and spheroids with a static external magnetic field applied along the axis of revolution [71]. Subsequently they were studied for a ferromagnetic slab of finite thickness [72], but by letting the thickness of the slab become infinite, solutions corresponding to a semi-infinite medium could also be obtained [73]. It was found that these surface waves, called Damon-Eshbach modes, are nonreciprocal. In the case of a semi-infinite medium, the frequency of a Damon-Eshbach mode is independent of the magnitude of the wave vector $\vec{k}_{\parallel}$, but does depend on its direction. Characteristically, their propagation is limited to a restricted set of directions. In the case of a slab, Damon-Eshbach modes on the upper and lower faces propagate in opposite directions.

The surface magnetostatic modes described in the preceding discussion were obtained with a static, external magnetic field applied parallel to the surface of the ferromagnet. When the applied magnetic field is applied perpendicular to the surface, no magnetostatic surface modes occur [74].

Surface modes that are guided by the interface between two magnetic materials [75, 76], or which are guided by two magnetic media separated by a gap [77, 78] have also been studied.

In the opposite limit, where the exchange interactions dominate the dipolar interactions, surface spin waves have been found to exist on the surface of a semi-infinite ferromagnet [79–81]. In contrast with magnetostatic surface waves, surface spin waves are reciprocal.

The situation in which both the dipolar and the exchange interactions have to be taken into account simultaneously in obtaining the frequencies and associated fields of surface waves on a ferromagnet has been treated in considerable detail by Wolfram and de Wames [82, 83], and by Mills and Benson [84]. These authors have considered the effects on these modes of changing the orientation of the externally imposed, static magnetic field, and of different boundary conditions for the magnetization at the surface, for samples in the form of a slab, as well as for semi-infinite media. Among the results they obtain is that when the applied magnetic field is oriented perpendicular to the surface of a ferromagnet a new type of surface mode appears: the dipole-exchange surface mode. This mode has no analogue in either the dipolar-coupled or exchange-coupled limits, and exists only in the transition region where the dipolar and exchange energies are comparable [85].

Surface modes on the surfaces of ferromagnetic media have recently been studied experimentally both by the technique of Brillouin scattering [86–94] and by ferromagnetic resonance measurements [95]. It appears as if the exchange interaction has to be included in the theory, along with the dipolar interaction, to obtain a good agreement with the experimental results [96–99].

The last type of surface wave on a ferromagnetic surface to be studied is the magnetoelastic surface wave. This is a surface wave that results from the coupling of Rayleigh waves to bulk spin waves and to the Damon-Eshbach surface waves through the magnetoelastic coupling of strain to the direction of magnetization in a magnetic medium. There has been a great deal of interest in such surface acoustic waves in recent years [100–111]. This interest arises from the possibility of using magnetoelastic surface wave propagation to study the bulk spin wave mode structure and the Damon-Eshbach surface waves with a probe of short wavelength, in contrast with microwave resonance studies, which probe only geometrical resonances of the entire sample [112–116]. This is due to the fact that the wavelength of Rayleigh waves, and hence its penetration depth, can readily be made small in comparison to the sample dimensions. Thus the Rayleigh wave can probe the nature of other elementary excitations near the surface of a crystal under circumstances where the medium can be regarded as semi-infinite. Also, if the Rayleigh wave couples to other low frequency excitations, e.g. bulk spin waves, its properties can then be altered, and thus manipulated (e.g. the attenuation rate, or dispersion, of the wave may be altered through the variation of external parameters, such as an external, static magnetic field).

The equations of motion of a semi-infinite magnetoelastic solid ($x_2 > 0$) are several in number. If a static, external magnetic field H_0 is applied along the x_3-axis, and the saturation magnetization M_s is also along this axis, the components of the transverse magnetization vector $\vec{m}$ obey the pair of equations

$$\left(\frac{\partial}{\partial t} + \frac{1}{\tau}\right) m_1 = \omega_H m_2 + \omega_M \frac{\partial \phi}{\partial x_2} - D\nabla^2 m_2 + \omega_M \sum_{\mu\nu} b_{2\mu\nu} \frac{\partial u_\mu}{\partial x_\nu} \tag{15a}$$

$$\left(\frac{\partial}{\partial t} + \frac{1}{\tau}\right) m_2 = -\omega_H m_1 - \omega_M \frac{\partial \phi}{\partial x_2} + D\nabla^2 m_1 - \omega_M \sum_{\mu\nu} b_{1\mu\nu} \frac{\partial u_\mu}{\partial x_\nu}, \tag{15b}$$

where τ is the transverse relaxation time for the spins in the system; $\omega_H = \gamma H_0$, $\omega_M = \gamma M_s$, with γ (< 0) the gyromagnetic ratio; ϕ is the magnetic scalar potential, from which the demagnetizing field $\vec{h}$ associated with spin motion is obtained according to $\vec{h} = -\nabla\phi$; D is the exchange constant (for a cubic medium); and the $\{b_{\alpha\mu\nu}\}$ are elements of the magnetoelastic coupling tensor. The equations of motion of the elastic subsystem are again given by Eq. (1), where the stress tensor $T_{\alpha\beta}$ now takes the form

$$T_{\alpha\beta} = \sum_{\mu\nu} C_{\alpha\beta\mu\nu} \frac{\partial u_\mu}{\partial x_\nu} + \sum_{\mu} b_{\mu\alpha\beta} m_\mu. \qquad (16)$$

Finally, one has the Maxwell equation

$$\nabla \cdot \vec{b} = 0, \qquad (17)$$

where $\vec{b} = h + 4\pi\vec{m}$ is the dynamic magnetic induction in the medium. In the vacuum outside the solid ($x_2 < 0$) only ϕ is nonzero, and it satisfies the Laplace equation $\nabla^2\phi = 0$. The boundary conditions at the surface $x_2 = 0$ are that ϕ must be continuous across this surface, together with b_2; the stress components $T_{\alpha 2}$ ($\alpha = 1, 2, 3$) must vanish at the surface; finally, the components of the transverse magnetization satisfy the conditions $((\partial/\partial x_2) - \lambda)\, m_\alpha = 0$ ($\alpha = 1, 2$) at $x_2 = 0$. Here λ is a phenomenological parameter that provides a measure of the strength of the surface pinning (magnetic) field that acts on the spins at the surface of a ferromagnetic medium and inhibits their motion. The solution of these equations of motion and the boundary conditions in general require the use of a computer.

Several types of magnetoelastic surface wave have been found on a semi-infinite substrate. In general, one has a Rayleigh wave somewhat modified due to its coupling to the spin subsystem through the magnetoelastic interaction. However, under special conditions one can obtain magnetoelastic surface waves that have no counterpart in a purely elastic system, and which owe their existence to the presence of the magnetoelastic interaction. For example, for propagation perpendicular to the saturation magnetization in the medium, and neglecting the exchange interaction (D = 0), a magnetoelastic wave polarized perpendicular to the sagittal plane can propagate [100, 101, 109], in addition to a Rayleigh wave that is uncoupled from the spin system. This wave is a magnetic analogue of the Bleustein-Gulyaev wave [61, 62] on the surface of a piezoelectric medium. The magnetoelastic surface wave is nonreciprocal, however.

Recently, it has been shown that a surface acoustic wave, polarized perpendicular to the sagittal plane, can be guided by the planar interface between two semi-infinite media in contact, either when one of the media is piezoelectric and the other is magnetoelastic [117], or when both are magnetoelastic [118]. The latter wave is the magnetoelastic analogue of the Maerfeld-Tournois wave [69] at the interface between two piezoelectric media. In both cases, the direction of propagation is perpendicular to the direction of the saturation magnetization in the system, and the waves are nonreciprocal. The interest in such waves is due to the fact that many nonreciprocal circuits [119] can be realized through their use.

Dispersion relations for surface waves of Rayleigh type in ferromagnetic, and antiferromagnetic, crystals with an easy-plane anisotropy have been determined by Gerus and Tarasenko [120]. It was assumed that the anisotropy axis is parallel to the crystal surface, and that the waves propagate in the basal plane. The magnetoelastic coupling in such crystals is found to decrease greatly the speed of these surface waves from that of standard Rayleigh waves. The amplification of these waves in a magnetic piezoelectric crystal adjacent to a semiconductor carrying a current is shown to be more easily achievable than the amplification of surface waves in a nonmagnetic piezoelectric crystal by the same method [121].

Surface magnetoelastic waves in a ferromagnetic film on a nonmagnetic substrate have been investigated by Camley [111], who finds Rayleigh waves and Love waves modified by the magnetoelastic interaction.

2.3.2 Antiferromagnetic Surfaces

Shortly after the discovery of surface spin waves on the surface of exchange-coupled ferromagnets [79, 80] surface spin waves (reciprocal) were discovered on the surface of two-sublattice, exchange-coupled antiferromagnets [122], and these modes were subsequently the objects of intensive studies [74].

An interesting way of studying properties of such modes was proposed by Maekawa [123]. It is based on the idea that a surface acoustic wave propagating along a surface deforms the crystal in the vincinity of the surface. This deformation can be written in terms of the symmetric (pure strains) and antisymmetric (pure rotations) parts of the displacement gradient tensor $v_{\alpha\beta} = \partial u_\alpha / \partial x_\beta$. The rotational part of the latter tensor couples more strongly to the spin subsystem in the magnetic crystal than does the pure strain part, because the strain components are small at the surface due to the stress-free boundary conditions. Thus, the surface acoustic wave can excite a surface magnon with the same wave vector, and be attenuated thereby. This attenuation is found to be observable in the vincinity of the surface spin-flop transition predicted by Mills and Saslow [122] in an antiferromagnet of the type of MnF_2, which affords a way of observing that transition.

Quite recently, magnetostatic surface waves on an antiferromagnet in which dipolar and exchange interactions coexist were discovered [124]. In this work the exchange interaction was treated in a mean field approximation, and the dipolar interaction in the magnetostatic approximation. These modes are nonreciprocal.

2.3.3 Paramagnetic Surfaces

Very recently surface acoustic waves on a new class of magnetic surface have been studied both theoretically and experimentally. These are surface acoustic waves on paramagnetic surfaces [125, 126]. Paramagnetic crystals are interesting in this context because their elastic moduli are strongly temperature dependent, and can be changed significantly by the application of an external, static magnetic field. This in turn gives rise to strong magnetic field and temperature dependences of the speeds of surface acoustic waves on such surfaces. These dependences arise from the magnetoelastic coupling between the elastic deformations in the crystal and the spins of the paramagnetic ions, and from the Zeeman interaction of these ions with the external magnetic field. In addition, in paramagnetic crystals the part of the crystal Hamiltonian that describes the magnetoelastic interaction contains *both* the symmetric and the antisymmetric components of the displacement gradient tensor (i. e. the pure strains and rotations, respectively), in order that the crystal Hamiltonian be rotationally invariant. Camley and Fulde [126] obtained the contribution of the magnetoelastic Hamiltonian to the Helmholtz free energy of the crystal by second order perturbation theory. From this result the stresses were obtained by differentiation with respect to the displacement gradients, and the equations of motion and boundary conditions follow from Eqs. (1) and (2), respectively. The solution of the resulting equations for the speeds of Rayleigh waves was carried out according to standard methods.

The presence of terms including rotations in the crystal Hamiltonian gives rise to interesting effects on the propagation of both bulk acoustic [127–130] and surface acoustic waves [125, 126]. For example, if the stress-free surface is the plane $x_2 = 0$, the speed of surface acoustic waves propagating in the x_3-direction with a magnetic field along the x_2-direction would be the same as the speed of surface waves propagating in the x_3-direction with the magnetic field along the x_3-direction, if the terms containing the pure rotations were omitted from the magnetoelastic Hamiltonian. The two speeds are different when the rotations are included in the Hamiltonian, and this difference has been observed in experiments on $CeAl_2$ [125] (Fig. 4). In Ref. [125] the temperature dependence of the speed of surface acoustic waves on $CeAl_2$ and SmSb was also measured and found to be in good agreement with theoretical predictions. The modes should be nonreciprocal, but the nonreciprocity should only become observable at surface acoustic wave frequencies higher than those for which the theory and experiments were carried out.

3 Power Flow in Surface Acoustic Waves

A practically important characteristic of surface acoustic waves is the power flow associated with them. This is given in terms of the energy flux vector $\vec{P}$ that has the dimensions of (energy/area × time), and in electromagnetic theory is called the *Poynting vector* [131]. In terms of this vector the differential form of the law of conservation of energy is given by

$$\frac{\partial U}{\partial t} + \nabla \cdot \vec{P} = -F, \tag{18}$$

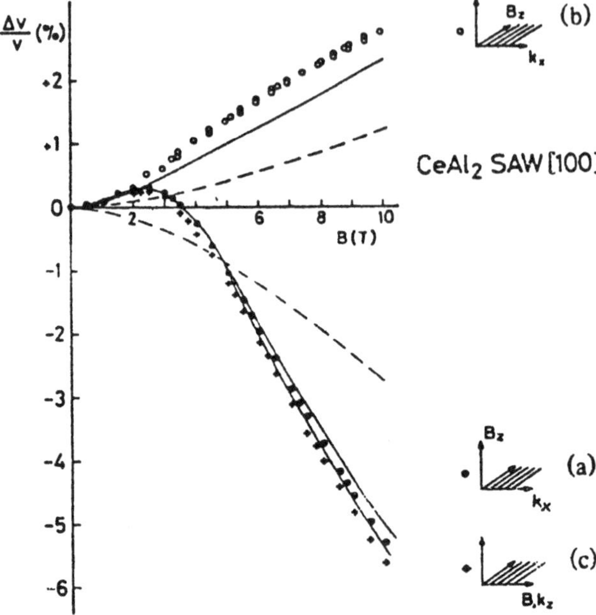

Fig. 4 The magnetic field dependence of the speed of Rayleigh waves propagating in the (100) direction on a (001) surface of $CeAl_2$. The difference between the speeds of Rayleigh waves in case (a), in which the external magnetic field is perpendicular to the surface and to the direction of propagation, and in case (b), in which the external magnetic field is parallel to the direction of propagation, is due to the terms containing pure rotations in the magnetoelastic interaction Hamiltonian for this crystal [125].

where U is the energy density in the wave and F is the rate at which energy is absorbed by dissipative processes in the system supporting the wave. The magnitude of $\vec{P}$ thus gives the energy crossing unit area perpendicular to the direction of $\vec{P}$ per unit time. A knowledge of $\vec{P}$ is useful in calculating cross sections for the scattering of bulk [132] and surface [133] elastic waves from point defects in solids, in the calculation of the efficiencies of second harmonic generation of Rayleigh waves [134], and in obtaining reflection and transmission coefficients for Rayleigh waves incident on geometrical and/or material discontinuities [135].

For purely elastic media $\vec{P}$ is given by [136]

$$P_\alpha = -\sum_\beta \dot{u}_\beta T_{\beta\alpha} \qquad \alpha = 1, 2, 3, \tag{19}$$

where the stress tensor $T_{\alpha\beta}$ is given by Eq. (2). For piezoelectric media $\vec{P}$ is given by [137]

$$P_\alpha = -\sum_\beta \dot{u}_\beta T_{\beta\alpha} - \dot{\phi} D_\alpha \quad \alpha = 1, 2, 3, \tag{20}$$

where the stress tensor is now given by Eq. (9), and the displacement vector $\vec{D}$ is given by Eq. (10). Finally, for magnetoelastic media, $\vec{P}$ is given by [138]

$$P_\alpha = -\sum_\beta \dot{u}_\beta T_{\beta\alpha} - \frac{D}{\omega_M} \sum_\beta \dot{m}_\beta \frac{\partial m_\beta}{\partial x_\alpha} - \frac{1}{4\pi} \dot{\phi} b_\alpha, \tag{21}$$

where the stress tensor in this case is given by Eq. (16).

The time averaged power flow is conveniently expressed by

$$\langle P_\alpha \rangle = \mathrm{Re} P_\alpha^c, \tag{22}$$

where $\vec{P}^c$ is the complex energy flux vector. It is obtained by multiplying each of Eq. (19)–(21) by $\frac{1}{2}$, and replacing each of the time derivatives by its complex conjugate.

The preceding results apply to bulk waves as well as to surface waves. For surface wave applications, however, it is usually not the power flow per unit area that is of greatest interest. Instead, because the wave is localized to the surface, it is the time average of the total power carried parallel to the surface by the surface wave, per unit width normal to the direction of propagation. (There is no power flow normal to the surface for non-leaky surface waves). Thus, if the surface is the plane $x_3 = 0$, and propagation is in the x_1-direction, the latter quantity is given by

$$W_1 = \int_{-\infty}^{\infty} \langle P_1 \rangle \, dx_3 . \tag{23}$$

It is sometimes of interest to write W_1 as a sum of the contributions from the medium ($x_3 > 0$) and from the vacuum in contact with it ($x_3 < 0$), as in the case of surface acoustic waves on semi-infinite piezoelectric and magnetoelastic media:

$$W_1 = \int_0^{\infty} \langle P_1 \rangle \, dx_3 + \int_{-\infty}^{0} \langle P_1 \rangle \, dx_3 \equiv W_1^{>} + W_1^{<}. \tag{24}$$

The relative amounts of energy carried in the medium and in the vacuum can be determined from $W_1^{>}$ and $W_1^{<}$.

4 Attenuation of Rayleigh Waves

As Rayleigh waves propagate along solid surface they are attenuated. The two general types of mechanisms responsible for this attenuation can be described as *intrinsic* and *extrinsic.*

4.1 Intrinsic Attenuation Mechanisms

In insulating crystals the principal intrinsic mechanism limiting the mean free path of Rayleigh waves is anharmonicity, particularly that due to the cubic terms in the expansion of the crystal's potential energy in powers of the displacements of the atoms from their equilibrium positions which, in the continuum limit, correspond to the leading (cubic) nonlinear terms in the potential energy density of an elastic medium. In the limit $\omega_R \tau \gg 1$, where ω_R is the frequency of the Rayleigh wave and τ is a characteristic relaxation time for the thermal phonons in the solid, this mechanism has been studied from the lattice point of view by Maradudin and Mills [139]. They found the attenuation rate for Rayleigh waves to be proportional to $\omega_R T^4$, where T is the absolute temperature. This frequency and temperature dependence was confirmed in subsequent theoretical work, by King and Sheard [140], and by Sakuma and Nakayama [141], and the latter authors have obtained the leading corrections to the $\omega_R T^4$ dependence found in Refs. [139] and [140]. This dependence is in accord with the measurements of Salzmann *et al.* [142] and of Budreau and Carr [143], who verified the $\omega_R T^4$ dependence at temperatures below 40 K. Above 60 K the attenuation rate becomes independent of temperature and varies as ω_R^2. The latter dependence can be understood on the basis of the viscosity theories of Maris [140] and of King and Sheard [145], that are valid in the limit $\omega_R \tau \ll 1$. Thus, at least in insulating crystals, it appears as if one has a good understanding of the intrinsic mechanism responsible for the attenuation of Rayleigh waves.

A very simple approach to Rayleigh wave attenuation has been taken by Press and Healy [146]. They start from the equation (Eq. (6)) expressing the speed of Rayleigh waves for an isotropic medium in terms of the speeds of bulk longitudinal and transverse waves, c_l and c_t, respectively, and introduce dissipation by letting each speed become complex, $c_j \to c_j(1 + i\delta_j)$ ($j = R, l, t$). Then defining $\alpha_j = \omega\delta_j/2c_j$, where ω is the frequency of the Rayleigh wave, they obtain a relation between the attenuation coefficient for Rayleigh waves, α_R, and those for the longitudinal and transverse bulk waves, α_l and α_t, in the form

$$\alpha_R = (A_l\alpha_l + A_t\alpha_t)/D, \tag{25}$$

where

$$A = 4(1 - (c_R^2/c_t^2))\,(c_t^2/c_l^2)\,(c_l/c_R),$$
$$A_t = [4(1 - (c_R^2/c_l^2)) - (2 - (c_R^2/c_t^2))^3]\,(c_t/c_R) \quad \text{and}$$
$$D = 4(1 - (c_R^2/c_t^2))\,(c_t^2/c_l^2) + 4(1 - (c_R^2/c_l^2)) - (2 - (c_R^2/c_t^2))^3.$$

The relation given by Eq. (25) seems to be reasonably well obeyed by many non-crystalline solids that in first approximation are isotropic.

4.2 Extrinsic Attenuation Mechanisms

Rayleigh waves can also be attenuated by extrinsic mechanisms. Since a Rayleigh wave is localized in the vicinity of the solid surface over which it propagates, a particular interest attaches to extrinsic attenuation mechanisms specific to the near vicinity of the surface. The damping of Rayleigh waves by density fluctuations in the surface of a solid has been studied by Nakayama and Sakuma [147], and their damping by point defects near the surface was first investigated by Steg und Klemens [148]. The latter authors found that the attenuation rate varies as ω_R^5. Subsequent studies of the attenuation of Rayleigh waves by point defects at or near the surface of a solid have been carried out by Sakuma [149], by Dacol and Zimerman [150], and by Wallis *et al.* [133]. Unfortunately, the calculations of Sakuma and of Dacol and Zimerman contains errors that invalidate some of their results [133, 151].

A more important extrinsic mechanism for the attenuation of Rayleigh waves is random surface roughness, which is found in varying degrees on all solid surfaces, even carefully prepared ones. The first study of this attenuation mechanism was carried out by Urazakov and Fal'kovskii [152]. In a subsequent work, Maradudin and Mills [153] used a rather different approach to solve this problem. Both sets of authors find that the attenuation rate is proportional to ω_R^5, in the limit that the wavelength of the Rayleigh wave is longer than the mean distance between consecutive peaks and valleys on the surface. The frequency variation of the attenuation rate then becomes much slower than ω_R^5 as the wavelength of the Rayleigh waves becomes comparable to, or shorter than, this characteristic distance. The most important attenuation mechanism associated with surface roughness turns out to be the roughness induced scattering of the Rayleigh wave into other Rayleigh waves with the same frequency, but with different wave vectors $k_{\parallel}$: scattering of the Rayleigh wave into bulk waves proves to be a less important attenuation mechanism. This problem has been further studied in two recent papers by Eguiluz and Maradudin [154, 155], on the basis of new computational methods.

To my knowledge, the ω_R^5 dependence of the attenuation rate for Rayleigh waves due to surface roughness has not yet been verified experimentally.

Not all rough surfaces are randomly rough, however. When surface roughness is due to a periodic grating ruled on the surface, the attenuation rate of a Rayleigh wave due to its scattering into bulk waves was first calculated by Brekhovskikh [156], whose result was subsequently corrected by Sabine [157]. The corrected results are in quite good agreement with the experimental results of Rischbieter [158] for the attenuation of Rayleigh waves by a symmetric sawtooth profile on the surface of aluminium (Fig. 5). Additional experimental results have been presented by Humphryes and Ash [159].

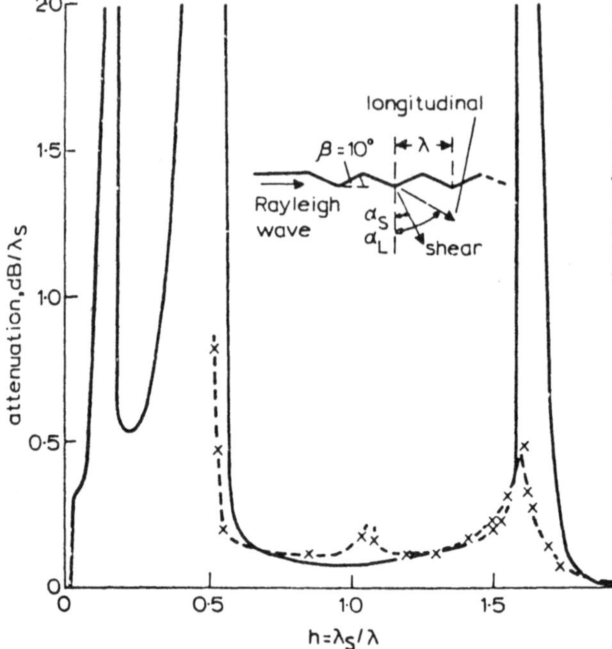

Fig. 5
Experimental (crosses) and theoretical (full line) results for the attenuation of Rayleigh waves propagating across a sawtooth profile grating ruled on an aluminium surface. λ and λ_s are the period of the grating and the wavelength of shear waves, respectively [157].

5 Propagation of Rayleigh Waves on Curved Surfaces

Up to now we have considered the propagation of surface acoustic waves only along planar surfaces or interfaces. Interesting results are obtained when the restriction to planar surfaces is lifted, and the surface is allowed to be curved. In particular, the frequencies of Rayleigh waves propagating over curved surfaces become dispersive. However, in addition to Rayleigh waves, a curved (convex) boundary can support other types of waves, localized to the curved boundary. Such surface waves were first studied, in the context of a problem in acoustics, by Lord Rayleigh [160], and today are sometimes called "whispering gallery" waves.

Subsequently, Rayleigh waves on the surface of a cylinder and of a sphere have been studied by several authors [161], since these are the only structures, apart from a semi-infinite medium, for which the equations of motion of elasticity, and the appropriate boundary conditions can be solved exactly. To obtain results for other surfaces one has to resort to approximate methods of solution, but perhaps more interesting results are obtained. In this section I consider surface acoustic wave propagation for three types of curved surfaces: smoothly curved surfaces in one or two dimensions; randomly rough surfaces; and surfaces with periodic, one-dimensional gratings ruled on them.

5.1 Cylindrical and Spherical Surfaces

The propagation of Rayleigh waves has been studied on general, smoothly curved surfaces, at every point of which there are either one or two principal radii of curvature. The former I will call *cylindrical surfaces*, the latter I will call *spherical surfaces.*

The propagation of high frequency elastic surface waves down a homogeneous, isotropic cylinder of circular cross section has been studied by Hudson [162]. Subsequently, Rosenfeld and Keller [163] studied the propagation of such waves down homogeneous, isotropic rods of arbitrary cross section. They considered modes whose displacements are confined to a boundary layer near the surface of the rod, and found that to lowest order these waves travel along helices with the speed of Rayleigh waves on a flat surface. They obtain the leading correction to this speed, which is of the order of the reciprocal of the wavelength of the wave.

The propagation of high frequency elastic surface waves down a homogeneous, isotropic cylinder that has a cross section boundary of nonconstant curvature, that may be open or closed, has been studied by Wilson and Morrison [164]. The boundary surface is stress-free, and Rayleigh wave-type solutions are sought. They find that in the case of an open boundary, for which the curvature reaches its algebraic maximum at a single point, modes exist in which the displacement field is essentially confined to the neighborhood of the point of maximum curvature, as well as to the neighborhood of the surface. These waves are quite similar to the elastic and piezoelectric edge modes that will be discussed in section 8. In the case of a closed boundary for which the curvature reaches its algebraic maximum at a single point it is argued that surface wave modes exist that differ by only exponentially small quantities from those obtained for the corresponding open boundary case. For closed boundaries that are symmetric and for which the curvature achieves its algebraic maximum at two points, two Rayleigh type modes are found.

The case of Rayleign waves propagating circumferentially around a cylinder of circular cross-section of radius R with a stress-free boundary has been investigated by several authors. Viktorov [165] and Rulf [166] have investigated the propagation of Rayleigh waves on convex and concave cylindrical surfaces in detail. It is found that Rayleigh-type surface modes exist on a convex cylindrical surface. These waves are dispersive, i.e. their speed depends on the radius of the cylinder. They go over into Rayleigh waves on a semi-infinite medium bounded by a planar surface when the radius of the cylinder becomes infinite.

Rulf [166] shows, however, that there is a critical or cut off wavelength above which no proper Rayleigh wave can exist. Thus, a Rayleigh wave that propagates on a surface of variable curvature toward a point of minimum radius of curvature may be scattered and partially transformed into bulk waves.

In the case of a concave cylindrical surface, the Rayleigh-type of surface wave is a leaky wave i.e. it radiates energy radially into the medium, and therefore propagates with attenuation. The leakage decreases with increasing R. The absorption coefficient γ due to surface wave radiation into the medium over the Rayleigh wavelength λ_R on a concave cylindrical surface has been calculated by Viktorov [167] as a function of R/λ_R and found to be in good agreement with his experimental results for Dural [168].

Circumferential Rayleigh waves propagating over the surface of a convex circular cylinder surrounded by a liquid have been studied by Grace and Goodman [169], by an extension of Viktorov's method.

In contrast with these Rayleigh-type circumferential surface waves on a cylindrical surface are the analogues of the whispering gallery waves that have been studied by several authors [170–172]. The discussion of these waves by Brekhovskikh [173] is particularly lucid. The simplest case consists of a wave polarized parallel to the generator of a right circular cylinder that propagates circumferentially around it. The resulting wave motion can be described in terms of rays propagating along the boundary and suffering successive reflections from it (Fig. 6a). An infinite number of such modes exist, differing from each other by the values of the glancing angle their rays make with the boundary. These modes are exactly equivalent to surface waves on a *flat* surface in which the speed of bulk transverse waves c_t decreases linearly in the vicinity of the surface to a value smaller at the surface than it has in the bulk (Fig. 6b). In this case they are essentially Love waves [39]. Similar, but more complicated, whispering gallery modes are found for circumferential waves polarized in the plane perpendicular to the generator of the cylinder [173]. Such modes also exist on a flat surface with speeds of bulk transverse and longitudinal waves that vary with the coordinate normal to the surface. In this case they are essentially Sezawa waves [47]. The modes for both polarization have frequencies that are dispersive, i.e. depend on the radius of the cylinder, and are not leaky modes.

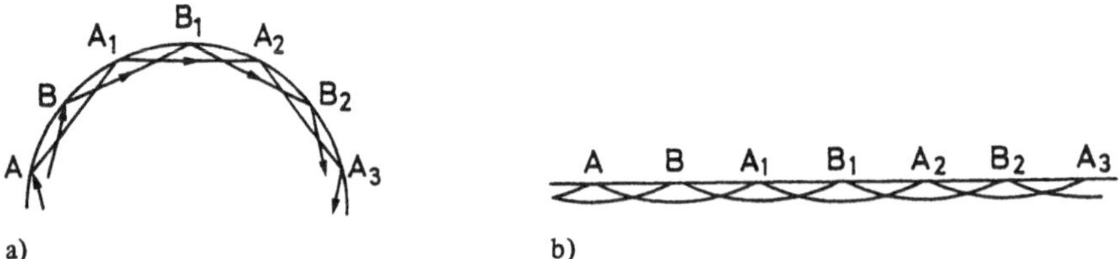

Fig. 6 (a) A schematic depiction of the propagation, circumferentially around a cylinder, of whispering gallery waves polarized perpendicular to the sagittal plane. Drawn are waves propagating along the boundary and suffering successive reflections from it. (b) The ray pattern for the analogues of whispering gallery waves polarized parallel to the sagittal plane and propagating along the plane surface of a medium in which the speed of transverse (shear) wave increases linearly with increasing distance into the medium from the surface [173].

Rayleigh-type waves on a spherical surface have been studied by Sezawa [174] and Petrashen' [175], and a discrete spectrum of such waves found. Whispering gallery modes in a convex spherical medium have been investigated by Brekhovskikh [173]. The main characteristics of these waves prove to be the same in the spherical case as in the cylindrical.

The similarity of the elastic whispering gallery modes to the electronic orbits that skip along the surface of a metal by periodic specular reflection in the presence of a magnetic field parallel to the surface [176–185] should be noted.

5.2 Randomly Rough Surfaces

It has been pointed out in section 4.2 that when a Rayleigh wave propagates along a randomly rough planar surface it is attentuated due to roughness-induced scattering into bulk waves and into other Rayleigh waves. However, a Rayleigh wave propagating along a randomly rough surface not only is attenuated but its frequency changes as well. This was pointed out by Urazakov and Fal'kovskii [152], but these authors did not study the frequency shift, nor was it considered by Maradudin and Mills [153]. The frequency shift has recently been obtained by two different methods by Eguiluz and Maradudin [154, 155] to $O(\delta^2/a^2)$, where δ is the root mean square departure from flatness, and a is the mean distance between consecutive peaks and valleys on the surface. One approach is based on Ryleigh's method [186], in which the surface localized elastic displacement field in the solid obtained from the equations of motion in the region beyond the maximum amplitude of the surface profile is continued in to the rough surface itself, and used in satisfying the stress-free boundary conditions along this surface. In the second [155] the stress-free boundary conditions on the displacement field at the actual rough surface are replaced by effective boundary conditions at the plane $x_3 = 0$ for the displacement field averaged over the ensemble of realizations of the surface profile. Both approaches ultimately yield the same results. The frequency shift is small, for reasonable values of δ/a ($\lesssim 0.05$), but should be measurable with present-day experimental capabilities.

5.3 Gratings

The propagation of Rayleigh waves across periodically corrugated surfaces has been the object of a great deal of theoretical and experimental interest. The reasons for this are several. Periodically corrugated surfaces, because they are easier to treat than randomly rough surfaces, are sometimes used to obtain physical insights about properties of the latter that are difficult to obtain otherwise [187]. A more practical reason for this interest can be found in the possibility of device applications for Rayleigh waves propagating across periodic structures. The periodicity of the surface structure causes the conversion of some of the energy in the Rayleigh wave into longitudinal and transverse bulk waves, in a controllable fashion, so that such structures can serve as surface-bulk wave transducers. Also, it is a characteristic of periodic structures in general that they possess stop bands for the waves that propagate through them. Consequently, periodically corrugated surfaces can be used in the design of Rayleigh wave filters.

Datta and Hunsinger [188] have investigated theoretically the propagation of Rayleigh waves across a periodic array of metallic strips overlaid on an elastic substrate, and this paper contains references to earlier work on this type of problem, or on the propagation of Rayleigh waves across a periodic array of shallow grooves.

The studies just mentioned were all carried out on the basis of perturbation theory, similar to that used when the surface is randomly rough. Recently, Glass *et al.* [189] have studied the propagation of Rayleigh waves across large amplitude gratings on an isotropic substrate using both Rayleigh's method, and the extinction theorem form of Green's theorem [190]. They have compared their results with those obtained by first order perturbation theory and have determined where the latter breaks down. Gaps are found in the Rayleigh wave dispersion curve at values of its wave vector corresponding to the boundaries of the one-dimensional Brillouin zones defined by the periodicity of the grating (Fig. 7).

Surface acoustic waves in periodic, layered piezoelectric media have been studied by a novel method by Datta and Hunsinger [191]. The boundary conditions are incorporated into the field equations by the use of position-dependent material con-

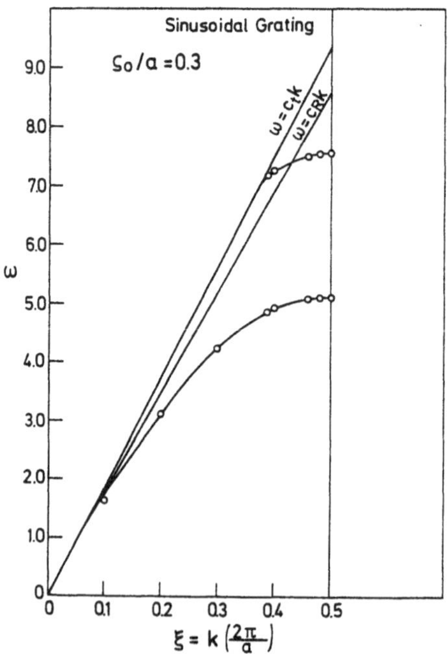

Fig. 7

The dispersion curves in the one-dimensional first Brillouin-zone for Rayleigh waves propagating perpendicular to the grooves of a sinusoidal grating whose profile is given by $\zeta(x_1) = \zeta_0 \cos(2\pi x_1/a)$, with $\zeta_0/a =$ $= 0.3$. The values of the speeds of bulk longitudinal and transverse waves used are $c_\ell = 5 \times 10^3$ m sec^{-1} and $c_t = 3 \times 10^3$ m sec^{-1}, respectively, and ω is measured in units of $(a \times 10^3$ rad m sec$^{-1}.)^{-1}$ [189].

stants [192–199]. The field quantities (elastic displacement components and electrostatic potential) are expanded in an orthonormal series of Laguerre functions in the coordinate perpendicular to the surface, and in a Fourier series in the coordinate along the direction of propagation. The results obtained are in good agreement with experimental data.

Surface acoustic waves polarized perpendicular to the sagittal plane do not exist on a planar elastic surface. Such waves can exist, however, on a grating. They were predicted and observed recently by Auld *et al.* [200] for a flat surface on which deep grooves has been cut periodically (Fig. 8). It will be of interest to study such grating-trapped surface acoustic waves on gratings with other profiles.

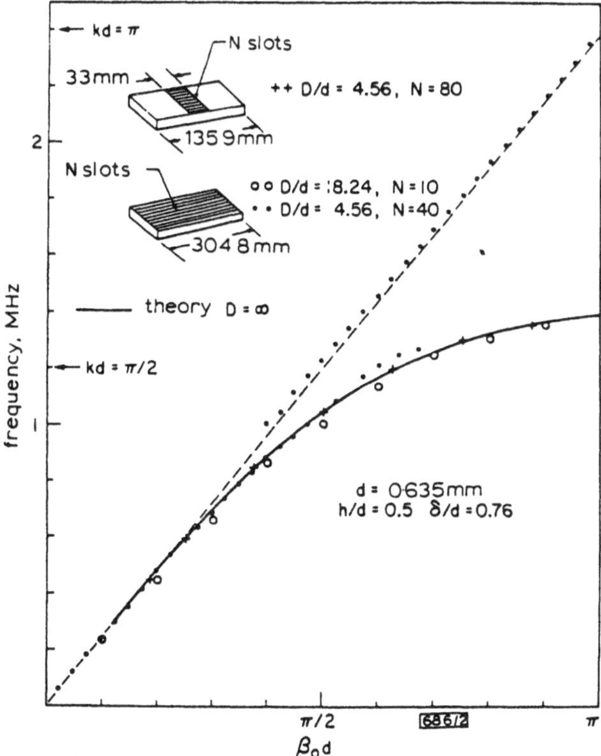

Fig. 8 Comparison of theory and experiment for the dispersion curve of surface acoustic waves polarized perpendicular to the sagittal plane and propagating perpendicular to the grooves of a grating obtained by cutting equally spaced, parallel slots into the surface of an aluminium plate [200].

6 Refraction of Surface Acoustic Waves by Geometrical and Material Discontinuities

The reflection and transmission of Rayleigh waves at a 90° corner, or at the edge of a plate is a problem of interest both in geophysics and, more recently, surface acoustic wave technology. The case of a single corner has been treated by a variety of approximate methods [201–204]. In practice, however, surface acoustic wave devices consist of plates with thicknesses in the 10–100 acoustic wavelength range. The reflection of a Rayleigh wave, incident along the top surface, from the end of an isotropic plate of finite thickness has recently been studied by an expansion of the elastic displacement field in a semi-infinite plate with a stress-free boundary in terms of the normal modes of an infinite extended plate of the same thickness [135]. The coefficients in the expansion were determined from the requirement that the end of the plate be stress-free. This condition was satisfied at a finite number of points on the end of the plate, whose number was increased until energy was conserved to within 1 %. Satisfaction of the boundary conditions at the end of the plate required the inclusion of both propagating and evancescent (decaying along the $-x_1$-direction, away from the plate end) modes. This means that part of the incident Rayleigh wave is converted into reflected "bulk" plate modes, while the remaining surface wave energy travels around both corners and leaves along the bottom surface (Fig. 9).

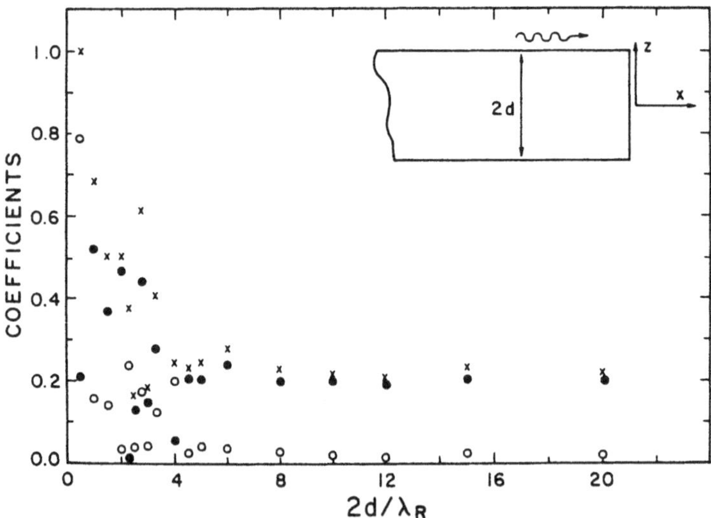

Fig. 9 Reflection (o), transmission (•) and sum of reflection and transmission (x) coefficients, for the interaction of a Rayleigh wave with the end of a plate, plotted as a function of the thickness of the plate. λ_R is the wavelength of the Rayleigh wave. The reflection coefficient is defined as the fraction of the energy in the incident Rayleigh wave that returns in the form of a Rayleigh wave along the upper surface; the transmission coefficient is the fraction of the energy that returns in the form of a Rayleigh wave along the lower surface [135].

Experimental results for the reflection of Rayleigh waves from 12–170° corners and from rounded corners have been presented by Viktorov [205, 206]. Rounding the corner decreases the energy in the incident Rayleigh wave converted into bulk waves.

The reflection, transmission, and bulk wave scattering of Bleustein-Gulyaev waves by an electrode of finite thickness and finite length, deposited on a piezoelectric surface has been determined exactly by Curtis and Redwood [207]. The surface region outside the electrode is assumed to be metallized. Subsequently, these authors have presented an approximate method for obtaining the reflection coefficient of a surface acoustic wave from a surface step, and have applied it to the reflection of Rayleigh and Bleustein-Gulyaev waves from a step [208].

7 Nonlinear Interactions of Rayleigh Waves

All elastic solids possess elastic nonlinearities to some degree. These manifest themselves through terms of third and higher order in the expansion of the elastic strain energy density in powers of the finite strain parameters. These nonlinear terms lead to interactions between the otherwise independent solutions of the equations of motion of a solid that are obtained when only the linear stress-strain relation, Eq. (2), is employed. We have already seen one consequence of these interactions for surface acoustic waves, in section 4.1: the attenuation of Rayleigh waves. Additional nonlinear effects that have their origin in the nonlinear terms in the stress-strain relation are second, third, ... harmonic generation of Rayleigh waves, and nonlinear mixing of two, or more, Rayleigh waves.

Interest in nonlinear effects associated with surface acoustic waves arises for several reasons. Because the acoustic energy density is high in these surface waves, nonlinear effects are important even at the relatively small acoustic power levels that can be realized in practice. A quantitative knowledge of surface wave nonlinearities is necessary for the design of surface wave devices [209]. At the high power density encountered in signal processing applications, harmonic generation leads to increased propagation losses [210], and sets limits to the efficiency of surface wave delay lines. Furthermore, nonlinear mixing of surface waves has been proposed for the switching and steering of surface acoustic waves [211], and also for high frequency convolution and correlation [212, 213]. Thus the calculation of surface wave nonlinearities is important for the complete evaluation of devices used in surface acoustic wave technology.

The second harmonic generation of Rayleigh waves was first discussed theoretically by Viktorov [214]. Experiments on second harmonic generation of surface acoustic waves were first reported by Rischbieter [158, 215] for aluminium and steel surfaces. A number of other experimentalists have subsequently studied harmonic generation and nonlinear parametric mixing of surface acoustic waves, in a variety of materials, mostly piezoelectric [216–232].

In this section we confine our attention to second harmonic generation of Rayleigh waves caused by the third order elastic nonlinearities, in a semi-infinite elastic medium, with a planar, stress-free boundary.

The problem to be solved can be formulated in the following way. The equations of motion of an elastic medium occupying the upper half-space ($x_3 > 0$), including the leading nonlinear terms can be written in the form [233]

$$\rho_0 \frac{\partial^2 u_\alpha}{\partial t^2} = \sum_{\beta\mu\nu} C_{\beta\alpha\mu\nu} \frac{\partial^2 u_\mu}{\partial x_\nu \partial x_\beta} + \sum_{\beta\mu\nu\rho\sigma} C^{(f)}_{\beta\alpha\mu\nu\rho\sigma} \frac{\partial^2 u_\mu}{\partial x_\nu \partial x_\beta} \frac{\partial u_\rho}{\partial x_\sigma}, \tag{26}$$

where ρ_0 is the mass density of the undeformed medium, and

$$C^{(f)}_{\beta\alpha\mu\nu\rho\sigma} = C_{\beta\alpha\mu\nu\rho\sigma} + C_{\beta\nu\rho\sigma}\delta_{\alpha\mu} + C_{\beta\alpha\sigma\nu}\delta_{\mu\rho} + C_{\beta\sigma\mu\nu}\delta_{\alpha\rho}. \tag{27}$$

The $\{C_{\beta\alpha\mu\nu\rho\sigma}\}$ are the third order elastic moduli. In addition to the decaying wave boundary conditions at $x_3 = +\infty$, the displacement field satisfies stress-free boundary conditions at the plane $x_3 = 0$. These can be written inthe form

$$T_{3\beta}|_{x_3=0} = 0 \quad \beta = 1, 2, 3, \tag{28}$$

where

$$T_{\alpha\beta} = \sum_{\mu\nu} C_{\alpha\beta\mu\nu} \frac{\partial u_\mu}{\partial x_\nu} + \sum_{\mu\nu\rho\sigma} C^{(s)}_{\alpha\beta\mu\nu\rho\sigma} \frac{\partial u_\mu}{\partial x_\nu} \frac{\partial u_\rho}{\partial x_\sigma}, \tag{29}$$

with

$$C^{(s)}_{\alpha\beta\mu\nu\rho\sigma} = \frac{1}{2} C_{\alpha\beta\mu\nu\rho\sigma} + \frac{1}{2} C_{\alpha\beta\sigma\nu}\delta_{\mu\rho} + C_{\sigma\beta\mu\nu}\delta_{\alpha\rho} + C_{\alpha\sigma\mu\nu}\delta_{\beta\rho} - C_{\alpha\beta\mu\nu}\delta_{\rho\sigma}. \tag{30}$$

It is the fact that the boundary conditions are nonlinear, as well as the equations of motion, that introduces the difficulties into the determination of the second harmonic wave.

There have been several theoretical studies of second harmonic generation of surface acoustic waves since the work of Viktorov. Løpen [219] analyzed his data on second harmonic generation in α-quartz in terms of theoretical calculations based on determining the solution for the second harmonic outside the nonlinear forcing region and identifying it with the solution just inside the interaction region. He obtained reasonable agreement between his experimental and theoretical results. Tiersten and Baumhauer [234] treat the nonlinear interaction directly, by an iterative approach, solving the iterated nonlinear equations of motion and boundary conditions asymptotically for large x_1 (propagation distance). Vella *et al.* [233] studied the problem by two different approaches; first, by evaluating the work done per unit distance by the nonlinear forces and surface stresses on a normal mode surface acoustic wave, and assuming that the work is totally converted into surface wave energy; and second by an iterative solution of the nonlinear equations of motion and the associated boundary conditions. Ljamov *et al.* [235] treated the nonlinearities by means of the secular equation, but did not include the nonlinear effects of the second order (linear) elastic moduli. Portz [236] used a Green's function approach to the problem, but his work is now known to contain errors. Several other workers [221, 227, 237] have presented treatments that are phenomenological in nature; however, their phenomenological parameters have not been related yet to the appropriate elastic moduli.

On the assumption that the initial Rayleigh wave is propagating in the x_1-direction, each of the preceding theoretical approaches yields the results that the amplitude of the generated second harmonic surface acoustic wave grows linearly with x_1. This is a consequence of the fact that the nonlinear interaction giving rise to second harmonic generation is phase-matched, due to the linear dispersion relation for Rayleigh waves, $\omega = c_R k$.

In recent work Normandin *et al.* [238, 239] have re-examined the problem of the second harmonic generation of surface acoustic waves and have shown that the complete solution for the second harmonic waves includes, in addition to a contribution that grows linearly with x_1, a contribution that grows linearly with x_3. The same conclusion has been reaches by Camley *et al.* [240] on the basis of a Green's function treatment that corrects the work of Portz [236].

The efficiency of second harmonic generation (essentially the ratio of the power in the second harmonic wave to the square of the power in the incident Rayleigh wave both per unit width of the Rayleigh wave) has been tabulated by Normandin *et al.* [238] for a large number of crystal surfaces.

8 Edge Modes

Just as significant advances in the study of elastic waves in continuous media accompanied the reduction in the dimensionality of the wave vector characterizing their propagation from three to two, so have advances accompanied the reduction of the dimensionality of the wave vector from two to one.

In 1972, edge modes were discovered independently and virtually simultaneously by Lagasse [241] and by Maradudin *et al.* [193]. These modes are vibrational modes that propagate in a wavelike fashion along the apex of a wedge formed by the intersection of two plane faces of an elastic, piezoelectric, or magnetic medium, and whose displacement amplitudes decay exponentially with increasing distance into the wedge from its apex and faces (Fig. 10). (In the case of piezoelectric wedges the amplitudes of the components of the accompanying electric field also decay exponentially with increasing distance from the apex and faces into both the wedge and the region outside it. The same is true for the magnetic field in the case of a dipole-coupled ferromagnet). The speed of propagation of these modes is slower than that of Rayleigh surface waves on the faces of the medium. Moreover, be-

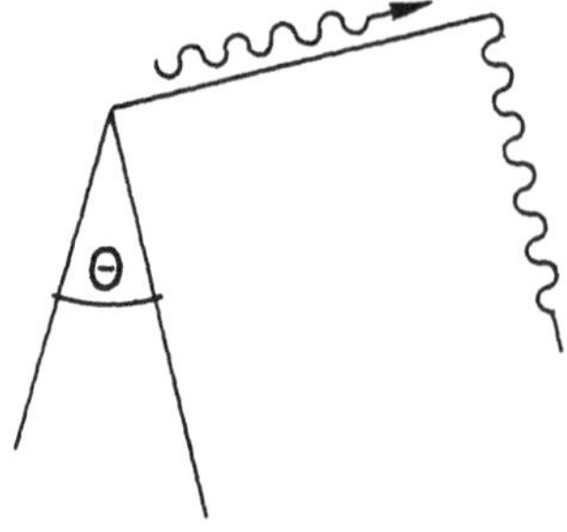

Fig. 10
A schematic depiction of a wedge-shaped topographic wave guide along which edge modes can propagate.

cause there is no characteristic length associated with a wedge, edge modes are dispersionless, i.e. their speed of propagation is independent of the one-dimensional wave vector q characterizing their propagation.

The material edge that supports edge modes is a particular type of wavequide for surface acoustic waves – a wedge topographic wave guide. Waveguides for surface acoustic waves have been discussed in some detail in a recent review by Oliner [242], and edge modes were reviewed somewhat earlier [197, 243, 244], so the emphasis here will be on more recent work.

The interest in waveguides derives from several considerations. Their use in devices overcomes the most important difficulty associated with wide surface wave beams [242], which is beam spreading, that causes cross talk between neighboring beams. This is because the field cross section associated with the waveguide remains constant as the wave progresses. Waveguides offer the possibility of guiding acoustic waves along the plane of the surface of a medium on which the waveguide has been constructed, without lateral diffraction, even along curved paths. The guiding action in the case of a wedge guide comes about because by decreasing the apex angle of the wedge the frequencies of the associated edge modes can be made much lower than those of Rayleigh surface waves on the medium which, in turn, are lower than those of bulk waves. Thus, in the absence of anharmonic coupling, edge modes cannot leak their energy into other vibrational modes of the system. In addition, the acoustic fields are strongly confined to the apex of the wedge, and so extend very little into the substrate on either side of the wedge. The minimum radius of curvature through which an acoustic ridge guide fabricated from a wedge can turn for a given radiation loss is determinated by a combination of the ratio of the speeds of edge and Rayleigh modes, and the mechanical isolation between the wedge shaped ridge and the substrate [245]. A small value of the former and a large degree of the latter ensure strong guiding, and both are achievable through the use of edge modes.

Moreover, because of the high acoustic field density near the tip of a small angle wedge, associated with the edge modes in such a structure, the absence of dispersion from these modes, and their low speeds, such a structure is a good candidate for devices that employ nonlinear interactions of acoustic waves [143, 246].

Several methods have been used to obtain the frequencies of edge modes and the associated displacements. The finite element method [247] used by Lagassė [241] has the desirable feature that the effects of the substrate on which the wedge is constructed can be incorporated into these calculations. More accurate results can be obtained by the mapped orthogonal functions method [193–197] which, however, has been applied only to the ideal infinite wedge. For a wedge on a substrate almost all the energy resides within the wedge portion at higher frequencies, and the structure behaves as an ideal infinite wedge [248]. The dispersion curve is then a constant, independent of wave vector. As the frequency is lowered, the mode becomes less localized, some of the energy spreads into the substrate and the speed increases. This behavior has been confirmed theoretically and experimentally [249]. Plate theory [250] and variational methods [251] have also been employed in the study of this problem. All of these calculations predict a multimode structure for the solutions, with more and more modes appearing as the angle of the wedge is decreased, whose speeds are below the speed of Rayleigh waves. For $\theta \gtrsim 60°$ only one edge mode exists for an isotropic wedge; there are six edge modes for a 20° wedge, and the number has increased to nine for a 15° wedge [194, 195] (Fig. 11). An analytic solution has been claimed for a 90° isotropic wedge [252]. However, the

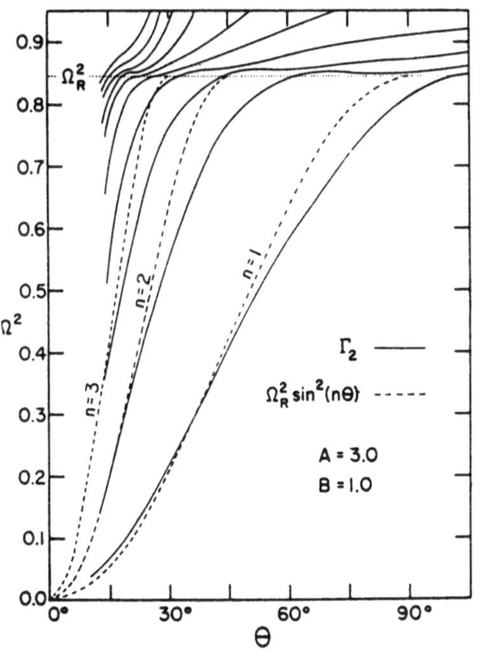

Fig. 11
The square of the speed of propagating of edge modes (in units of the speed of bulk transverse acoustic waves) of Γ_2 symmetry for an elastically isotropic wedge as a function of the wedge angle Θ. Also shown is the empirical relation $\Omega^2 = \Omega_R^2 \sin^2 n\Theta$, where Ω_R is the speed corresponding to Rayleigh waves [194].

speed obtained for the so-called Poisson case, $c_l = \sqrt{3}\,c_t$, viz. $c_E = 0.7923\,c_t$ is considerably smaller than the value $c_E = 0.9013\,c_t$ obtained in Ref. [193].

The empirical dependence of the speed of the nth edge mode on the wedge angle θ,

$$c_E = c_R \sin n\theta \qquad n = 1, 2, \ldots\ (n\,\theta < 90^\circ) \tag{31}$$

where c_R is the speed of Rayleigh waves, has been proposed by Lagasse *et al.* [253]. The real speeds, however, have been found to lie below those given by Eq. (31), and in fact this expression becomes quite inaccurate for all but the smallest values of n.

Lagasse *et al.* [254] have studied theoretically the effects on edge modes of rounding the edge. This leads to cylindrical surfaces that are like the ones studied by Wilson and Marrison [164]. It is found that the modes become dispersive, i.e. their speed becomes a function of q, but the change in speed from that for an ideal wedge is only by a few per cent, if the radius of curvature of the tip is not too large.

The propagation of waves along topographic wave-guides of other shapes than wedges has also been studied, and the reader is referred to the review of Oliner [242] for a discussion of this work. Here, we consider only one such study, because it relates to edge modes. The propagation of elastic waves that are localized at the top of a semi-infinite rectangular ridge made out of an isotropic elastic medium was studied by Subbaswamy and Maradudin [255] (Fig. 12). The frequencies of these waves were obtained as functions of qd, where q is the one-dimensional wave vector describing their propagation along the ridge, and d is the half-width of the ridge. It was found that the two modes of lowest frequency, although nondegenerate for small values of qd, become degenerate in the limit of large qd with a frequency equal to that of an edge mode associated with an isolated right-angle wedge made of the same material as the ridge (Fig. 13). Thus, these modes are localized not only to the top of the ridge, but to its edges as well.

Edge modes have also been studied on piezoelectric wedges. Wedges in which the surfaces of the wedges have been metallized to force the electrostatic potential inside the medium to vanish on the surfaces have been investigated by Sharon [196]. Lagasse [241] studied the case of wedges whose surfaces have not been grounded,

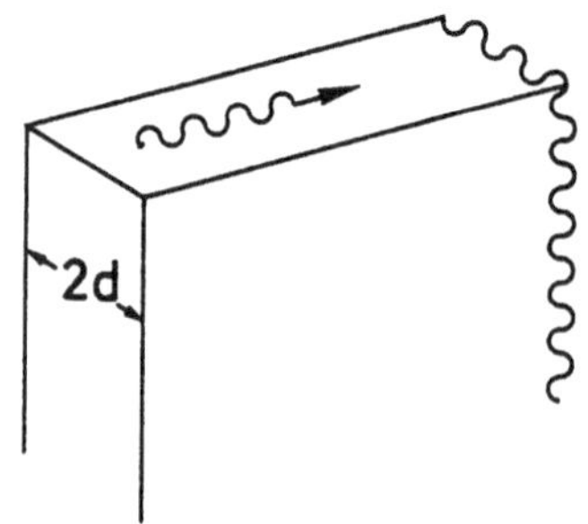

Fig. 12

Schematic depiction of a rectangular ridge topographic wave guide.

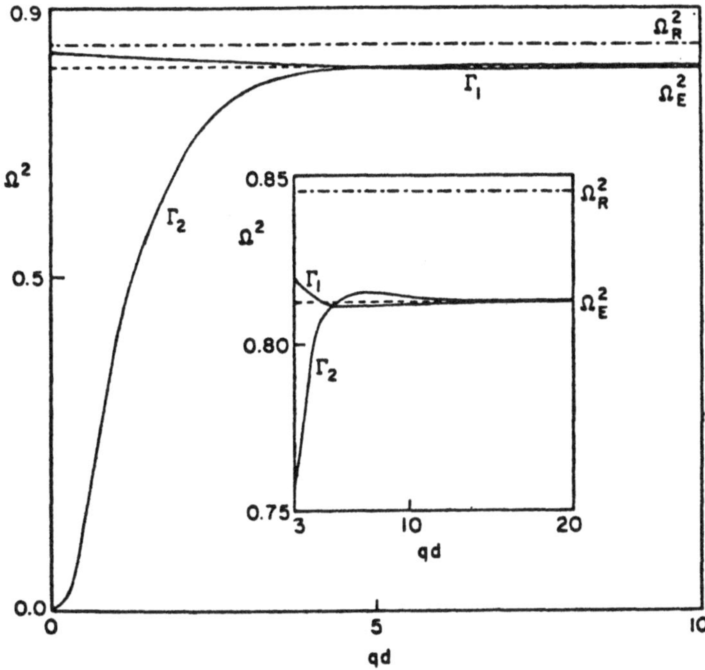

Fig. 13 The squares of speeds of the lowest frequency vibration modes of Γ_1 and Γ_2 symmetry (in units of the speed of bulk transverse modes) for an isotropic rectangular ridge. Ω_E is the speed of edge modes along a right angle wedge of the same material, while Ω_R is the speed of Rayleigh waves on the same material. The inset shows the asymptotic approach of the lowest eigenfrequencies to the edge mode value [225].

but he neglected the electric field outside the wedge. Apart from the presence of the electric field, which could be useful in device applications, the results for the frequencies and displacement fields of edge modes in piezoelectric crystals are qualitatively and quantitatively similar to those for edge modes in purely elastic wedges.

A detailed investigation of edge modes on an arbitrarily oriented, triclinic piezoelectric wedge has been carried out by Datta and Hunsinger [256] by means of the method used earlier by Sharon [196]. In addition to the usual stress-free boundary conditions, they have used the electrical boundary conditions that either the normal component of the electric displacement vector vanishes on the wedge surface, or the electrostatic potential does. In both cases the electric field vanishes outside the wedge. Numerical calculations are performed for a 40° wedge of $LiNbO_3$.

Edge modes have also been studied in right angle wedges of an exchange coupled Heisenberg ferromagnet (edge magnons). The dispersion curve was obtained in the form $\omega(k) = D(qa)^2 - E(qa)^4$ in the long wavelength limit by a continuum approach [257], where q is the one-dimensional wave vector of the edge mode and a is the lattice parameter of the ferromagnet. It is of interest to note that the coefficient D in this dispersion relation is the same as that found for surface spin waves and bulk spin waves [79, 80] for the same value of the wave vector. The coefficient E, however, is smaller for magnon edge modes than for surface spin waves, and the latter, in turn, is smaller than the coefficient E for bulk spin waves. Subsequently, from a lattice approach the dispersion relation for edge magnons was obtained for all values of q in the one-dimensional first Brillouin zone for the crystal [258].

Other types of edge modes will be discussed in Part II of this review.

Part II. Surface Polaritons

1 Introduction

A polariton is an electromagnetic wave that is linearly coupled to an electric or magnetic dipole active elementary excitation in a condensed medium, i.e. it is a photon coupled to a plasmon, photon, magnon, etc. A *surface polariton* is a polariton whose associated electromagnetic field is localized at the surface of the medium [73, 259–267].

In this Part of this review we are concerned with surface polaritons in a variety of physical systems and with their properties. Unlike surface acoustic waves, surface polaritons have yet to be utilized in technological contexts. Consequently the investigations of these modes and their properties up to the present time have been carried out primarily from the standpoint of elucidating the basic physics of these excitations, and this orientation is reflected in the material presented here. We begin with an introductory description of surface polaritons in the simplest system that can display them, viz. the planar interface between two different media, one of which commonly, but not necessarily, is vacuum. The remainder of this Part is then devoted to a survey of properties of surface polaritons in more complicated structures, how they can be studied experimentally, and effects in which they play a role.

2 Surface Polaritons in Various Two-Layer Systems

The bulk of the discussion in the second Part of these lectures will be concerned with various properties of surface polaritons at the planar interface between two different dielectric media. For the most part these two media will be characterized by isotropic dielectric tensors, and one of the two media will be vacuum. We therefore preface the subsequent discussion by describing in this section properties of surface polaritons in various two-layer systems.

2.1 Two Isotropic Media

The simplest structure that supports surface polaritons is the plane interface between two semi-infinite, optically isotropic dielectric media. This is the case we consider in the present subsection.

The physical system we consider is the following. The region $x_3 < 0$ is filled by a dielectric medium (medium a) characterized by an isotropic, frequency-dependent dielectric constant $\epsilon_a(\omega)$, that we assume is real. The region $x_3 > 0$ is filled by a second dielectric medium (medium b) characterized by a real dielectric constant $\epsilon_b(\omega)$. We assume that both media are nonmagnetic, so that the corresponding magnetic permeabilities are equal to unity. The solution of Maxwell's equations that describes a p-polarized (TM) wave propagating (with no loss of generality due to the isotropy of both media) in the x_1-direction but whose amplitudes decay exponentially with increasing distance into each medium from the interface are

$$\vec{E}(\vec{x}, t) = A\left(1, 0, \frac{ik_\parallel}{\alpha_b}\right) e^{ik_\parallel x_1 - \alpha_b x_3 - i\omega t} \qquad x_3 > 0 \tag{32a}$$

$$= B\left(1, 0, \frac{-ik_\parallel}{\alpha_a}\right) e^{ik_\parallel x_1 + \alpha_a x_3 - i\omega t} \qquad x_3 < 0 \tag{32b}$$

$$\vec{H}(\vec{x}, t) = A\left(0, \frac{-i\omega\epsilon_b}{c\alpha_b}, 0\right) e^{ik_\parallel x_1 - \alpha_b x_3 - i\omega t} \qquad x_3 > 0 \tag{33a}$$

$$= B\left(0, \frac{i\omega\epsilon_a}{c\alpha_a}, 0\right) e^{ik_\parallel x_1 + \alpha_a x_3 - i\omega t} \qquad x_3 < 0 \tag{33b}$$

where $\alpha_{a,b}(k_\parallel \omega) = (k_\parallel^2 - \epsilon_{a,b}(\omega)(\omega/c)^2)^{1/2}$. Both of these constants must be real and positive in order that Eqs. (32)–(33) describe electromagnetic waves localized to the interface $x_3 = 0$.

If we now require that the tangential components of $\vec{E}(x, t)$ and $\vec{H}(x, t)$ be continuous across this interface, we find that $A = B$, and obtain as the condition that a p-polarized wave exist

$$\frac{\epsilon_a(\omega)}{\alpha_a(k_\parallel \omega)} + \frac{\epsilon_b(\omega)}{\alpha_b(k_\parallel \omega)} = 0. \tag{34}$$

Since α_a and α_b are both required to be positive, we see from this equation that a necessary condition that it have a solution is that either $\epsilon_a(\omega)$ and $\epsilon_b(\omega)$ (but not both) must be negative at the frequency ω of the surface wave. Of the two dielectric media in contact along the interface $x_3 = 0$ in the structure under consideration here, the one whose dielectric constant is negative at the surface wave frequency is termed the *surface active medium*; the one whose dielectric constant is positive is termed the *surface inactive medium*.

The electromagnetic surface wave whose field vectors are given by Eqs. (32) and (32) (with A = B), and whose frequency is obtained from the dispersion relation, Eq. (34), is called a *surface polariton.*

It can be shown [73] that an s-polarized (TE) surface polariton cannot exist in this structure.

With the use of the definitions of $\alpha_a(k_\parallel \omega)$ and $\alpha_b(k_\parallel \omega)$ the dispersion relation can be rearranged into the form

$$k_\parallel^2 = \frac{\omega^2}{c^2} \frac{\epsilon_a(\omega)\, \epsilon_b(\omega)}{\epsilon_a(\omega) + \epsilon_b(\omega)} . \tag{35}$$

This gives an explicit expression for the magnitude of the wave vector of the surface polariton as a function of its frequency, a relation that can be inverted trivially to yield ω as a function of $k_\parallel$. However, because the method used to obtain Eq. (35) can introduce spurious solutions, in order that any solution of Eq. (35) corresponds to a surface polariton it must satisfy the conditions that $\alpha_a(k_\parallel \omega)$ and $\alpha_b(k_\parallel \omega)$ both be positive, and that either $\epsilon_a(\omega)$ or $\epsilon_b(\omega)$ be negative at that frequency.

If, instead of working in the regime where the effects of retardation are important, we turn to the case where they are unimportant, i.e. to the electrostatic limit, it is Laplace's equation for the scalar potential that we must solve. The solution of this equation that is wavelike in directions in the plane $x_3 = 0$, but decays exponentially as $|x_3| \to \infty$, is

$$\phi(\vec{x}, t) = A e^{ik_\parallel x_1 - |k_\parallel| x_3 - i\omega t} \qquad x_3 > 0 \tag{36a}$$

$$= B e^{ik_\parallel x_1 + |k_\parallel| x_3 - i\omega t} \qquad x_3 < 0. \tag{36b}$$

When we impose the boundary conditions that ϕ and $\epsilon\, \partial\phi/\partial x_3$ be continuous across the interface $x_3 = 0$, we find that A = B, and obtain as the condition that such a wave exist

$$\epsilon_a(\omega) + \epsilon_b(\omega) = 0. \tag{37}$$

The surface excitation whose potential is given by Eq. (36) (with A = B) and whose frequency is obtained from Eq. (37) is called an *unretarded surface polariton.* We again see that in order that such an excitation exist one of the dielectric constants must be negative.

If we compare Eqs. (35) and (37) we see that the frequency of the unretarded surface polariton is not only the limiting frequency of a surface polariton as $c \to \infty$, it is also the limiting frequency of a surface polariton as $k_\parallel \to \infty$.

We also note that the frequency that is the solution of Eq. (37) is independent of the wave vector $\vec{k}_\parallel$, unlike the solution of Eq. (34) or of Eq. (35). The reason for this ultimately is that in the electrostatic limit there is no characteristic length in the problem. In the electromagnetic case there is one, connected with the finite speed of light. It can be taken to be the vacuum wavelength of the elementary excitation in the dielectric medium that contributes a zero at a frequency Ω_0 to its dielectric constant. This is defined by $\lambda_0 = 2\pi/k_0$ where $k_0 = \Omega_0/c$. In the common case that medium b ($x_3 > 0$) is vacuum, $\epsilon_b(\omega) = 1$, and medium a ($x_3 < 0$) is a dielectric whose dielectric constant is $\epsilon_a(\omega) = \epsilon(\omega)$, the surface active medium is clearly the dielectric medium. The surface polariton dispersion curve for such a system, according to the definitions of α_a and α_b, must lie to the right of the "light line", $\omega = ck_\parallel$, and to the right of the bulk polariton dispersion curve defined by the equation $k_\parallel^2 = \epsilon(\omega)(\omega^2/c^2)$.

2.2 An Anisotropic Medium in Contact With an Isotropic Medium

There are many crystals whose optical properties are markedly anisotropic. This can be due either to the crystal class to which they belong, or to the action of external forces. It is therefore of interest to consider the situation where medium b ($x_3 > 0$) is isotropic, with dielectric constant $\epsilon_b > 0$, while the second medium, medium a ($x_3 < 0$), is an anisotropic dielectric.

A simple case of an anisotropic dielectric medium in contact with an isotropic dielectric medium along a plane interface is that in which the interface is perpendicular to a principal axis of the anisotropic dielectric tensor, which we take to be the x_3-axis, and the wave vector $\vec{k}_\parallel$ is parallel to a second principal axis, which we take to be the x_1-axis. We denote by $\epsilon_\parallel(\omega)$ and $\epsilon_\perp(\omega)$ the principal values of the dielectric tensor along the axes parallel to $\vec{k}_\parallel$ and perpendicular to the surface, respectively.

We consider a p-polarized surface polariton propagating along the x_1-direction. Nothing is assumed to vary with x_2. Then proceeding as in section 2.1 we find that the dispersion relation for such a polariton is

$$\frac{\epsilon_\parallel(\omega)}{\epsilon_b} = -\frac{\alpha_a}{\alpha_b} \tag{38}$$

where now $\alpha_a(k_\parallel\omega) = [\epsilon_\parallel(\omega)/\epsilon_\perp(\omega)]^{1/2}\,[k_\parallel^2 - \epsilon_\perp(\omega)(\omega^2/c^2)]^{1/2}$, and $\alpha_b(k_\parallel\omega) = [k_\parallel^2 - \epsilon_b(\omega^2/c^2)]^{1/2}$ are real and positive. Equation (38) is equivalent to

$$k_\parallel^2 = \frac{\omega^2}{c^2}\,\frac{\epsilon_b\epsilon_\perp(\omega)\,[\epsilon_b - \epsilon_\parallel(\omega)]}{\epsilon_b^2 - \epsilon_\parallel(\omega)\,\epsilon_\perp(\omega)}\,. \tag{39}$$

Solutions to Eq. (39) correspond to surface polaritons only if either of the following two conditions are satisfied:

(i) $\epsilon_\parallel(\omega)$ and $\epsilon_\perp(\omega)$ are both negative, $k_\parallel > \frac{\omega}{c}\sqrt{\epsilon_b}$;

(ii) $\epsilon_\parallel(\omega) < 0$, $\epsilon_\perp(\omega) > 0$, $\frac{\omega}{c}\sqrt{\epsilon_b} < k_\parallel < \frac{\omega}{c}\sqrt{\epsilon_\perp(\omega)}$.

Case (i) represents the generalization of the surface polaritons discussed in Section 2.1 to anisotropic materials. Such surface polaritons have been designated "type I" surface polaritons by Bryksin *et al.* [268]. Case (ii), on theother hand, presents a new type of surface polariton, that is designated "Type II" by Bryksin *et al.* It exists only if $\epsilon_\perp(\omega) > \epsilon_b$ and only for a limited range of values of $k_\parallel$. In particular, it does not exist in the unretarded limit, $c \to \infty$. The type II surface polariton is an example of what is sometimes called a "photon-induced" or "virtual" surface polariton (Fig. 14).

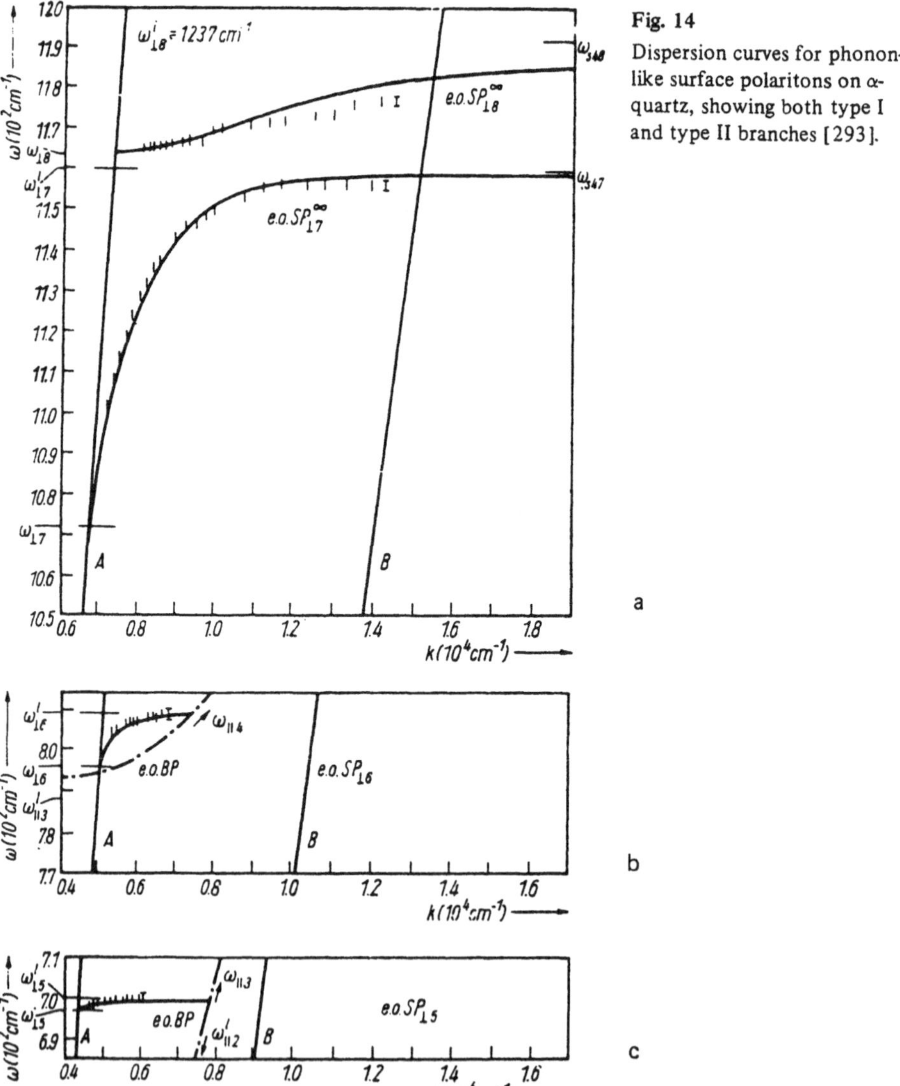

Fig. 14
Dispersion curves for phonon-like surface polaritons on α-quartz, showing both type I and type II branches [293].

A second type of anisotropic dielectric medium is a gyrodielectric medium. It is characterized by a dielectric tensor with off-diagonal elements that are complex conjugates of each other.

A common way of achieving this type of medium is through the application of a static, external magnetic field to a semiconductor containing free charge carriers. It is known that in this situation the semiconductor exhibits a variety of magneto-optical effects including cyclotron resonance, magnetoplasma resonance, Faraday effect, and Voigt effect. If a free surface is present, the magnetic field will also modify the surface polaritons associated with surface plasmons. The resulting surface waves are referred to as surface magnetoplasmon polaritons.

The dispersion curves for surface magnetoplasmon polaritons have been worked out for a semiconductor, such as n-type InSb, in which the energy band is simple and spherical, to a good approximation [269]. Medium b was assumed to be vacuum. When the external magnetic field is parallel to the surface, and the wave vector of the surface polariton is perpendicular to the magnetic field, the resulting dispersion curves are nonreciprocal. When the wave vector is oriented parallel to the magnetic field, the electromagnetic fields in the medium are characterized by two decay constants. In this case it is found that the dispersion curve has a portion that corresponds to the existence of pseudosurface waves (one decay constant real and positive, the second pure imaginary), and a portion that corresponds to generalized surface waves (the two decay constants are complex conjugates of each other). When the magnetic field is perpendicular to the surface, both pseudosurface waves and generalized surface waves are found again. This geometry, however, possesses cylindrical symmetry about the external magnetic field, so the results are independent of the direction of propagation of the wave in the surface.

The effects on the dispersion curves of surface magnetoplasmon polaritons due to their coupling to surface optical phonons have also been worked out for n-type InSb [270].

Many semiconductors are characterized by conduction or valence bands that have several equivalent extrema rather than a single extremum, and that have ellipsoidal, rather than spherical, constant energy surfaces. In general, free carriers associated with different extrema will have different effective masses and hence different cyclotron frequencies. This leads to additional surface polariton modes.

An example of such a crystal is p-type PbTe, which has valence band maxima in the ⟨111⟩ directions in wave vector space. The dispersion curves for surface magnetoplasmon polaritons on a (001) surface of this semiconductor have been worked out for the case that the external magnetic field is parallel to the surface [271]. When this field is in the [100] direction, and $\vec{k}_{\parallel}$ is perpendicular to it, the dispersion curve for p-type PbTe has two more branches, one for $k_{\parallel} > 0$ and one for $k_{\parallel} < 0$, than occur in n-type InSb. This behavior is typical of the multiple carrier case and becomes even more pronounced when the external magnetic field does not make equal angles with the major axes of all the constant energy ellipsoids.

Thus, when $\vec{H}_0$ is in the [101] direction, p-type PbTe exhibits even more surface polariton branches.

Another example of multiple carriers is provided by the electron-hole plasma that arises if a semiconductor is maintained at a sufficiently high temperature. Here again one finds additional surface polariton modes [272].

2.3 Surface Polaritons on Magnetic Media

Up to this point in the present section we have considered only nonmagnetic, dielectric media. In this subsection we turn to a consideration of surface polaritons in magnetic media, and relax our earlier assumption that the magnetic permeability $\mu(\omega)$ is unity.

That surface polaritons should exist at the interface between two magnetic media or between a magnetic and a dielectric medium follows from the results of the preceding subsections and the fact that in the absence of free charges Maxwell's equations are invariant under the interchange of $\vec{E}$ and $\vec{H}$, of $\overleftrightarrow{\epsilon}(\omega)$ and $\overleftrightarrow{\mu}(\omega)$, and of ω and $-\omega$. A necessary condition for the existence of such surface polaritons is that one of the magnetic media be surface active in the case that two magnetic media are in contact across a plane surface, and that the magnetic medium be surface active in the case that it is in contact with a dielectric medium.

The theory of surface polaritons propagating along the interface between two magnetic media can be carried out in complete analogy to the case of two dielectrics [273]. Just as in the gyrodielectric case, the dispersion relation exhibits the nonequivalent propagation in the $+\vec{k}_\parallel$ and $-\vec{k}_\parallel$ directions. The magnetic vector of the surface polariton lies in the sagittal plane, and the mode is a TE mode. In Ref. [273] surface polariton dispersion curves at a vacuum-ferromagnetic interface are worked out for both the case of a semi-infinite ferromagnetic insulator, where the dielectric tensor of the magnetic medium, as well as of the vacuum, is assumed to be the three-dimensional unit tensor, and for a ferromagnetic metal, where the effects of the dielectric tensor of the magnetic medium are taken into account.

2.4 Power Flow in a Surface Polariton

The energy in an electromagnetic wave crossing unit area normal to the direction of propagation per unit time is given by the Poynting vector [131] $\vec{S} = (c/4\pi)\vec{E} \times \vec{H}$, where the field $\vec{E}$ and $\vec{H}$ in this expression are required to be real. The time average of the power flow is given by the real part of the complex Poynting vector [274], $\langle\vec{S}\rangle = \mathrm{Re}\,\vec{S}^c$, where $\vec{S}^c = (c/8\pi)\vec{E} \times \vec{H}^*$. When the electromagnetic fields Eqs. (32)–(33) associated with a surface polariton propagating in the x_1-direction along the planar interface between vacuum ($x_3 > 0$, $\epsilon_b(\omega) \equiv 1$) and an isotropic dielectric medium ($x_3 < 0$, $\epsilon_a(\omega) \equiv \epsilon(\omega)$) are substituted into these expressions, the time averaged Poynting vector is found to be

$$\langle\vec{S}\rangle = \hat{x}_1 \frac{|A|^2}{8\pi} \frac{k_\parallel \omega}{\alpha_b^2} e^{-2\alpha_b x_3} \qquad x_3 > 0 \tag{40a}$$

$$= -\hat{x}_1 \frac{|A|^2}{8\pi} \frac{k_\parallel \omega}{\alpha_a \alpha_b} e^{2\alpha_a x_3} \quad x_3 < 0. \tag{40b}$$

Since $k_\parallel$, ω, α_a, α_b are all assumed to be positive, we see that the power flow in the vacuum is in the direction of propagation of the wave, but the power flow in the dielectric medium is in the direction opposite to the direction of propagation of the wave.

In view of the localized nature of the wave, more useful than the time averaged Poynting vector itself is the total power flow in the surface polariton, given by

$$\vec{P} = \vec{P}^> + \vec{P}^< \equiv L_2 \int_0^\infty \langle\vec{S}\rangle \, dx_3 + L_2 \int_{-\infty}^0 \langle\vec{S}\rangle \, dx_3, \tag{41}$$

where L_2 is the width of the beam. In the present case we find that

$$\vec{P}^> = \hat{x}_1 L_2 \frac{|A|^2}{16\pi} \frac{k_\parallel \omega}{\alpha_b^3}, \quad \vec{P}^< = -\hat{x}_1 L_2 \frac{|A|^2}{16\pi} \frac{k_\parallel \omega}{\alpha_a^2 \alpha_b}. \tag{42}$$

The ratio of the magnitude of the power flow in the vacuum to that in the dielectric medium is

$$\frac{|\vec{P}^>|}{|\vec{P}^<|} = \frac{\alpha_a^2}{\alpha_b^2} = \epsilon(\omega)^2. \tag{43}$$

This ratio can be quite large. For example, for a simple free electron metal, for which $\epsilon(\omega) = 1 - (\omega_p^2/\omega^2)$ and ω is small compared to the plasma frequency ω_p e.g. for ω in the infrared, we find that $|\vec{P}^>|/|\vec{P}^<| \cong \omega_p^4/\omega^4$. In the case of copper, for which $\hbar\omega_p \cong 10$ eV, if ω is the frequency that corresponds to a wavelength of 10.6 μm, then this ratio is $\sim 0.5 \times 10^8$, and for all practical purpose we can assume that all of the power flow takes place in the vacuum.

2.5 Surface Polaritons With Damping

It has been assumed that the dielectric tensors of the two dielectric media in contact across a plane interface are real. Thus, dissipative processes in the two media have been neglected. If this assumption is relaxed, and the dielectric tensor of medium a is assumed to be both isotropic and complex, $\epsilon_a(\omega) = \epsilon_a^{(1)}(\omega) + i\epsilon_a^{(2)}(\omega)$, while medium b is assumed to be vacuum, the attenuation length L of a surface polariton at such an interface can be obtained directly. This quantity is the distance required for the energy density of the wave to decrease to $1/e$ of its initial value.

The dispersion relation for surface polaritons in this case is still given by Eq. (35). If we assume that ω is real and $k_\parallel$ is complex, $k_\parallel = k_\parallel^{(1)} + ik_\parallel^{(2)}$, then the expressions for $k_\parallel^{(1)}$ and $k_\parallel^{(2)}$ in the case of a very lossy medium for which $\epsilon_a^{(2)}(\omega) \gg \epsilon_a^{(1)}(\omega)$ and $\epsilon_a^{(2)}(\omega) \gg 1$, take the simple forms

$$k_\parallel^{(1)} = \frac{\omega}{c}\left[1 - \frac{\epsilon_a^{(1)}(\omega) + \frac{3}{4}}{2\,\epsilon_a^{(2)}(\omega)}\right], \quad k_\parallel^{(2)} = \frac{\omega}{2c}\,\frac{1}{\epsilon_a^{(2)}(\omega)}. \tag{44}$$

(Expressions for $k_\parallel^{(1,2)}$ when these assumptions are not made can be found in Ref. [73].) The existence of an imaginary part of $k_\parallel$ shows that the amplitude of the surface polariton is exponentially damped, according to the factor $\exp(-k_\parallel^{(2)} x_1)$, as it propagates in, say, the x_1-direction. The attenuation length L is thus found to be given by

$$L = \frac{1}{2\,k_\parallel^{(2)}} = \frac{c}{\omega}\,\epsilon_a^{(2)}(\omega), \tag{45}$$

which is very long compared with the wavelength of the surface polariton. Experimental results for L [275] are in good agreement with the result given by Eq. (45).

The attenuation of surface polaritons just discussed is due to intrinsic damping mechanisms present in medium a. Damping of surface polaritons by an important extrinsic mechanism is discussed in Section 8.3.

We see from Eq. (44) that if $\epsilon_a^{(1)}(\omega) < -3/4$, $k_\parallel^{(1)}$ lies to the right of the light line, and we have a damped surface polariton. If, on the other hand, $\epsilon_a^{(1)}(\omega) > -3/4$, $k_\parallel^{(1)}$ lies to the left of the light line, and we have a new type of surface polariton known as a *Zenneck mode* [276]. Zenneck modes can exist even when the real part of $\epsilon_a(\omega)$ is positive, provided that the imaginary part of $\epsilon_a(\omega)$ is large enough to bind them to the surface, i.e., to give rise to decay constants α_a and α_b with positive real parts. With the assumption made above we have that

$$\alpha_a = \frac{\omega}{c}\left(\frac{\epsilon_a^{(2)}(\omega)}{2}\right)^{\frac{1}{2}}(1 - i), \quad \alpha_b = \frac{\omega}{c}\,(2\,\epsilon_a^{(2)}(\omega))^{-\frac{1}{2}}(1 + i). \tag{46}$$

Since α_a and α_b are complex quantities, Zenneck modes are "generalized" surface waves.

3 Observation of Surface Polaritons

Several experimental methods have been used up to the present time for the observation of surface polaritons.

One of the earlier techniques used, which is particularly applicable to the investigation of surface polaritons in the unretarded regime, is the observation of energy loss by electrons transmitted through [277] or reflected by [278] the sample. A peak in the energy loss spectrum can occur as a result of the excitation of surface plasmons. The penetration of the electric field of the surface polariton into the vacuum outside the sample provides a mechanism for strong coupling of these excitations to the incident electron.

It has also been observed that surface polaritons can radiate under certain conditions if they are excited by electrons that graze the surface, in the case of a rough surface [279], or are incident normally, in the case of a grating [280]. Indeed, Teng and Stern [280] obtained the dispersion curve for surface plasmons on an aluminium surface by bombarding an optical grating with normally incident 10 keV electrons and observing the peak in the angular distribution of the emitted light.

Conventional optical absorption or reflection measurements cannot be used for the direct observation of surface polaritons. This is due to the following circumstance. Conservation of energy requires that the frequency of the incident light (assumed incident from the vacuum) must equal the frequency of the surface polariton: $c(k_\parallel^2 + k_3^2)^{1/2} = \omega_{sp}(k_\parallel)$. The fact that the same value of $k_\parallel$ appears on both sides of this equation is due to the fact that, due to infinitesimal translational invariance, the component of the wave vector parallel to the surface is conserved across a planar surface. On the other hand, the requirement that the electromagnetic fields of the surface polariton decay into the vacuum with increasing distance from the vacuum-crystal interface imposes the condition that $ck_\parallel > \omega_{sp}(k_\parallel)$. These two conditions require that $k_3^2 < 0$, so the incident light must be attenuated in the x_3-direction in the vacuum above the crystal surface.

One can achieve such a field by the method of attenuated total reflection (ATR). This method was developed by Otto [281] for the study of surface polaritons on metal surfaces. It application to the study of surface polaritons associated with infrared active phonons in the infrared range of frequencies was proposed by Otto [282] and Ruppin [283]. In this method a prism of dielectric constant ϵ_p is placed above the crystal and is separated from it by a gap of thickness d and dielectric constant ϵ_g. Both of these dielectric constants are assumed to be positive and frequency independent in the frequency range in which surface polaritons occur. Light with its electric vector in the $x_1 x_3$-plane is incident on the interface between the prism and the gap, and makes an angle θ with the x_3-axis, where $\theta > \theta_c = \sin^{-1}(\epsilon_g/\epsilon_p)$. The angle θ_c is the critical angle for total internal reflection at the prism-gap interface, so the incident light undergoes total internal reflection within the prism. At the same time it gives rise to an electric field in the gap that varies exponentially with x_3, i.e. to a field with $k_3^2 < 0$. One can now have the frequency of the incident light equal to the frequency of the surface polariton for some value of $k_\parallel$, and when this happens there is a dip in the reflectivity of the system as a function of the frequency of the incident light.

The ATR method has been used to measure the dispersion curves of surface polaritons due to optical phonons [284–289], and due to coupling between optical phonons and surface plasmons [290, 291], in isotropic media. It has also been used to obtain the dispersion curves of surface polaritons at the interface between vacuum and an anisotropic dielectric medium [292–296], and of surface polaritons associated with surface magnetoplasmons in n-type InSb [297, 298].

A variant of the ATR method was used by Raether and coworkers [299–303] to study the dispersion curves of surface polaritons on metal surfaces. In this method a thin metal film is plated on the bottom of a quartz prism. Light is incident on the quartz-metal interface from within the quartz prism. The angle of incidence is greater than that for total internal reflection at the quartz-metal interface. The exponentially decaying electric field created thereby can penetrate the metal film and excite a surface polariton at the metal-vacuum boundary, which will radiate into the vacuum if the metal-vacuum surface is rough or a grating. The intensity of the radiated light will be greater for those frequencies of the incident light that satisfy the dispersion relation for these surface polaritons. At the same time, the reflectivity of this system will display a dip at that frequency.

Another experimental method used for the study of surface polaritons is the observation of the reflectivity of a surface upon which a grating has been ruled. The presence of the grating relaxes the conservation condition on the component of the wavevector of the incident light parallel to the surface, $\vec{k}_{\parallel}$. The effective values of $k_{\parallel}$ are $k_{\parallel}^{(n)} = (\omega/c) \sin\theta + (2\pi n/d)$, where ω is the frequency of the incident light, θ is the angle of incidence, d is the period of the grating, and n is an integer. When d is sufficiently small, one can have $ck_{\parallel}^{(n)} > \omega$ even when n is a small integer. The reflectivity exhibits a dip whenever the surface polariton frequency $\omega_{sp}(k_{\parallel}^{(n)})$ equals ω for some value of n, corresponding to the excitation of the surface polariton by the incident field. Because a randomly rough surface can be regarded as the superposition of many diffraction gratings, each with its own orientation and period, the preceding argument also indicates how an incident electromagnetic wave can couple to surface polaritons on a rough surface. This method has been used by several authors to study the surface polariton dispersion curve on metal surfaces [280, 304–306]. In particular, Ritchie *et al.* [305] succeeded in observing by this method the gaps opened up on the dispersion curve of surface polaritons on Au at wave vectors corresponding to the boundaries of the one-dimensional Brillouin zones defined by the period of the grating. Marschall, Fischer, and Queisser [307, 308] used this method to obtain the first complete surface polariton dispersion curve, for n-type InSb.

Recently, the techniques of Raman scattering have been applied successfully to the study of surface polaritons in thin films (2 500 Å thick) of GaAs deposited on a sapphire substrate [309]. The 4 880 Å line of an argon ion laser was used, together with near forward scattering at normal incidence. The thinness of the film made possible the study of the scattered radiation in transmission. The surface polariton dispersion curve has two branches in this structure, the higher (lower) frequency one corresponding to surface polaritons at the vacuum (substrate) film interface. Only the latter was observed in the initial experiments, but both branches have been seen in subsequent experiments [310, 311]. The theory of such experiments has been worked out by several authors [312–319].

The newest experimental method for measuring the dispersion curve of a surface polariton consists of a crystal with a prism above its surface at a distance d, exactly as in an ATR experiment, but with no external source of electromagnetic radiation. Instead, the entire system is heated to a temperature of 150 °C. The surface of the heated crystal emits radiation that includes an exponentially decaying (in x_3) component associated with surface polaritons. The latter component is converted into outgoing radiation by the prism and appears as a peak in the emissivity of the crystal as a function of frequency at fixed emission angles, or as a function of angle at fixed frequencies. This method, due to Vinogradov and Zhizhin [320], has subsequently been used in additional such studies [321–323].

Trasmission radiation is radiation emitted when a charged particle moving with constant velocity passes from a medium with one dielectric constant into a second medium with a different dielectric constant [324]. Since the interface between the two media is essential for the occurence of transition radiation, it is natural to think that surface polaritons play a role in the formation of the transition radiation. This is indeed the case [325], and a spectral analysis of the transition radiation in the appropriate frequency range could yield information concerning the dispersion curve for surface polaritons. These waves have to be made radiative in order that they can be observed and studied. Two different experimental geometries have been suggested for achieving this, both of which employ a prism, as in an ATR experiment, and the theory of each worked out [326, 327]. However, no experimental results employing transition radiation for the study of surface polaritons have been reported up to now, as far as I am aware.

4 Surface Polaritons in Layered Media

Only surface polaritons in two layer systems have been discussed thus far. Systems of physical interest exist, however, that consist of a larger number of layers. These include three layer systems consisting, for example, of a dielectric substrate separated from vacuum by an oxide layer, or a thin film of one semiconductor deposited on a substrate consisting of a different semiconductor. Surface polaritons can exist in such structures under suitable conditions.

In a structure in which the region $x_3 > d$ is filled by a dielectric characterized by the dielectric constant $\epsilon_1(\omega)$, the region $0 < x_3 < d$ is filled with a dielectric of dielectric constant $\epsilon_2(\omega)$, while the region $x_3 < 0$ is filled by a dielectric with a dielectric constant $\epsilon_3(\omega)$, the dispersion relation for p-polarized surface polaritons is

$$\left[\frac{\epsilon_2(\omega)}{\epsilon_1(\omega)}\frac{\alpha_1}{\alpha_2}+1\right]\left[\frac{\epsilon_2(\omega)}{\epsilon_3(\omega)}\frac{\alpha_3}{\alpha_2}+1\right]=\left[\frac{\epsilon_2(\omega)}{\epsilon_1(\omega)}\frac{\alpha_1}{\alpha_2}-1\right]\left[\frac{\epsilon_2(\omega)}{\epsilon_3(\omega)}\frac{\alpha_3}{\alpha_2}-1\right]e^{-2\alpha_2 d}, \tag{47}$$

where $\alpha_i = (k^2 - \epsilon_i(\omega)(\omega^2/c^2))^{1/2}$ (i = 1, 2, 3) is real and positive. Here k is the wave vector of the mode. In the limit as $d \to \infty$, Eq. (47) yields the two equations

$$\frac{\epsilon_2(\omega)}{\epsilon_1(\omega)}\frac{\alpha_1}{\alpha_2} + 1 = 0, \quad \frac{\epsilon_2(\omega)}{\epsilon_3(\omega)}\frac{\alpha_3}{\alpha_2} + 1 = 0, \tag{48}$$

that are the dispersion relations for surface polaritons at the interfaces $x_3 = d$ and $x_3 = 0$, respectively. In order to have surface polaritons associated with both interaces, one of the two dielectric media in contact across each interface must be surface active and the other surface inactive. This means that either both $\epsilon_1(\omega)$ and $\epsilon_3(\omega)$ are positive while $\epsilon_2(\omega)$ is negative or *vice versa*. If both $\epsilon_1(\omega)$ and $\epsilon_2(\omega)$ are positive while $\epsilon_3(\omega)$ is negative, or *vice versa*, only one surface polariton is expected in this structure. In structures in which a polariton bound to each surface can exist, for values of k such that $\alpha d \approx 1$, the surface polariton at one interface "feels" the existence of the surface polariton at the other interface, a "repulsion of levels" occurs, and the dispersion curves for the surface polaritons localized at each of the two interfaces become distorted by the interaction between these waves.

It was in such a three layer system, consisting of a GaAs film in contact with vacuum on one side and with a sapphire substrate on the other, that the first experimental observations of surface polaritons by the technique of Raman scattering were carried out, by Ushioda and his colleagues [309–310]. The theory of such experiments is given in Refs. [312–319]. (In fact, sapphire is optically anisotropic, and cannot be characterized by an isotropic dielectric constant $\epsilon_3(\omega)$. The dispersion relation that replaces Eq. (47) in this case can be found in Ref. [328].)

Multilayer structures consisting of prism, gap, and crystal are necessarily used in calculations of attenuated total reflection spectra. In some cases the crystal itself is modeled by a two or more layer system, to represent the fact that its optical properties in the vicinity of the surface are different from those in the bulk. This is the case, for example, in the investigations of surface polaritons on the surface of an organic solid, crystalline anthracene [329, 330], in which the surface polariton arises from coupling of photons to the electronic and vibronic transitions in the crystal. The latter can be different in the surface from what they are in the bulk. Surface polaritons in other organic crystals have also been studied by the ATR technique recently [331, 332].

5 Guided Wave Polaritons

In addition to the surface polaritons discussed in section 4, three layer systems can support a second type of wave that is localized to the middle layer, if not to each of the two interfaces, when the dielectric constant of the middle layer is larger than those of both of the surrounding media. In such modes, called *guided wave polaritons,* the electromagnetic fields decay exponentially with increasing distance along the outward normal to each of the two surfaces of the middle layer, but are standing waves within the middle layer itself. These are the modes that form the basis for the new integrated optics technology [333]. Integrated optics devices, however, use guided waves with frequencies in the visible, while recent investigations of these modes by physicists have been for frequencies in the infrared, where the dielectric constant $\epsilon(\omega)$ of the film shows strong frequency dispersion, in polar materials.

The three layer system considered in section 4 can support guided wave polaritons of both p- and s-polarization. The simplest case to consider is that in which the film is surrounded by vacuum ($\epsilon_1(\omega) = \epsilon_3(\omega) = 1$, $\epsilon_2(\omega) = \epsilon(\omega)$), so that the film has a plane of reflection symmetry at $x_3 = \frac{1}{2}$ d. The dispersion relations for p-polarized guided wave polaritons in such a film are

$$\epsilon(\omega)\frac{\alpha_0}{\hat{\alpha}} = \tan\frac{1}{2}\hat{\alpha}d, \quad \epsilon(\omega)\frac{\alpha_0}{\hat{\alpha}} = -\cot\frac{1}{2}\hat{\alpha}d, \tag{49}$$

where $\alpha_0 = (k^2 - (\omega^2/c^2))^{1/2}$ and $\hat{\alpha} = (\epsilon(\omega)(\omega^2/c^2) - k^2)^{1/2}$ are real and positive. Thus we are concerned with values of ω for which $\epsilon(\omega) > 0$, and with values of the wave vector k for which $(\omega/c) < k < \epsilon^{1/2}(\omega)(\omega/c)$. The first (second) of Eqs. (49) gives the frequencies of modes whose magnetic vector is even (odd) in the variable $x_3 - \frac{1}{2}$ d. The corresponding equations for s-polarized guided wave polaritons are

$$\frac{\hat{\alpha}}{\alpha_0} = \cot\frac{1}{2}\hat{\alpha}d, \quad \frac{\hat{\alpha}}{\alpha_0} = -\tan\frac{1}{2}\hat{\alpha}d, \tag{50}$$

where the first (second) equation gives the frequencies of modes in which the electric vector is even (odd) in $x_3 - \frac{1}{2}$ d. Consequently, for each polarization there is an infinite number of guided wave polaritons possible.

Dispersion relations for guided wave polaritons in asymmetric three layer structures can be found in Ref. [73].

Guided wave polaritons have now been studied by Raman spectroscopy [334] (Fig. 15). The theory of these scattering experiments has been worked out by Subbaswamy and Mills [335].

For additional information concerning guided wave polaritons the reader is referred to a recent review by Subbaswamy and Mills [336].

6 Refraction of a Surface Polariton by Geometrical and Material Discontinuities

A great deal is known about the properties of surface polaritons in a wide range of systems of physical interest. With this information in hand one can now envision optical circuits impressed on surfaces over which surface polaritons propagate, rather similar in nature to those employed in the field of integrated optics. To assess the effectiveness of such a circuit a study of one basic phenomenon is required: the refraction of a surface polariton by a geometrical and/or material discontinuity.

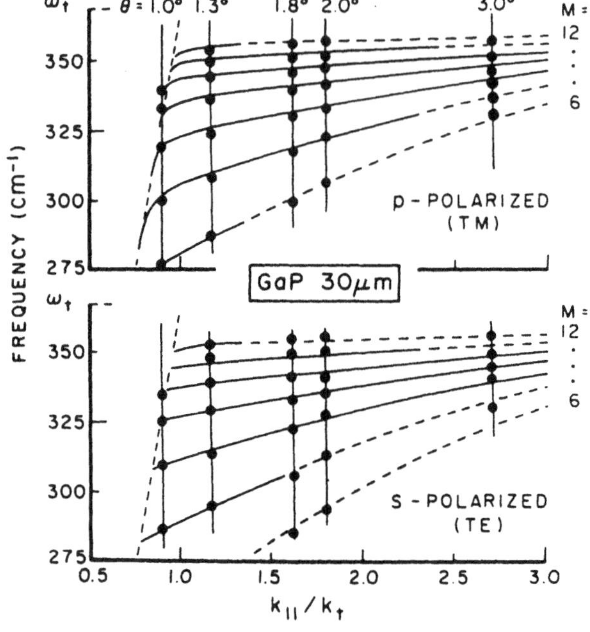

Fig. 15
Experimental (dots) and calculated (curves) dispersion of guided wave polaritons in a 30 μm thick GaP film. These modes are labeled by their mode number M, for $6 \leqslant M \leqslant 12$. The solid portions of the calculated curves indicate the regions where the normal components of the wave vectors are conserved to within $2\pi/d$ in a Raman scattering experiment, where d is the film thickness.

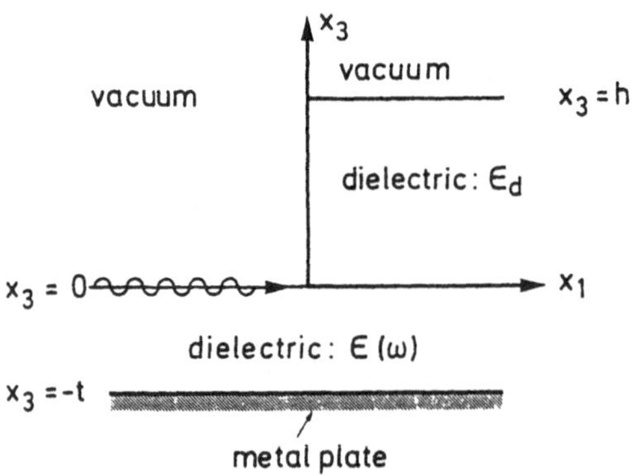

Fig. 16
A dielectric obtstacle from which surface polaritons can be refracted and scattered [337].

The refraction of a surface polariton incident normally on a step discontinuity caused by a dielectric (ϵ_d) overlayer of thickness h that covers half ($x_1 > 0$) of a dielectric substrate ($\epsilon(\omega)$, $x_3 < 0$) has been studied by Mahmoud and Beal [337] (Fig. 16). The dielectric constant ϵ_d is positive and assumed to be frequency-independent. Both the interface between the dielectric substrate and vacuum and the interface between the substrate and the dielectric overlayer can support a surface polariton of frequency ω. Mahmoud and Beal expanded the electromagnetic fields in the regions $x_1 < 0$ and $x_1 > 0$ in terms of the normal modes for each region, in order to satisfy the boundary conditions at the surface $x_1 = 0$. The complete spectrum of these modes includes a discrete part (the surface polariton) and a continuous part (radiative modes), and were obtained by a method due to Shevchenko [338]. The fields in the region $x_1 < 0$ consisted of an incident and reflected surface polariton and reflected radiative waves; in the region $x_1 > 0$ they consisted of a transmitted surface polariton and transmitted radiative waves. The radiative modes are required because it is not possible to satisfy the boundary conditions with the surface polaritons alone. From the electromagnetic fields obtained in this way, the power in each of the surface waves and in the reflected and transmitted radiative waves could be calculated, and from them the reflection coefficient for the surface polaritons, and the two components of the transmitted power. The reflection coefficient was found to increase nonmonotonically with increasing height of the dielectric abstacle h and to be larger for a larger fraction of the total incident power in the vacuum region for $x_1 < 0$.

Somewhat similar calculations were carried out for surface polaritons on a plane surface ($x_3 = 0$) that cross from a region with dielectric constant $\epsilon_1(\omega)$ ($x_1 < 0$) into a region with dielectric constant $\epsilon_2(\omega)$ [339] (Fig. 17). Reflection and transmission coefficients for the incident surface polariton were calculated, as well as the power flows in the reflected and transmitted waves, for different choices of $\epsilon_1(\omega)$ and $\epsilon_2(\omega)$ (Fig. 18).

A different version of the latter approach had been used earlier by Clarricoats and Slinn [340] in studies of the effects of geometrical and/or material discontinuities in closed wave guides.

By the use of these methods one can now begin to study the effects of non-normal incidence and of other types of discontinuities on the propagation characteristics of surface polaritons.

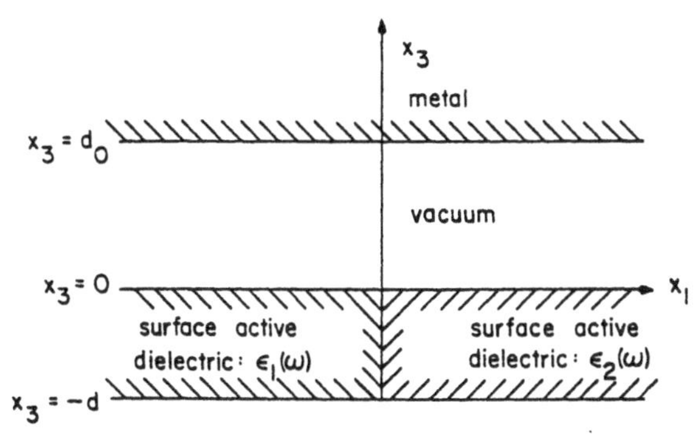

Fig. 17
A dielectric structure from which surface polaritons can be refracted and scattered [339].

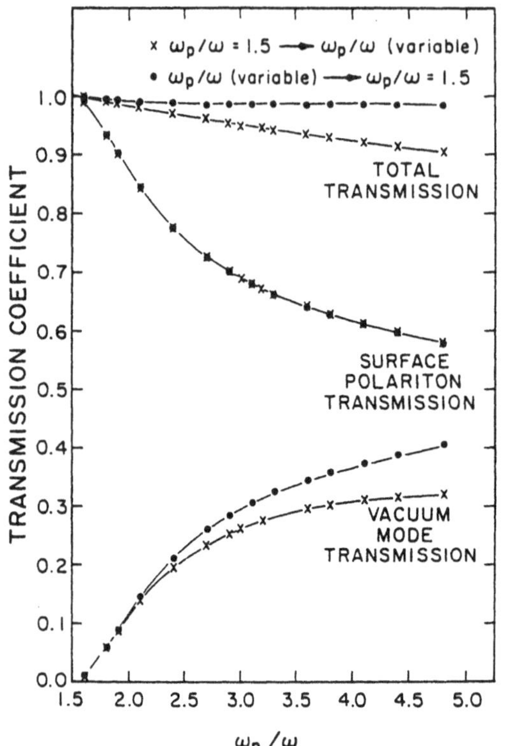

Fig. 18
Energy transmission curves for surface polaritons and vacuum modes in the structure depicted in Fig. 17 for $d = d_0 = 2\pi c/\omega$. A free electron dielectric constant of the form of $\epsilon(\omega) = 1 - (\omega_p^2/\omega^2)$ was chosen for each of the two dielectric media constituting the structure. The surface polariton frequency ω is fixed, and the two media are distinguished by the values of the corresponding bulk plasma frequencies ω_p. Curves are presented both for the propagation of a surface polariton from a medium with $\omega_p/\omega = 1.5$ into a medium with variable ω_p/ω (x) and from a medium with a variable ω_p/ω to a medium with $\omega_p/\omega = 1.5$ (•). Note that the transmission coefficient for surface polaritons is independent of the direction of propagation of the incident surface polariton. This may be due to the fact that the bulk of the energy flow in a surface polariton is carried by the vacuum, where there is no material discontinuity [339].

7 Spatial Dispersion

We have so far assumed that the relation between the electric displacement vector and the macroscopic electric field in a dielectric medium is local in space, although nonlocal in time. However, in principle the most general linear relation between $\vec{D}(\vec{x}, t)$ and $\vec{E}(\vec{x}, t)$ is nonlocal in space as well as in time. For a system described by a time-independent Hamiltonian the most general form of this relation is

$$D_\alpha(\vec{x}, t) = \sum_\beta \int d^3x' \int dt' \epsilon_{\alpha\beta}(\vec{x}, \vec{x}'; t - t')\, E_\beta(\vec{x}', t'), \tag{51}$$

where the integration over $\vec{x}'$ is over the volume of the dielectric medium. Causality requires that the nonlocal dielectric tensor $\epsilon_{\alpha\beta}(\vec{x}, \vec{x}'; t-t')$ vanish for $t' > t$.
If the dielectric medium is spatially homogeneous $\epsilon_{\alpha\beta}(\vec{x}, \vec{x}'; t-t')$ depends on $\vec{x}$ and $\vec{x}'$ only through their difference. In this case we can introduce the Fourier representation

$$\epsilon_{\alpha\beta}(\vec{x}, \vec{x}'; t - t') = \int \frac{d^3q}{(2\pi)^3} \int \frac{d\Omega}{2\pi}\, \epsilon_{\alpha\beta}(\vec{q}, \Omega)\, e^{i\vec{q}\cdot(\vec{x} - \vec{x}')}\, e^{-i\Omega(t-t')}. \tag{52}$$

If, in addition we assume a plane wave representation for the electromagnetic fields, e.g. $E_\alpha(\vec{x}, t) = \hat{E}_\alpha(\vec{k}, \omega) \exp(i\vec{k}\cdot\vec{x} - i\omega t)$, the relation between the amplitudes $\hat{E}_\alpha(\vec{k}, \omega)$ and $\hat{D}_\alpha(\vec{k}, \omega)$ is found to be

$$\hat{D}_\alpha(\vec{k}, \omega) = \sum_\beta \epsilon_{\alpha\beta}(\vec{k}, \omega)\, E_\beta(\vec{k}, \omega). \tag{53}$$

The fact that the dielectric tensor of a medium is a function of the wave vector of an electromagnetic wave propagating though it, as well as of its frequency, is referred to as *spatial dispersion.*

In many instances, when the incident electromagnetic radiation is in the visible or infrared portions of the spectrum, the dependence of the dielectric tensor $\epsilon_{\alpha\beta}(\vec{k}, \omega)$ on $\vec{k}$ can be safely ignored, since the wavelength of the radiation is very long compared to the interatomic separation.

However, there are many instances where the wave vector dependence of the dielectric tensor produces striking effects in the reflectivity, even when the wavelength of the radiation is long compared to the interatomic separation. An example of this is provided by the experimental and theoretical studies of the optical properties of CdS carried out by Hopfield and his collaborators [341]. In this case structure in the reflectivity, arising from three sharp exciton series and observed at low temperatures, differs markedly from that expected from a model that incorporates only the frequency dependence of the dielectric constant in the exciton regime. A quantitative interpretation of the data can be obtained from a model that includes the wave vector dependence of the dielectric tensor of the crystal as an essential feature [341].

The surface of a bounded medium plays an essential role in any theory of the optical properties of such a solid. The incident radiation must cross the boundary to enter the medium, and reflected and scattered radiation must cross it to leave the medium. Refraction of the radiation occurs in each of its interactions with the boundary, for non-normal incidence. And it is the surface of the solid that presents almost all the difficulties when the optical properties of the solid are calculated in the presence of spatial dispersion.

7.1 Theory

At the present time several approaches exist to the study of the optical properties of bounded media in the presence of spatial dispersion. The first is due to Pekar [342], and has been developed by several authors subsequently [341–343]. In this approach one reasons as follows. We do not in general know the nonlocal dielectric tensor of a bounded medium, particularly its dependence on $\vec{x}$ and $\vec{x}\,'$ in the vicinity of the surface, where it is expected to deviate the most from its form when $\vec{x}$ and $\vec{x}\,'$ are in the bulk of the medium. We do know, or are able to model, the dielectric tensor $\epsilon_{\alpha\beta}(\vec{k}, \omega)$ appropriate to the bulk of the crystal. We therefore solve the differential form of Maxwell's equations for the electromagnetic fields everywhere in the medium, as if it were unbounded with the use of $\epsilon_{\alpha\beta}(\vec{k}, \omega)$. Thus modifications of these fields due to the fact that the true dielectric tensor differs from its bulk form in the vicinity of the surface are not taken into account. When we do this we find that there are more electromagnetic waves in the medium than there are in the absence of spatial dispersion, enough more, in fact, that the usual Maxwell boundary conditions (continuity of the tangential components of $\vec{E}$ and $\vec{H}$) are insufficient in number to match all the waves in the medium to the waves in the vacuum outside it. *Additional boundary conditions* are required.

Despite attempts to provide microscopic justifications for them [343], in practice these additional boundary conditions are usually introduced in a somewhat *ad hoc* fashion. They generally take one of three forms: (1) the polarization associated with the excitation in the medium that contributes the dominant pole to the dielectric tensor in the frequency range of interest is assumed to vanish at the boundary [342]; (2) the normal derivative of this polarization is assumed to vanish at the boundary [344, 345]; (3) a linear combination of this polarization and its normal derivative is assumed to vanish at the boundary [346]. The number of additional boundary conditions assumed is the number needed in conjunction with the usual Maxwell boundary conditions to determine the amplitudes of all the waves in the problem.

It should be noted that Zeyher [347] has recently shown for the case of Wannier excitons that if the nonlocal dielectric tensor is constructed out of electron-hole wave functions calculated for a semi-infinite crystal with account taken of the Coulomb interaction between electron and hole, then the Pekar additional boundary condition, (1) above,is the correct one to use in a macroscopic formulation of spatial dispersion as a boundary value problem.

In the case of metal-metal interfaces, Forstmann and his coworkers [348–350] have argued that the conservation of the normal component of the energy flow vector across the boundary must be invoked as a new, additional boundary condition. This condition has also been used by Monreal *et al.* [351], where references to related work can be found, in the study of the optical properties of metals, when the electrons are assumed to be reflected diffusely from the surface. However, Boardman and Ruppin [352] have pointed out that the continuity of the normal component of the energy flow vector across a boundary is automatically ensured by the usual boundary conditions of plasma theory [353, 354], in particular by the continuity of the normal component of the electron's velocity. Consequently, in a correct formulation of a macroscopic hydrodynamic model of the conduction electrons in a metal continuity of energy does not add a new boundary condition. It is true, however, that the set of boundary conditions used by Boardman and Ruppin [352] admits the possibility of a mass flow across the boundary, but they argue that the consequences of this are second order in magnitude or are confined to a transition layer in the immediate vicinity of the surface.

Recently, a second approach to the study of the optical properties of bounded, spatially dispersive media has been advanced in work by Birman and Sein [355], Agarwal *et al.* [356], and Maradudin and Mills [357]. The last named authors have also obtained the dispersion relation for surface exciton polaritons by this approach (see section 8.2). In this work the emphasis is placed on the nonlocal dielectric tensor of a bounded medium, in particular of a semi-infinite dielectric medium occupying the half-space $x_3 > 0$. If the medium is homogeneous, then this dielectric tensor $\epsilon_{\alpha\beta}(\vec{x}, \vec{x}\,'; t - t')$ can depend on the coordinates $\vec{x}_\parallel$ and $\vec{x}_\parallel'$ only through their difference. It is therefore convenient to introduce the Fourier representation

$$\epsilon_{\alpha\beta}(\vec{x}, \vec{x}\,'; t - t') = \int \frac{d^2 q_\parallel}{(2\pi)^2} \int \frac{d\Omega}{2\pi}\, \epsilon_{\alpha\beta}(\vec{q}_\parallel \Omega | x_3 x_3')\, e^{i\vec{q}_\parallel \cdot (\vec{x}_\parallel - \vec{x}_\parallel') - i\Omega(t-t')} \tag{54}$$

that exploits this property. Then, if we write the fields in the medium in the form $E_\alpha(\vec{x}, t) = \hat{E}_\alpha(\vec{k}_\parallel \omega | x_3) \exp(i\vec{k}_\parallel \cdot \vec{x}_\parallel - i\omega t)$, the following integro-differential equation for $\hat{E}_\alpha(\vec{k}_\parallel \omega | x_3)$ is obtained:

$$\sum_\beta{}' \left\{ \left[(1 - \delta_{\alpha 3})\, ik_\alpha + \delta_{\alpha 3} \frac{d}{dx_3} \right] \left[(1 - \delta_{\beta 3})\, ik_\beta + \delta_{\beta 3} \frac{d}{dx_3} \right] + \right.$$
$$\left. + \delta_{\alpha\beta} \left(k_\parallel^2 - \frac{d^2}{dx_3^2} \right) \right\} \hat{E}_\beta(\vec{k}_\parallel \omega | x_3) \tag{55}$$
$$= \frac{\omega^2}{c^2} \sum_\beta{}' \int_0^\infty dx_3'\, \epsilon_{\alpha\beta}(\vec{k}_\parallel \omega | x_3 x_3')\, \hat{E}_\beta(\vec{k}_\parallel \omega | x_3').$$

The attractiveness of this approach lies in the fact that when $\epsilon_{\alpha\beta}(\vec{k}_\parallel \omega | x_3 x_3')$ is given, Eqs. (55) together with Maxwell's equations in the vacuum ($x_3 < 0$) and the usual Maxwell boundary conditions, contain all the information required for a unique solution of whatever problem is being considered, e.g. the determination of the reflection coefficient or the dispersion relation for surface polaritons. Any additional boundary conditions that are required for this solution to be obtained are contained in the integro-differential form of Maxwell's equations, and there is no reason in principle that requires one to resort to a microscopic theory to deduce new boundary conditions to supplement the ones that follow from the macroscopic theory. Two recent papers in which this approach was taken are those of Garcia-Moliner and Flores [358] and Mukhopadhyay and Lundqvist [359].

In addition to eliminating the need for introducing *ad hoc* additional boundary conditions this approach focuses attention on the part of the problem where all the physics resides, viz. the determination of the nonlocal dielectric tensor $\epsilon_{\alpha\beta}(\vec{k}_\parallel \omega | x_3 x_3')$. There now exist several efforts to calculate this function [347, 360–364] and more can be expected in the future.

A third approach [365–369] to the study of optical properties of bounded, spatially dispersive media does away with the nonlocal dielectric tensor altogether. In this approach one tries to solve Maxwell's equations for a reasonably realistic model of the crystal. The model used in most of this work consists of a semi-infinite crystal, or a crystal of finite thickness, whose sites are occupied by point dipoles that, in isolation, possess frequency dependent polarizabilities. In the solution of Maxwell's equations for electromagnetic fields in such a system additional boundary conditions never arise.

The last approach to this problem that will be discussed here involves obtaining a dispersion relation for surface polaritons on a spatially dispersive surface in terms of the dielectric tensor of the medium, with the aid of physically motivated assumptions about the nature of the elementary excitations in the medium.

The starting point for this approach was the derivation by Ritchie and Marusak [370] of a dispersion relation for surface plasmons in the presence of spatial dispersion, on the basis of the crucial assumption that the electrons in a semi-infinite dielectric medium are reflected specularly from the surface. Their equation has the simple form

$$1 + \frac{k_{\parallel}}{\pi} \int_{-\infty}^{\infty} \frac{dk_3}{\epsilon(\vec{k}, \omega)} = 0. \tag{56}$$

Equation (56) was subsequently generalized in two ways by Kliewer and Fuchs [344]. First, they included the effects of retardation, and second, they allowed for an anisotropic bulk dielectric tensor.

The dispersion relations obtained on the basis of the assumption that the electrons in a medium are specularly reflected from the boundary can be shown to be the same as those obtained on the assumption of the vanishing at the boundary of the normal derivative of the polarization associated with the excitation in the medium that contributes the dominant pole to the dielectric tensor in the frequency range of interest [345]. The dispersion relation analogous to the Kliewer-Fuchs equation, but with the assumption that it is the polarization, and not its normal derivative, that vanishes on the boundary (the Pekar [342] condition) has been obtained in effect by Rimbey and Mahan [345]. These authors in fact obtain the surface impedance $Z_p(\vec{k}_{\parallel}\, \omega)$ for p-polarized light. However, the dispersion relation for p-polarized surface polaritons is given in terms of the surface impedance by

$$\left(k_{\parallel}^2 - \frac{\omega^2}{c^2}\right)^{\frac{1}{2}} = i\,\frac{\omega}{c}\, Z_p(\vec{k}_{\parallel}\, \omega). \tag{57}$$

More recently in an effort to get away from the specular reflection assumption, Maradudin [371] obtained a generalization of the Ritchie-Marusak dispersion relation for a nonlocal dielectric tensor of the form $\epsilon_{\alpha\beta}(\vec{k}_{\parallel}\,\omega|x_3x_3') = \delta_{\alpha\beta}\,\epsilon(\vec{k}_{\parallel}\,\omega|x_3x_3')$, where $\epsilon(\vec{k}_{\parallel}\,\omega|x_3x_3')$ is assumed to be symmetric in x_3 and x_3'. This result was generalized in turn by Pattanayak and Birman [372], who included the effects of retardation, and assumed a uniaxial dielectric tensor defined by $\epsilon^{-1}_{\alpha\beta}(\vec{k}_{\parallel}\,\omega|x_3x_3') = \delta_{\alpha\beta}\,\epsilon^{-1}_{\alpha\alpha}(\vec{k}_{\parallel}\,\omega|x_3x_3')$, where $\epsilon^{-1}_{11} = \epsilon^{-1}_{22} \neq \epsilon^{-1}_{33}$. The nonzero elements $\epsilon^{-1}_{\alpha\alpha}(\vec{k}_{\parallel}\,\omega|x_3x_3')$ are assumed to be symmetric in x_3 and x_3'.

No additional boundary conditions are required when these dispersion relations are used.

Generalization of the latter types of dispersion relations to nonlocal dielectric tensors with off-diagonal elements would be welcome.

7.2 Experimental Results

There is a hope that experiments can help decide between different possible additional boundary conditions. At the present time, however, that hope remains unfulfilled. The use of reflectivity measurements, or measurements of optical transmission through thin films has failed to provide the desired information, because the fairly strong attenuation of surface polariton modes in the frequency range of interest tends to wash out the differences between the predictions obtained by the use of different additional boundary conditions.

In principle, the use of Brillouin scattering to study the exciton-polariton dispersion curve could discriminate among various additional boundary conditions. The theory of such scattering due to Brenig *et al.* [373] yields different predictions for the scattering cross section and the variation of linewidths and intensities as functions of the frequency of the incident light, for different choices of the additional boundary conditions. The experiments carried out so far, however, have lacked the resolution meeded to give information about the additional boundary conditions [374–376].

The experimental observation of the surface exciton polariton [377–384] now offers the possibility of testing macroscopic and microscopic models describing the behavior of excitations near a semiconductor surface. Surface exciton polaritons are elementary excitations at the surface of a crystal whose corresponding bulk excitation is a Wannier (i.e. delocalized) exciton. The correct description of bulk exciton polaritons requires that the effects of spatial dispersion be taken into account, and this is also true for their surface counterparts. The theory of surface exciton polaritons and the experiments that have been carried out to study their properties are well and thoroughly discussed in several recent review articles by Fischer and Lagois [380, 381, 384], so that little more will be said about them here, except to note a very interesting experimental study of them carried out recently by Lagois [382, 383]. In this work he compared the reflectivity spectra and ATR spectra taken on the same crystal surface (ZnO) in the frequency range where

A and B exciton polaritons exist. Both types of spectra were fitted using a bulk dielectric tensor of the form

$$\epsilon_{\alpha\beta}(\vec{k}, \omega) = \delta_{\alpha\beta}\epsilon_\infty \left\{1 + \frac{\Omega_A^2}{\omega_{AT}^2 + \beta_A k^2 - \omega^2 - i\omega\Gamma} + \right. \\ \left. + \frac{\Omega_B^2}{\omega_{BT}^2 + \beta_B k^2 - \omega^2 - i\omega\Gamma}\right\} \tag{58}$$

together with an exciton-free layer of thickness d_0 at the surface. In Eq. (58) ϵ_∞ is the optical frequency dielectric constant, ω_{AT} and ω_{BT} are the A and B exciton eigenenergies at $\vec{k} = 0$, and Γ is an empirical damping constant. The parameters β_A and β_B are related to the effective masses of the corresponding excitons. The vanishing of the exciton contribution to the macroscopic polarization at the interface between the exciton-free layer and the bulk was the additional boundary condition used. The best fits to both sets of experimental data showed that the empirical damping constant Γ appearing in Eq. (58) required to fit the ATR data is about three times larger than the value required to fit the reflectivity data. In addition, the excitonic eigenenergies obtained by fitting the ATR spectra are considerably higher than those obtained by fitting the reflectivity spectra (Fig. 19). Lagois shows further that the reflectivity and ATR spectra are determined by the two different branches of the excitonic polariton spectrum in the crystal, with the result that the incident light probes distances of up to 100 nm into the crystal in the former experiment, but distances of only about 25 nm in the latter. Using a multilayer model in which the excitonic eigenenergies and the damping constant are allowed to decrease toward their bulk values with increasing distance into the crystal from the surface (the former nonmonotonically, the latter monotonically), Lagois was able not only to fit the reflectivity and ATR spectra with the same dielectric function, but was also able to improve the fit to the experimental reflectivity over what was

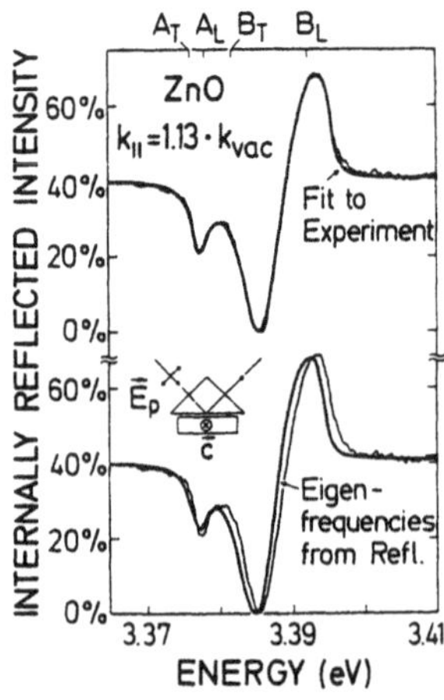

Fig. 19

Attenuation total reflection (ATR) spectra of the $A_{n=1}$ and $B_{n=1}$ excitonic polaritons in ZnO. The wave vector $\mathbf{k}_\parallel$ of the excitonic surface polariton is given by the angle of incidence and the prism's refractive index. Upper part: Experiment (weak line) and theory with constant exciton energy (heavy line) calculated with parameters obtained from a fit to the experimental ATR spectrum. The upper scale gives the transverse and longitudinal exciton eigenenergies. Lower part: Theory with constant eigenenergies obtained from fit to the experimental *reflection* spectrum [383].

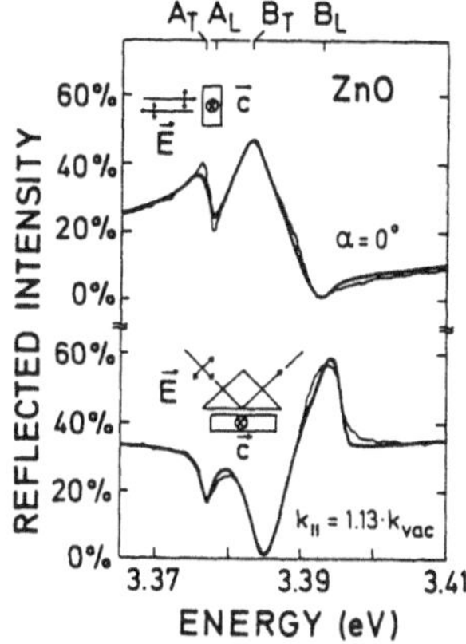

Fig. 20

Normal incidence reflection (upper part) and attenuated reflection (lower part) of the $A_{n=1}$ and $B_{n=1}$ excitonic polaritons in ZnO. The latter experimental spectrum is the same as in Fig. 19. The theoretical spectra (heavy lines) are calculated with the same set of parameters for reflection and ATR, including a depth dependence of the excitonic eigenenergies and damping [383].

possible with a dielectric constant with position-independent excitonic eigenenergies and damping constants (Fig. 20). These experimental results of Lagois, and his interpretation of them, pose a challenge to theorists, viz. to explain the position dependence of the dielectric constant on the basis of a consistent, microscopic model.

8 Surface Polaritons on Curved Surfaces

Like surface acoustic waves surface polaritons can propagate on curved surfaces. On such surfaces they can possess properties that are absent from surface electromagnetic waves propagating over a flat surface.

8.1 Cylindrical, Spherical, and Gently Bent Surfaces

The frequencies and associated electromagnetic fields of polaritons in spherical samples and propagating parallel to the axis of a cylinder of circular cross section have been worked out, e.g. by Englman and Ruppin, both with the neglect of retardation [385] and with its inclusion [386, 387]. The choices of these geometries are dictated by the fact that these are geometries for which Laplace's and, particularly, Maxwell's equations can be solved exactly. Problems of this kind are also discussed in the review articles by Economou and Ngai [261] and by Kliewer and Fuchs [262].

A theory of surface polaritons propagating over gently bent surfaces has been presented by Berry [388], who means by "gently bent" that the radii of curvature are large in comparison with the vacuum wavelength of the surface polariton, so that the surface is sufficiently locally flat for the surface polariton to be almost perfectly bound to it. However, over large distances such surfaces may depart arbitrarily far from any chosen plane and so cannot be considered as perturbations of a plane.

Berry considers propagation of a surface polariton circumferentially around a cylinder of radius R, constructed from a dielectric material with dielectric constant $\epsilon(\omega)$, surrounded by vacuum. In this case he finds that as the wave propagates around the cylinder it radiates energy into the vacuum, and consequently it is attenuated. It is therefore a pseudosurface, or leaky, surface wave. The attenuation length of the surface wave is determined. When the cylindrical surface is concave toward the vacuum a true surface wave is possible which is rigorously nonradiative. These results are then used to infer the properties of surface waves on gently bent surfaces with one and two principal radii of curvature. It is shown that a p-polarized surface wave can exist on any gently bent surface.

I am unaware of any experiments that test these predictions.

8.2 Randomly Rough Surfaces

The propagation of p-polarized surface polaritons over a randomly rough planar surface has been studied theoretically by several authors by a variety of approaches based on Maxwell's and Laplace's equations, in both the retarded [389–392] and unretarded cases [393, 394]. The equivalence of the dispersion relations obtained by the several groups who have studied this problem is now established [395]. These calculations share the feature that they are all carried out to the lowest non-vanishing order in the surface roughness profile function. In the retarded limit, it is found that the surface polariton is damped, due to roughness-induced scattering of the surface polariton into bulk electromagnetic waves and into other surface polaritons of the same frequency but with different (two-dimensional) wave vectors. Indeed, Mills [396] has calculated the inverse attenuation length of surface polaritons on a randomly rough surface by treating such attenuation as a scattering problem and using an energy conservation argument to identify the attenuation length. In addition, it is found that the frequency of a surface polariton is lowered by surface roughness, and this is in agreement with experimental results [397].

A dispersion relation for s-polarized surface polaritons propagating across a randomly rough surface has been obtained by Maradudin [73, 395], and its solution has been obtained numerically [398]. Since such waves cannot exist on a flat surface, the solution represents a roughness-induced, or trapped, surface polariton.

In the unretarded limit each branch of the surface polariton dispersion curve is found to be split into two branches by surface roughness [393–395], a result that has been verified experimentally [399, 400] (Fig. 21). This spliting arises for the following reason [393]. In a periodic structure a gap opens up in the dispersion curve for the corresponding structure without periodicity at a value of the wave vector for which the associated frequency is degenerate with the frequency at a second wave vector differing from the first by a translation vector of the lattice reciprocal to that defined by the periodicity of the structure. The surface polariton curve for a flat surface in the unretarded limit is flat, i.e. it is independent of the wave vector $\vec{k}_{\parallel}$. The frequency at one wave vector $\vec{k}_{\parallel}$ is therefore degenerate with the frequency at any other value of $\vec{k}_{\parallel}$. At the same time, the rough surface can be regarded as the superposition of gratings with all possible periods. Consequently, every point of the flat surface dispersion curve is connected by some reciprocal lattice vector to another point degenerate with it, so that a gap opens up at each point of the curve. It has been speculated that if the calculations were extended to higher order than first in the surface roughness profile function, each branch of the flat surface dispersion curve would split into many more branches [401]. There is no roughness-induced damping of the surface polariton in the electrostatic limit, because the radiative processes that attenuate the surface polariton when retardation is taken into account are not possible in the electrostatic limit.

There is also no inhomogeneous broadening of the surface polariton by surface roughness with or without retardation, because the frequencies of the surface polariton on a flat surface are independent of any parameter specifying the position of the surface [402]. This would not be the case for surface polaritons on a thin film where in the absence of roughness, the surface polariton frequency is a function of the film thickness.

The effect of surface roughness on the frequencies of unretarded surface polaritons in a sphere has been investigated theoretically in the small roughness limit [403]. It is found that each of the plasma frequencies in a smooth sphere [404] is split into two by surface roughness. There is again no broadening of these levels, because their frequencies are independent of the radius of the sphere in the absence of roughness.

Additional references to work on the effects of of random roughness on surface polaritons can be found in Refs. [395] and [405–407].

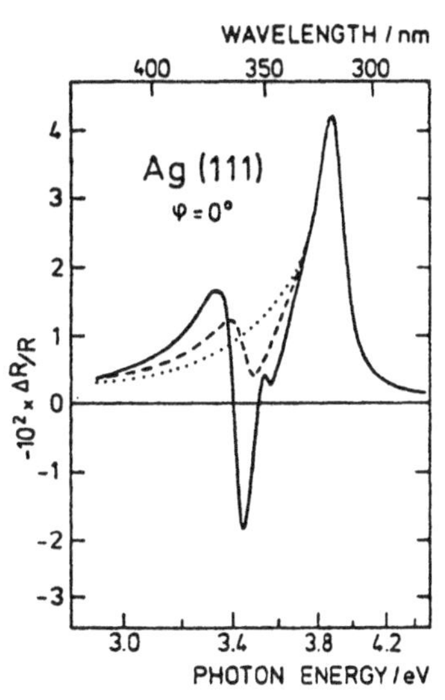

Fig. 21

Spectral dependence of the normal incidence electroreflectance (ER) signal for a smooth (. . .), a slightly rough, and a rough Ag (111) surface. The dashed curve gives the ER signal for a surface slightly roughened by electrochemically dissolving and redepositing 30 monolayers of surface. The solid curve corresponds to a surface prepared by dissolving and redepositing 120 monolayers [400].

8.3 Gratings

The dispersion relation for p-polarized surface polariton propagating perpendicular to the grooves of a large amplitude grating in contact with vacuum has been calculated recently, in both the retarded [408] and unretarded [401] cases, for sinusoidal and symmetric sawtooth grating profiles. Both Rayleigh's method [186] and the extinction theorem form of Green's theorem [190] were used in these calculations. Rayleigh's metod consists of obtaining the solutions of Maxwell's equations, or Laplace's equation, outside the maximum and minimum values of the grating profile function, and then continuing these solutions in to the interface itself, where they are used to satisfy the boundary conditions. It is known in the case of the scattering of a scalar plane wave from a sinusoidally corrugated hard wall, whose profile is given by $\zeta(x_1) = \zeta_0 \cos(2\pi x_1/a)$, that Rayleigh's method is valid as long as $\zeta_0/a < 0.072$ [409–411]. By a theorem due to Millar [412], Rayleigh's theorem is known to be invalid for profiles that contain sharp corners, e.g. the symmetric sawtooth profile function. Calculations based on Green's theorem are, in principle, free from such limitations. However, it was found in both the retarded and unretarded cases that convergent results could be obtained by both approaches for values of ζ_0/a of up to 0.6 in the former case, and of up to 0.25–0.30 in the latter. These results lend support to the conclusion of Toigo *et al.* [391] that the equations obtained by Rayleigh's method are generally valid, and that it is because one usually solves them by simple expansions, e.g. in Fourier series, that convergence for all surface profiles is not obtained. However, because just such simple expansions were used in the work of Refs. [401] and [408], it is possible that the convergence obtained is asymptotic in nature. That is, the inclusion of more plane waves results in surface polariton frequencies that are steadily improving until a critical number of terms is reached; the inclusion of more plane waves in the expansion progressively worsens the value of the frequency obtained. The possibility of this kind of behavior had been noted earlier by Hill and Celli [411].

The results of Ref. [408] show that for surface polaritons on a grating gaps open up in the dispersion curve for surface polaritons on a flat surface at values of the wave vector equal to the boundaries of the one-dimensional Brillouin zones defined by the period of the grating. The widths of these gaps, as functions of ζ_0/a for a sinusoidal grating, were compared with the predictions of first order perturbation theory [413], and the limits of validity of the latter assessed. For example, the calculated gap at $k_{\parallel} = \pi/a$ when $\zeta_0/a = 0.1$ for the sinusoidal profile is $\sim 5\,\%$ smaller than the prediction of first order perturbation theory, while it is $\sim 20\,\%$ too small for $\zeta_0/a = 0.2$. From the results of calculations like these one can estimate quantitatively the effect on the surface polariton frequency of ruling a grating on the surface of a solid to measure the surface polariton dispersion curve (Section 3) (Fig. 22).

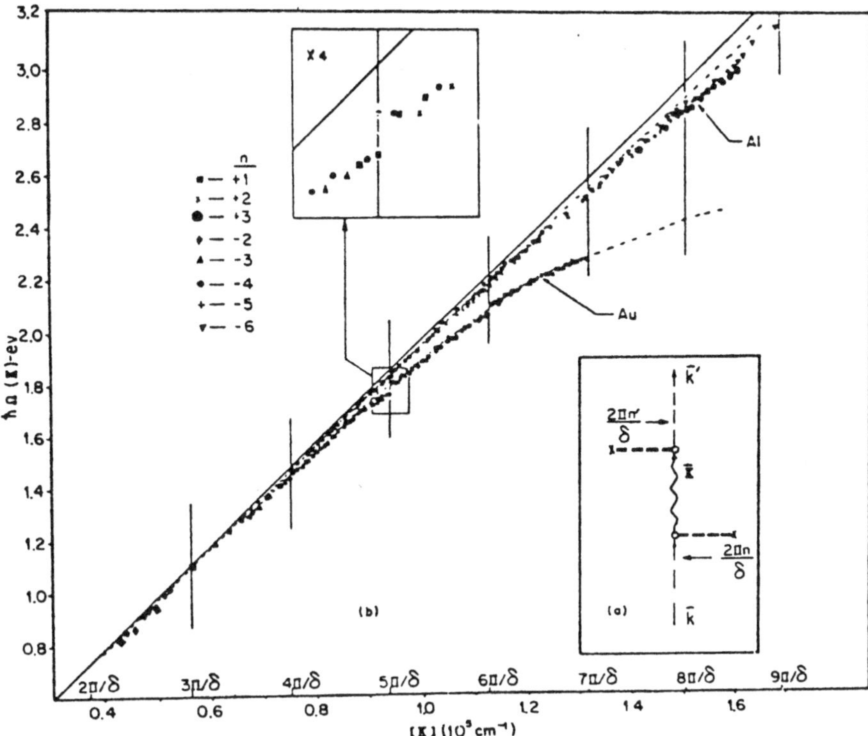

Fig. 22 Dispersion curve for surface polaritons propagating perpendicular to the grooves of a grating ruled on Al and Au surfaces, measured through the reflection of p-polarized light from such a surface. Note that gaps opened up in the Au dispersion curve by the grating at values of $k_{\|} = 5\ \pi/\delta$ and $6\ \pi/\delta$, where δ is the periode of the grating [305].

Discussions of earlier work on the propagation of surface polaritons across a grating, that employed rather special forms of the grating profile can be found in Refs. [413–418].

A calculation of the effects on a surface polariton of propagation over a moving grating, caused by the propagation of a Rayleigh wave across the surface of a solid has been carried out by Mills [419].

Extensive reviews of both the theoretical [395] and experimental [406] aspects of the propagation of surface polaritons across gratings have recently been written, and the interested reader is referred to them for additional information on this subject.

9 Surface Polaritons in Inhomogeneous Media

In the determination of surface polariton dispersion curves in Section 2 it was assumed that the bulk dielectric tensor, or magnetic permeability tensor, of the surface active medium remains uniform right up to the surface itself. However, it is known that, for example, depletion, accumulation, and inversion layers can exist at semiconductor surfaces, under appropriate conditions, with the consequence that the electron number density can vary markedly with distance into the semiconductor from its surface before reaching its limiting bulk value. A considerable interest also attaches to the question of the sensitivity of the dispersion curve for surface plasmons or surface polaritons at a metal surface to the electron number density profile at the surface. Finally, the presence of a thin transition layer at the interface between two media is known to have a considerable effect on the dispersion relation for surface polaritons [420]. The dielectric constant of the transition layer is assumed to vary with the coordinate normal to the plane surfaces bounding the layer. In this section we review the work that has been done on these and other problems in which the effects on surface polaritons of inhomogeneities in the properties of the medium in the vicinity of the surface over which they propagate are important.

9.1 Depletion, Accumulation and Inversion Layers on Semiconductor Surfaces

The earliest theoretical calculation of surface polariton dispersion curves in the presence of a depletion layer at a semiconductor surface was carried out by Wallis *et al.* [421]. These authors used the assumption that the free carrier concentration was zero in the immediate vicinity of the surface, and jumped discontinuously to the bulk value at a distance d from the surface. They found that the dispersion curve in the presence of a depletion layer falls below the dispersion curve obtained with no depletion layer, in qualitative agreement with experimental results of Marschall *et al.* [307, 308] on n-type InSb (Fig. 23).

In subsequent work on this problem the discontinuously varying electron number density was replaced by one that varied continuously with the coordinate normal to the surface. Conwell [422] considered the guided wave polaritons that can propagate in a semi-infinite dielectric medium whose dielectric constant in the medium ($x_3 < 0$) had the form $\epsilon(\omega;\, x_3) = \epsilon_b(\omega) + \Delta\epsilon(\omega)\, f(x_3)$, with $f(x_3) = \mathrm{erfc}(x_3/s)$ or $f(x_3) = \exp(-x_3^2/t^2)$.

A spatially varying dielectric constant of the form

$$\epsilon(\omega;\, x_3) = \epsilon_b(\omega) + \Delta\epsilon(\omega) \exp(x_3/d) \qquad x_3 < 0 \tag{59}$$

was subsequently used by Conwell [423] and by Guidotti *et al.* [424, 425]; the former in the context of a graded index of refraction arising from a composition gradient in a polar material, for example, from the in or out diffusion of impurities or one of the constituents during crystal growth; the latter in connection with ac-

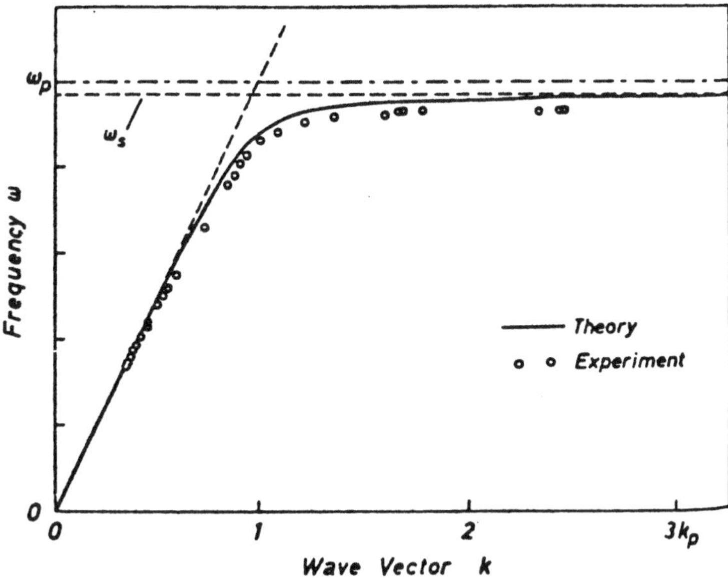

Fig. 23 Experimental and theoretical results for the surface polariton dispersion curve for InSb [307].

cumulation or depletion layers on metal surfaces. Solutions of Maxwell's equation were obtained only in the case that $|\Delta\epsilon/\epsilon_b| \ll 1$. It was found [423] that in the case of an accumulation (depletion) layer the surface polariton dispersion curve lies above (below) the curve obtained without the accumulation (depletion) layer. In their work Guidotti *et al.* [424, 425] found an extra branch in the dispersion curve in the presence of either a depletion or accumulation layer whose existence, however, was questioned by Conwell [423] on the grounds that the assumption that $|\Delta\epsilon_b/\epsilon_b| \ll 1$ was not satisfied by frequencies on this branch. Several additional articles were devoted to this question [426, 427].

The first mathematically rigorous treatment of the effects of depletion, accumulation and inversion layers on surface polariton dispersion curves was carried out by Cunningham *et al.* [428] who used a dielectric constant of the form $\epsilon(\omega; x_3) = \epsilon_\infty[1 - \omega_p^2(x_3)/\omega^2]$ in the semiconductor, where the square of the local plasma frequency, $\omega_p^2(x_3)$, was approximated by a sequence of straight line segments. The fact that $\epsilon(\omega; x_3)$ can vanish for certain values of x_3 in the semiconductor was taken into account in these calculations. No extra branch of the type found by Guidotti *et al.* [424, 425] was found in these calculations. The work of Cunningham *et al.* [428] was subsequently improved on by Kao and Conwell [429, 430], using the same mathematical techniques, but for the exponentially varying form of the dielectric constant given by Eq. (59). They included damping in this dielectric constant, which has the effect of eliminating the singularities in the electric field that are obtained when a real, local dielectric constant is used, and which validates the use of a local dielectric constant. They also found that extra branches can exist, both for depletion and accumulation layers, but they have different properties from those found by Guidotti *et al.* [424, 425]. In the case of depletion layers, where the extra branches have the character of guided wave polaritons, their existence requires larger values of the decay length than were considered by Guidotti *et al.* [424, 425].

A comparison of surface polariton dispersion curves calculated on the basis of step function, linear, and exponential surface electron profiles, but without damping has been carried out by Aers *et al.* [431] using exact and multilayer techniques. They conclude that all of these profiles lead to qualitatively similar dispersion curves.

All of the dielectric constants considered so far have been local. An elaborate calculation, that includes spatial dispersion (on the basis of a fluid model [432, 433]) as well as a double step function electron density profile has been carried out for depletion and accumulation layers by Aers and Boardman [434]. A static, externally applied magnetic field parallel to the surface and perpendicular to the direction of propagation has also been included in their treatment. In the presence of spatial dispersion complex frequencies for real wave vectors are found, in the absence of any intrinsic damping in the system. The modes are non reciprocal, and fall into two types. The first type of mode exists for all wave vectors, while the second exists for only a limited range of wave vectors. The theoretical results are at least in qualitative agreement with the existing experimental results [435, 436]. A similar calculation, but in the electrostatic limit has been carried out by Eguiluz and Quinn [437].

A somewhat similar calculation has recently been carried out, in the nonretarded limit, and in the absence of an external magnetic field, by Das Sarma *et al.* [438]. They consider a degenerate semiconductor that has a low carrier concentration surface layer. Spatial dispersion effects are taken into account on the basis of a hydrodynamical model [439, 440, 441] and the effects of the optical phonons are included. In addition to the usual coupled surface plasmon-surface optical phonon modes [442], which are known to be independent of the electron density profile in the long wavelength unretarded limit, an infinite number of higher frequency modes, called higher multipole excitations [439, 440], are found. These are identified as bulk plasmons of the low density surface region that cannot propagate into the high density region and, hence, are localized near the surface. The experimental observation of such modes has not been reported up to the present time.

The surface polaritons discussed up to now in this section have all been p-polarized waves. S-polarized surface polaritons can also exist on semiconductor surfaces described by the dielectric constant of Eq. (59). Their properties are discussed in Ref. [443].

9.2 Metal Surfaces

There is an extensive literature devoted to the study of the effects of spatial dispersion and a diffuse surface electron density profile on the dispersion curve for surface plasmons at a metal surface. In particular, if the frequency of the surface plasmon is written in the collisionless form

$$\omega_{sp}(k) = \frac{\omega_p}{\sqrt{2}}\left[1 + (a_1 + ia_2)\frac{v_F k}{\omega_p} + (a_3 + ia_4)\frac{v_F^2 k^2}{\omega_p^2} + \ldots\right] \qquad (60)$$

where v_F is the Fermi velocity and ω_p is the bulk plasma frequency, the question is what are the (real) coefficients $a_1, a_2 \ldots$? Most of the attention has focused on a_1 and a_3. The coefficient a_2 in the absence of collisions is due to Landau damping.

On the experimental side, three metals have been studied: Mg [444, 445], Al [445–449] and In [449]. For Mg, Kunz [444] finds that $a_1 < 0$ and $a_3 > 0$. The data of Kloos and Raether [445] show a rather weaker dependence of $\omega_{sp}(k)$ on k, but are not inconsistent with those of Kunz. In the case of Al, the results of Kloos and Raether [445] indicate that $a_1 \approx 0$, $a_3 \approx 0$, while the experiments of Porteus and Fath [447], as analysed by Duke and Landman [448], yield the result that $a_1 > 0$, $a_3 \approx 0$. The discrepancy between the two results for a_1 may be due in part to the very complex analysis required to extract $\omega_{sp}(k)$ from the experimental data of Porteus and Fath [447]. The data of Krane and Raether [449] for Al suggest that $a_1 \gtrsim 0$ and $a_2 > 0$. The result of Refs. [447–449], however, are in disagreement with those of Bagchi *et al.* [446]. The data for In [449] indicate that $a_1 > 0$.

Most theoretical studies of the surface plasmon dispersion curve have been carried out on the basis of two different approaches. The first is a microscopic, quantum mechanical approach based on the random phase approximation (RPA) [438, 450–457]. The second is an essentially classical approach based on a hydrodynamical description of the conduction electrons in a metal [458–462]. The results obtained depend strongly on the assumptions made. For example, in the RPA calculations of Feibelman [451] the static electron surface density profile was assumed to be a step function, and the result that $a_1 = 0$ was obtained (see however, Refs. [452, 463]). In subsequent work, in which the electron surface density profile was assumed to be smoothly varying, Feibelman [445] found that $a_1 \neq 0$, and depended both on the distance over which the surface density profile changes and on its shape. Indeed, Harris and Griffin [453] have pointed out that with the RPA a nonzero value of a_1 can only arise as a consequence of the surface structure, and that in the high density RPA limit the use of either a sharp (e.g. step function) electron surface density profile or an infinite potential barrier for the electrons is not justified (however, see Ref. [454]).

Similar results are found in the hydrodynamic approach. Ritchie [458], using a step function electron density profile, found that $a_1 > 0$. Bennett [459] using an electron density profile that fell to zero linearly over a distance of a few angstroms showed that a_1 could be negative. In addition, Bennett found a number of modes at higher frequencies than $\omega_{sp}(k)$, and these have come to be called "multipole excitations". They were not obtained in the work of Feibelman [451, 455] and of Beck and Celli [454], and have yet to be seen experimentally, however. Boardman *et al.* [462] have used a double step electron surface density profile to show that $\omega_{sp}(k) \to \omega_p/\sqrt{2}$ as $k \to 0$, independently of the shape of the surface density profile, [451], and that the dispersion curve is also insensitive to the shape of this profile. However, the work of Boardman *et al.* [462] has been criticized by Forstmann *et al.* [348–350] on the ground that the boundary conditions they used were inconsistent because they violated the condition that the normal component of the energy current be continuous across boundaries. The effect of this on the conclusions of Boardman *et al.* [462] has not been determined.

The work just described was all carried out in the electrostatic limit. Recently, the effects of retardation have been included in the theory of the dispersion curve of surface plasmons in metals. Eguiluz and Quinn [464] and Boardman *et al.* [465] used a double step electron surface density profile in the hydrodynamic approach, while subsequently the former authors repeated their calculations with smoothly varying profiles [466]. In all these calculations higher frequency modes are found, but Boardman *et al.* [465] argue that their association with multipoles of the fluctuation charge density, as suggested by Eguiluz and Quinn [464] cannot be rigorous. The criticism by Forstmann *et al.* [348–350] of the work in Ref. [462] also applies to that of Refs. [464] and [465], however.

9.3 Transition Layer

The presence of a transition layer at the interface between two media (one of which can be vacuum) can have a significant effect on the surface polariton dispersion curve associated with the interface [420]. By a transition layer is meant a very thin layer occupying the region $0 < x_3 < d$, whose thickness d is very much smaller than λ, the surface polariton wavelength, in which the dielectric properties differ from those of the two media on both sides of it. In some cases d can be as small as ~ 1 Å. Much of the recent work concerning the optical properties of transition layers has been concerned with exploiting the thinness of the layer to obtain effective boundary conditions for the electromagnetic fields in the media on both sides of the layer. The properties of the layer in this approach thus appear only in these boundary conditions.

This has been done by expanding the electromagnetic field components within the transition layer in a series in the Dirac delta function and its derivatives, $\delta^{(n)}(x_3)$ $(n = 0, 1, 2, \ldots)$ and using the theory of generalized functions together with Maxwell's equations to obtain the discontinuities across the layer of the electromagnetic field components in the media on both sides of it [467]. The results, however, are not presented as a systematic expansion in a small parameter, such as d/λ, for example, so that it is not easy to determine the significance of truncating the expansion in $\delta^{(n)}(x_3)$ after some small number of terms.

A more transparent derivation of such boundary conditions has been given by Agranovich and his coworkers on the assumption that the dielectric tensor of the transition layer, $\epsilon_{\alpha\beta}(\omega; x_3)$, is a function of the coordinate normal to the planes bounding it. In the case of an isotropic layer Agranovich and Yudson [468] eliminated the layer, and consequently the boundary conditions at the interface $x_3 = d$ in favor of modified boundary conditions at the interface $x_3 = 0$ that incorporate the effects of the layer exactly to $O(d/\lambda)$. These effective boundary conditions are obtained from the true boundary conditions and Maxwell's equations for the components in much the same way as this has been done for thin elastic plates, either on a semi-infinite substrate or sandwiched between two semi-infinite media [51, 52]. In the case that the layer is anisotropic the corresponding boundary conditions have been obtained by Agranovich *et al.* [469]. The parameters in these effective boundary conditions can be determined exactly if $\epsilon_{\alpha\beta}(\omega; x_3)$ is known exactly; otherwise they are treated as parameters of the theory, to be obtained from experimental results. Using this approach Agranovich and his coworkers have investigated several interesting problems involving thin films on substrates or transition layers on the surface of a semi-infinite medium [470].

An alternative approach to the problem of a transition layer has been taken by Brodsky and Urbakh [471, 472]. They are concerned with obtaining electromagnetic fields in the vicinity of a transition layer when the latter, together with the media on both sides, are spatially dispersive, and the transition layer is inhomogeneous. These fields are sought with an accuracy up to terms of $O(a\omega/c)$, where a is a distance of atomic dimensions and ω is the frequency of the electromagnetic field. Unfortunately, in applying their approach to the determination of the surface polariton dispersion curve all the calculations are carried out in a local approximation for the dielectric constants of the media on both sides of the transition layer. The spatial dispersion of the layer, however, is taken into account. Their starting points is Eq. (55), in which the nonlocal dielectric tensor is written in the form $\epsilon_{\alpha\beta}(\vec{k}_\parallel \omega | x_3 x_3') = \epsilon^{(0)}_{\alpha\beta}(\vec{k}_\parallel \omega | x_3 x_3') + \delta\epsilon_{\alpha\beta}(\vec{k}_\parallel \omega | x_3 x_3')$, where the first term alone gives rise to an exactly solvable problem, while the second is sensibly nonzero only for x_3 and x_3' in the transition layer. Equations (55) are then rearranged to yield a system of equations for only those components of the electromagnetic field that have no discontibuity across the surface, in which $\delta\epsilon_{\alpha\beta}(\vec{k}_\parallel \omega | x_3 x_3')$ plays the role of a perturbation [473]. This procedure allows all terms of first order in $(a\omega/c)$ to be obtained systematically. The results are quite cumbersome, and have been used only in the case that $\epsilon^{(0)}_{\alpha\beta}(\vec{k}_\parallel \omega | x_3 x_3')$ is replaced by a local function.

In addition to its limitation to transition layers for which $d/\lambda \ll 1$ or $(a\omega/c) \ll 1$, the theory of the effects of transition layers on surface polaritons just described differs from the work discussed in section 9.1 in a fundamental manner. This is that the dielectric tensor of the layer is (at least implicitly) assumed not to vanish for any x_3 inside the layer. This is in contrast with the behavior of the $\epsilon(\omega; x_3)$ used in the study of depletion, accumulation, and inversion layers on semiconductor surfaces. All the difficulties in theories of surface polaritons in inhomogeneous systems of the latter type in fact arise from the fact that $\epsilon(\omega; x_3)$ can vanish within these layers.

10 Nonlinear Effects Associated with Surface Polaritons

Surface polaritons are solutions of Maxwell's equations when a linear relation is assumed between the polarization in each medium and the corresponding macroscopic electric field. However, it is well known that all solids possess optical nonlinearities to a greater or lesser extent. These nonlinearities can be expressed by writing the polarization in a medium as the sum of a linear and a nonlinear contribution, $\vec{P} = \vec{P}^L + \vec{P}^{NL}$. The linear polarization can be written formally as $\vec{P}^L = \overleftrightarrow{\chi}^{(1)} \vec{E}$, where $\overleftrightarrow{\chi}^{(1)}$ is the (linear) susceptibility tensor of the medium. The nonlinear polarization takes different forms depending on the nature of the medium being considered, but has the characteristic of containing the components of the electromagnetic fields in the medium to higher than first order.

The introduction of the nonlinear polarization into Maxwell's equations results in several new effects, absent from the linear theory. In particular, it can result in a coupling, or interaction, between the otherwise independent solutions of this equation that are obtained when only the linear approximation to the polarization is employed. It can also give rise to new types of surface polaritons that are essentially nonlinear in nature. We consider both effects in turn.

10.1 Nonlinear Generation and Interaction of Surface Polaritons

Among the nonlinear optical effects that have their origins in the nonlinear polarization are harmonic generation, nonlinear mixing, and optical rectification. The advent of the laser has provided the sources of the powerful electric fields that are required for the experimental study of these and other nonlinear optical effects, and a large body of work now exists devoted to these effects in bulk samples [474].

The nonlinear terms in Maxwell's equations can also give rise to the analogues for surface polaritons of all the nonlinear optical processes that exist for bulk excitations. The earliest theoretical study of nonlinear interactions of surface polaritons is found in unpublished work by Maddox and Mills [475]. These authors studied second harmonic generation and nonlinear mixing of surface polaritons, as well as the generation of volume electromagnetic waves, in the infrared region of the spectrum, by the nonlinear interactions occuring at the plane interface between vacuum and a compound semiconductor of the zincblende structure caused by the second order susceptibility of the latter. In this work the nonlinear electromagnetic fields generated by the dielectric nonlinearities in the semiconductor were obtained by a direct, iterative solution of the nonlinear Maxwell equations, together with the appropriate boundary conditions at the interface and at infinity. Subsequently, these same nonlinear effects, together with the optical rectification of a surface polariton, were studied by Bonsall and Maradudin [476], who used a Green's function formulation of the problem, and concluded that these effects, although weak, might still be observable.

The second harmonic generation of surface polaritons on a metal surface was first studied theoretically by Mills [477]. This is a particularly interesting effect because for frequencies well below the plasma frequency ω_p of the conduction electrons, e.g. in the infrared, the surface polariton curve is nearly linear. Thus the driven output wave of frequency $2\,\omega$ and wave vector $2k_\parallel$, where ω and $k_\parallel$ are the frequency and wave vector of the incident surface polariton, has a wave vector very close to that of a freely propagating surface polariton of frequency $2\,\omega$. Thus the nonlinear interaction occurs under conditions of near phase matching, leading to an enhancement of the efficiency of second harmonic generation. Mills used a Green's function approach in his calculation of the second harmonic fields. On applying his results to the second harmonic generation of a surface polariton on the surface of copper at the frequency of a CO_2 laser, he found "the second harmonic output to be modest, but of detectable magnitude".

Subsequently Fukui and Stegeman [478] examined the same problem but from a different point of view. They argued that it is unrealistic to assume that the amplitude of the second harmonic wave assumes its full value at the point where it starts to be generated, which is the form for the second harmonic wave obtained by a direct application of the Green's function method [476, 477]. Instead, it should increase from zero with increasing distance of propagation of the wave from the point at which it starts to be generated. Fukui and Stegeman obtained the second harmonic fields by an interative solution of the nonlinear Maxwell equation, and then by subtracting appropriate multiples of the fields of a freely propagating surface polariton of frequency $2\,\omega$ they obtained the growing wave fields sought. They conclude, however, that in the case of surface polaritons on silver their mean free path is too short for the growth characteristics of the second harmonic power to be observed under present day experimental conditions.

It has been shown recently [193, 479] how the Green's function approach to second harmonic generation of surface polaritons can be modified to yield the exact growing wave solutions. A simpler, but approximate, method for obtaining such solutions has been presented by Sipe [480].

The nonlinear mixing of two oppositely travelling surface polaritons on a semi-infinite metal has been analyzed by Fukui *et al.* [481, 482]. They show that free space radiation fields and transverse bulk waves in the metal can be generated at the sum frequency of the two surface polaritons.

The nonlinear mixing of two volume electromagnetic waves to produce a surface polariton was suggested by DeMartini and Shen [483], and the idea was implemented experimentally by DeMartini *et al.* [484, 485]. This approach makes possible the generation of surface polaritons of appreciable intensity in the infrared (100–500 cm^{-1}), whose properties can then be studied in a variety of ways.

Other nonlinear interactions in which surface polaritons play a role have been studied experimentally by Simon *et al.* [486, 487].

A review of nonlinear interactions of surface polaritons has recently been written by Stegeman and Fukui [488], to which the reader is referred for a fuller treatment of this subject.

10.2 Nonlinear Surface Polaritons

In view of the current interest in various static and dynamic properties of strongly nonlinear systems [489], it is perhaps not surprising that investigations should be undertaken of surface polaritons on optically nonlinear media.

The first such study seems to have been carried out by Tomlinson [490], who considered s-polarized surface polaritons at the planar interface between two dielectric media, one of which is isotropic (dielectric constant $\epsilon^I(\omega)$) and the other of which is uniaxial and is characterized by a diagonal dielectric tensor whose elements depend on the amplitude of the electric field according to

$$\begin{aligned} \epsilon_{11}(\omega, |\vec{E}|^2) = \epsilon_{22}(\omega, |\vec{E}|^2) &= \epsilon_0(\omega) + a(\omega)(|E_1|^2 + |E_2|^2) \\ \epsilon_{33}(\omega) &= \epsilon(\omega). \end{aligned} \tag{61}$$

He demonstrated the existence of such waves, which have no counterpart in a linear system ($a(\omega) \equiv 0$), and studied their propagation characteristics (e.g. power flow) in the case that $a(\omega) > 0$. These modes were rediscovered independently by Maradudin [491], who obtained their dispersion curves and the associated electric fields for both $a(\omega) > 0$ and $a(\omega) < 0$, on the assumption that the frequency dependence of $a(\omega)$ can be neglected in the frequency range where surface polaritons exist, and that the linear dielectric medium is vacuum. The frequencies of these waves are functions of the amplitude of the electric field of the surface polariton, but are independent of its wave vector. The former is characteristic of nonlinear problems; the latter may be a reflection of the fact that s-polarized surface polaritons do not exist in the linear limit of the system studied.

P-polarized surface polaritons in the same system were discovered by Agranovich [492], also independently of the work of Tomlinson. The forms of the electric field components were obtained in the case that $a(\omega) < 0$, and the dispersion relation for these waves was determined. No solutions of the latter were obtained; however, it was pointed out that p-polarized surface polaritons can exist in the presence of the nonlinear terms on the right hand side of Eq. (61) even when $\epsilon^I(\omega)$, $\epsilon_0(\omega)$, and $\epsilon(\omega)$ are all positive, which is not possible for linear surface polaritons.

One can look forward to additional studies of these nonlinear surface polaritons, both theoretical and experimental. It would, e.g. be of interest to obtain the form of the ATR spectrum associated with such modes.

11 Edge Modes

I conclude this survey of the properties of surface polaritons with a discussion of electromagnetic excitations localized at the apex of a dielectric wedge. These modes are wavelike in the direction parallel to the edge of the wedge, and decay in amplitude with increasing distance from the dielectric-vacuum interface both into the medium and into the vacuum, as well as from the apex of the edge. Such modes were first studied in the electrostatic limit, but recently the first results in which the effects of retardation have been taken into account have been reported.

11.1 Electrostatic Edge Modes

The earliest study of electrostatic edge modes was carried out for a dielectric wedge formed by the intersection of two plane surfaces making an internal angle $2\alpha (r > 0, 0 < \theta < 2\alpha, -\infty < z < \infty)$, and characterized by an isotropic, frequency-dependent dielectric constant $\epsilon(\omega)$ [493]. The complementary space was assumed to be vacuum. Laplace's equation was solved by the method of separation of variables to obtain a scalar potential proportional to $\exp(iqz)$, and localized about the apex of the wedge. The dispersion curves obtained from the boundary conditions possessed some unusual properties. The frequencies of the modes were independent of the wave vector q. This is not surprising, since there is no characteristic length in the structure. However, in addition to depending on the wedge angle 2α, they turned out to be functions of a continuously varying quantum number, which was the separation constant arising from the solution of the differential equation. In addition [494], the electric fields associated with these edge modes were singular, sufficiently so that the field energy was infinite in the neighborhood of the wedge. Hence the fields violate the Meixner criterion [495] and cannot be normalized.

The singular features of these solutions for a sharp wedge are removed if the edge of the wedge is rounded. The first investigation of electrostatic edge modes of a dielectric wedge whose edge has been rounded was carried out by Davis [494] for a wedge whose boundary is a hyperbolic cylinder. The solutions in this case had to be carried out numerically. Subsequently the problem was solved analytically for a wedge whose boundary is a parabolic cylinder [496]. The results of both calculations showed that the singularities in the electric fields of edge modes are removed by rounding the edge of the wedge, and the spectrum of the modes becomes discrete, consisting of an infinite number of branches, and their frequencies are now functions of the wave vector q. The latter effect arises because there is now a characteristic length in the structure, viz. the radius of curvature of the tip of the wedge.

Recently, the dispersion relation has been obtained for surface plasmons propagating parallel to the grooves of a diffraction grating [497]. Such a structure can be regarded as an assembly of an infinite number of parallel wedges. The results for a sinusoidal grating are similar to those for edge modes on a parabolic wedge for large values of q, when the resulting scalar potential is localized near the apex of each wedge forming the grating. For smaller values of q, however, the wave localized to each wedge of the grating becomes more extended and detects the presence of neighboring grooves. The dispersion curves are modified accordingly.

Electrostatic edge modes in other geometries have been studied by Lucas and his coworkers [498].

11.2 Retarded Edge Modes

When the effects of retardation are taken into account, it is the full set of Maxwell equations that has to be solved for the electromagnetic fields and dispersion relation of edge modes at the apex of a dielectric wedge, and not the simpler scalar Laplace equation. Very recently, Boardman *et al.* [499] have succeeded in obtaining solutions of Maxwell's equations that represent retarded edge modes for a wedge whose boundary is a parabolic cylinder. The dispersion relation is obtained in the form of the vanishing of the product of two infinite determinants, where the zeros of each determinant yield the frequencies of modes with a definite parity with respect to the symmetry plane of the wedge. The dispersion relation was solved numerically, and an infinite number of branches of the edge mode spectrum is obtained, with q-dependent frequencies. They are well described by the electrostatic edge mode dispersion curves except in the immediate vicinity of the light line, where the retarded curves bend over and go to zero along the light line, while the unretarded curves tend to finite limits as $q \to 0$ (Fig. 24).

The results of such studies provide a quantitative, theoretical basis for the use of dielectric wedges as wave guides for electromagnetic surface waves.

Acknowledgements

It is a pleasure to thank Dr. R. E. Camley and Dr. B. Fischer for their critical reading of various parts of this manuscript. I am grateful to Professor A. D. Boardman for sending me a preprint copy of his work on retarded edge modes.

I am indebted to Fr. G. Keck, Fr. M. B. Maradudin, Fr. E. Fowler and to Fr. A. Vierhaus for a superb job typing a long manuscript under severe time pressure.

This work was carried out under the auspices of an Alexander von Humboldt Senior U.S. Scientist Award, whose support is gratefully acknowledged. I would like to thank Professor H. Bilz and Professor M. Cardona for the hospitality of the Max-Planck-Institut für Festkörperforschung, where this work was done.

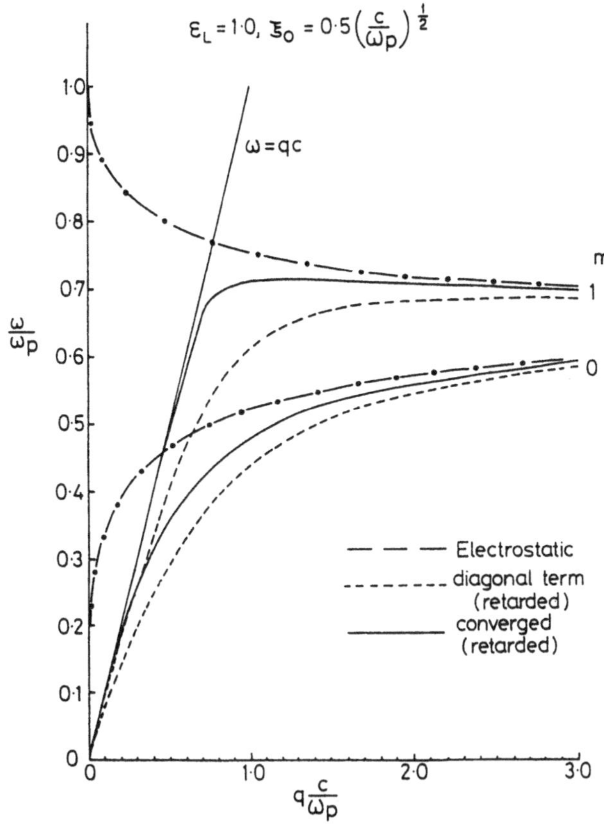

Fig. 24 Dispersion curves for the two lowest order retarded edge polaritons propagating along a metallic parabolic wedge defined by $x_1 = \frac{1}{2}((x_2^2/\xi_0^2) - \xi_0^2)$, compared with the corresponding curves obtained in the electrostatic approximation. The narrow dashed curve is the result of an approximate calculation in which retardation is taken into account. ϵ_L is the background dielectric constant of the metal, and ω_p is its plasma frequency [499].

References

[1] *Lord Rayleigh*, Proc. Lond. Math. Soc. 17, 4 (1887).

[2] *B. A. Auld, Acoustic Fields and Waves in Solids*, vol. II (John Wiley and Sons, New York, 1973), p. 95.

[3] *J. D. Maines* and *E. G. S. Paige*, Proc. IEEE, 64, 639 (1976); *A. A. Oliner*, in *Acoustic Surface Waves*, ed. *A. A. Oliner* (Springer-Verlag, New York, 1978) p. 1.

[4] *F. A. Firestone* and *J. R. Frederick*, J. Acoust. Soc. Am. 18, 200 (1946); see also, the discussion in I. A. Viktorov, *Rayleigh and Lamb Waves* (Plenum, New York, 1967), Chap. III.

[5] See, for example, *L. D. Landau* and *E. M. Lifshitz, Theory of Elasticity* (Pergamon Press, New York, 1970).

[6] *T. C. Lim* and *G. W. Farnell*, J. Appl. Phys. **39**, 4319 (1968); J. Acoust. Soc. Am. **45**, 845 (1969); See also *R. G. Pratt* and *T. C. Lim*, Appl. Phys. Lett. **15**, 403 (1969).

[7] *F. R. Rollins, T. C. Lim*, and *G. W. Farnell*, Appl. Phys. Lett. **12**, 236 (1968).

[8] *R. Stoneley*, Proc. Roy. Soc. (London) **A232**, 447 (1955).

[9] *R. F. S. Hearmon, An Introduction to Applied Anisotropic Elasticity* (Oxford University Press, Oxford, 1961), pp. 5 and 8.

[10] *L. Dobrzynski* and *A. A. Maradudin*, Phys. Rev. **B14**, 2200 (1976).

[11] *G. W. Farnell*, in *Physical Acoustics*, vol. VI, eds. *W. P. Mason* and *R. N. Thurston* (Academic Press, New York, 1970), p. 109.

[12] Ref. [4], pp. 7–25.

[13] *R. M. White* and *F. W. Voltmer*, Appl. Phys. Lett. **8**, 40 (1966).

[14] *W. S. Goruk* and *G. I. Stegeman*, Phys. Lett. **51A**, 419 (1975).

[15] *S. Mishra, G. D. Holah*, and *R. Bray, Proc. 3rd Int. Conf. Light Scattering in Solids* (Flammarion, Paris, 1975), p. 198.

[16] *J. R. Sandercock*, Solid State Comm. **26**, 547 (1978).

[17] *D. L. Mills, A. A. Maradudin*, and *E. Burstein*, Phys. Rev. Lett. **21**, 1178 (1968); in *Light Scattering Spectra of Solids*, ed. *George B. Wright* (Springer-Verlag, New York, 1969), p. 399; Ann. Phys. (N.Y.) **56**, 504 (1970).

[18] *B. I. Bennett, A. A. Maradudin*, and *L. R. Swanson*, in *Proceedings of the Second International Conference on Light Scattering in Solids*, ed. *M. Balkanski* (Flammarion, Paris, 1971), p. 443; Ann. Phys. (N.Y.) **71**, 357 (1972).

[19] *J. R. Sandercock*, Phys. Rev. Lett. **28**, 237 (1972); Phys. Rev. Lett. **29**, 1735 (1972).

[20] *G. Dresselhaus* and *A. S. Pine*, Solid State Comm. **16**, 1001 (1975).

[21] *A. Dervisch* and *R. Loudon*, J. Phys. C **9**, L669 (1976).

[22] *J. Dil* and *E. Brody*, Phys. Rev. **B14**, 5218 (1976).

[23] *R. Loudon*, J. Phys. C **11**, 403 (1978).

[24] *K. R. Subbaswamy* and *A. A. Maradudin*, Indian J. Pure Appl. Phys. **16**, 282 (1978).

[25] *A. Dervisch* and *R. Loudon*, J. Phys. C **11**, L291 (1978).

[26] *R. Loudon*, Phys. Rev. Lett. **40**, 581 (1978).

[27] *R. Loudon*, J. Phys. C **11**, 2623 (1978).

[28] *N. L. Rowell* and *G. I. Stegeman*, Phys. Rev. **B18**, 2598 (1978).

[29] *N. L. Rowell* and *G. I. Stegeman*, Phys. Rev. Lett. **41**, 970 (1978).

[30] *K. R. Subbaswamy* and *A. A. Maradudin*, Phys. Rev. **B18**, 4181 (1978).

[31] *V. Bortolani, F. Nizzoli*, and *G. Santoro*, Phys. Rev. Lett. **41**, 39 (1978); J. Phys. F **8**, L215 (1978); Phys. Lett. **68A**, 342 (1978).

[32] *V. Bortolani, F. Nizzoli, G. Santoro*, and *E. Tosatti*, Solid State Comm. **26**, 507 (1978).

[33] *A. M. Marvin, V. Bortolani, F. Nizzoli*, and *G. Santoro*, J. Phys. C **13**, 299 (1980); J. Phys. C **13**, 1607 (1980).

[34] *L. Bjerkan* and *K. Fossheim*, Solid State Comm. **21**, 1147 (1977).

[35] *E. Krätzig*, Solid State Comm. **9**, 1205 (1971).

[36] *J. Poliquen, M. Depoorter*, and *A. Defebvre, Proc. IVth International Conference on Solid Surfaces*, Cannes, 1980, LeVide, Les Couches Minces, 201, (Suppl. vol. 1), 116 (1980).

[37] *M. Borisov*, Proceedings of the Autumn School on Surface Physics, Varna, Bulgaria, October, 1980 (to be published).

[38] *G. W. Farnell* and *E. L. Adler*, in *Physical Acoustics*, vol. IX, eds. *W. P. Mason* and *R. N. Thurston* (Academic Press, New York, 1972), p. 35.

[39] *A. E. H. Love*, *Some Problems of Geodynamics* (Cambridge University Press, London, 1911).

[40] Ref. [2], p. 97.

[41] Ref. [2], p. 100.

[42] Ref. [2], p. 102.

[43] *H. Lamb*, Phil. Trans. Roy. Soc. (London) **A203**, 1 (1904).

[44] *W. M. Ewing, W. S. Jardetzky*, and *F. Press*, *Elastic Waves in Layered Media* (McGraw-Hill, New York, 1957), pp. 281–288.

[45] Ref. [4], Chap II.

[46] Ref. [2], p. 77.

[47] *K. Sezawa*, Bull. Earthquake Res. Inst. **3**, 1 (1927).

[48] *R. Stoneley*, Proc. Roy. Soc. (London) **A106**, 416 (1924).

[49] *K. Sezawa* and *K. Kania*, Tokyo Bull. Earthquake Res. Inst. **17**, 1 (1939).

[50] *J. G. Scholte*, Roy. Astron. Soc. London, Monthly Notices Geophys. Suppl. **5**, 120 (1947).

[51] *V. R. Velasco, B. Djafari-Rouhani, L. Dobrzynski*, and *F. Garcia-Moliner*, *Proc. IVth International Conference on Solid Surfaces*, Cannes, 1980, Le Vide, Les Couches Minces, **201** (Suppl. vol. II), 774 (1980).

[52] *B. Djafari-Rouhani, L. Dobrzynski, V. R. Velasco*, and *F. Garcia-Moliner* (preprint, 1981).

[53] *H. F. Tiersten*, J. Appl. Phys. **40**, 770 (1969).

[54] *I. Murdoch*, J. Mech. Phys. Solids **24**, 137 (1976).

[55] *V. R. Velasco* and *F. Garcia-Moliner*, Physica Scripta **20**, 111 (1979).

[56] *V. Bortolani, F. Nizzoli, G. Santoro, A. Marvin*, and *J. R. Sandercock*, Phys. Rev. Letters **43**, 224 (1979).

[57] *J. R. Sandercock, F. Nizzoli, V. Bortolani, G. Santoro*, and *A. M. Marvin*, Proc. IVth International Conference on Solid Surfaces, Cannes, 1980, Le Vide, Les Couches Minces, **201** (Suppl. vol. II) (1980).

[58] *D. C. Tseng* and *R. M. White*, J. Appl. Phys. **38**, 4274 (1967).

[59] *C. C. Tseng*, J. Appl. Phys. **38**, 4281 (1967).

[60] *M. I. Kaganov* and *I. L. Sklovskaya*, Fiz. Tver. Tela **8**, 3840 (1966) (Soviet Physics-Solid State **8**, 2789 (1967)).

[61] *J. L. Bleustein*, Appl. Phys. Lett. **13**, 412 (1968).

[62] *Yu. V. Gulyaev*, *Zh. Eksper. i Teor. Fiz, Pis. v. Red.* **9**, 63 (1969) (Soviet Physics-JETP Letters **9**, 37 (1969)).

[63] *C. Maerfeld, F. Gires*, and *P. Tournois*, Appl. Phys. Lett. **18**, 269 (1971).

[64] *R. G. Curtis* and *M. Redwood*, J. Appl. Phys. **46**, 2406 (1975).

[65] *G. Koerber* and *R. F. Vogel*, IEEE Trans. Sonic Ultrason. SU-**19**, 3 (1972).

[66] *C. C. Tseng*, Appl. Phys. Lett. **16**, 253 (1970).

[67] *G. G. Kessenikh, V. N. Lyubimov*, and *D. G. Sannikov*, Kristallografiya **17**, 591 (1972) (Soviet Physics-Crystallography **17**, 512 (1972)).

[68] *G. G. Kessenikh* and *L. A. Shuvalov*, Kristallografiya **21**, 5 (1976) (Soviet Physics-Crystallography **21**, 1 (1976)).

[69] *C. Maerfeld* and *P. Tournois*, Appl. Phys. Lett. **19**, 117 (1971).

[70] *J. Mercereau* and *R. P. Feynman*, Phys. Rev. **104**, 63 (1956).

[71] *L. R. Walker*, Phys. Rev. **105**, 390 (1957).

[72] *R. W. Damon* and *J. R. Eshbach*, J. Phys. Chem. Solids **19**, 308 (1961).

[73] A derivation of the dispersion relation for Damon-Eshbach modes on a semi-infinite ferromagnet can be found, e.g. in *A. A. Maradudin*, Proceedings of the Autumn School on Surface Physics, Varna, Bulgaria, October, 1980 (to be published).

[74] *T. Wolfram* and *R. E. DeWames*, Progr. Surf. Sci., ed. S. Davison, **2**, 233 (1972).

[75] *T. Wolfram*, J. Appl. Phys. **41**, 4748 (1970).

[76] *R. E. Camley* and *A. A. Maradudin*, Solid State Comm. (to be published).

[77] *H. Pfeiffer*, phys. stat. solidi **A18**, K53 (1973); **A19**, K85 (1973).

[78] *P. Grünberg*, J. Appl. Phys. **51**, 4338 (1980).

[79] *I. P. Ipatova, A. A. Klochikhin, A. A. Maradudin*, and *R. F. Wallis*, in *Elementary Excitations in Solids*, eds. *A. A. Maradudin* and *G. F. Nardelli* (Plenum, New York, 1969), p. 476.

[80] *R. F. Wallis, A. A. Maradudin, I. P. Ipatova*, and *A. A. Klochikhin*, Solid State Comm. **5**, 89 (1967).

[81] *B. N. Fillipov*, Fiz. Tver. Tela **9**, 1339 (1967) (Soviet Physics-Solid State **9**, 1048 (1967).

[82] *R. E. DeWames* and *T. Wolfram*, J. Appl. Phys. **41**, 987 (1970); Solid State Comm. **7**, 1451 (1969); Appl. Phys. Lett. **15**, 297 (1969).

[83] *T. Wolfram* and *R. E. DeWames*, Phys. Rev. **B1**, 4358 (1970); Phys. Rev. Lett. **24**, 1489 (1970); Solid State Comm. **8**, 191 (1970).

[84] *D. L. Mills*, in *Localized Excitations in Solids*, ed. *R. F. Wallis* (Plenum, New York, 1968), p. 426; *H. Benson* and *D. L. Mills*, Phys. Rev. **178**, 839 (1969); **188**, 849 (1969).

[85] *T. Wolfram* and *R. E. DeWames*, Physics Letters **30A**, 2 (1969); J. de Physique **32**, C1–1171 (1971).

[86] *P. Grünberg* and *F. Metawe*, Phys. Rev. Letters **39**, 1561 (1977).

[87] *F. Metawe* and *P. Grünberg*, J. Magn. Magn. Mater. **9**, 80 (1978).

[88] *P. H. Chang, A. P. Malozemoff, M. Grimsditch, W. Senn*, and *G. Winterling*, Solid State Comm. **27**, 617 (1978).

[89] *J. R. Sandercock* and *W. Wettling*, IEEE Trans. Magnetics **14**, 442 (1978).

[90] *A. P. Malozemoff, M. Grimsditch, J. Aboaf*, and *A. Brunsch*, J. Appl. Phys. **50**, 5885 (1979).

[91] *A. P. Malozemoff, P. H. Chang*, and *M. Grimsditch*, J. Appl. Phys. **50**, 5896 (1979).

[92] *J. R. Sandercock* and *W. Wettling*, J. Appl. Phys. **50**, 7784 (1979).

[93] *M. Grimsditch, A. P. Malozemoff*, and *A. Brunsch*, Phys. Rev. Lett. **43**, 711 (1979).

[94] *M. Grimsditch, A. P. Malozemoff, A. Brunsch*, and *G. Suran*, in *Proceedings of the International Conference on Magnetism* 1979, (in press).

[95] *R. E. Camley* and *M. Grimsditch*, Phys. Rev. **B22**, 5420 (1980).

[96] *R. E. Camley* and *D. L. Mills*, Phys. Rev. **B18**, 4821 (1978).

[97] *R. E. Camley* and *D. L. Mills*, Solid State Comm. **28**, 321 (1979).

[98] *M. G. Cottam*, J. Phys. **C12**, 1709 (1979).

[99] *R. E. Camley, Talat S. Rahman*, and *D. L. Mills*, Phys. Rev. **B23**, 1226 (1981).

[100] *M. Matthews* and *H. van de Vaart*, Appl. Phys. Lett. **15**, 373 (1969).

[101] *J. P. Parekh*, Electron. Lett. **5**, 323 (1969).

[102] *J. P. Parekh*, Electron. Lett. **6**, 430 (1970).

[103] *J. P. Parekh*, in *Proc. 1972 IEEE Ultrasonics Symposium* (IEEE, Boston, 1972), p. 333.

[104] *J. P. Parekh* and *H. L. Bertoni*, Appl. Phys. Lett. **20**, 362 (1972).

[105] *S. Shen, J. P. Parekh*, and *G. Thomas*, in *Proc. 1974 IEEE Ultrasonics Symposium* (IEEE, Boston, 1974), p. 483.

[106] *J. P. Parekh* and *H. L. Bertoni*, J. Appl. Phys. **45**, 434 (1974).

[107] *P. Emtage*, Phys. Rev. **B13**, 3063 (1976).

[108] *R. Q. Scott* and *D. L. Mills*, Solid State Comm. **18**, 7 (1976).

[109] *R. Q. Scott* and *D. L. Mills*, Phys. Rev. **B15**, 3545 (1977).

[110] *R. E. Camley* and *R. Q. Scott*, Phys. Rev. **B17**, 4327 (1978).

[111] *R. E. Camley*, J. Appl. Phys. **50**, 5272 (1972).

[112] *M. Sparks*, Phys. Rev. **B1**, 3831 (1970).

[113] *M. Sparks*, Phys. Rev. **B1**, 3856 (1970).

[114] *M. Sparks*, Phys. Rev. **B1**, 3809 (1970).

[115] *T. Wolfram* and *R. E. DeWames*, Phys. Rev. **B4**, 3125 (1971).

[116] *R. Henry, S. D. Brown, P. E. Wigen, P. J. Besser*, Phys. Rev. Lett. **28**, 1272 (1972).

[117] *Makoto Tsutsumi, Tushar Bhattacharyya*, and *Nobukaki Kumagai*, J. Appl. Phys. **46**, 5072 (1975).

[118] *R. E. Camley* and *A. A. Maradudin*, Appl. Phys. Lett. (to appear).

[119] *M. F. Lewis* and *E. Patterson*, Appl. Phys. Lett. **20**, 276 (1972).

[120] *S. V. Gerus* and *V. V. Tarasenko*, Fiz. Tver. Tela **17**, 2247 (1975) (Soviet Physics – Solid State **17**, 1487 (1976)).

[121] *Yu. V. Gulyaev, A. Yu. Karabanov, A. M. Kmita, A. V. Medved'*, and *Sh. S. Tursunov*, Fiz. Tver. Tela **12**, 2595 (1970) (Soviet Physics – Solid State **12**, 2085 (1971)).

[122] *D. L. Mills* and *W. M. Saslow*, Phys. Rev. **171**, 488 (1968).

[123] *S. Maekawa*, *Proc. Conf. on Magnetism and Magnetic Materials*, eds. *J. Becker, G. H. Lander*, and *J. J. Rhyne* (AIP, New York, 1975), p. 542.

[124] *R. E. Camley*, Phys. Rev. Lett. **45**, 283 (1980).

[125] *C. Lingner* and *B. Lüthi*, Phys. Rev. **B23**, 256 (1981).

[126] *R. E. Camley* and *P. Fulde*, Phys. Rev. **B** (to appear).

[127] *R. L. Melcher*, Phys. Rev. Lett. **25**, 1201 (1970).

[128] *V. Dohm* and *P. Fulde*, Z. Phys. **B21**, 369 (1975).

[129] *P. Fulde*, in *Handbook on the Physics and Chemistry of Rare Earths*, eds. *K. A. Gschneider* and *L. Eyring* (North-Holland, Amsterdam, 1979), Chap. 17.

[130] *P. S. Wang, B. Lüthi* and *C. Lingner*, Z. Phys. **B34**, 157 (1979); *P. S. Wang* and *B. Lüthi*, Phys. Rev. **B15**, 2718 (1977).

[131] See, for example, *J. D. Jackson*, *Classical Electrodynamics* (John Wiley and Sons, New York, 1962), p. 190.

[132] *W. M. Visscher*, J. Appl. Phys. **51**, 825 (1980); **51**, 835 (1980).

[133] *R. F. Wallis, D. L. Mills,* and *A. A. Maradudin,* Phys. Rev. **B19**, 3981 (1979).

[134] See, for example, *R. Normandin, M. Fukui,* and *G. I. Stegeman,* J. Appl. Phys. **50**, 81 (1979).

[135] *K. Portz, G. I. Stegeman,* and *A. A. Maradudin,* Appl. Phys. Lett. (to be published).

[136] See, for example, *A. E. H. Love, A Treatise on the Mathematical Theory of Elasticity* (Dover, New York, 1944), p. 177.

[137] Ref. [2], inside front cover.

[138] *R. E. Camley* and *A. A. Maradudin,* Phys. Rev. B (to be published).

[139] *A. A. Maradudin* and *D. L. Mills,* Phys. Rev. **173**, 881 (1968).

[140] *P. J. King* and *F. W. Sheard,* Proc. Roy. Soc. (London) **A320**, 175 (1970).

[141] *T. Sakuma* and *T. Nakayama,* Appl. Phys. Lett. **25**, 176 (1974).

[142] *E. Salzmann, T. Plieninger,* and *K. Dransfeld,* Appl. Phys. Lett. **13**, 14 (1968).

[143] *A. J. Budreau* and *P. H. Carr,* Appl. Phys. Lett. **18**, 239 (1971).

[144] *H. J. Maris,* Phys. Rev. **188**, 1308 (1969).

[145] *P. J. King* and *F. W. Sheard,* J. Appl. Phys. **40**, 5189 (1969).

[146] *F. Press* and *I. Healy,* J. Appl. Phys. **28**, 1323 (1957).

[147] *T. Nakayama* and *T. Sakuma,* J. Appl. Phys. **46**, 2445 (1975).

[148] *R. B. Steg* and *P. G. Klemens,* Phys. Rev. Lett. **24**, 381 (1970).

[149] *T. Sakuma,* Phys. Rev. **B8**, 1433 (1973).

[150] *D. K. Dacol* and *A. H. Zimerman,* Phys. Rev. **B11**, 974 (1975).

[151] *D. K. Dacol,* private communication.

[152] *E. I. Urazakov* and *L. A. Fal'kovskii,* Zhur. Eksper. i Teor. Fiz. **63**, 2297 (1972) (Soviet Physics – JETP **36**, 1214 (1973)).

[153] *A. A. Maradudin* and *D. L. Mills,* Ann. Phys. (N.Y.) **100**, 262 (1976).

[154] *A. Eguiluz* and *A. A. Maradudin,* Technical Report No. 80–85 from the Department of Physics, University of California, Irvine.

[155] *A. Eguiluz* and *A. A. Maradudin,* Technical Report No. 81–11 from the Department of Physics, University of California, Irvine.

[156] *L. M. Brekhovskikh,* Soviet Physics – Acoustics **5**, 288 (1960).

[157] *P. V. H. Sabine,* Electronics Lett. **6**, 149 (1970).

[158] *F. Rischbieter,* Acustica **16**, 75 (1965).

[159] *R. F. Humphryes* and *E. A. Ash,* Electronics Lett. **5**, 175 (1969).

[160] *Lord Rayleigh, The Theory of Sound,* vol. II, 2nd Ed. (Dover, New York, 1945), Section 287.

[161] Reference [44] contains a good bibliography of work on this subject.

[162] *G. E. Hudson,* Phys. Rev. **63**, 46 (1943).

[163] *Gerald Rosenfeld* and *Joseph B. Keller,* J. Acoust. Soc. Am. **55**, 555 (1974).

[164] *L. O. Wilson* and *J. A. Morrison,* J. Math. Phys. **16**, 1795 (1975).

[165] *I. A. Viktorov,* Akust. Zh. **4**, 131 (1958) (Soviet Physics – Acoustics **4**, 131 (1958)). See also, Ref. [4], pp. 29–42.

[166] *Benjamin Rulf,* J. Acoust. Soc. Am. **45**, 493 (1969).

[167] *I. A. Viktorov,* Ref. [4], p. 37.

[168] *I. A. Viktorov,* Akust. Zhur. **7**, 21 (1961).

[169] *O. D. Grace* and *R. R. Goodman*, J. Acoust. Soc. Am. **39**, 173 (1966).

[170] *A. I. Lapin*, Abstracts of Papers for the Third All-Union Symposium on Wave Diffraction (in Russian) (Nauka, Moscow, 1964), pp. 108–109.

[171] *V. S. Buldyrev* and *A. I. Lapin*, in *Numerical Methods of Solving Problems in Mathematical Physics* (in Russian) (Nauka, Moscow, 1966), pp. 131–143.

[172] *V. S. Buldyrev*, Vestn. Leningrad Univ. Ser. fiz. Khim. **22**, 38 (1965).

[173] *L. M. Brekhovskikh*, Akust. Zh. **13**, 541 (1967) (Soviet Physics – Acoustics **13**, 462 (1968).

[274] *K. Sezawa*, Bull. Earthquake Res. Inst. (Tokyo) **2**, 21 (1927).

[175] *G. I. Petrashen'*, Dokl. Akad. Nauk USSR **52**, 763 (1946).

[176] *M. S. Khaikin*, Zh. Eksper. i Teor. Fiz. Pis'ma v Red. **4**, 164 (1966) (Soviet Physics – JETP Letters **4**, 113 (1966)).

[177] *T. W. Nee* and *R. E. Prange*, Physics Letters **25A**, 582 (1967).

[178] *R. E. Prange* and *Tsu-Wei Nee*, Phys. Rev. **168**, 779 (1968).

[179] *T. W. Nee*, *J. F. Koch*, and *R. E. Prange*, Phys. Rev. **174**, 758 (1968).

[180] *J. F. Koch* and *J. D. Jensen*, Phys. Rev. **184**, 643 (1969).

[181] *J. F. Koch*, Physik Kondensierten Materie **9**, 148 (1969).

[182] *H. J. Fischbeck* and *J. Mertsching*, phys. stat. solidi **27**, 345 (1968).

[183] *E. A. Kaner*, *N. M. Makarov*, and *I. M. Fuks*, Zh. Eksper. i Teor. Fiz. **55**, 931 (1968) (Soviet Physics – JETP **28**, 483 (1969)).

[184] *N. M. Makarov* and *I. M. Fuks*, Zh. Eksper. i Teor. Fiz. **60**, 806 (1971) (Soviet Physics – JETP **33**, 436 (1971)).

[185] *J. Mertsching* and *H. J. Fischbeck*, phys. stat. solidi **41**, 45 (1970).

[186] *Lord Rayleigh*, Phil Mag. **14**, 70 (1907); *Theory of Sound* vol. II, 2nd Ed. (Dover, New York, 1945), p. 89.

[187] See, for example, *P. Beckmann* and *A. Spizzichino*, *The Scattering of Electromagnetic Waves From Rough Surfaces* (Pergamon, Oxford, 1963).

[188] *Supriyo Datta* and *Bill J. Hunsinger*, *Proc. 1979 Ultrasonics Symposium*, 673 (1979).

[189] *N. E. Glass*, *R. Loudon*, and *A. A. Maradudin* (unpublished work).

[190] *E. Wolf*, in *Coherence and Quantum Optics*, eds. *L. Mandel* and *E. Wolf* (Plenum, New York, 1973), p. 339.

[191] *Supriyo Datta* and *Bill J. Hunsinger*, J. Appl. Phys. **50**, 3370 (1979).

[192] *W. Ludwig* and *B. Lengeler*, Solid State Comm. **2**, 83 (1964).

[193] *A. A. Maradudin*, *R. F. Wallis*, *D. L. Mills*, and *R. L. Ballard*, Phys. Rev. **B6**, 1106 (1972).

[194] *S. L. Moss*, *A. A. Maradudin*, and *S. L. Cunningham*, Phys. Rev. **B8**, 2999 (1973).

[195] *T. M. Sharon*, *A. A. Maradudin*, and *S. L. Cunningham*, Phys. Rev. **B8**, 6024 (1973).

[196] *T. M. Sharon*, *Proc. IEEE Ultrasonics Symposium*, 1973 (IEEE, New York, 1973), p. 126.

[197] *A. A. Maradudin*, Japan. Journ. Appl. Phys. Suppl. **2**, 871 (1974).

[198] *Supriyo Datta* and *Bill J. Hunsinger*, J. Appl. Phys. **49**, 475 (1978).

[199] *Supriyo Datta* and *Bill J. Hunsinger*, Phys. Rev. **B16**, 4224 (1977).

[200] *B. A. Auld*, *J. J. Gagenpain*, and *M. Tan*, Electron Lett. **12**, 650 (1976).

[201] *E. R. Lapwood*, Geophys. J. Roy. Astron. Soc. **4**, 174 (1961); *J. Kane* and *J. Spence*, Geophysics **28**, 715 (1963).

[202] *J. A. Hudson* and *L. Knopoff,* J. Geophys. Res. 69, 281 (1964); *K. Viswanathan, J. T. Kuo,* and *E. R. Lapwood,* J. Geophys. Roy. Astron. Soc. **24**, 401 (1971).

[203] *L. E. Alsop* and *A. S. Goodman,* I.B.M. J. Res. Develop. **16**, 365 (1972).

[204] *M. Munasinghe* and *G. W. Farnell,* J. of Geophys. Res. 78, 2454 (1973); *M. Munasinghe,* Ultrasonics **14**, 9 (1976).

[205] *I. A. Viktorov,* Dokl. Akad. Nauk SSSR, **119**, 463 (1958). See also, *I. Cl. de Bremaecker,* Geophysics **23**, 253 (1958).

[206] *I. A. Viktorov,* Akust. Zhur. 7, 90 (1961).

[207] *R. G. Curtis* and *M. Redwood,* J. Appl. Phys. **46**, 2406 (1975).

[208] *R. G. Curtis* and *M. Redwood,* J. Appl. Phys. **46**, 4627 (1975).

[209] *R. M. White,* Proc. IEEE **58**, 1238 (1970).

[210] *E. G. H. Lean, C. C. Tseng,* and *C. T. Powell,* Appl. Phys. Lett. **16**, 32 (1970); *A. Slobodnik,* J. Acoust. Soc. Am. **48**, 203 (1970).

[211] *V. L. Newhouse, C. L. Chen,* and *K. L. Davis,* J. Appl. Phys. **43**, 2603 (1972).

[212] *M. Luukkala* and *G. S. Kino,* Appl. Phys. Lett. **18**, 393 (1971).

[213] *M. Luukkala* and *J. S. Surakka,* J. Appl. Phys. **43**, 2510 (1972).

[214] *I. A. Viktorov,* Akust. Zhur. **9**, 296 (1963) (Soviet Physics – Acoustics **9**, 242 (1964)).

[215] *F. Rischbieter,* Acustica **18**, 109 (1967).

[216] *A. Hikata, B. Chick,* and *C. Elbaum,* J. Appl. Phys. **36**, 229 (1965).

[217] *P. H. Carr,* IEEE Trans. Sonics and Ultrasonics SU-**13**, 103 (1966).

[218] *V. A. Krasilnikov* and *L. K. Zarembo,* IEEE Trans. Sonics and Ultrasonics SU-**14**, 12 (1967).

[219] *P. O. Løpen,* J. Appl. Phys. **39**, 5400 (1968).

[220] *E. G. Lean, C. T. Powell,* and *L. Kuhn,* Appl. Phys. Lett. **15**, 10 (1969).

[221] *E. G. Lean* and *C. C. Tseng,* J. Appl. Phys. **41**, 3912 (1970).

[222] *A. J. Slobodnik,* J. Acoust. Soc. Am. **48**, 203 (1970).

[223] *W. J. Tanski* and *W. G. Mayer,* Appl. Phys. Lett. **19**, 148 (1971).

[224] *L. R. Adkins,* in *Proc. Ultrasonics Symposium Boston, 1972* (IEEE, New York, 1972), p. 292.

[225] *Y. Nakagawa, K. Yamanouchi,* and *K. Shibayama,* Electron. Comm. Japan. **55A**, 1 (1972).

[226] *W. J. Tanski* and *W. G. Mayer,* J. Appl. Phys. **44**, 506 (1973).

[227] *E. L. Adler, E. Bridoux, G. Coussot,* and *E. Dieulesaint,* IEEE Trans. Sonics and Ultrasonics SU-**20**, 13 (1973).

[228] *J. W. Gibson* and *P. H. E. Meijer,* J. Appl. Phys. **45**, 3288 (1974).

[229] *K. L. Davis* and *V. Newhouse,* IEEE Trans. Sonics and Ultrasonics SU-**22**, 33 (1975).

[230] *P. G. Ivanov̆,* Soviet Physics – Solid State **16**, 2234 (1975).

[231] *P. J. Vella, G. I. Stegeman,* and *V. M. Ristic,* J. Appl. Phys. **48**, 82 (1977).

[232] *A. Alippi, A. Palma, L. Polmieri,* and *G. Socino,* J. Appl. Phys. **48**, 2182 (1977).

[233] *P. J. Vella, T. C. Padmore, G. I. Stegeman,* and *V. M. Ristic,* J. Appl. Phys. **45**, 1993 (1974).

[234] *H. F. Tiersten* and *J. C. Baumhauer,* J. Appl. Phys. **45**, 4272 (1974).

[235] *V. E. Lyamov, T. H. Hsu,* and *R. M. White,* J. Appl. Phys. **43**, 800 (1972).

[236] *K. G. Portz,* Ph. D. Dissertation, University of California, Irvine, 1979 (unpublished).

[237] *R. Torguet, C. Carles, E. Bridoux,* and *M. Moriamez,* Proc. IEEE Ultrasonics Symposium, Boston, 1972 (IEEE, New York, 1972), pp. 147–150.

[238] *R. Normandin, M. Fukui,* and *G. I. Stegeman,* J. Appl. Phys. **50**, 81 (1979).

[239] *R. Normandin, M. Fukui,* and *G. I. Stegeman,* J. Appl. Phys. **50**, 1131 (1979).

[240] *R. E. Camley, A. A. Maradudin,* and *G. I. Stegeman* (unpublished work).

[241] *P. E. Lagasse,* Electronics Lett. **8**, 372 (1972).

[242] *A. A. Oliner,* in *Acoustic Surface Waves,* ed. *A. A. Oliner* (Springer-Verlag, New York, 1978), p. 187.

[243] *P. E. Lagasse, I. M. Mason,* and *E. A. Ash,* IEEE Trans. Micro. Theory Tech. MTT-**21**, 225 (1973).

[244] *B. Aalami,* IEEE Trans. on Sonics and Ultrasonics SU-**20**, 252 (1973).

[245] *E. A. J. Marcatilli,* Bell Syst. Tech. J. **48**, 2161 (1969).

[246] *I. M. Mason, M. D. Motz,* and *J. Chambers, Proc. 1972 IEEE Ultrasonics Symposium* (IEEE, Boston, 1972), p. 314.

[247] See, for example, *O. C. Zienkiewicz, The Finite Element Method in Engineering Science* (McGraw-Hill, New York, 1971).

[248] Ref. [242], p. 211.

[249] See Ref. [243].

[250] *G. D. Boyd, J. McKenna, R. V. Schmidt,* and *R. N. Thurston, Proc. 1973 Ultrasonics Symposium* (IEEE, New York, 1973).

[251] *R. S. Wagers, Proc. 1973 IEEE Ultrasonics Symposium* (IEEE, New York, 1973).

[252] *V. S. Bondarenko, V. F. Dubovitskii,* and *A. P. Krutov,* Fiz. Tver. Tela 17, 3032 (1975) (Soviet Physics – Solid State 17, 2010 (1976)).

[253] See Ref. [243].

[254] *P. E. Lagasse, M. Cabus, M. Verplanhen, Proc. 1974 IEEE Ultrasonics Symposium* (IEEE, New York, 1974).

[255] *K. R. Subbaswamy* and *A. A. Maradudin,* J. Appl. Phys. **48**, 3410 (1977).

[256] *Supriyo S. Datta* and *Bill J. Hunsinger,* Phys. Rev. **B16**, 4224 (1977).

[257] *T. M. Sharon* and *A. A. Maradudin,* Solid State Comm. **13**, 187 (1973).

[258] *A. A. Maradudin, S. L. Moss,* and *S. L. Cunningham,* Phys. Rev. **B15**, 4490 (1977)

[259] *D. L. Mills* and *E. Burstein,* Rep. Prog. Phys. **37**, 817 (1974); *E. Burstein, A. Hartstein, J. Schoenwald, A. A. Maradudin, D. L. Mills,* and *R. F. Wallis,* in *Polaritons,* eds. *E. Burstein* and *F. de Martini* (Pergamon Press, New York, 1974), p. 89.

[260] *V. V. Bryksin, D. N. Mirlin,* and *Yu. A. Firsov,* Usp. Fiz. Nauk **113**, 29 (1974) (Soviet Physics-Uspekhi **17**, 305 (1974).

[261] *E. N. Economou* and *K. L. Ngai,* Adv. Chem. Phys. **29**, 265 (1974).

[262] *K. L. Kliewer* and *R. Fuchs,* Adv. Chem. Phys. **27**, 355 (1974).

[263] *A. Otto,* in *Advances in Solid State Physics,* Festkörperprobleme **XIV**, 1 (1974); *A. Otto,* in *Optical Properties of Solids New Developments,* ed. *B. O. Seraphin* (North Holland, Amsterdam, 1976), p. 677.

[264] *R. J. Bell, R. W. Alexander, C. A. Ward,* and *I. L. Tyler,* Surface Science **48**, 253 (1975).

[265] *J. Nkoma, R. Loudon,* and *D. R. Tilley,* J. Phys. **C7**, 3547 (1976).

[266] *A. A. Maradudin, R. F. Wallis,* and *L. Dobrzynski, Surface Phonons and Polaritons* (Garland STPM Press, New York, 1980).

[267] *S. Ushioda*, in *Progress in Optics*, ed. *E. Wolf* (to appear).
[268] *V. V. Bryksin, D. N. Mirlin*, and *I. I. Reshina*, Zh. Eksper. i Teor. Fiz. Pis. v Red. **16**, 445 (1972) (Soviet Physics-JETP Lett. **16**, 315 (1972)).
[269] *J. J. Brion, R. F. Wallis, A. Hartstein*, and *E. Burstein*, Phys. Rev. Lett. **28**, 1455 (1972).
[270] *J. J. Brion, R. F. Wallis, A. Hartstein*, and *E. Burstein*, Surf. Sci. **34**, 73 (1973).
[271] Ref. [266], p. 297–299.
[272] *P. G. Flahive* and *J. J. Quinn*, Phys. Rev. Lett. **31**, 586 (1973).
[273] *A. Hartstein, E. Burstein, A. A. Maradudin, R. Brewer*, and *R. F. Wallis*, J. Phys. C **6**, 1266 (1973).
[274] Ref. [131], p. 205.
[275] *J. Schoenwald, E. Burstein*, and *J. M. Elson*, Solid State Comm. **14**, 1223 (1974).
[276] *J. Zenneck*, Ann. Phys. (Leipzig) **23**, 846 (1907).
[277] *H. Boersch, J. Geiger*, and *W. Stickel*, Phys. Rev. Lett. **17**, 379 (1966).
[278] *H. Ibach*, Phys. Rev. Lett. **24**, 1416 (1970).
[279] *H. Boersch* and *G. Sauerbrey*, in *Optical Properties and Electronic Structure of Metals and Alloys*, ed. *F. Abelès* (North-Holland, Amsterdam, 1966), p. 386.
[280] *Ye-Yung Teng* and *E. A. Stern*, Phys. Rev. Lett. **19**, 511 (1967).
[281] *A. Otto*, Z. Phys. **216**, 398 (1968).
[282] *A. Otto*, Z. angew. Phys. **224**, 65 (1969); phys. stat. solidi **42**, K37 (1970).
[283] *R. Ruppin*, Solid State Comm. **8**, 1129 (1970).
[284] *V. V. Bryksin, Yu. M. Gerbshtein*, and *D. N. Mirlin*, Fiz. Tver. Tela **13**, 2125 (1972) (Soviet Physics-Solid State **13**, 779 (1972)).
[285] *V. V. Bryksin, Yu. M. Gerbshtein*, and *D. N. Mirlin*, Fiz. Tver. Tela **14**, 543, 3368 (1972) (Soviet Physics-Solid State, **14**, 453, 2849 (1972)).
[286] *N. Marschall* and *B. Fischer*, Phys. Rev. Lett. **28**, 511 (1972).
[287] *A. S. Barker*, Phys. Rev. Lett. **28**, 892 (1972).
[288] *B. Fischer, I. L. Tyler*, and *R. J. Bell*, in *Polaritons*, eds. *E. Burstein* and *F. de Martini* (Pergamon, Oxford, 1974), p. 123.
[289] *B. Fischer, D. Bäuerle*, and *W. J. Buckel*, Solid State Comm. **14**, 291 (1974).
[290] *V. V. Bryksin, D. N. Mirlin*, and *I. I. Reshina*, Solid State Comm. **11**, 695 (1972).
[291] *I. I. Reshina, Yu. M. Gerbshtein*, and *D. N. Mirlin*, Fiz. Tver. Tela **14**, 1280 (1972) (Soviet Physics-Solid State **14**, 1104 (1972)).
[292] *V. V. Bryksin, D. N. Mirlin*, and *I. I. Reshina*, Zh. Eksp. i Teor. Fiz. Pis. v Red. **16**, 445 (1972) (Soviet Physics-JETP Lett. **16**, 315 (1972)).
[293] *H. J. Falge* and *A. Otto*, phys. stat. solidi (b) **56**, 523 (1973).
[294] *B. Fischer, W. J. Buckel*, and *D. Bäuerle*, Solid State Comm. **15**, 1801 (1974).
[295] *C. H. Perry, B. Fischer*, and *W. J. Buckel*, Solid State Comm. **13**, 1261 (1973).
[296] *V. V. Bryksin, D. N. Mirlin*, and *I. I. Reshina*, Fiz. Tver. Tela **15**, 118 (1973) (Soviet Physics-Solid State **15**, 760 (1973)).
[297] *E. D. Palik, R. Kaplan, R. W. Gammon, H. Kaplan, J. J. Quinn*, and *R. F. Wallis*, Phys. Lett. **45A**, 143 (1973).
[298] *A. Harstein* and *E. Burstein*, Solid State Comm. **14**, 2223 (1974).
[299] *E. Kretschmann* and *H. Raether*, Z. Naturforsch. **23a**, 2135 (1968).

[300] *R. Bruns* and *H. Raether*, H. Physik **237**, 98 (1970).

[301] *E. Kretschmann*, Optics Comm. **6**, 185 (1972).

[302] *E. Kretschmann*, Optics Comm. **10**, 353 (1974).

[303] *I. Pockrand* and *H. Raether*, Optics Comm. **18**, 395 (1976).

[304] *J. E. Stewart* and *W. S. Gallaway*, Appl. Opt. **1**, 421 (1962).

[305] *R. H. Ritchie, E. T. Arakawa, J. J. Cowan*, and *R. N. Hamm*, Phys. Rev. Lett. **21**, 1530 (1968).

[306] *I. Pockrand* and *H. Raether*, Appl. Opt. **16**, 1784 (1977).

[307] *N. Marschall, B. Fischer*, and *H. J. Queisser*, Phys. Rev. Lett. **27**, 95 (1971).

[308] *B. Fischer, N. Marschall*, and *H. J. Queisser*, Surface Sci. **34**, 50 (1973).

[309] *D. J. Evans, S. Ushioda*, and *J. D. McMullen*, Phys. Rev. Lett. **31**, 369 (1973).

[310] *J.-Y. Prieur* and *S. Ushioda*, Phys. Rev. Lett. **34**, 1012 (1975).

[311] *J. B. Valdez* and *S. Ushioda*, Phys. Rev. Lett. **38**, 1098 (1977).

[312] *Y. J. Chen, E. Burstein*, and *D. L. Mills*, Phys. Rev. Lett. **34**, 1516 (1975).

[313] *J. S. Nkoma, Proceedings of the Third International Conference on Light Scattering in Solids*, Campinas, Brazil, 1975.

[314] *J. S. Nkoma* and *R. Loudon*, J. Phys. C **8**, 1950 (1975).

[315] *Y. R. Chen, E. Burstein*, and *D. L. Mills, Proceedings of the Third International Conference on Light Scattering in Solids*, Campinas, Brazil, 1975.

[316] *Y. Chen, E. Burstein, D. L. Mills*, and *S. Ushioda, Proceedings of the Fourth International Conference on Raman Spectroscopy*, Brunswick, Maine, 1975.

[317] *D. L. Mills, Y. J. Chen*, and *E. Burstein*, Phys. Rev. **B13**, 4419 (1976).

[318] *V. M. Agranovich* and *T. A. Leskova*, Fiz. Tver. Tela **17**, 1367 (1975) (Soviet Physics-Solid State **17**, 880 (1975)).

[319] *V. M. Agranovich* and *T. A. Leskova*, Fiz. Tver. Tela **19**, 804 (1977) (Soviet Physics-Solid State **19**, 465 (1977)).

[320] *E. A. Vinogradov* and *G. N. Zhizhin*, Zhur. Eksper. i Teor. Fiz. Pis. v Red. **24**, 84 (1976) (Soviet Physics-JETP Lett. **24**, 71 (1977).

[321] *E. A. Vinogradov, G. N. Zhizhin, A. G. Mal'shukov*, and *V. I. Yudson*, Solid State Comm. **23**, 915 (1977).

[322] *E. A. Vinogradov, G. N. Zhizhin*, and *A. G. Mal'shukov*, Zh. Eksper. i Teor. Fiz. **73**, 1480 (1977) (Soviet Physics-JETP **46**, 778 (1977)).

[323] *E. A. Vinogradov, V. I. Yudson*, Zh. Eksper. i Teor. Fiz. **75**, 1919 (1978) (Soviet Physics-JETP **48**, 967 (1978)).

[324] *V. L. Ginzburg* and *I. M. Frank*, J. Phys. USSR **9**, 353 (1945).

[325] *V. E. Pafomov*, in *Nuclear Physics and Interaction of Particles with Matter; Proc. (Trudy) of the P. N. Lebedev Physics Institute*, vol. 44, Chapter 6, and references therein, ed. *A. D. V. Skobel'tsyn* (Consultants Bureau, New York, 1971).

[326] *E. Kröger* and *H. Raether*, Z. Physik **244**, 1 (1971).

[327] *M. F. Bishop* and *A. A. Maradudin*, Solid State Comm. **12**, 1225 (1973).

[328] *D. L. Mills* and *A. A. Maradudin*, Phys. Rev. Lett. **31**, 372 (1973).

[329] *Kenichi Tomioka, Mark G. Sceats*, and *Stuart A. Rice*, J. Chem. Phys. **66**, 2984 (1977).

[330] *M. G. Sceats, K. Tomioka*, and *Stuart A. Rice*, J. Chem. Phys. **66**, 4486 (1977).

[331] *A. Brillante, M. R. Philpott*, and *I. Pockrand*, J. Chem. Phys. **70**, 5739 (1979).

[332] *Michael R. Philpott, Iven Pockrand, Aldo Brillante,* and *J. D. Swalen,* J. Chem. Phys. **72**, 2774 (1980).
[333] *P. K. Tien,* Revs. Mod. Phys. 49, 361 (1977). See also, *R. A. Collin, Field Theory of Guided Waves* (McGraw-Hill, New York, 1960).
[334] *J. B. Valdez, G. Mattei,* and *S. Ushioda,* Solid State Comm. **27**, 1089 (1978).
[335] *K. R. Subbaswamy* and *D. L. Mills,* Solid State Comm. **27**, 1985 (1978).
[336] *K. R. Subbaswamy* and *D. L. Mills,* in *Progress in Optics,* ed. *E. Wolf* (to appear).
[337] *Samir F. Mahmond* and *John C. Beal,* IEEE Transactions on Microwave Theory and Techniques, **MTT-23**, 193 (1975).
[338] *V. V. Shevchenko, Continuous Transitions in Open Waveguides* (Golem, Boulder Colo., 1971).
[339] *G. I. Stegeman, A. A. Maradudin,* and *Talat S. Rahman,* Phys. Rev. B (to appear).
[340] *P. J. Clarricoats* and *K. R. Slinn,* Proc. Inst. Elect. Eng. **114**, 878 (1967).
[341] *J. J. Hopfield* and *D. G. Thomas,* Phys. Rev. **122**, 35 (1961); *G. D. Mahan* and *J. J. Hopfield,* Phys. Rev. **135**, A 428 (1964).
[342] *S. I. Pekar,* Zh. Eksper. i Teor. Fiz. **33**, 1022 (1957) (Soviet Physics-JETP **6**, 785 (1958)); J. Phys. Chem. Solids **5**, 11 (1958).
[343] *V. M. Agranovich* and *V. L. Ginzburg, Spatial Dispersion in Crystal Optics and the Theory of Excitons* (Interscience, London, 1966).
[344] *K. L. Kliewer* and *R. Fuchs,* Phys. Rev. **172**, 607 (1968).
[345] *P. R. Rimbey* and *G. D. Mahan,* Solid State Comm. **15**, 35 (1974).
[346] *J. J. Hopfield,* J. Phys. Soc. Japan Suppl. **21**, 77 (1966).
[347] *R. Zeyher,* in Proc. 1980 Annual Conf. Condensed Matter Division EPS, Antwerpen (Belgium) April, 1980, ed. *J. T. Devreese* (Plenum, New York) (to appear).
[348] *F. Forstmann* and *H. Stenschke,* Phys. Rev. Lett. **38**, 1365 (1977).
[349] *F. Forstmann* and *H. Stenschke,* Phys. Rev. **B17**, 1489 (1978).
[350] *F. Forstmann,* Z. Physik **32**, 385 (1979).
[351] *R. Monreal, F. Garcia-Moliner,* and *F. Flores,* Physica Scripta **22**, 155 (1980).
[352] *A. D. Boardman* and *R. Ruppin* (preprint, 1981).
[353] *M. H. Cohen,* Phys. Rev. **126**, 389 (1962).
[354] *T. H. Stix, The Theory of Plasma Waves* (McGraw-Hill, New York, 1962), p. 73.
[355] *J. L. Birman* and *J. J. Sein,* Phys. Rev. **B6**, 2482 (1972).
[356] *G. Agarwal, D. N. Pattanayak,* and *E. Wolf,* Opt. Comm. **4**, 225 (1971); Opt. Comm. **4**, 260 (1971); Phys. Rev. Lett. **27**, 1022 (1971).
[357] *A. A. Maradudin* and *D. L. Mills,* Phys. Rev. **B7**, 2787 (1973).
[358] *F. Garcia-Moliner* and *F. Flores,* J. Physique **38**, 851 (1977).
[359] *G. Mukhopadhyay* and *S. Lundqvist,* Physica Scripta **17**, 69 (1978).
[360] *D. C. Mattis* and *G. Beni,* Phys. Rev. **B18**, 3816 (1978).
[361] *V. V. Hyzhnyakov, A. A. Maradudin,* and *D. L. Mills,* Phys. Rev. **B11**, 3149 (1975).
[362] *S. Sakoda,* J. Phys. Soc. Japan **40**, 152 (1976).
[363] *A. D'Andrea* and *R. DelSole,* Solid State Comm. **30**, 145 (1979).
[364] *A. Eguiluz* and *A. A. Maradudin,* Ann. Phys. (N.Y.) **113**, 29 (1978).
[365] *C. W. Deutsche* and *C. A. Mead,* Phys. Rev. **138**, A63 (1965).
[366] *G. D. Mahan* and *G. Obermair,* Phys. Rev. **183**, 834 (1969).

[367] *M. R. Philpott*, J. Chem. Phys. **60**, 1410 (1974); **60**, 2530 (1974); Phys. Rev. **B14**, 3471 (1976).

[368] *C. A. Mead* and *M. R. Philpott*, Phys. Rev. **B17**, 914 (1978).

[369] *C. A. Mead*, Phys. Rev. **B17**, 4644 (1978).

[370] *R. H. Ritchie* and *A. L. Marusak*, Surf. Sci. **4**, 243 (1966).

[371] *A. A. Maradudin*, Surf. Sci. **81**, 13 (1979).

[372] *D. N. Pattanayak* and *J. L. Birman*, Solid State Comm. **28**, 927 (1978).

[373] *W. Brenig, R. Zeyher*, and *J. L. Birman*, Phys. Rev. **B6**, 4617 (1972).

[374] *R. G. Ulbrich* and *C. Weisbuch*, Phys. Rev. Lett. **38**, 865 (1977).

[375] *G. Winterling* and *E. Koteles*, Solid State Comm. **23**, 95 (1977).

[376] *C. Hermann* and *P. Y. Yu*, Solid State Comm. **28**, 313 (1978).

[377] *J. Lagois* and *B. Fischer*, Phys. Rev. Lett. **36**, 680 (1976); Solid State Comm. **18**, 1519 (1976); in Proc. 13th Intern. Conf. Phys. Semiconductors, ed. *F. G. Fumi* (North-Holland, Amsterdam, 1976), p. 788.

[378] *I. Hirabayashi, T. Koda, Y. Tokura, J. Murata*, and *Y. Kaneko*, J. Phys. Soc. Japan **40**, 1215 (1976); J. Phys. Soc. Japan **43**, 173 (1977).

[379] *Y. Tokura, I. Hirabayashi*, and *T. Koda*, J. Phys. Soc. Japan **42**, 1071 (1977).

[380] *J. Lagois* and *B. Fischer*, Introduction to Surface Exciton Polaritons, in *Festkörperprobleme – Advances in Solid State Physics*, ed. *J. Treusch* (Vieweg, Braunschweig, 1978), p. 197.

[381] *B. Fischer* and *J. Lagois*, Surface Exciton Polaritons, in *Excitons*, Topics in Current Physics, vol. 14, ed. *K. Cho* (Springer-Verlag, Heidelberg, 1979), Ch. 4.

[382] *J. Lagois*, in Proc. 15th Intern. Conf. on Phys. Semicond., Kyoto, 1980, ed. *S. Tanaka* (to be published).

[383] *J. Lagois*, Phys. Rev. B (to appear).

[384] *J. Lagois* and *B. Fischer*, in *Surface Polaritons*, eds. *V. M. Agranovich* and *D. L. Mills* (North-Holland, Amsterdam, 1981) (to appear).

[385] *R. Englman* and *R. Ruppin*, J. Phys. C **1**, 614 (1968).

[386] *R. Ruppin* and *R. Englman*, J. Phys. C **1**, 630 (1968).

[387] *R. Englman* and *R. Ruppin*, J. Phys. C **1**, 1515 (1968).

[388] *M. V. Berry*, J. Phys. A **8**, 1952 (1975).

[389] *E. Kröger* and *E. Kretschmann*, phys. stat. solidi **76**, 515 (1976).

[390] *A. A. Maradudin* and *W. Zierau*, Phys. Rev. **B14**, 484 (1976).

[391] *F. Toigo, A. Marvin, V. Celli*, and *N. R. Hill*, Phys. Rev. **B15**, 5618 (1977).

[392] *N. R. Hill*, Ph. D. dissertation, University of Virginia (1978) (unpublished).

[393] *E. Kretschmann, T. L. Ferrell*, and *J. C. Ashley*, Phys. Rev. Lett. **42**, 1312 (1979).

[394] *Talat S. Rahman* and *A. A. Maradudin*, Phys. Rev. **B21**, 2137 (1980).

[395] See the discussion of this point by *A. A. Maradudin* in *Surface Polaritons*, eds. *V. M. Agranovich* and *D. L. Mills* (North-Holland, Amsterdam, 1981) (to appear).

[396] *D. L. Mills*, Phys. Rev. **B12**, 4036 (1975).

[397] *S. Ushioda, A. Aziza, J. B. Valdez*, and *G. Mattei*, Phys. Rev. **B19**, 4012 (1979).

[398] *A. A. Maradudin* and *W. Zierau* (unpublished work).

[399] *R. E. Palmer* and *S. E. Schnatterly*, Phys. Rev. **B4**, 2329 (1971).

[400] *R. Kötz, H. J. Lewerenz*, and *E. Kretschmann*, Phys. Lett. **70A**, 452 (1979).

[401] *N. E. Glass* and *A. A. Maradudin*, Phys. Rev. B (to appear).

[402] *A. A. Maradudin* (unpublished work).

[403] *R. G. Lee, Jr., A. A. Maradudin*, and *B. Mühlschlegel* (unpublished work).

[404] See, for example, Ref. [385].

[405] *H. Raether*, in *Physics of Thin Films*, vol. 9 (Academic Press, New York, 1977), p. 145.

[406] *H. Raether*, in *Surface Polaritons*, eds. *V. M. Agranovich* and *D. L. Mills* (North-Holland, Amsterdam, 1981) (to appear).

[407] *S. Ushioda*, in *Progress in Optics*, ed. *E. Wolf* (to appear).

[408] *B. Laks, D. L. Mills*, and *A. A. Maradudin*, Phys. Rev. B (to appear).

[409] *R. Petit* and *M. Cadilhac*, C. R. Acad. Sci. **B262**, 468 (1966).

[410] *R. F. Millar*, Proc. Camb. Phil. Soc. **69**, 175, 217 (1971); Radio Sci. **8**, 785 (1973).

[411] *N. R. Hill* and *V. Celli*, Phys. Rev. **B17**, 2478 (1978).

[412] *R. F. Millar*, Proc. Camb. Phil. Soc. **65**, 773 (1969).

[413] *R. S. Elliot*, IRE Trans. **AP-2**, 71 (1954).

[414] *R. A. Hurd*, Can. J. Phys. **32**, 727 (1954).

[415] *Dean A. Watkins*, *Topics in Electromagmetic Theory* (Wiley, New York, 1958), pp. 14–19.

[416] *R. A. Collin*, *Field Theory of Guided Waves* (McGraw-Hill, New York, 1960), pp. 465–469.

[417] *R. M. Bevensee*, *Electromagmetic Slow Wave Systems* (Wiley, New York, 1964), pp. 57–61.

[418] *R. E. Collin* and *F. J. Zucker*, *Antenna Theory – Part 2* (McGraw-Hill, New York, 1969), pp. 229–232.

[419] *D. L. Mills*, Phys. Rev. **B15**, 3097 (1977).

[420] *V. M. Agranovich* and *A. G. Mal'shukov*, Opt. Comm. **11**, 169 (1974).

[421] *R. F. Wallis, J. J. Brion, E. Burstein*, and *A. Hartstein*, in *Proceedings of the Eleventh International Conference on the Physics of Semiconductors* (PWN – Polish Scientific Publishers, Warsaw, 1972), p. 1448.

[422] *E. Conwell*, Solid State Comm. **14**, 915 (1974).

[423] *E. Conwell*, Phys. Rev. **B11**, 1508 (1975).

[424] *Daniel Guidotti, Stuart A. Rice*, and *Howard L. Lemberg*, Solid State Comm. **15**, 113 (1974).

[425] *S. A. Rice, D. Guidotti, H. L. Lemberg, W. C. Murphy*, and *A. N. Bloch*, in *Advances in Chemical Physics*, eds. *I. Prigogine* and *S. A. Rice* (Wiley, New York, 1974), vol. XXVII.

[426] *E. M. Conwell*, Phys. Rev. **B14**, 5515 (1976).

[427] *D. Guidotti* and *S. A. Rice*, Phys. Rev. **B14**, 5518 (1976).

[428] *S. L. Cunningham, A. A. Maradudin*, and *R. F. Wallis*, Phys. Rev. **B10**, 3342 (1974).

[429] *E. M. Conwell* and *C. C. Kao*, Solid State Comm. **18**,1123 (1976).

[430] *Cheng C. Kao* and *Esther M. Conwell*, Phys. Rev. **B14**, 2464 (1976).

[431] *G. C. Aers, A. D. Boardman*, and *P. A. Young*, J. Phys. C. **10**, 2437 (1977).

[432] *M. C. Steele* and *B. Vural*, *Wave Interactions in Solid State Plasmas* (McGraw-Hill, New York, 1969), pp. 39–66.

[433] *W. P. Allis, S. J. Buchsbaum,* and *A. Bers, Waves in Anisotropic Plasmas* (MIT Press, Cambridge, Mass., 1963), pp. 248–252.

[434] *G. C. Aers* and *A. D. Boardman,* J. Phys. C **11**, 945 (1978).

[435] *A. Hartstein* and *E. Burstein,* Solid State Comm. **14**, 1223 (1974).

[436] *A. Hartstein, E. Burstein, E. D. Palik, R. W. Gammon,* and *B. W. Henvis,* Phys. Rev. **B12**, 3186 (1975).

[437] *Adolfo Eguiluz* and *J. J. Quinn,* Phys. Rev. **B13**, 4299 (1976).

[438] *S. das Sarma, J. J. Quinn,* and *Adolfo Eguiluz* (preprint, 1980).

[439] *A. Eguiluz, S. C. Ying,* and *J. J. Quinn,* Phys. Rev. **B11**, 2118 (1975).

[440] *A. Eguiluz* and *J. J. Quinn,* Phys. Rev. **B14**, 1347 (1976).

[441] *S. das Sarma* and *J. J. Quinn,* Phys. Rev. **B20**, 4872 (1979).

[442] *K. W. Chui* and *J. J. Quinn,* Phys. Lett. **35A**, 469 (1971).

[443] *E. M. Conwell* and *C. C. Kao,* Optics Comm. **17**, 98 (1976).

[444] *C. Kunz,* Z. Physik, **196**, 311 (1966).

[445] *T. Kloos* and *H. Raether,* Phys. Lett. **44A**, 157 (1973).

[446] *A. Bagchi, C. B. Duke, P. J. Feibelman,* and *J. O. Porteus,* Phys. Rev. Lett. **27**, 998 (1971).

[447] *J. O. Porteus* and *W. N. Fath,* Phys. Rev. **B8**, 491 (1973).

[448] *C. B. Duke* and *U. Landman,* Phys. Rev. **B8**, 505 (1973).

[449] *K. J. Krane* and *H. Raether,* Phys. Rev. Lett. **37**, 1355 (1965).

[450] *H. Kanazawa,* Progr. Theor. Phys. **26**, 851 (1961).

[451] *P. J. Feibelman,* Phys. Rev. **176**, 551 (1976); Phys. Rev. **B3**, 220 (1971).

[452] *D. E. Beck,* Phys. Rev. **B4**, 1555 (1971).

[453] *J. Harris* and *A. Griffin,* Can. J. Phys. **48**, 2592 (1970); Phys. Rev. **B3**, 749 (1971); Phys. Lett. **34A**, 51 (1971); Phys. Lett. **37A**, 387 (1971).

[454] *D. E. Beck* and *V. Celli,* Phys. Rev. Lett. **28**, 1124 (1972); Surf. Sci. **37**, 48 (1973).

[455] *P. J. Feibelman,* Phys. Rev. Lett. **30**, 975 (1973).

[456] *J. E. Inglesfield* and *E. Wikborg,* J. Phys. C **6**, L158 (1973).

[457] *R. H. Ritchie,* Surf. Sci. **34**, 1 (1973).

[458] *R. H. Ritchie,* Progr. Theor. Phys. **29**, 607 (1963).

[459] *A. J. Bennett,* Phys. Rev. **B1**, 203 (1970).

[460] *J. Heinrichs,* Phys. Rev. **B7**, 3487 (1973).

[461] *L. Kleinman,* Phys. Rev. **B7**, 2288 (1973).

[462] *A. D. Boardman, B. V. Paranjape,* and *R. Teshima,* Surf. Sci. **49**, 275 (1975).

[463] *D. M. Newns,* Phys. Rev. **B1**, 3304 (1970).

[464] *A. Eguiluz* and *J. J. Quinn,* Phys. Lett. **53A**, 151 (1975).

[465] *A. D. Boardman, B. V. Paranjape,* and *Y. O. Nakamura,* phys. stat. solidi (b) **75**, 347 (1976).

[466] *A. Eguiluz* and *J. J. Quinn,* Phys. Rev. **B14**, 1347 (1976).

[467] *S. F. Timashev* and *M. A. Drykin,* phys. stats. solidi (b) **76**, 67 (1976).

[468] *V. M. Agramovich* and *V. I. Yudson,* Opt. Comm. **15**, 422 (1972).

[469] *V. M. Agranovich, S. A. Darmanyan,* and *A. G. Mal'chukov,* Opt. Comm. **33**, 234 (1980).

[470] A summary of this work is contained in *V. M. Agranovich*, in *Light Scattering in Solids*, eds. *J. L. Birman, H. Z. Cummins*, and *K. K. Rebane* (Plenum, New York, 1979), p. 113.

[471] *A. M. Brodsky* and *M. I. Urbach*, Zh Eksper. i Teor. Fiz. 77, 656 (1979) (Soviet Physics-JETP **50**, 330 (1979)).

[472] *A. M. Brodsky* and *M. I. Urbach*, Surf. Sci. **94**, 369 (1980).

[473] *A. Bagchi, R. G. Barrera*, and *A. K. Rajagopal*, Phys. Rev. **B20**, 4824 (1979).

[474] See, for example, *N. Bloembergen, Nonlinear Optics* (W. A. Benjamin, New York, 1965).

[475] A description of this work is contained in *R. Maddox*, Ph. D. dissertation, University of California, Irvine (1975) (unpublished).

[476] *L. Bonsall* and *A. A. Maradudin*, J. Appl. Phys. **49**, 253 (1978).

[477] *D. L. Mills*, Solid State Comm. **24**, 669 (1977).

[478] *M. Fukui* and *G. I. Stegeman*, Solid State Comm. **26**, 239 (1978).

[479] *G. I. Stegeman* and *A. A. Maradudin* (unpublished work).

[480] *J. E. Sipe*, Solid State Comm. (to appear).

[481] *M. Fukui, J. E. Sipe, V. C. Y. So*, and *G. I. Stegeman*, Solid State Comm. **27**, 1265 (1978).

[482] *M. Fukui, V. C. Y. So, J. E. Sipe*, and *G. I. Stegeman*, J. Phys. Chem. Solids **40**, 523 (1979).

[483] *F. de Martini* and *Y. R. Shen*, Phys. Rev. Lett. **36**, 216 (1976).

[484] *F. de Martini, G. Giuliani, P. Mataloni, E. Palange*, and *Y. R. Shen*, Phys. Rev. Lett. **37**, 440 (1976).

[485] *F. de Martini, M. Colocci, S. E. Kohn*, and *Y. R. Shen*, Bull. Am. Phys. Soc. **22**, 279 (1977).

[486] *H. J. Simon, D. E. Mitchell*, and *J. G. Watson*, Phys. Rev. Lett. **33**, 1531 (1974).

[487] *H. J. Simon, R. E. Benner*, and *J. G. Rako*, Optics Comm. **23**, 245 (1977).

[488] *G. I. Stegeman* and *M. Fukui* (unpublished work).

[489] See, for example, the review by *A. R. Bishop, J. A. Krumhansl*, and *S. E. Trullinger*, Physica **1D**, 1 (1980).

[490] *W. J. Tomlinson*, Optics Lett. **5**, 323 (1980).

[491] *A. A. Maradudin*, Z. Physik B (to appear).

[492] *V. M. Agranovich*, Zh. Eksper. i Teor. Fiz. Pis. v Red. **32**, 532 (1980).

[493] *L. Dobrzynski* and *A. A. Maradudin*, Phys. Rev. **B6**, 3810 (1972).

[494] *L. C. Davis*, Phys. Rev. **B14**, 5523 (1976).

[495] *Josef Meixner*, IEEE Trans. **AP-20**, 442 (1972).

[496] *A. Eguiluz* and *A. A. Maradudin*, Phys. Rev. **B14**, 5526 (1976).

[497] *N. E. Glass* and *A. A. Maradudin* (unpublished work).

[498] *A. Moussiaux, A. Ronveaux*, and *A. Lucas*, Canad. J. Phys. **55**, 1423 (1977).

[499] *A. D. Boardman, G. C. Aers*, and *R. Teshima* (preprint 1981).

High Resolution Electron Energy Loss Spectroscopy on Semiconductor Surfaces

Hans Lüth

2. Physikalisches Institut der Rheinisch-Westfälischen Technischen Hochschule Aachen, Aachen, Germany.

Summary: The application of inelastic electron scattering in reflection to the study of semiconductor surfaces is duscussed. The special type of experimental technique called high resolution electron energy loss spectroscopy (HRELS) is characterized by the use of primary energies between 1 and 10 eV and by an energy resolution better than 15 meV. The method is particularly useful for the investigation of low energy surface excitations with quantum energies below 0.5 eV. Results are reported on well defined surfaces of Si, GaAs and ZnO. On clean surfaces scattering by surface phonons, surface plasmons and free carriers is considered. Furthermore, results for some simple adsorption systems are reported.

1 Introduction

Electron energy loss spectroscopy – recently called ELS or EELS – is a well established electron scattering technique which has been applied successfully in the past to the study of elementary excitations of solids [1, 2, 3, 4]. Because of the "strong interaction" of electrons with solids in particular thin solid films have been studied in transmission and high energy electrons (5–50 KeV) have been applied as probe.

In surface physics a modern type of this technique is used with low energy electrons (1–200 eV) as probe and the experiments are performed in reflection on well defined surfaces prepared in ultrahigh vacuum (UHV). The present paper is concerned with a special version of this technique, sometimes called high resolution electron energy loss spectroscopy (HRELS or HREELS). The probe in this approach is a monoenergetic electron beam with primary energies between 1 and 10 eV and an energy width between 4 and 15 meV, typically. This energetic resolution is achieved by a first monochromator, the electron beam then is projected at the well defined surface – clean or with adsorbate – and the reflected beam is collected by a second monochromator (analyser) where the energy analysis of the backscattered electrons is performed.

The angular resolution of the systems used ranges between 0.5° and 3°; measurements are mainly performed with the analyser looking at the specular beam direction. But recently also some studies have been made about angular distributions of the inelastically scattered electrons [5].

The information which is obtained by this technique resembles that of optical absorption, reflection and Raman spectroscopy, i.e. transition energies between occupied and empty states and excitation energies of collective modes, but in a spatial region close to the surface, are detected. In contrast to optical spectroscopies [6] (information depth 500 – 1000 Å at high absorption) HRELS is more surface sensitive, i.e. the method is of great importance in surface physics, in order to study excitations within the topmost atomic layers of a solid.

Except the first studies of Ibach on ZnO [7, 8] (Ch. 3.3) and on Si surfaces [9, 10] (Ch. 3.1) HRELS has mainly been applied up to now to the investigation of adsorbate vibrations on metal surfaces; a great deal of important information about adsorption and surface reaction mechanisms even with larger molecules like hydrocarbons could be obtained [11]. Also transitions between electronic surface states at energies higher than ~ 0.5 eV have been studied on some well defined semiconductor surfaces [12]. This latter work was not concerned with the low loss energy range below 0.5 eV and therefore the energetic resolution being used was rather poor (100 – 250 meV). Reviews about those studies have been given by Ibach [13] and Froitzheim [14]. A review with particular emphasis on adsorbates on metals and on scattering into directions away from the specular direction has been published by Willis [5].

If one concentrates on semiconductor surfaces, in particular, on the low energy regime of losses between 0 and 0.5 eV, a variety of interesting phenomena might be expected to occur in HRELS studies. Beside collective excitations like surface phonons [8, 15] adsorbate vibrations like on metal surfaces can be studied. Such studies of adsorbates and surface reactions might, in particular, be useful for a better understanding of expitaxial growth of III–V compounds and more generally for all technological processes concerning semiconductor devices. Moreover, low energy electronic excitations being connected with the free carriers of a semiconductor might contribute to the effects being observable in HRELS. In the bulk these effects have already been studied by Raman effect: e.g. collective excitation of conduction band plasmons, single particle scattering on free carriers etc. [16–18]. Similar effects, but now related to the surface have recently been investigated by Matz and Lüth [19, 20] on GaAs by HRELS.

The purpose of the present paper is to review results which have been obtained by HRELS on semiconductor surfaces in the loss energy range 0–0.5 eV. Within the limited number of studies electronic surface state transitions in this energy range have not been found so far. This topic, therefore, is not treated. On the other hand electronic effects being related to free carriers in semiconductors seem to be specific for these materials and are considered in some detail. On these materials a mixture of several phenomena, vibronic, electronic ones and in some cases coupling between those might be taken into account. As far as adsorption systems have been studied on semiconductors, they are also included; the particular case of hydrogen on GaAs shows the intimate relation between adsorption and electronic properties of the substrate space charge layer [21].

The nature of the physical effects rather than experimental details are emphasized in this review. The experimental results are arranged according to the different classes of materials on which they have been obtained. A short discussion is also given on theoretical approaches to the scattering mechanism as far as they are used for the analysis of data on semiconductors.

2 Theoretical Approaches to the Scattering Mechanism

Essentially two different scattering mechanisms might be taken into account in inelastic scattering of low energy electrons on solid surfaces: (i) The electrons interact with the spatially extended dipole field of a particular surface excitation, e.g. a surface phonon, a plasmon or molecular adsorbate vibrations; (ii) the electrons are scattered inelastically on the localized atomic potential of surface atoms (substrate or adsorbate). There is only little work about the latter, so called impact scattering mechanism [5]. In particular, scattering into directions away from the specular beam on adsorbate vibrations of H adsorbed on W (100) [22] and of OH on NiO layers [24] have been interpreted in terms of impact scattering. Most results having been obtained by HRELS on semiconductor surfaces so far, are described in terms of dipole scattering on long range charge density fluctuations. In the simplest picture the solid is regarded as a continuum with a frequency and wavevector dependent dielectric function $\epsilon(\omega, \underline{q}) = \epsilon_1 + i\epsilon_2$. The energy transfer from the primary electron to the solid is ascribed to the damping of the electric field $\underline{\mathcal{E}}$ of the electron within the solid; i.e. according to classical electrodynamics the energy transfer for the Fourier component at frequency ω (quantum energy $\hbar\omega$) and wavevector $\underline{q}$ of a particular elementary excitation of the solid is proportional to $\underline{\mathcal{E}}(\omega, \underline{q})\, \underline{\dot{D}}(\omega, \underline{q})$. The damping of the field is proportional to $\epsilon_2(\omega, \underline{q})$. The field inside the solid is screened by $|\epsilon|^{-1}$ when the electron itself is moving inside the solid and by $|\epsilon + 1|^{-1}$ when the electron considered as a point charge is moving in front of a semiinfinitive dielectric, respectively. The energy loss probability results, therefore, as being proportional to $\epsilon_2/|\epsilon|^2 = -\mathrm{Im}\,\epsilon^{-1}$ for bulk and to $-\mathrm{Im}\,(\epsilon + 1)^{-1}$ for surface scattering. This simple classical "dielectric theory" was originally developed by Fermi [25], Hubbard [26], Fröhlich and Pelzer [27]. Reviews have been given by Raether [1] and Geiger [2].

In reflection experiments like in surface physics in prinicple both scattering mechanisms, bulk scattering (scattering process inside the solid) and surface scattering (scattering outside the solid) appear simultaneously. All experiments having been performed so far in the low energy range ($1\ \mathrm{eV} < E_0 < 10\ \mathrm{eV}$) with high resolution ($\Delta E < 10\ \mathrm{meV}$), however, have been interpreted successfully by taking into account only surface scattering. A main reason might be, that in surface scattering the scattered electrons (loss energy $\hbar\omega_s$) are more sharply peaked around the specular direction (angular halfwidth $\psi_E = \hbar\omega_s/2E_0$) then in bulk scattering. For the geometry used with the common high resolution spectrometers (angular resolution typically 2°) then the amount of bulk scattering related to surface scattering can be estimated to be in the order of 10^{-2}. A more extensive discussion of these problems has been given by Froitzheim [14].

The first semiclassical theoretical description of inelastic scattering in reflection has been given by Lucas and Šunjić [28]. In their treatment the electron is considered as a particle and the crystal surface is treated as an infinitely high repulsive barrier with reflection coefficient of unity. By this theory Lucas and Šunjić could already describe Ibachs experiments on ZnO [7] (see Ch. 3.3) quantitatively.

Dielectric theory for reflection has been treated quantum-mechanically by Evans and Mills [29, 30], and by Mills [31, 32] and later on by Šokčević et al. [33]. A review of some essential points of this theory is included in a paper by Froitzheim [14]. The requirement for elementary excitations giving rise to inelastic scattering consists in the assumption that long-range electric field fluctuations are set up outside the crystal. Dipole interaction is taken into account and the scattering is treated in time dependent Born approximation. Scattering of electrons, which penetrate into the solid, is completely neglected. The conservation of energy

$$E'(\underline{k}') - E_0(\underline{k}_0) = \pm \hbar\omega_s \tag{1}$$

and of wavevector parallel to the surface

$$\underline{k}'_{\parallel} - \underline{k}_{0\parallel} = \underline{q}_{\parallel} \pm \underline{G}_{\parallel} \tag{2}$$

emerges naturally from this theory of surface scattering because of the two-dimensional periodicity of the solid surface. Surface scattering from long-range field fluctuations gives rise to a relatively sharp angle distribution around the direction of specular reflection: The angular half width given by $\psi_E = \hbar\omega_s/2E_0$ results from the fact, that maximum coupling of the primary electron to the field fluctuation is possible at glancing incidence when the electron velocity parallel to the surface resembles the group velocity $\omega_s/q_{\parallel}$ of the surface excitation. This condition allows an estimate of the wavevector transfer parallel to the surface $q_{\parallel} \approx k_0\ (\hbar\omega_s/2E_0)$ (k_0 primary wave vector). The exact expression for the wavevector transfer $\underline{q}_{\parallel} = (q_{\parallel x}, q_{\parallel y}, 0)$ derived from energy and wavevector conservation (Equs. 1 and 2) is [29]:

$$\underline{k}' - \underline{k}_0 = \begin{pmatrix} q_{\parallel x} \\ q_{\parallel y} \\ \Delta k_z \end{pmatrix} = k_0 \begin{pmatrix} [\hbar\omega_s/2E_0]\ \sin\theta_I - \psi\cos\theta_I \cos\varphi \\ \psi \sin\varphi \\ [\hbar\omega_s/2E_0]\ \cos\theta_I + \psi \sin\theta_I \cos\varphi \end{pmatrix}. \tag{3}$$

θ_I is the angle of incidence measured with respect to the surface normal. The angles ψ and φ determine the $\underline{k}'$ direction of the inelastically scattered electron: ψ denotes the angle between $\underline{k}'$ and the direction of specular reflection, whereas φ is the azimuthal angle of $\underline{k}'$ around the specular direction; $\underline{k}'(\varphi = 0)$ lies in the plane of incidence between the surface normal and the specular direction.

For primary energies in the 10 eV and loss energies in the 20 meV range $q_{\parallel}$ is estimated to be in the order of 10^{-3} Å^{-1}, i.e. collective excitations with $\underline{q}$ near the Γ point of the Brillouin zone are excited only. In most cases, therefore, the wave-vector dependent dielectric function $\epsilon(\omega, \underline{q})$ can be substituted by $\epsilon(\omega, q = 0)$ in dielectric theory.

Taking into account only single inelastic scattering events (Born approximation) the processes shown in Fig. 1 contribute to the inelastic scattering cross section. With the assumptions $\hbar\omega_S/2E_0 \ll 1$, i.e. also $k_z' \approx k_z \gg q$ for small angle scattering as discussed above, only the two-step processes of Fig. 1 (loss-diffraction (a) and diffraction-loss (b)) have to be taken into account and the inelastic scattering cross section is obtained as [29, 31]:

$$\frac{d^2S}{d(\hbar\omega)\, d\Omega} = \frac{m^2 e^2 v_\perp^2}{8\pi^2 \epsilon_0 \hbar^4 \cos\theta_I} \left(\frac{k'}{k_0}\right) \times$$

$$\times \frac{|v_\perp q_\parallel (R_s + R_I) + i(R_I - R_s)(\omega - \underline{v}_\parallel \cdot \underline{q}_\parallel)|^2}{q_\parallel^2 [v_\perp^2 q_\parallel^2 + (\omega - \underline{v}_\parallel \cdot \underline{q}_\parallel)^2]^2} \times P(\underline{q}_\parallel, \omega); \quad (4)$$

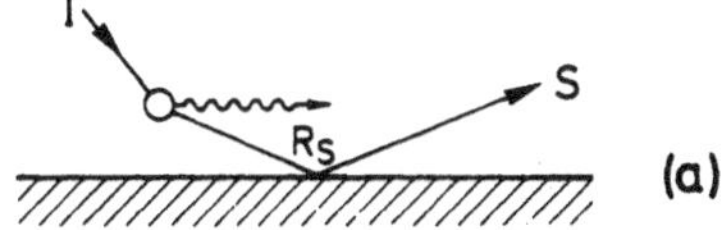

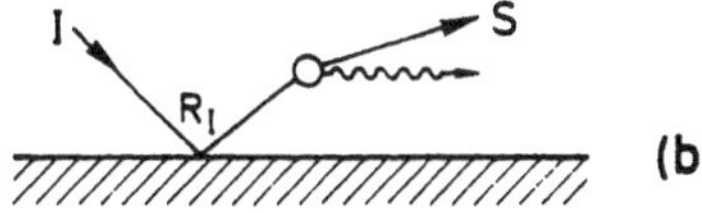

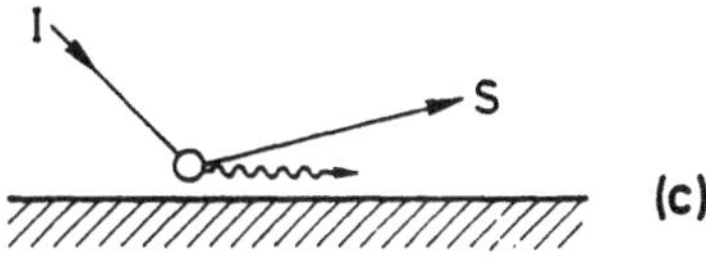

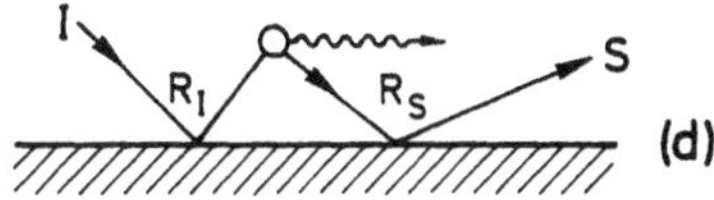

Fig. 1
Schematic representation of the four processes by which an incident electron (I) is scattered (S) into a direction close to specular reflection. The circles denote the point where the electron is scattered by electric field fluctuations in the vacuum above the surface. For small angle inelastic scattering only the processes (a); loss-diffraction (reflection coefficient R_s) and (b): diffraction-loss (reflection coefficient R_I) have to be taken into account.

$v_{\parallel}$ and $v_{\perp}$ are the components of the electron velocity parallel and normal to the surface, respectively; R_I and R_s are the complex reflectivity coefficients of the electron before and after suffering an energy loss $\hbar\omega$. For an energy loss due to scattering on an isotropic semiinfinite half space the function $P(\underline{q}_{\parallel}, \omega)$ is obtained as

$$P(\underline{q}_{\parallel}, \omega) = [1 + N(\omega)] \frac{2q_{\parallel}}{\pi} \operatorname{Im} \left(\frac{-1}{\epsilon(\omega, \underline{q}_{\parallel}) + 1} \right); \quad (5)$$

i.e., it essentially represents the classic surface loss function. $N(\omega)$ is the Bose occupation probability.

There are important cases, where near the surface a spatially varying dielectric constant is given: e.g. the electronic properties near the surface might be modified by electronic surface states localized in the topmost atomic layers or the vibronic properties near the surface might be different from those of the bulk. These problems can be treated conveniently by introducing a surface dielectric function $\epsilon_s(\omega)$ [12] within a certain surface layer as is done in the analysis of optical spectroscopy data [6]. Under certain conditions the essential structures of a loss spectrum are then determined by Im $\epsilon_s(\omega)$ [12]. The same procedure would be possible if an adsorbate being described as an array of independent oscillating dipoles on the surface forms the topmost atomic layer. In this case, however, an atomistic treatment as has been presented by Evans and Mills gives more interesting information [29]. In particular, the orientation quasi-selection rule is derived, that vibrations with a dynamic dipole moment parallel to the surface scatter electrons less by a factor $|\epsilon|^{-2}$ than those with dipole moment perpendicular to the surface. The underlying physical reason for this rule is the image dipole within the substrate material, which partially compensates the effect of a parallel dipole and enhances the action of a dipole normal to the surface (Fig. 2). For most semiconductors $|\epsilon|$ is in the order of 10, i.e. the orientation selection rule should be applied and might give interesting information about the orientation of adsorbed molecules.

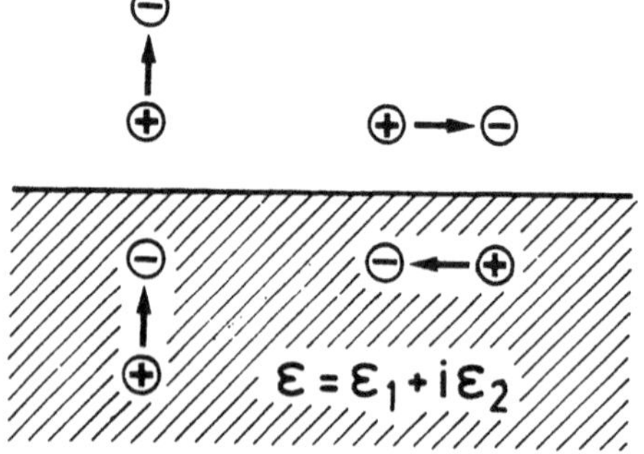

Fig. 2
Qualitative explanation of the orientation quasi-selection rule for dipole surface scattering: The image dipoles within the substrate (dielectric function $\epsilon = \epsilon_1 + i\,\epsilon_2$) enhance (normal orientation) or compensate (parallel orientation) the effect of dipoles being located in the vacuum above the surface.

In optical reflection spectroscopy with IR-light being reflected under grazing incidence the dipole orientation selection rule is strictly valid, since both adsorbed dipole and image dipole "see" essentially the same electric field of the light ($\lambda \sim 10^4$ Å). On the other hand, the limitations of this selection rule in HRELS might be seen from a comparison of the relevant electron wavelengths being in the order of 1–5 Å with the dipole/image dipole distance. For larger adsorbed molecules and for larger chemisorption bond lengths the adsorbed dipole and its image dipole can not be considered as one unity and the dipole orientation selection rule might break down.

Newns has recently considered theoretically inelastic surface scattering of electrons on only one single molecule adsorbed on a metal surface [34, 35]. Dipole interaction between the long-range electrostatic potential of the electron and the time-varying dipole moment of the adsorbed molecule is the only scattering mechanism being taken into account. He arrives at essentially the same conclusions concerning the orientation selection rule, peaking of scattering intensity around specular direction etc. as do Evans and Mills in their treatment of an independent array of adsorbed molecules. A thorough quantum-mechanical treatment including image charge effects of the scattered primary electron has also been given by Persson [36]. A further discussion of these theoretical approaches is beyond the scope of this paper, in particular, since they have not been applied up to now to studies on semiconductor surfaces.

3 Experimental Results

3.1 Silicon Surfaces

Within the class of elemental semiconductors the Si (111) surface has been studied by Ibach [9] and Ibach et al. [10] by means of HRELS in the loss energy range below 0.5 eV. Electronic transitions between surface states at energies higher then 0.5 eV not being the subject of this paper have been investigated both on Si (111) and on Ge (111) [12].

3.1.1 Surface Phonons on Si (111)

Fig. 3 shows loss spectra measured on a Si (111) surface with (2x1) superstructure prepared by cleavage in UHV. The spectra are measured each time after exposures to increasing dosages of oxygen. The prominent loss at 56 meV is the only one, which is observed on the clean surface. A corresponding gain peak not shown in Fig. 3 occurs at – 56 meV. The intensity of the gain peak is related to the loss intensity by the Boltzmann factor exp $(\hbar\omega_{56}/k_B T)$ at room temperature. At 5 eV primary energy the intensity of the loss peak I_{56} is smaller than that of the zero loss line I_0 of the elastically scattered electrons by about 7×10^{-3}. According to Fig. 3 this loss at 56 meV sensitively responds to adsorbate coverages: Adsorbed oxygen decreases its intensity and makes it finally vanish. From these observations

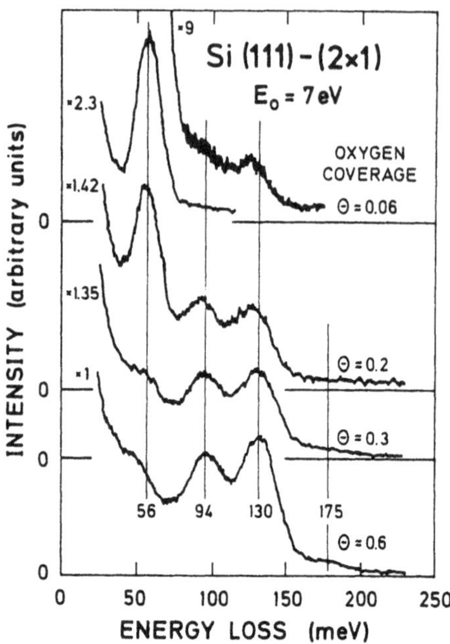

Fig. 3

Energy loss spectra of a Si (111) surface with (2 × 1) superstructure prepared by cleavage in UHV. The spectra are measured after exposures to oygen. Θ = 1 means monolayer coverage. The amplification factors are calculated taking into account the elastic intensity I_0 and a variation of the acceptance angle (After H. Ibach et al. [10]).

Ibach already interpreted the loss as due to surface scattering on optical surface phonons which are accompanied by a long-range dipole field [9]. Due to the cubic symmetry optical bulk phonons in Si (at 65 meV at Γ point) are not infrared-active. At the surface, however, the symmetry is lowered – the two atoms of the unit mesh on (111) are in different symmetry positions – and a vibration of the two sublattices may set up a long-range electric field derived from an effective dynamic charge in the topmost atomic layers.

Further arguments for this interpretation are derived from the experimentally observed angular dependence of the 56 meV loss. Strong peaking around the specular direction is found: The observed angular half width corresponds well to the theoretical values expected for long-range Coulomb scattering [8, 9]. Furthermore, according to Fig. 4 the $(\cos \theta_I)^{-1}$ angular dependence expected for surface scattering (Equ. (4)) is nearly found [14]. Small deviations are ascribed by Froitzheim to bulk scattering contributions or to surface roughness effects [14].

Since in surface scattering only the wave vector $k_{\|}$ parallel to the surface is conserved, also the excitation of bulk phonons with $q_{\|}$ near Γ but $q_{\perp}$ from throughout the whole Brillouin zone might be taken into account in the interpretation. According to their dispersion such optical bulk phonons with $\underline{q}$ perpendicular to (111) exist at energies between 65 meV and 52 meV [37]. There is no reason why these phonons should not be excited with comparable probability, i.e. one would expect in this

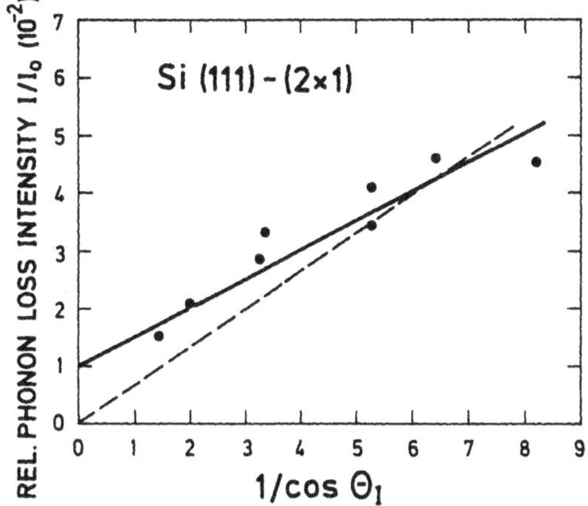

Fig. 4
Dependence of the relative phonon loss ($\hbar\omega = 56$ meV) intensity on the angle of incidence Θ_I.
I_0 ist the intensity of the elastic peak. The $(\cos\Theta_I)^{-1}$ dependence expected for surface scattering is plotted in broken line (After H. Froitzheim [14]).

case a broad structure ranging from 52 to 65 meV and folded with the experimental halfwidth of about 15 meV. In the measured spectrum, however, the phonon line shows a halfwidth which is nearly equal to that of the elastically scattered electron beam. Excitation of bulk phonons, therefore, can be ruled out.

Based on these arguments Evans and Mills [29] have assumed in a microscopic model, that the loss peak at 56 meV is due to surface scattering on a two-dimensional wave-like dipole layer which is set up by vibrations of the two sublattices against each other in the topmost atomic layer. According to the arguments given in Ch. 2 (orientation selection rule) only the dipole moment of vibrations perpendicular to the surface is taken into account in the calculation. For the scattering cross section with $q_{\parallel} \approx 0$ Evans and Mills [29] derive

$$\frac{dS}{d\Omega} = \frac{4mn_0e^2e_\perp^{*2}|R|^2\epsilon^2}{(4\pi\epsilon_0)^2\,\hbar\omega_sME_0\cos\theta_I(1+\epsilon)^2} \times \frac{1}{1-\exp(-\hbar\omega_s/k_BT)}$$
$$\times\frac{(\psi\cos\varphi\cos\theta_I-\psi_E\sin\theta_I)^2+\psi^2\sin^2\varphi}{(\psi^2+\psi_E^2)^2}; \qquad (6)$$

$e_\perp^*$ is the effective dipole charge for phonon vibrations perpendicular to the surface, E_0 is the primary energy; $\psi_E = \hbar\omega_s/2\,E_0$ with $\hbar\omega_s$ as loss energy, M is the Si atomic mass and n_0 is the two-dimensional density of dipole active surface unit cells. The angles θ_I, ψ and φ have already been explained in connection with Equ. (3). The reflection coefficient R is assumed to be a smooth function of energy, i.e. $R_I \approx R_s = R$. Assuming for simplicity a circular aperture of the analyser rather than a rectangular slit like in the experiment integration of the inelastic intensity over the acceptance cone yields an effective dipole charge $e_\perp^* = 0.7e$ for the surface phonon on Si (111).

Based on the experimental data of Ibach [9] Thoma and Zimmermann [38] and Zimmermann [39] have calculated in a simple lattice dynamical model dispersion curves of surface phonons on a non-reconstructed Si (111) surface. From these calculations also a weakening of the force constants in or between the first two topmost atomic layers in the order of 10 % is inferred.

The connection of this type of surface phonon on Si (111) with the particular (2 X 1) superstructure does not seem to be clear at present. In preliminary experiments on the (7 X 7) superstructure obtained by annealing Ibach [8] could not observe the loss at 56 meV. Because of a finite background in the experimental data, however, surface phonon excitation with considerably lower effective dipole charge $e_\perp^*$ on the (7 X 7) superstructure can not be excluded.

3.1.2 Oxygen Adsorption on Si (111)

The effect of oxygen adsorption on the loss spectrum of a well defined Si (111) surface with (2 X 1) superstructure is also shown in Fig. 3 [10]: For coverages lower than $\theta \approx 0.2$ beside the surface phonon loss new oxygen characteristic vibration losses appear at 90 and 125 meV. For higher coverages the two new losses shift to 94 and 130 meV and a small additional shoulder at 175 meV appears.

The measurements have been performed with the analyser in the direction of specular reflection and the interpretation has been given in terms of dielectric theory for surface scattering, i.e. on top of the Si substrate the adsorbed oxygen forms a layer in which molecular vibrations with dynamic dipole moment give rise to long-range electric fields outside the crystal. These fields are responsible for surface scattering. The orientation quasi-selection rule must be applied, that preferentially vibrations with dipole moment perpendicular to the surface are observed. Oxygen must, therefore, be chemisorbed in a configuration which enables three vibration modes with dipole moment perpendicular to the surface. Since the highest excitation energy of 175 meV resembles the O–O stretching frequency (195 meV) of the free O_2 molecule, Ibach et al. [10] originally suggested two possible adsorption models:

(i) an upright bridge-bonded O–O$<^{\text{Si}}_{\text{Si}}$ configuration, or

(ii) a perioxidic Si$\diagup$O–O$\diagdown$Si arrangement. In Model (i) the weak high energy band arises from a weak dipole moment being induced in the originally nonpolar O_2 by the chemisorption bonds to the surface; in model (ii) the O–O stretch vibration parallel to the surface is also accompanied by a weak vibration component perpendicular to the surface. A more detailed quantitative discussion for this case has been presented by Froitzheim [14]. Adsorption models in which the oxygen – even dissociated atomic oxygen – is bonded in a configuration with lower symmetry can also yield three normal modes with dipole moment perpendicular to the surface. From cluster calculations Goddard et al. [41, 42] conclude that the most favourable configuration would be an asymmetric Si$\diagup$O–O$\cdot\cdot$Si molecular configuration.

With regard to the observed vibration frequencies the HREIS results are compatible with all three molecular types of adsorption. Even a dissociative adsorption on a site of lower symmetry can not be ruled out completely but seems less probable with regard to the observed vibration modes, in particular that at 175 meV.. It is worthwhile to mention, however, that from other experimental results [143, 144] and from theoretical calculations [45] a dissociative adsorption is concluded. An unequivocal, final answer to the question of oxygen adsorption in the submonolayer range on Si (111) cannot be given at present.

3.1.3 Hydrogen Adsorption on Si

The adsorption of atomic hydrogen on Si (111) surfaces has first been studied by Froitzheim et al. [46]. The surfaces had been prepared by cleavage in UHV, i.e. the (2 X 1) superstructure was given on the clean surface. Only one vibrational loss at 257 meV due to adsorbed hydrogen was found. A relatively high background signal up to loss energies of about 150 meV, however, might have prevented the detection of further low energy vibrational losses. The loss at 257 meV has been attributed by the authors to the Si-H stretching vibration, the H atoms being bonded on top of the surface Si atoms. A strong argument for the interpretation is derived from a self-consistent first principle calculation of the electronic structure of the Si-H system by Appelbaum and Hamann [47]: For H adsorbed in on-top sites a force constant for the Si-H bond is derived which agrees well with the value obtained from the experimental vibration frequency [46].

Results of recent measurements by Wagner et al. [40] with a better signal to background ratio are shown in Fig. 5. On Si (111) with (7 X 7) superstructure prepared by ion bombardment and annealing the same loss at 257 meV like on Si (111)–(2 X 1) is observed. But additional peaks occur at 78 meV and 108 meV. Due to a burning W filament the exposure temperature of the substrate was about 100 °C. With increasing H coverage a (1 X 1) reconstruction develops [40]. All three

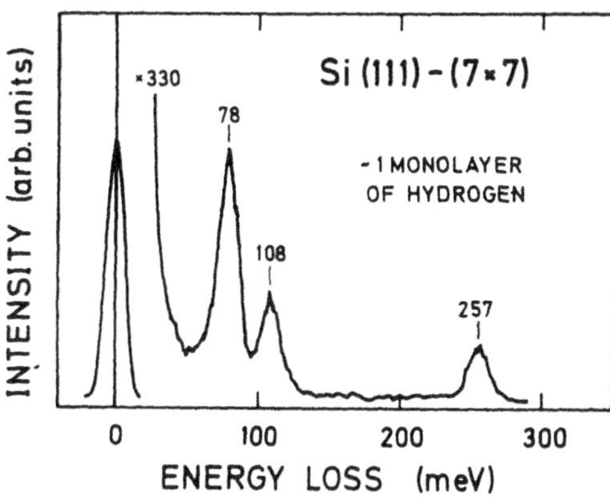

Fig. 5

Loss spectrum of a Si (111) – (7 × 7) surface after exposure to atomic hydrogen [(1 × 1) reconstruction]; measurement under angle of specular reflection (After H. Wagner et al. [40]).

hydrogen losses in Fig. 5 are observed in the whole exposure range, from low coverages up to saturation. Essentially the same results like on Si (111)–(7 × 7) have also been found on Si (100)–(2 × 1) prepared by annealing in UHV [40]. From this observation on Si (100) and from a comparison with IR data the authors were lead to the following interpretation: Like on Si (111)–(2 × 1) the 257 meV loss is due to the stretching vibration of H-atoms being bonded in on-top sites as Si-H. The corresponding Si-H bending mode might partially contribute to the 78 meV loss. The 10 meV loss is ascribed to the wagging and scisser modes of SiH_2 complexes. Their rocking vibration, then, should significantly contribute also to the loss at 78 meV. Meanwhile Wagner et al. [40] could support this interpretation by further experiments also for H-adsorption on the Si (111)–(7 × 7) surface (Fig. 5). The interpretation implies that even on Si (111)–(7 × 7) SiH_2 complexes exist from low coverages up to saturation, i.e. adsorbing H breaks Si-Si bonds or the adsorbing atoms find enough broken Si-bonds to form the SiH_2 complexes. The results of Wagner et al. [40] do not seem to be consistent with adsorption sites which have been suggested by other authors [47–49].

3.2 The Gallium Arsenide (110) Surface

Within the class of III-V compound semiconductors only GaAs (110) surfaces prepared by cleavage in UHV have been investigated by Matz and Lüth [19, 20] recently in the energy loss range below 0.5 eV. Previous work of Froitzheim and Ibach with worse resolution (~ 100 meV) was concerned only with electronic transitions at energies higher than 1 eV [50].

3.2.1 Surface Phonons and Plasmons

In contrast to the elemental semiconductors like Si (Ch. 3.1) the existence of IR-active bulk TO-phonons in compound semiconductors leads to the theoretical prediction of surface phonons being observable in HRELS. Born-Huang dynamical equations for the two "ionic" sublattices (long wavelength limit: $\underline{q} \approx \underline{0}$) together with Maxwell equations have solutions with both div $\underline{P} = \underline{0}$ and curl $\underline{P} = \underline{0}$ for a semiinfinite halfspace ($\underline{P}$ is polarisation). The solutions are surface phonons, usually called Fuchs-Kliewer phonons; they are accompanied by a long-range electric field:

$$\mathcal{E}_x = \mathcal{E}_0 \sin (q_{\parallel} x - \omega t) \exp (- q_{\parallel} |z|), \tag{7a}$$

$$\mathcal{E}_z = \mathcal{E}_0 \cos (q_{\parallel} x - \omega t) \exp (- q_{\parallel} |z|) \operatorname{sgn} (z), \tag{7b}$$

where the surface phonon wavevector $\underline{q}_{\parallel}$ is parallel to the surface in x-direction. The crystal described by its dielectric function $\epsilon(\omega) = \epsilon_1(\omega) + i\epsilon_2(\omega)$ fills the halfspace with $z < 0$.

The mathematical form (Equs. (7a, b)) and Fig. 6 show that the character of these phonons is mixed: transverse – longitudinal, and that the electric field decays into the crystal and into the vacuum with a penetration depth of their inverse wavenumber $1/q_{\parallel}$.

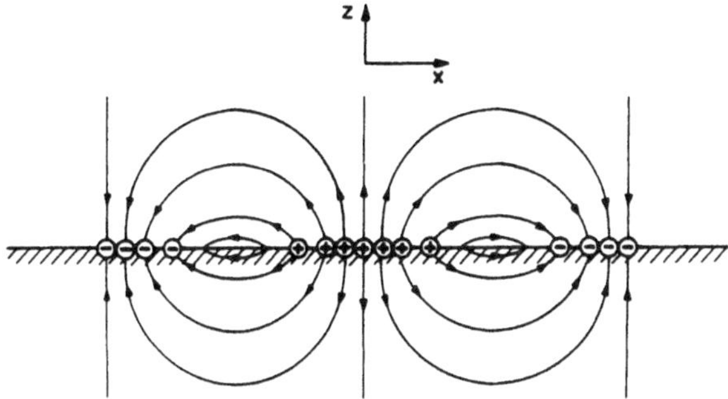

Fig. 6 Electric field produced by a long wavelength surface charge fluctuation, given e. g. with a Fuchs-Kliewer surface phonon. Inside the material ($z < 0$) the arrows indicate the direction of the polarisation field.

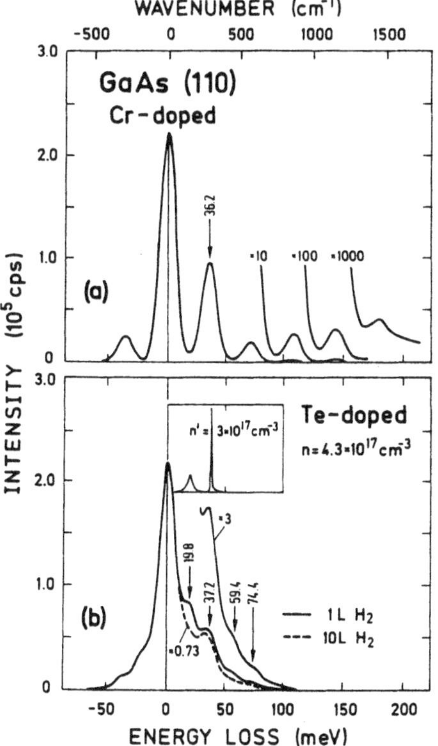

Fig. 7

(a) Loss spectrum of a clean cleaved GaAs (110) surface of semiinsulating material (angle of incidence 8o°).

(b) Loss spectra measured on an n-type sample after exposures to atomic hydrogen (angle of incidence 70°; H dosages unknown). Inset: Calculated surface loss function $-\operatorname{Im}(1+\epsilon)^{-1}$ in arbitrary units; $\epsilon(\omega)$ contains contributions from the TO lattice oscillator and from the free electron gas (density $n' = 3 \times 10^{17}\ cm^{-3}$) (Equ. (12)) (After R. Matz and H. Lüth [20]).

The excitation exists at only one frequency

$$\omega = \omega_{TO} \left(\frac{\epsilon(0)+1}{\epsilon(\infty)+1} \right)^{1/2}, \tag{8}$$

for which the condition

$$\epsilon(\omega) = -1 \tag{9}$$

is satified, as is discussed in detail elsewhere [2, 4, 15]. The condition of Equ. (9) for the frequency of the Fuchs-Kliewer surface phonons is essentially identical with the requirement that the surface loss function $-\text{Im}\,[1+\epsilon(\omega)]^{-1}$ has a maximum in a spectral range of negligible Im $\epsilon(\omega)$. In surface scattering Fuchs-Kliewer surface phonons, therefore, are expected to cause the predominant spectral structures. On ZnO these phonons have been observed experimentally by Ibach in HRELS [7] (see Ch. 3.3) and by Lüth in surface photoconductivity [51].

Results of HRELS measurements on clean GaAs (110) surfaces of high resistivity, Cr-doped material with negligible free carrier concentration are shown in Fig. 7a [19, 20]. The series of energetically equidistant gain and loss peaks indicates multiple scattering on one and the same excitation, whose excitation energy is derived from the energetic distance as being (36.2 ± 0.2) meV. With the well known optical IR-data [52] Equ. (8) yields a quantum energy of 36.6 meV for the corresponding Fuchs-Kliewer surface phonon. This numerical agreement and the observation of a sharply peaked scattering intensity around the direction of specular reflection strongly supports the interpretation of Fig. 7a in terms of excitation of Fuchs-Kliewer surface phonons. The intensity distribution of the multiphonon losses I_m obeys a Poisson distribution law

$$P(m) = I_m / \sum_{\nu} I_{\nu} = (m!)^{-1} Q^m \exp(-Q) \tag{10}$$

(Fig. 8a) as is expected from theory [28, 30]. In a simple quantum-mechanical picture this Poisson distribution results for the transition probability of the lattice oscillator from the ground state into the m-th excited state if the passing electron is described as a time dependent perturbation. Q then is essentially the Fourier transform of the time dependent perturbation. As is expected for elementary Bose-type excitations in thermal equilibrium "Stokes and Anti-Stokes" line intensities should be related by a Boltzmann factor, i.e. the intensity ratio of the gain (I_{-m}) and loss peaks (I_m) should be

$$I_{-m}/I_m = \exp(-m\hbar\omega/k_B T). \tag{11}$$

According to Fig. 8b this relation is found for the experimental data. Using Evans' and Mills' theoretical expression for the inelastic scattering cross section [29] Matz and Lüth have calculated from the experimental data in Fig. 7a an effective ionic charge e* being responsible for the dipole field of the Fuchs-Kliewer surface

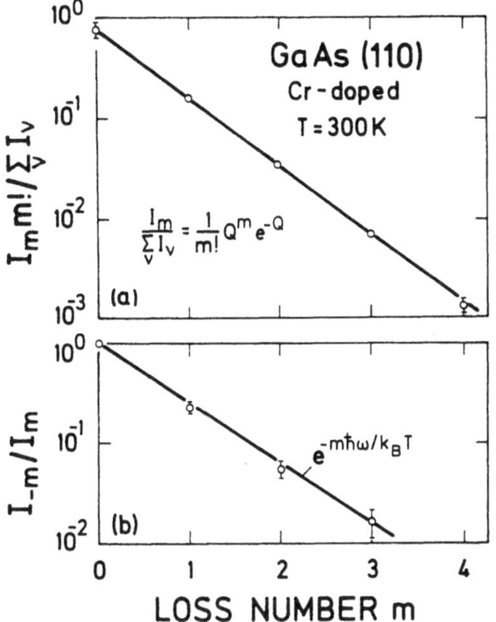

Fig. 8
(a) Poisson distribution of the loss intensities I_m measured on a clean semiinsultaling GaAs surface (Fig. 7a).
(b) Intensity ratio of the m-th phonon gain and the m-th phonon loss versus loss number m. The straight line is calculated with $\hbar\omega$ = 36.0 meV. (After R. Matz and H. Lüth [19]).

phonons [19]. In agreement with the Szigeti charge from IR optical data (0.51 e [52]) this effective charge results as (0.53 ± 0.03)e. The agreement is not surprising, since the penetration depth of the Fuchs-Kliewer surface phonons can be estimated by means of Equ. (7) from the wavevector transfer parallel to the surface $q_{\parallel} \approx k_0(\hbar\omega/2E_0) \sin \theta_I$ (see Equ. 3) to be larger than 300 Å. Bulk properties should, therefore, be relevant for the dynamics of the surface phonon.

On (111) surfaces of n-type GaAs with carrier concentrations between 10^{17} and 10^{18} cm^{-3} Matz and Lüth measured spectra with a different spectral shape [20] (Fig. 7b). On clean cleaved surfaces and also after exposure to small dosages of dissociated hydrogen (up to 1 L) or to residual gas a series of loss and gain peaks is observed at energies $\hbar\omega_+$ resembling those of the Fuchs-Kliewer surface phonon. Additional gains and losses (including multiples) emerge with a considerably smaller quantum energy $\hbar\omega_-$. The energetic position of these losses $\hbar\omega_-$ appeared to be sensitive to bulk doping and to band bending being varied by gas treatment. The authors, therefore, interpreted these losses as being derived from surface plasmon-like excitations of the "free" electron gas in the conduction band of GaAs. In this case of "conducting" GaAs crystals the dielectric function $\epsilon(\omega)$ in the spectral range of the Reststrahlen-band is influenced by the free carriers. In the simplest approximation this contribution can be described by the Drude model. Thus, $\epsilon(\omega)$ is assumed to be

$$\epsilon(\omega) = 1 + \chi_{VE} + \chi_{PH}(\omega) + \chi_{FE}(\omega),$$

with

$$\chi_{VE} = \epsilon(\infty) - 1,$$

$$\chi_{PH}(\omega) = [\epsilon(0) - \epsilon(\infty)] \frac{\omega_{TO}^2}{\omega_{TO}^2 - \omega^2 - i\omega\gamma},$$

$$\chi_{FE}(\omega) = -\left(\frac{\omega_p}{\omega}\right)^2 \frac{1}{1 - 1/i\omega\tau}, \tag{12}$$

where the first term describes the constant contribution of the valence electrons, the second one the TO-phonon and the third one the Drude free electron contribution. $\epsilon(0)$ and $\epsilon(\infty)$ are the static and the electronic dielectric constants, ω_{TO} the frequency of the TO-phonon and γ the damping constant taken from IR-data [52]. The bulk plasmon frequency

$$\omega_p^2 = ne^2/\epsilon_0 m_n^* \tag{13a}$$

and the Drude relaxation time

$$\tau = m_n^* \mu/e \tag{13b}$$

can be determined from transport data [53] (m_n^* effective electron mass, μ mobility). The free electron concentration n, then, is the only free parameter. Within the framework of dielectric theory the surface loss function $-\text{Im}\,[1 + \epsilon(\omega)]^{-1}$ allows the calculation of the essential spectral structures in HRELS (only single losses). The inset in Fig. 7b shows the surface loss function calculated by means of $\epsilon(\omega)$ from Equ. (12). With a carrier concentration of $n' = 3 \times 10^{17}\ \text{cm}^{-3}$ the peak positions of the spectrum measured after a 1 L hydrogen exposure are well described. The concentration n' does not agree with the bulk concentration $n = 4.3 \times 10^{17}\ \text{cm}^{-3}$. This disagreement results from the band bending which is induced by treatment with atomic H: The formation of a depletion layer reduces the carrier concentration in the space charge layer within a depth of several hundred Ångstroms below the surface. The carrier concentration within this region is relevant for the plasmon, since its long range electric field has the same spatial structure like that of the Fuchs-Kliewer surface phonon (Fig. 6). In Fig. 9 experimentally determined loss peak positions $\hbar\omega_+$ and $\hbar\omega_-$ are compared with theoretical predictions from the dielectric theory [20]. The peak positions $\hbar\omega_+$ and $\hbar\omega_-$ (curves in full line) are calculated from the surface loss function using the dielectric function $\epsilon(\omega)$ of Equ. (12) with an effective carrier concentration n' as parameter. The lower branch $\hbar\omega_-$ has surface plasmon-like character for small n' whereas $\hbar\omega_+$ is surface phonon-like. Near $n' \approx 10^{18}\ \text{cm}^{-3}$ the two branches interchange their character thus indicating a coupling between the two modes via their long range electric fields.

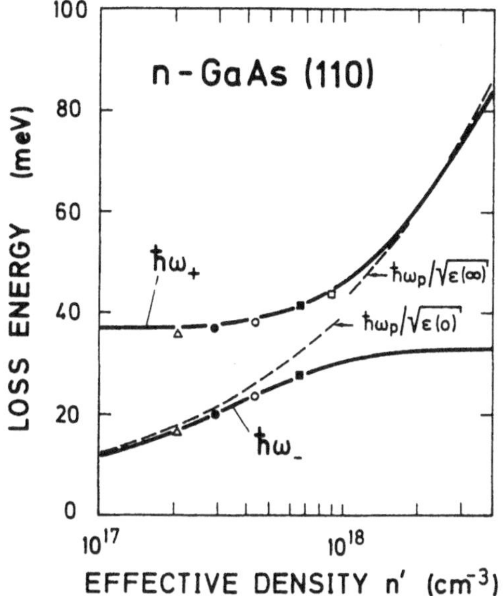

Fig. 9
Loss peak positions $\hbar\omega_+$ and $\hbar\omega_-$ calculated from the maxima of the surface loss funktion with $\epsilon(\omega)$ according to Equ. (12) (curves in full Line). Broken line: plasmon frequency without coupling to surface phonon. The experimental points are measured on samples with different doping:

a) Te-doped (bulk density: n = 9×10^{17} cm^{-3}): □ clean, ■ after exposure to 1 L residual gas.
b) Te-doped (bulk density: n = 4.3×10^{17} cm^{-3}): ○ clean, ● after exposure to 1 L dissociated H_2.
c) Si-doped (bulk density: n = 3×10^{17} cm^{-3}: △clean.

(After R. Matz and H. Lüth [20]).

For Te-doped samples the experimentally observed losses after cleavage in UHV fit very well into the theoretical scheme, when the effective carrier concentration n′ is taken as that of the bulk (n) determined by Hall effect. As is well known already from other investigations [54] flat bands are found here (n′ = n). On Si-doped samples usually the bands are bent upwards to some extent after cleavage [54] which explains the deviation of n′ from the bulk concentration n in this case. Treatment with atomic hydrogen after cleavage in UHV obviously bends the bands also upwards because in Fig. 9 the effective surface carrier concentration n′ must be assumed to be reduced below the bulk values n in order to achieve a good fit to the theoretical curves. The measurement of the coupled surface phonon-plasmon losses, therefore, allows conclusions concerning the type of the space charge layer. In adsorption studies information about the direction of charge transfer between substrate and adsorbate can be obtained.

It is an interesting phenomenon, that inspite of the strongly inhomogeneous carrier distribution in the space charge layer along z the simple picture of the surface plasmon seems to hold. It would be interesting to investigate more in detail the correlation between band bending and plasmon shift. Model calculations to this problem have already been performed by Cunningham et al. [55]. A quantitative application to the present GaAs case, however, is difficult, since the GaAs crystals are probably compensated by unknown deep lying impurities which complicate a quantitative description of the space charge layer.

Surface phonon and plasmon are accompanied by electric fields of the same wavelength. With decreasing wavelength λ the screening of this field in the free electron gas becomes less effective and finally for $\lambda \ll \lambda_s$, the screening length, the collective response of the electron gas has disappeared; the collective modes at $\hbar\omega_+$ and $\hbar\omega_-$ decay into single particle excitations. Intraband electron-hole pair excitation is favoured. This additional "single particle" scattering might also be taken into account in the interpretation of spectra like in Fig. 7b since the screening wave vector ($q_s = 2\pi/\lambda_s$)

$$q_s^2 = m_n^* \, \omega_P^2 / \epsilon(\omega) \, k_B T \tag{14}$$

calculated for the conduction band Boltzmann gas in the Debye form is of the order of 10^6 cm^{-1}. This is just the order of magnitude of the $q_{\parallel}$-transfer at the excitation of 20 meV surface plasmons estimated from Equ. (3). Matz and Lüth [20] therefore, have also used Lindhard's dielectric function in the Boltzmann limit [18] with $\tau \to \infty$ instead of the Drude term χ_{FE} in Equ. (12):

$$\chi_{FE}(\omega, \underline{q}) = -\left(\frac{\omega_P}{\omega}\right)^2 + i \, \frac{\pi e^2 n \omega}{\epsilon_0 k_B T q^3} \left(\frac{m_n^*}{2\pi k_B T}\right)^{1/2} \exp\left(-\frac{m_n^* \omega^2}{2 k_B T q^2}\right), \tag{15}$$

to estimate the effect both of intraband scattering and collective plasmon and phonon excitation within the framework of dielectric theory. The appropriate expression for the scattering efficiency has been derived from Equ. (4) and (5) by Mills [31] as

$$\frac{1}{|R|^2} \, \frac{d^2 S}{d\Omega d(\hbar\omega)} = \frac{m^2 e^2 v_\perp^2}{\hbar^4 \pi^3 \epsilon_0 \cos\theta_I} \left(\frac{k'}{k_0}\right) [N(\omega) + 1] \times$$

$$\times \frac{v_\perp^2 q_{\parallel}}{v_\perp^2 q_{\parallel}^2 + (\omega - \underline{v}_{\parallel} \cdot \underline{q}_{\parallel})^2]^2} \; \mathrm{Im} \, \frac{-1}{1 + \epsilon(\omega, q_{\parallel})} \, . \tag{16}$$

The explanation of the symbols is like in Equs. (4), (5) and (6). Integration of Equ. (16) over the angular aperture of the spectrometer (assumed as a cone of 0.8° half angle width around specular direction) yields the loss intensity $dS/d(\hbar\omega)$ normalized to the elastically reflected intensity. Furthermore $dS/d(\hbar\omega)$ has been convoluted with an assumed Gaussian-shaped transmission function (FWHM: 12 meV) of the analyser to take into account the finite energy resolution of the equipment. Results of this calculation are shown in Fig. 10. Only single step excitations, of course, are obtained within this calculation according to Equ. (16); multiple losses as seen in the experimental data of Fig. 7b are not treated. The theoretical curve in Fig. 10 plotted in broken line does not contain decay of the coupled eigenmodes $\hbar\omega_+$ and $\hbar\omega_-$ into electron-hole pairs. The agreement of this curve with the experimental data (Fig. 7b) is poorer as if single particle scattering is taken into account (Fig. 10, curve in full line). The calculation suggests that

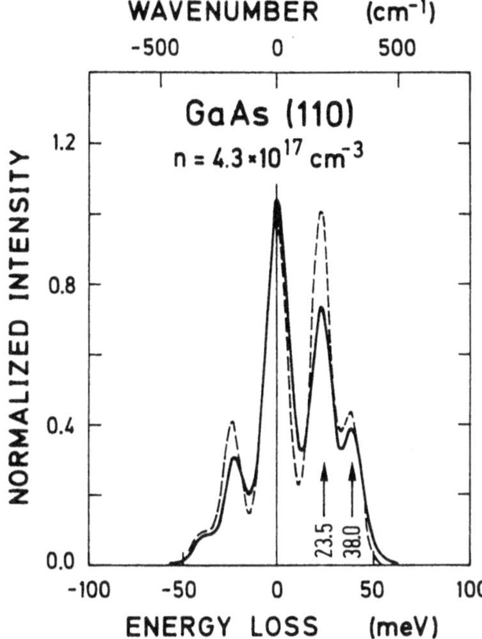

Fig. 10
Loss spectra calculated using Mill's theory for inelastic scattering [31] (Equ. (16) and Lindhard's electronic dielectric function ϵ (ω, $\underline{q}$) (Equ. 15) superimposed on the TO lattice oscillator of GaAs. Full line: single particle scattering included. Broken line: single particle scattering not included. Experimental peak positions (in meV) are given by arrows (After R. Matz and H. Lüth [20]).

scattering on electron-hole pairs also contributes to the spectra on n-type GaAs (110), eventhough the energetic resolution of the experimental technique is not sufficient to study details of the spectral shape being induced by intraband scattering.

In a better approximation further broadening of the calculated structures in Fig. 10 might occur due to plasmon dispersion and due to scattering of the conduction band electrons (finite electron mobility) not being taken into account in Equ. (15).

In conclusion, the positions $\hbar\omega_+$ and $\hbar\omega_-$ obtained in this more quantitative description via Lindhard's $\epsilon(\omega, \underline{q})$ coincide with those from Drude's $\epsilon(\omega)$ (Equ. (12), Fig. 9) within about 1 %.

3.2.2 Hydrogen Adsorption on GaAs (110)

Up to now only one adsorption system has been studied by HRELS within the class of III-V compounds: hydrogen on cleaved GaAs (110) surfaces [21]. Both on Cr-doped and on n-type material Lüth and Matz found two characteristic hydrogen induced vibration losses at 1890 cm^{-1} (234.3 meV) and 2150 cm^{-1} (266.6 meV). Adsorption of hydrogen could only be observed after predissociation of the molecular species H_2 in the gas phase, i.e. only atomic H adsorbes on GaAs (110) at 300 K (like on Si). Corresponding losses due to atomic deuterium D were observed at 1380 cm^{-1} and 1660 cm^{-1} (Fig. 11). Inspite of careful investigation no vibrational losses were detected near 4400 cm^{-1} where the stretching vibration of H_2 is expected.

The agreement of the measured loss energies with IR-vibrational frequencies of free molecules, in particular, of AsH_3 (AsD_3) [56] and some trimethylamine-halogenogallanes [57] shows that atomic H and D adsorb both on As and on Ga surface sites. The corresponding vibrational frequencies are 2150 cm^{-1} (1660 cm^{-1}) for the As-H (As-D) and 1890 cm^{-1} (1380 cm^{-1}) for the Ga-H (Ga-D) bond, respectively. It seems reasonable to assume that for saturated adsorption all Ga and As surface atoms (equal surface density) bond an H atom (monolayer coverage). Then, using Evans' and Mills' theory for the inelastic scattering cross section [29], the intensity ratios $I_{H\text{-}Ga}/I_0$ and $I_{H\text{-}As}/I_0$ of the Ga-H and the As-H peaks with respect to the integrated elastic peak intensity I_0 allow the determination of effective dynamic ionic charges $e_\perp^*$ of the adsorbed H : $e^*_{H\text{-}Ga} = 0.042$ e, $e^*_{H\text{-}As} = 0.038$ e (e is elementary charge). It should be emphasized that the net effect of H adsorption is the formation of a depletion layer (see Ch. 3.2.1), i.e. upon adsorption electronic charge is transferred from the substrate into the adsorbate layer.

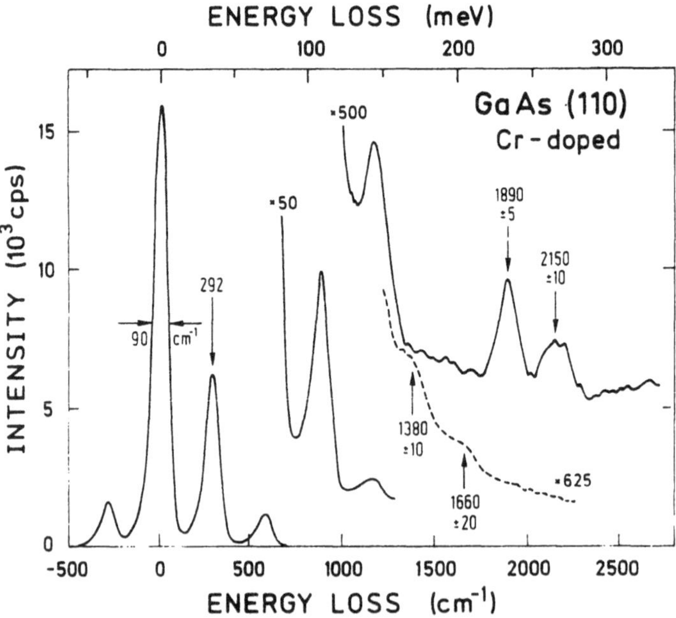

Fig. 11 Loss spectra measured on two different cleaved (110) surfaces of semiinsulating GaAs after adsorption of H (full line) and D (broken line).
H_2 and D_2 exposures 1000 L (H and D doses unknown). (After H. Lüth and R. Matz [21]).

From a comparison of the H and D vibrational frequencies Lüth and Matz have constructed the whole vibrational potentials for the Ga-H and As-H chemisorption bonds [21]: The H and D frequencies on the As and on the Ga site are related by factors 1.37 and 1.3, respectively, whereas the value $\sqrt{2}$ would be expected for harmonic potentials and a rigid GaAs surface. The deviation from $\sqrt{2}$ due to a non-rigid GaAs surface can be estimated to be at maximum 0.7 % by considering hypothetical As-H, As-D, Ga-H and Ga-D molecules. The much stronger deviation observed in the experimental results (Fig. 11) is therefore ascribed to the anharmonicity of the adsorption potentials. Because of their chemical identity the potentials for H and D are equal, only the H atoms are lighter and, therefore, their vibrational levels are more spaced in energy than those of the D atoms. The first excited level probes the anharmonicity of the potential more than the D level does; this level is therefore shifted down in energy as compared with a purely harmonic potential. The anharmonicity is related to the finite dissociation energy E_D which therefore can be determined by extrapolation from the anharmonic part of the potential. The quantitative analysis is performed in terms of a Morse potential approximation. This type of potential has been applied with great success to the quantitative description of free diatomic molecules [58, 59]. With a Morse potential (Fig. 12)

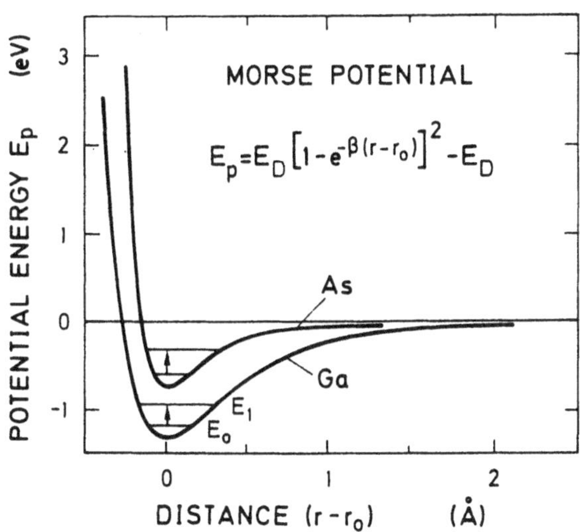

Fig. 12 Morse potentials $E_p\,(r - r_0)$ for hydrogen adsorbed on As and on Ga sites. The calculation is based on experimental data of the GaAs (110) surface (Fig. 11). E_D is the dissociation energy. The first two H-vibrational levels E_0 and E_1 and the experimentally observed transitions (arrows) are also plotted. The potentials are referred to the same equilibrium position r_0. (After H. Lüth and R. Matz [21]).

the vibrational wave equation is solved rigorously giving the following vibrational energy levels E_n [59]:

$$E_n = \hbar\omega_0 \,(n+\frac{1}{2}) - \hbar\omega_0 x_0 \,(n+\frac{1}{2})^2$$

$$= (2E_D \hbar^2 \beta^2/\mu)^{1/2} \,(n+\frac{1}{2}) - \frac{1}{2}(\hbar^2\beta^2/\mu)\,(n+\frac{1}{2})^2; \qquad (17)$$

μ is the reduced mass, here the mass of D and H, respectively. The experimentally determined excitation energies $\hbar\omega = E_1 - E_0$ yield by means of Equ. (17) the two parameters E_D and β for both adsorption sites Ga and As. The anharmonicity terms $\hbar\omega_0 x_0$ which are also obtained are in the order of magnitude being typical for molecules like HCl, HF etc. [59]. The dissociation energy on the Ga site E_D(Ga) is derived as (1.3 ± 0.3) eV, that on the As site emerges as (0.73 ± 0.1) eV. The corresponding Morse potentials are plotted in Fig. 12.

The potentials are obtained in dependence on the displacement $(r-r_0)$ from the equilibrium position r_0. But with the assumption that the repulsive part of the potential (steep slope at $r<r_0$) essentially determines the equilibrium position with respect to the Ga and As bonding partner, one might conclude that the Ga-H bond length is larger by about 0.15 Å than that of the As-H bond. The covalent bonding radii of Ga (1.26 Å) and of As (1.15 Å) together with the atomic radius of H (0.3 Å) indicate the same tendency in bonding length.

The different dissociation energies for H on Ga and on As sites are contrasted by the experimental result, that both adsorption sites are emptied upon annealing within the same temperature range (Fig. 13). The relatively low desorption temperature (~ 80 °C) is in agreement with the low dissociation energies 0.7 eV (As-H)

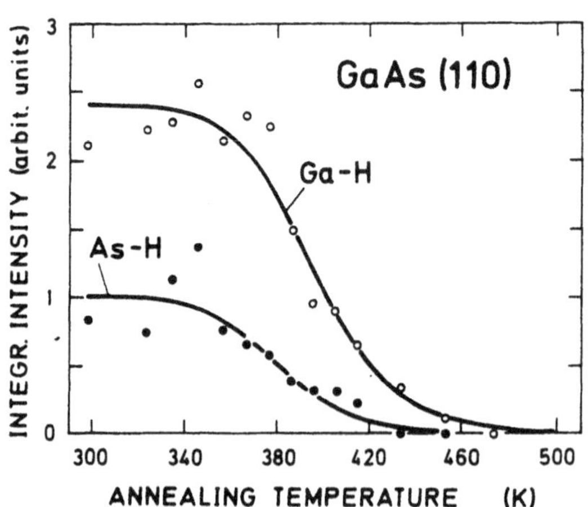

Fig. 13
Integrated intensity of the two H losses observed on the GaAs (110) surface at 1890 cm^{-1} (Ga-H) and 2150 cm^{-1} (As-H) versus annealing temperature. (After H. Lüth and R. Matz [21]).

and 1.3 eV (Ga-H). For an understanding of the desorption mechanism one has to take into account that H_2 desorbs rather than atomic H [60], i.e. upon desorption two H atoms from one and the same type of site or from two different sites (Ga and As) have to recombine and about the dissociation energy of H_2 (4.3 eV per H_2 molecule) is gained. The total desorption potential, therefore, consists of a superposition of the two atomic potentials (Fig. 12) and a "molecular" H_2 potential describing the approach of an H_2 molecule to the surface before dissociation. For the case that the desorbing H_2 is formed by H atoms originating from a Ga and from an As site, only one desorption energy between 0.7 and 1 eV is derived. When the H_2 molecule is formed by H atoms from sites of the same type, two different desorption energies exist, one for desorption from Ga sites, the other from As sites. A qualitative estimation of the intersection point of the "molecular" H_2 potential with the two atomic potentials in Fig. 12 suggests, however, that both desorption energies are similar. This is in agreement with the observation that both adsorption sites are emptied within the same temperature range (Fig. 13).

The interpretation of the desorption data implies that the mobility of the chemisorbed H is extremely low, such that exchange between the two types of sites can be neglected. This conclusion is supported by the quantiative description of the desorption data in Fig. 13. The curves in full line have been calculated under the assumption of second order kinetics. With an assumed desorption energy for the As site of 0.7 eV/atom a preexponential factor $k_2 = 2 \times 10^{-7}$ cm^2/s fits best. For the Ga site this k_2 yields a desorption energy of 0.73 eV/atom. The relatively small k_2 as compared with H on most metal surfaces [61] is indicative for a low surface mobility of the chemisorbed species and/or severe "steric hindrances" for the recombination and desorption of H_2. In this context it is interesting to mention that such a low preexponential factor k_2 in the order of 10^{-7} cm^2/s has also been found for second order desorption of D_2 from Cu (110) [62].

Recently Mokwa could also show by thermal desorption spectroscopy (mass spectrometry) that H_2 desorbes from cleaved GaAs (110) surfaces in a second order process with a desorption energy estimated to be smaller than 1 eV [60]. Thus, a quite consistent picture emerges for the adsorption of hydrogen on GaAs (110), essentially derived from spectroscopic data. It would be interesting to approach the question for the two different desorption mechanisms considered above by means of angular resolved desorption techniques.

3.3 ZnO Surfaces

ZnO surfaces, both the polar (0001) (Zn), $(000\bar{1})$ (O) and the nonpolar $(10\bar{1}0)$ ones, were the first, which have been investigated by Ibach by means of HRELS [7, 8]. After cleavage in UHV and after cleaning air cleaved crystals by annealing in UHV spectra like those in Fig. 14 were measured. The similarity of the results on semiinsulating GaAs (110) (Fig. 7a) to those on ZnO is obvious. The interpretation which was given by Ibach to the results is that, which has been discussed more in

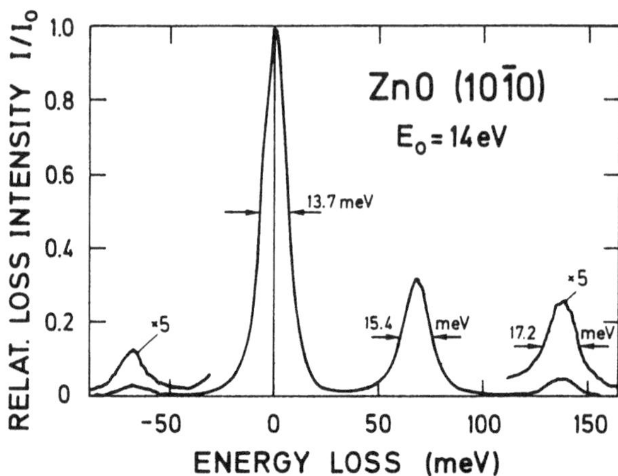

Fig. 14 Energy loss spectrum measured under specular reflection on a nonpolar Zn0 surface prepared by cleavage in UHV (angle of incidence 45°). (After H. Ibach [8]).

detail in Ch. 3.2.1 for GaAs. For the nonpolar and for the polar surface excitation energies of the Fuchs-Kliewer surface phonons of (68.6 ± 0.5) meV and (67.5 ± 0.5) meV, respectively, are obtained. These energies are well separated from those of the LO bulk phonons: 73.1 meV and 71.5 meV, respectively. From IR-data of ZnO, taking the dielectric functions $\epsilon_{\parallel}(\omega)$ and $\epsilon_{\perp}(\omega)$ parallel and normal to the hexagonal c-axis, Ibach has calculated the excitation energies for the corresponding Fuchs-Kliewer surface phonons from the condition $\mathrm{Re}\,\epsilon(\omega) = -1$ (Equ. (9)): The calculated values 69.0 meV (10$\bar{1}$0) and 67.3 meV (0001) are in good agreement with the measured excitation energies [7]. Furthermore, the intensity ratio of the one-phonon gain peak I_{-1} to the one-phonon loss I_1 versus temperature obyes a Boltzmann factor dependence $\exp(-\hbar\omega/k_B T)$. Like for GaAs (110) (Fig. 8a) the intensities I_m versus loss number obey a Poisson distribution law [7]. These results have been explained quantitatively already by the first theoretical approach of Lucas and Šunjić [28] and later on by Evans and Mills [29, 30] and Mills [31, 32]. Like in the case of GaAs an effective ionic charge (Szigeti charge) can be calculated from the inelastic scattering cross section. In agreement with calculations from bulk IR-data it results as about one elementary charge.

Ibach showed also that the scattering probability measured as the ratio I_1/I_0 versus primary energy E_0 (Fig. 15) is proportional to $E_0^{-1/2}$ [8]. This dependence can be understood qualitatively in terms of the interaction time $\tau \sim v_0^{-1} \sim E_0^{-1/2}$ between scattered electron and collective surface excitation. A careful study of the energetic halfwidth of the first phonon loss (15.4 meV in Fig. 14) as compared with that of the primary beam (13.7 meV) suggests that the surface phonon is damped to a

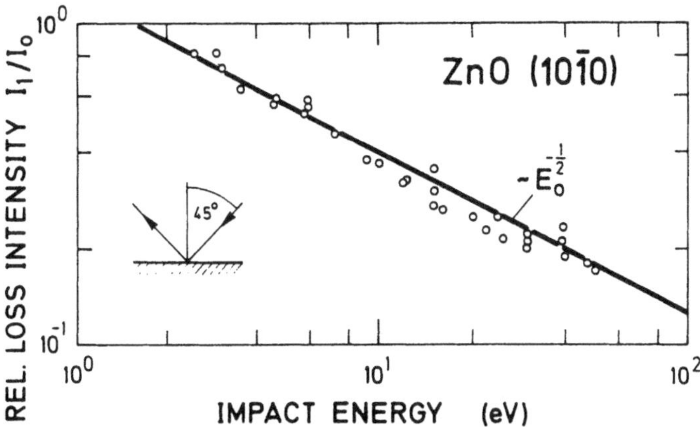

Fig. 15 Relative loss intensity of the first surface phonon loss on Zn0 (10$\bar{1}$0) (Fig. 14) versus impact energy (After H. Ibach [8]).

certain extent. This damping might partially be due to free electrons in the conduction band. To first approximation, however, the effect of free carriers can be neglected in these measurements as can be seen from a comparison with the results on GaAs (Fig. 7b).

ZnO surfaces (bulk material always n-type) treated with atomic hydrogen or exposed to excessive UV irradiation exhibit very strong accumulation layers [63–65]. Surface carrier concentrations in the space charge layer up to $\Delta N \approx 10^{14}$ cm^{-2} can be achieved [64, 66]. As is derived from space charge layer calculations [67], the thickness of those strong accumulation layers is in the order of 10 Å. Electrons within such a layer should be quantized along the z-direction normal to the surface and "free" only within the surface along x and y. HRELS results on those "conducting" ZnO surfaces have been published by Goldstein et al. [68]. Fig. 16 shows results obtained by those authors on surfaces treated with oxygen (depletion layer) and with atomic hydrogen. The results on the depleted surface resemble those of Ibach (Fig. 14), only the energetic resolution (~ 20 meV) is worse. After hydrogen treatment the valleys between the multiple phonon losses are filled and the phonon loss intensity is reduced. Similar results have been obtained already by Ibach [69] on hydrogen treated surfaces and also Matz and Lüth found the same effect due to an accumulation layer even with better spectrometer resolution (~ 10 meV) [70]. Pronounced peak-like structures like in the case of GaAs (110) as due to surface plasmons (Fig. 7b) could never be seen. Assuming a semiinfinite halfspace with carrier concentrations of 10^{18} cm^{-3} and 1×10^{19} cm^{-3} corresponding to surface concentrations of 10^{12} cm^{-2} and 5×10^{12} cm^{-2}, respectively, a simple calculation of the surface loss function like in the case of GaAs (Equ. (12)) yields surface plasmon frequencies of about 22 meV and 44.6 meV.

Furthermore, the phonon-like loss should be shifted in the latter case to about 113 meV. But neither plasmon-like structures like in Fig. 6b for GaAs nor large shifts of the surface phonon have been observed. This result is probably explained by the penetration depth of the space charge layer on ZnO. On GaAs the carriers being involved in the plasma oscillations are found in the whole bulk material, except for depletion layers, where the plasmon finally vanishes. On ZnO with accumulation layers the bulk carrier concentration is low, "free" carriers only exist within a depth of 5–100 Å. On the other hand, the penetration depth of the surface plasmon according to the wave vector transfer $q_{\|}$ would amount to at least a couple of hundred Ångstroms, i.e. within the thin quantized accumulation layers of ZnO such a "three-dimensional (3D)" surface plasmon like on GaAs can not exist as an eigenstate.

The description of the ZnO results in terms of a "two-dimensional (2D)" surface plasmon [71] with negligible extension in z-direction, therefore, seems much more reasonable. Such an interpretation has been presented by Gersten [72] and by Goldstein et al. [68] for the results in Fig. 16. In his semi-classical model Gersten assumes two-dimensional free charge density fluctuations of the type

$$\rho = [-e(N-N_0) + Q\exp(i\underline{q}_{\|}\cdot\underline{r}) + Q\exp(-i\underline{q}_{\|}\cdot\underline{r})]\delta(z) \qquad (18)$$

on top of an insulating infrared-active ZnO crystal. N is the instantaneous number of surface electrons per unit area and N_0 the mean of this quantity. Q is the wave amplitude of the plasma oscillation, $q_{\|}$ the wavevector parallel to the surface (z = 0).
A coupling of these 2D surface plasmons to the TO-lattice oscillator, i.e. the Fuchs-

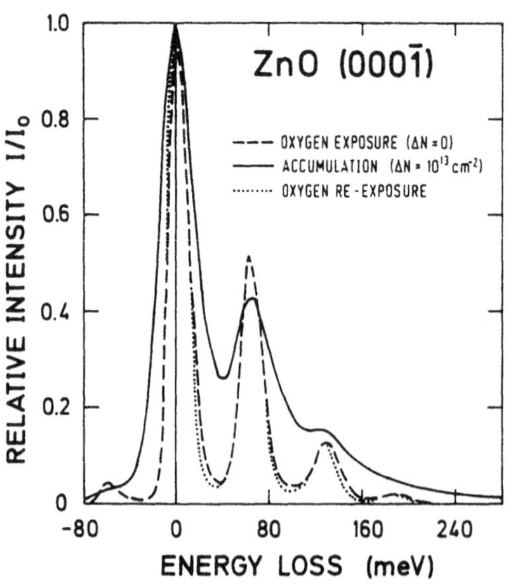

Fig. 16
Energy loss spectra measured under specular reflection on a ZnO polar O-face (primary energy E_0 = 3.7 eV, angle of incidence Θ_I = 45°). The surface was cleaned by Ar-bombardment and annealing. The accumulation layer was produced by treatment with atomic hydrogen. Δ N is the surface carrier concentration. (After Y. Goldstein et al. [68]).

Kliewer surface phonon, is taken into account. Two eigenmodes $\hbar\omega_+$ and $\hbar\omega_-$ of this coupled system – called plasmaron – are calculated in dependence of $q_{\parallel}$. Their dispersion curves are shown in Fig. 17 for a surface carrier concentration of $\Delta N = 2 \times 10^{13}$ cm^{-2}. The $\hbar\omega_-$ mode approaches zero for $q_{\parallel} \to 0$, since for a 2D plasmon the positive and negative charge can be imagined to lie on rows; thus for $\lambda \to \infty$ the restoring force vanishes. For a 3D plasmon the charges are arranged on parallel sheets and the restoring force remains finite with increasing distance between the sheets ($\lambda \to \infty$).

Gersten, furthermore, calculates inelastic scattering cross sections for HRELS on the plasmaron modes in the first Born approximation [72]. A comparison of his results with experimental data (Fig. 16) is shown in Fig. 18. The experimental curves (full line) are obtained by subtracting loss spectra of the clean ZnO surface with negligible surface carrier concentration from those measured on the hydrogen treated surface with accumulation layer. This subtraction eliminates spectral structures due to multiple excitation. Such multiple excitation processes are prominent in the experimental curves for losses $\hbar\omega > 100$ meV (Fig. 16); they are, however, not considered in the calculation. On the other hand, the main effect of the $\hbar\omega_+$ plasmon-like dispersion branch (Fig. 17) is observed in the spectra just for loss energies above 100 meV (Fig. 18). For the theoretical curves (broken line) a smoothing-out due to nonspecularities of the ZnO surface has been taken into account to achieve a better fit. Even for spectra measured at different angles (5°, 10°) away from the specular direction (0°) the agreement between theory and experiment seems reasonable. For lower surface carrier concentrations ($\Delta N < 10^{12}$ cm^{-2}) the agreement is poorer [68]. This might be explained by the fact that in this case the description in terms of a 2D surface plasmon breaks down. The space charge layer thickness for $\Delta N = 10^{12}$ cm^{-2} is in the order of 100 Å [67].

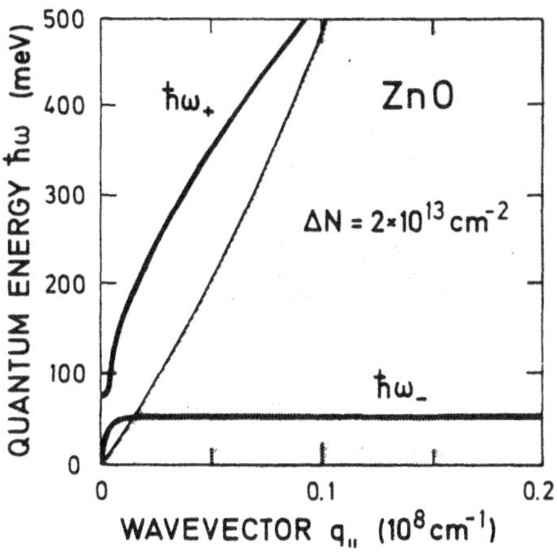

Fig. 17
Calculated dispersion branches of coupled Fuchs-Kliewer surface phonons and 2D surface plasmons. The lattice data of Zn0 are used and a two-dimensional free electron gas with a surface density $\Delta N = 2 \times 10^{13}$ cm^{-3} is assumed on the surface. The shaded area indicates the region of single particle excitation (decay of plasmarons into electron-hole pairs). (After J. I. Gersten [72]).

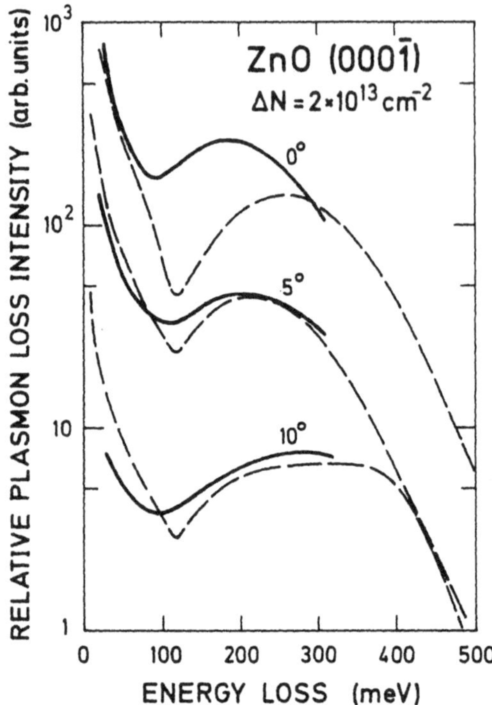

Fig. 18 Curves in full line:
Difference between energy loss spectra of ZnO (000$\bar{1}$) with an accumulation layer (surface carrier concentration $\Delta N = 2 \times 10^{13}$ cm^{-2}) and those of the clean surface without accumulation. The spectra are measured at different angles (0°, 5°, 10°) with respect to the specular direction.
Curves in broken line:
Calculated scattering intensities due to 2D surface plasmon excitation. (After Y. Goldstein et al. [68]).

The presented results show that the assumption of 2D plasmon excitation might explain at least part of the effect of accumulation layers on the HRELS data. It should, however, be emphasized that effects due to subbands resulting from the z-quantization perpendicular to the surface are not explicitely included in the theoretical description of the scattering probability. Single particle electron scattering and also transitions between occupied and empty subbands might yield further contributions to the spectra measured on semiconductors with strong accumulation layers. Because of the broad, washed out structures on ZnO a distinction between these differenct effects might be difficult. Raman scattering on 2D plasmons coupled to LO bulk phonons has been discussed theoretically by Burstein et al [73] taking into account also inter- and intra-subband transitions.

4 Concluding Remarks

The experimental results presented here are the only ones on semiconductors being known to the author so far. In spite of the limited number of data, nevertheless, problems which are characteristic for semiconductor surfaces might be realized, since examples from the main classes of semiconductors have been investigated. Thus, the different types of surface phonons being characteristic for infrared and non-infrared active materials have been discussed. Also different aspects of free carrier effects have been studied on GaAs and on ZnO. Comparable studies on elemental semiconductors are missing so far.

These carrier phenomena and the strong excitations being due to Fuchs-Kliewer surface phonons on infrared materials are characteristic for semiconductors in contrast with HRELS studies on metal surfaces. They have to be taken into account, in particular, in studies of adsorbate vibrations: The severe disadvantage emerges that low energy bending modes or vibrations of molecules against the substrate surface can scarcely be observed on the background of these collective excitations. On the other hand, as is shown for the case of GaAs, useful information about the space charge layer can be extracted from HRELS data, when electronic excitations are involved.

There is a wide field of possible studies on metal overlayers on semiconductors, about which nothing is known up to now as far as HRELS is concerned. In particular, in the submonolayer range HRELS should be applied also to these problems. Furthermore, nothing has been done up to now by means of this technique in the field of organic semiconductors. One might expect that at least on thin samples HRELS measurements might be possible inspite of the high resistivity of these materials.

Concerning special experimental problems in the application of HRELS to semiconductor surfaces, it seems, that work function inhomogeneities on semiconductor surfaces might be responsible for an energy broadening of the reflected electron beam. On metal surfaces usually the energy width of the primary electron beam determined by the spectrometer resolution is also found for the elastic beam after reflection on the sample. On the other hand, this is not true for cleaved GaAs (110) surfaces. Broadenings from 7 to 14 meV are observed, which might be due to work function inhomogeneities being induced by steps or other surface imperfections. These effects can complicate studies on semiconductor surfaces when high energy resolution is required.

Acknowledgement

I am thankful to Richard Matz for a number of useful discussions and for some computer calculations.

Experimental work on GaAs (110) presented in this paper has financially been supported by the Deutsche Forschungsgemeinschaft (SFB 56 "Festkörperelektronik").

References

[1] *H. Raether*: In Springer Tracts in Modern Physics, Vol. **38** (Springer Verlag Berlin, Heidelberg, New York 1965), p. 85.

[2] *J. Geiger*: "Elektronen und Festkörper", Vol. **128**, (Sammlung Vieweg, Braunschweig 1968).

[3] *J. Daniels, C. v. Festenberg, H. Raether, K. Zeppenfeld*: In Springer Tracts in Modern Physics, Vol. **54**, (Springer Berlin, Heidelberg, New York 1970) p. 76.

[4] *H. Raether*: "Excitation of Plasmons and Interband Transitions by Electrons", in Springer Tracts in Modern Physics, Vol. **88**, (Springer Berlin, Heidelberg, New York 1980).

[5] *R.F. Willis*: "Vibrational Spectroscopy of Adsorbates", in Springer Series in Chemical Physics, Vol. **15** (Springer Berlin, Heidelberg, New York 1980) p. 23.

[6] *H. Lüth*: Appl. Phys. **8**, 1 (1975).

[7] *H. Ibach*: Phys. Rev. Lett. **24**, 1416 (1970).

[8] *H. Ibach*: J. Vac. Sci. Technol. **9**, 713 (1972).

[9] *H. Ibach*: Phys. Rev. Lett. **27**, 253 (1971).

[10] *H. Ibach, K. Horn, R. Dorn* and *H. Lüth*: Surf. Science **38**, 433 (1973)

[11] *S. Lehwald* and *H. Ibach*: Surf. Science **89**, 425 (1979).

[12] *H. Froitzheim, H. Ibach* and *D. L. Mills*: Phys. Rev. **B 11**, 4980 (1975).

[13] *H. Ibach*: in Proceedings of the Twelfth International Conference on the Physics of Semiconductors, 1974 (unpublished).

[14] *H. Froitzheim*: "Electron Energy Loss Spectroscopy", in Topics in Current Physics, Vol. 4, "Electron Spectroscopy for Surface Analysis", Ed. H. Ibach (Springer Berlin, Heidelberg, New York 1977) p. 205.

[15] *H. Ibach*: "Dynamische Eigenschaften der Festkörperoberlfäche", in Festkörperprobleme/Advances in Solid State Physics, Vol. XI, Ed. O. Madelung, (Pergamon/Vieweg, Oxford/Braunschweig 1971) p. 135.

[16] *A. Mooradian*: in Advances in Solid State Physics, Vol. **9**, Ed. H.J. Queisser (Pergamon/Vieweg, Oxford/Braunschweig 1969) p. 74.

[17] *G. Abstreiter, R. Trommer* and *M. Cardona*: Solid State Commun. **30**, 703 (1979).

[18] *W. Hayes* and *R. Loudon*: "Scattering of Light by Crystals", (Wiley, New York 1978) p. 301.

[19] *R. Matz* and *H. Lüth*: Proceedings 3rd European Conference on Surface Science, Cannes 1980, Suppl. a la Revue "Le Vide, les Couches Minces", No. 201 (Soc. Française du Vide, 1980) p. 762.

[20] *R. Matz* and *H. Lüth*: Phys. Rev. Lett. **46**, 500 (1981).

[21] *H. Lüth* and *R. Matz*, to be published.

[22] *W. Ho, R.F. Willis, E.W. Plummer*: Phys. Rev. Lett. **40**, 1463 (1978).

[23] *R.F. Willis, W. Ho, E.W. Plummer*: Surf. Science **80**, 593 (1979).

[24] *S. Anderson* and *J.W. Davenport*: Solid State Commun. **28**, 677 (1978).

[25] *E. Fermi*: Phys. Rev. **57**, 485 (1940).

[26] *J. Hubbard*: Proc. Phys. Soc. (London), **A 68**, 976 (1955).

[27] *H. Fröhlich* and *H. Pelzer*: Proc. Phys. Soc. (London), **A 68**, 525 (1955).

[28] *A.A. Lucas* and *M. Šunjić*: Progress in Surf. Science, Vol. **2**, Ed. S.G. Davison (Pergamon Press 1972) p. 2.

[29] *E. Evans* and *D. L. Mills*: Phys. Rev. **B 5**, 4126 (1972).

[30] *E. Evans* and *D. L. Mills*: Phys. Rev. **B 7**, 853 (1973)

[31] *D. L. Mills*: Surf. Science **48**, 59 (1975).

[32] *D. L. Mills*: "Interaction of Low Energy Electrons with Surface Lattice Vibrations", in Surfaces and Interfaces, Vol. I, ed. L. Dobrzynski (Marcel Dekker, New York 1978).

[33] *D. Šokčević, Z. Lenac, R. Brako* and *M. Šunjić*: Z. Physik **B 28**, 273 (1977).

[34] *D.M. Newns*: Phys. Lett. **60 A**, 461 (1977).

[35] *D.M. Newns*: "Theory of Dipole Electron Scattering from Adsorbates", in Springer Series in Chemical Physics, Vol. **15**, Vibrational Spectroscopy of Adsorbates, ed. R.F. Willis (Springer Berlin, Heidelberg, New York 1980) p. 7.

[36] *B.N.J. Persson*: Surf. Science **92**, 265 (1980).

[37] *G. Dolling*: "Inelastic Scattering of Neutrons" (IAEA, Vienna 1965) p. 249.

[38] *K. Thoma* and *R. Zimmermann*: Z. Physik **252**, 168 (1972).

[39] *R. Zimmermann*: Appl. Phys. **3**, 235 (1974).

[40] *H. Wagner*: private communication, and *H. Wagner, R. Butz, U. Backes, D. Bruchmann*, to be published.

[41] *W.A. Goddard III, A. Redondo* and *T. C. McGill*: Solid State Commun. **18**, 981 (1976).

[42] *W.A. Goddard III, J.J. Barton, A. Redondo* and *T.C. McGill*: J. Vac. Sci. Technol. **15**, 1274 (1978).

[43] *R. Ludeke* and *A. Koma*: Phys. Rev. Lett. **34**, 1170 (1975).

[44] *G. Hollinger, Y. Jugnet, P. Pertosa, L. Porte* and *Tran Minh Duc*: Chem. Phys. Lett. **36**, 441 (1975).

[45] *M. Chen, I.P. Batra* and *C.R. Brundle*: J. Vac. Sci. Technol. **16**, 1216 (1979).

[46] *H. Froitzheim, H. Ibach* and *S. Lehwald*: Phys. Lett. **55 A**, 247 (1975).

[47] *J.A. Appelbaum* and *D.R. Hamann*: Phys. Rev. Lett. **34**, 806 (1975).

[48] *G. Schulze* and *M. Henzler*: Proc. 3rd European Conference on Surface Science, Cannes 1980, Suppl. a la Revue "Le Vide, les Couches Minces", No. 201 (Soc. Française du Vide, 1980) p. 967.

[49] *M. Seel* and *P.S. Bagus*: to be published.

[50] *H. Froitzheim* and *H. Ibach*: Surf. Science **47**, 713 (1975).

[51] *H. Lüth*: Phys. Rev. Lett. **29**, 1377 (1972).

[52] *M. Hass* and *B. W. Henvis*: J. Phys. Chem. Solids **23**, 1099 (1962).

[53] *R.K. Willardson* and *C.A. Beer*: "Semiconductors and Semimetals", Vol. **3** (Academic Press, New York 1967).

[54] *M. Liehr* and *H. Lüth*: J. Vac. Sci. Technol. **16**, 1200 (1979).

[55] *S.L. Cunningham, A.A. Maradudin* and *R.F. Wallis*: Phys. Rev. **B 10**, 3342 (1974).

[56] *V.M. Mc Conaghie* and *H.H. Nielsen*: Phys. Rev. **75**, 633 (1949).

[57] *N.N. Greenwood* and *A. Storr*: J. Chem. Soc. 1965, 3, 3426 (1965).

[58] *P.M. Morse*: Phys. Rev. **34**, 57 (1929).

[59] *G. Herzberg*: "Molecular Spectra and Molecular Structure, I. Spectra of Diatomic Molecules", 2nd Edition, (Van Nostrand Reinhold, New York, 1950).

[60] *W. Mokwa*: private communication and Ph.D. Thesis, RWTH Aachen, 1981.

[61] *D. Menzel*: "Desorption Phenomena", in Topics in Applied Physics, Vol. **4**, "Interactions on Metal Surfaces", Ed. R. Gomer (Springer, Berlin 1975), p. 101.

[62] *I.E. Wachs* and *R.J. Madix*: Surf. Science **84**, 375 (1979).

[63] *D. Kohl* and *G. Heiland*: Surf. Science **63**, 96 (1977).

[64] *Y. Grinshpan, M. Nitzan* and *Y. Goldstein*: Phys. Rev. **B 19**, 1098 (1979).

[65] *G. Heiland* and *H. Lüth*: "Adsorption on Oxides", in "The Chemical Physics of Solid Surfaces and Heterogeneous Catalysis", Vol. III, Eds. D.A. King and D.P. Woodruff, (Elsevier, Amsterdam), to be published.

[66] *D. Kohl* and *E. Veuhoff*: to be published in J. Physics C (1981).

[67] *D. Eger* and *Y. Goldstein*: Phys. Rev. **B 19**, 1089 (1979).

[68] *Y. Goldstein, A. Many, I. Wagner* and *J.I. Gersten*: Surf. Science **98**, 599 (1980).

[69] *H. Ibach*: private communication.

[70] *R. Matz* and *H. Lüth*: unpublished results.

[71] *F. Stern*: Phys. Rev. Lett. **18**, 546 (1967).

[72] *J.I. Gersten*: Surf. Science **92**, 579 (1980).

[73] *E. Burstein, A. Pinczuk* and *D.L. Mills*: Surf. Science **98**, 451 (1980).

Ternary Semiconductors

Angus MacKinnon
Referat für Theoretische Physik, Physikalisch-Technische Bundesanstalt, Braunschweig, Germany

Summary: The group of the ternary compounds, which includes the chalcopyrite and many other less well known ordered and disordered semiconductors, has a number of physical properties, which are not found in the simpler binary semiconductors. For example, the chalcopyrites have not only direct and indirect band gaps but also pseudodireect energy gaps, which are formally direct but act physically like indirect gaps. Electronic and dynamic properties of these complicated compounds can be understood in a few simpler cases, when relationships are established to the very well understood elementary semiconductors. In the transformation diamond-zincblende-chalcopyrite, for example, one atom of a sublattice is replaced at each step by two. In this manner the electronic and dynamical properties of chalcopyrite and related structures, like "Ordered Vacancy Compounds" are surveyed. Many, often even simple, properties of ternary semiconductors (eg. the structure) are on the other hand unknown or in dispute, so that a large number of theoretical and experimental problems are still waiting for their solution.

1 Introduction

In the development of semiconductor physics the elemental semiconductors Silicon and Germanium have always played a dominant role. Later the III-V and II-VI compounds, for example Gallium Arsenide and Zinc Selenide, took their place as important materials for the semiconductor industry. These materials are related to the type IV semiconductors in that in all cases the average valency of the elements is 4. This is the Grimm-Sommerfeld rule [1].

A simple extension of this rule involves replacing one sublattice of a III-V compound (eg. Ga As) by two atoms, giving a II-IV-V_2 compound (eg. $ZnGeAs_2$).

This new compound is related to its III-V parent in much the same way as the III-V to a type IV atom, Si or Ge. Similarly the II-VI compounds give I-III-VI_2 ternary semiconductors, like $AgGaSe_2$.

A further extension of this principle leads from a I-III-VI_2 to a 0-II-III_2-VI_4 compound. The atom with zero valency could be an inert gas or a vacancy. There are no reports of the former in the literature, whereas there are a large number of semiconductors with an empty sublattice. These are often referred to as defect tetrahedral or ordered vacancy compounds.

These and a large number of other systems which fulfil the Grimm-Sommerfeld rule are discussed in a monograph by Parthé [2]. The I-III-VI_2 and II-IV-V_2 compounds, referred to as chalcopyrites after the mineral $CuFeS_2$, are the subject of a book by Shay and Wernick [3].

In the following sections some of the theoretical principles will be discussed which allow us to develop a picture of the similarities and the differences between these compounds and the much better understood elemental and binary semiconductors.

2 Structure

In Fig. 1. The structure of a zincblende compound (eg. Ga As) is shown in a somewhat unusual manner. Here two zincblende cubes are drawn one above the other. Fig. 2a shows the chalcopyrite structure adopted by almost all $I\text{-}III\text{-}VI_2$ and $II\text{-}IV\text{-}V_2$ compounds. One sublattice of zincblende now contains an ordered arrangement of two atoms so that the Bravais lattice is body centred tetragonal (Space group $I\bar{4}2d$). Another possible way of distributing the two new atoms is shown in Fig. 2b. Here they are arranged in planes in an – ABAB – sequence along the z-axis. In the chalcopyrite case (Fig. 2a), these planes are the same and contain both atoms. This second structure (Fig. 2b) has yet to be found in nature but is useful for comparison.

Fig. 3 contains three possible structures for $0\text{-}II\text{-}III_2\text{-}VI_4$ compounds. The first (Fig. 3a) is adopted only by $CdIn_2 \square Se_4$ ($\square$ refers to the vacancy). In this case the space group is $P\bar{4}2m$ (i.e. simple tetragonal). The lattice constants however are equal within experimental error, so that the structure is sometimes referred to as cubic. This can cause confusion. The term *Pseudocubic* is therefore preferable. The structures shown in Fig. 3b and 3c, termed defect stannite and defect chalcopyrite (or thiogallate) respectively, are adopted by the other $0\text{-}II\text{-}III_2\text{-}VI_4$ compounds. They are both body centred tetragonal but with different distributions of the cations. Defect stannite, $I\bar{4}2m$ (Fig. 3b), has greater symmetry than defect chalcopyrite, $I\bar{4}$ (Fig. 3c). Examples of these are $HgIn_2 \square Te_4$ ($I\bar{4}2m$) and $CdGa_2 \square S_4$ ($I\bar{4}$). In many cases it is uncertain which of these two structures is adopted by a particular compound. Indeed it may well be possible to prepare some of them in both forms [5].

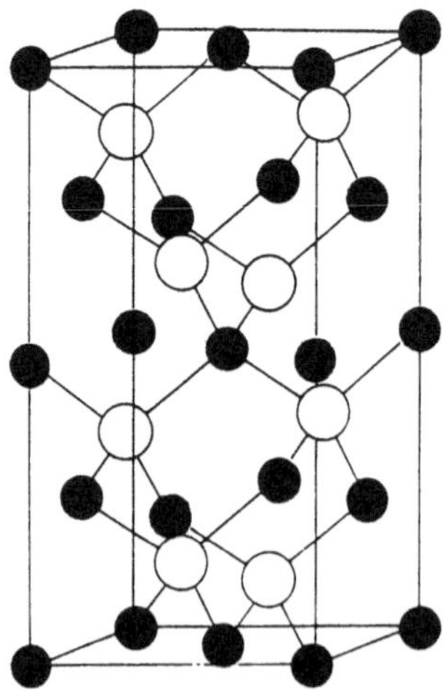

Fig. 1
Structure of zincblende. 2 cubic unit cells are shown (8 primitive cells).

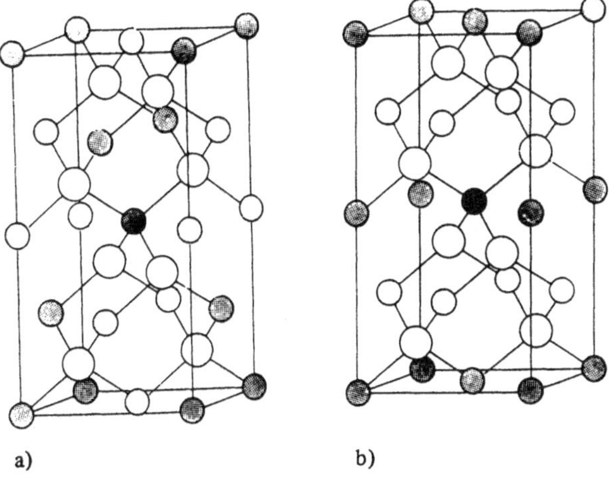

Fig. 2 (a) Chalcopyrite structure (b) alternative structure

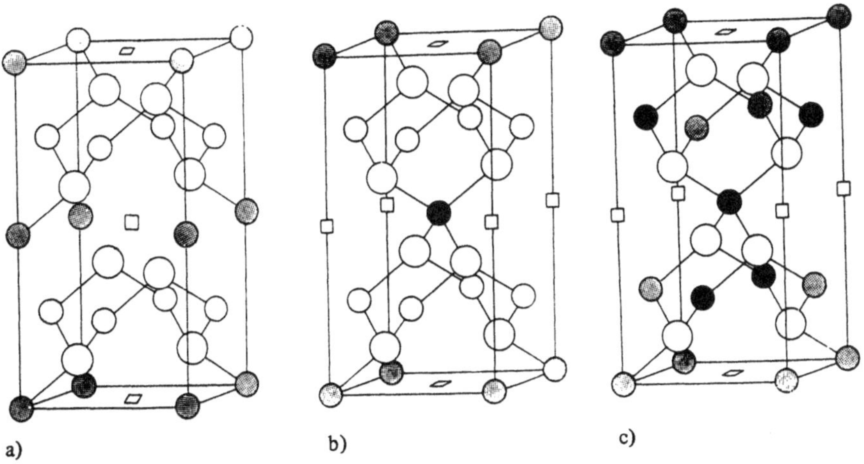

Fig. 3 (a) Cd In_2 □ Se_4 structure, (b) defect stannite, (c) defect chalcopyrite. □ refers to vacant sites.

3 The Folding Method

The similarity of the above structures to zincblende and diamond suggests that much can be learned by considering the differences from zincblende as perturbations. In order to illustrate the method which is employed, it is worth considering a simple one dimensional example. A one dimensional chain of atoms (Fig. 4a) with lattice constant „a", has the electronic energy band structure and phonon dispersion curves illustrated by the continuous lines in Fig. 5. When the single atom is replaced by two as in Fig. 4b the Brillouin zone is halved and the bands are "folded back" from the larger into the smaller zone. Where this leaves degenerate or almost degenerate states gaps open up giving the band structures and dispersion curves shown by the dashed lines in Fig. 5.

In 3 dimensions the situation is somewhat more complicated. Fig. 6 shows the Brillouin zone of the zincblende lattice (fine lines) together with two Brillouin zones of the body centred tetragonal (bct) lattice of chalcopyrite etc. (thicker lines). Four points in the zincblende zone are mapped onto each point in the bct zone. Thus Γ, X and two different W points of zincblende correspond to the Γ point of bct. Further symmetry considerations associate each irreducible representation from the zincblende group with one or more irreducible representations of the group of the ternary [6].

We are thus in a position to formulate a general rule for relating the electronic band structure and phonon dispersion curves of a ternary semiconductor to those of zincblende:

States which, when folded into the smaller Brillouin zone of the ternary compound, belong to the same irreducible representation repel each other.

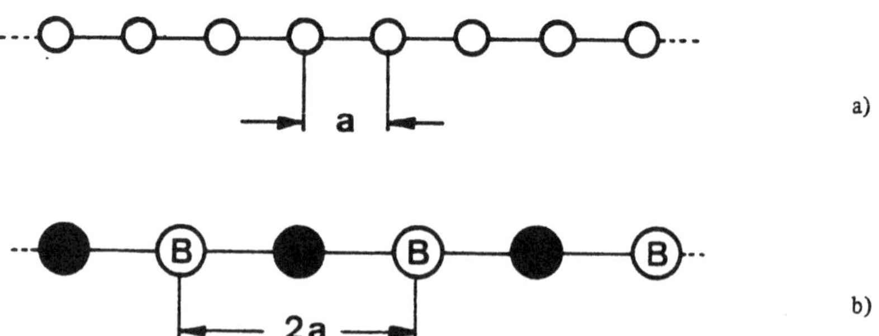

Fig. 4 (a) 1 D monatomic chain, (b) diatomic chain.

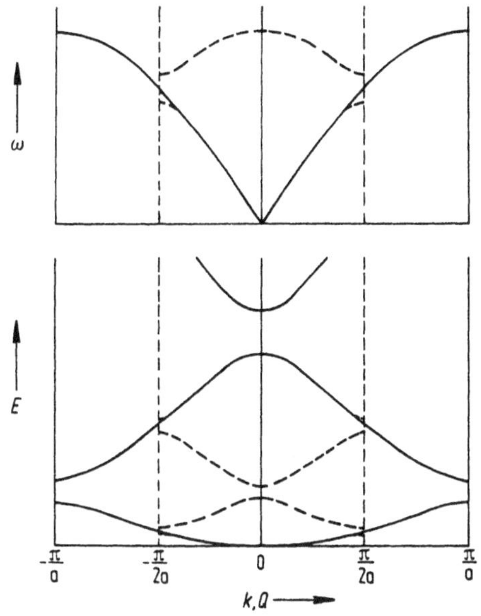

Fig. 5

Electronic energy band structure (lower diagramm) and phonon dispersion curves (upper diagramm) of a 1 D monatomic (continuous lines) and a diatomic (dashed lines) chain.

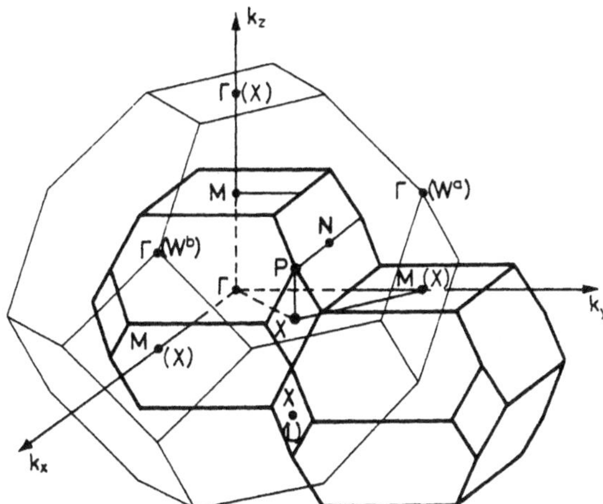

Fig. 6 Brillouin zone of zincblende (fcc) (thin lines) and two Brillouin zones of chalcopyrite (thicker lines). The letters in brackets refer to the symmetry points of the zincblende zone the others to chalcopyrite.

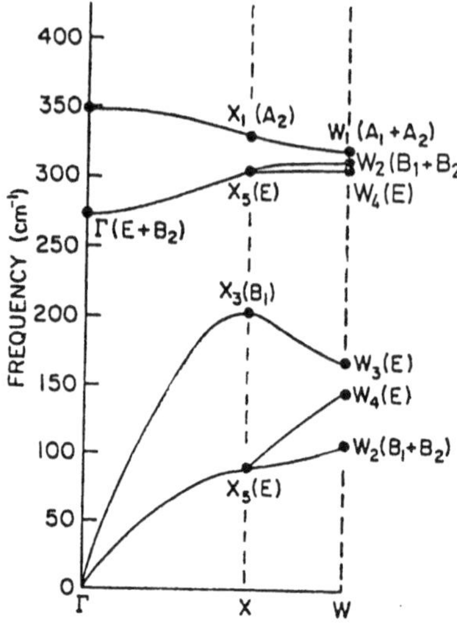

a)

Fig. 7

(a) Phonon dispersion curves of ZnS^{7}. Irreducible representations to which the phonons belong are indicated and the designations in parenthesis give the symmetry of these modes when they are folded back into the chalcopyrite structure.

(b) Real and imaginary parts of the dielectric constant of $CdGeAs_2$.

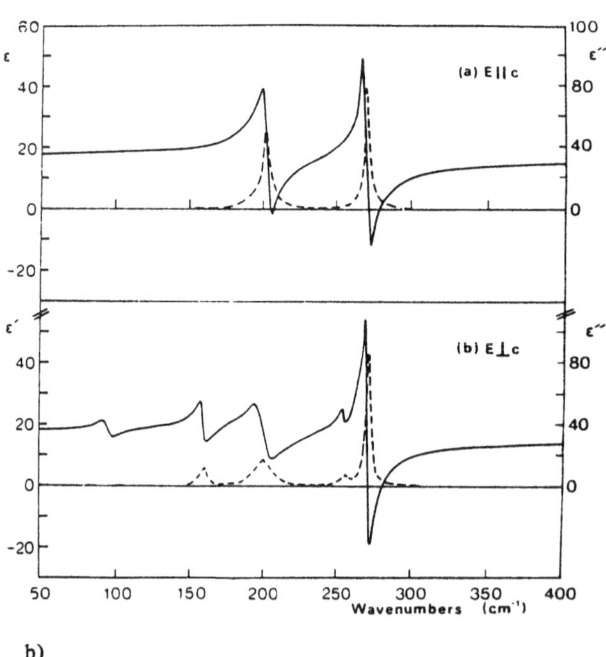

b)

This principle becomes much more useful when we consider the nature of the perturbations to zincblende and the strengths of the resulting physical processes.

The perburbations can be divided into 3 groups:

(a) Cationic Asymmetry. The existence of more than one atom on a zincblende sublattice.

(b) Anionic Displacement. The anions are no longer on sites with tetrahedral symmetry. They are therefore not constrained to remain on the ideal zincblende sites.

(c) Tetragonal Compression. In the tetragonal lattice the axial ratio c/a is smaller than the ideal zincblende value (c/a = 2).

Considering these perturbations singly and in combination leads to a hierarchy of strengths for physical processes such as absorption etc.:

Allowed-strong	allowed in zincblende
-weak	allowed in first order perturbation theory
-very weak	allowed in higher order perturbation theory.
Forbidden	forbidden in zincblende and ternary.

As an example of the application of these ideas Fig. 7a shows part of the phonon dispersion curves of a typical zincblende compound. The states at Γ, X and W are marked with the irreducible representations to which they correspond in chalcopyrite. In Fig. 7b an infrared spectrum of $CdGeAs_2$ is shown [7]. For $E \perp c$ there are one strong and several weaker lines. The strong line is the one which would be observed in zincblende, whereas the others are only seen because of the perturbations.

For $E \parallel c$ the two peaks have similar strengths. This can be understood by noticing that in Fig. 7a there are several near degeneracies among the higher states which tend to invalidate the simple perturbation theory.

4 Electronic Band Structure

The folding method discussed above allows us to make some predictions about the electronic band structure of ternary compounds. In the band structure of a typical zincblende semiconductor, GaP, (Fig. 8a) the two conduction band minima, at Γ and X, are almost equal in energy. When the band edge is at Γ the gap is direct and at X indirect. Both cases occur; Ge-direct, Si-indirect. Since both these points are folded back to Γ in chalcopyrite it is possible to obtain such crystals with two different types of direct gap, either from the Γ or from the X point of zincblende.

Figs. 8b and 8c show such band structures [3]. In Fig. 8b the conduction band edge is derived from Γ and is labelled Γ_1, whereas in Fig. 8c the edge is from X and is labelled Γ_3. In this case the gap is referred to as *pseudodirect*, since as we shall see, it is formally direct but looks in an experiment like indirect.

Fig. 9 shows absorption and electroreflectance on a normal direct gap material, $CdGeP_2$, for polarisations $E \parallel c$ and $E \perp c$.

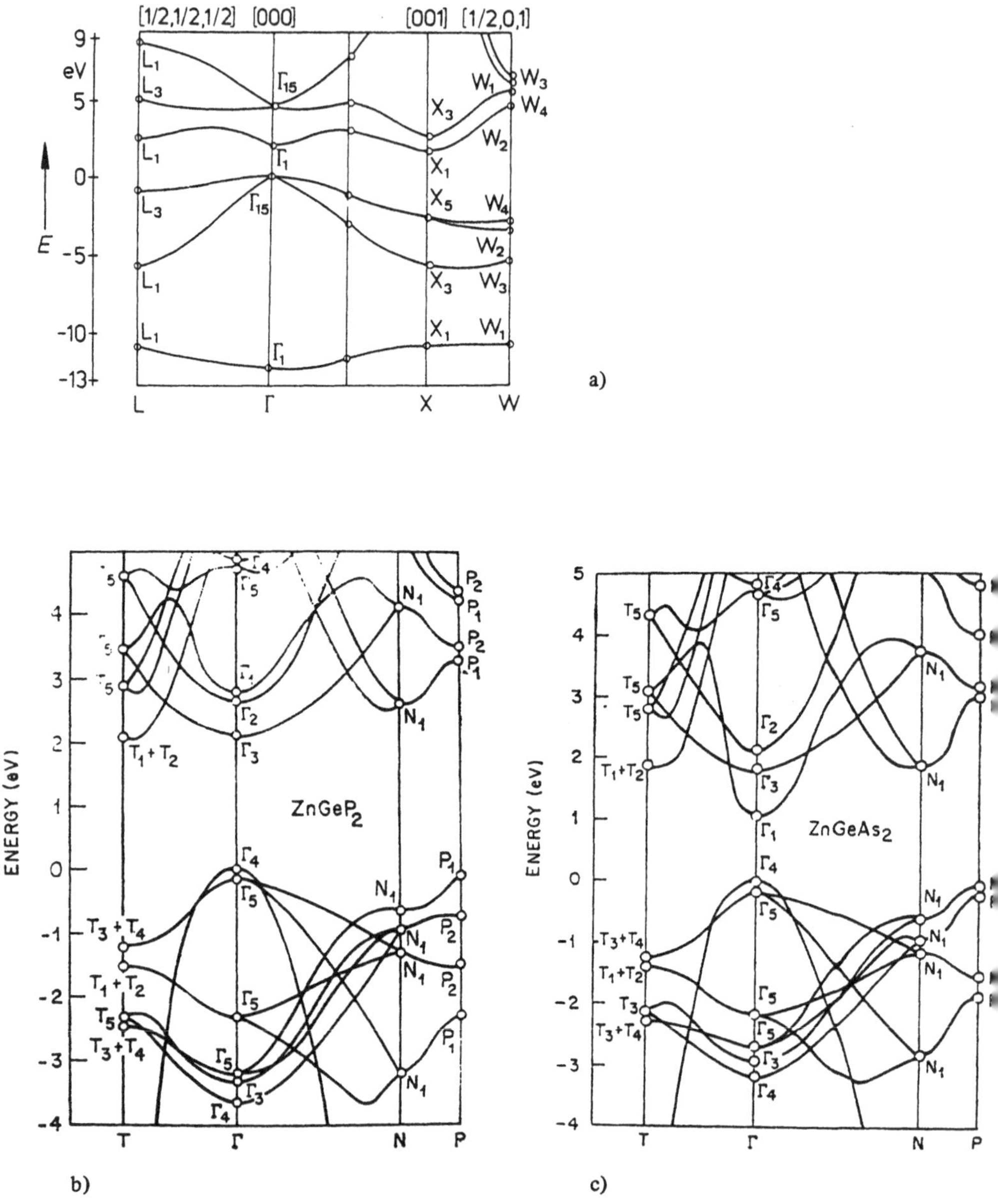

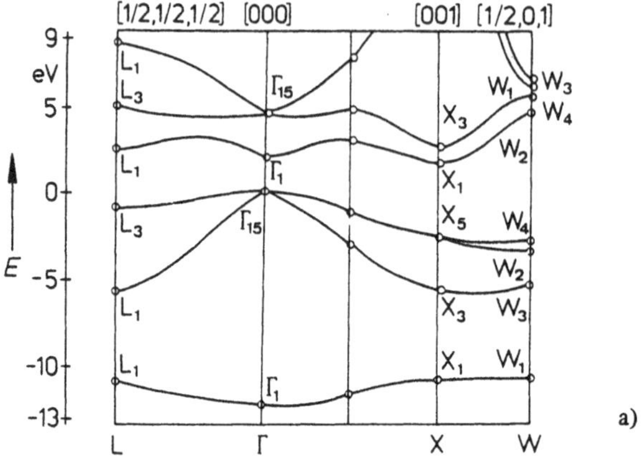

a)

b)

c)

Fig. 8 Electronic energy band structure of (a) GaP (b) $ZnGeAs_2$ (c) $ZnGeP_2$.

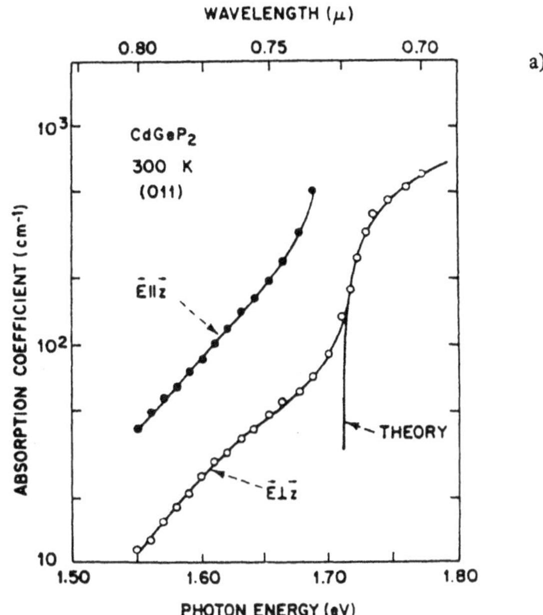

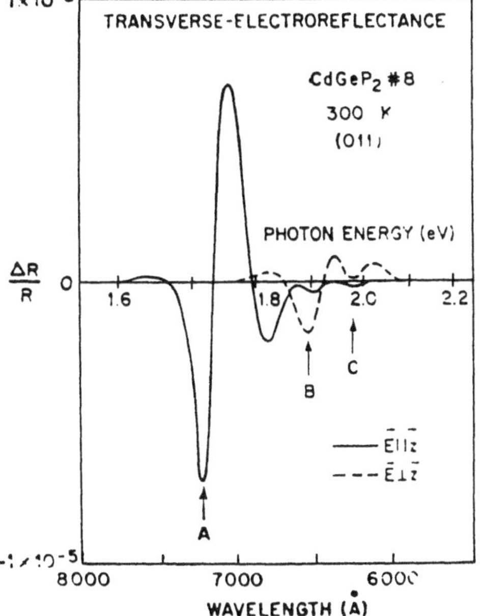

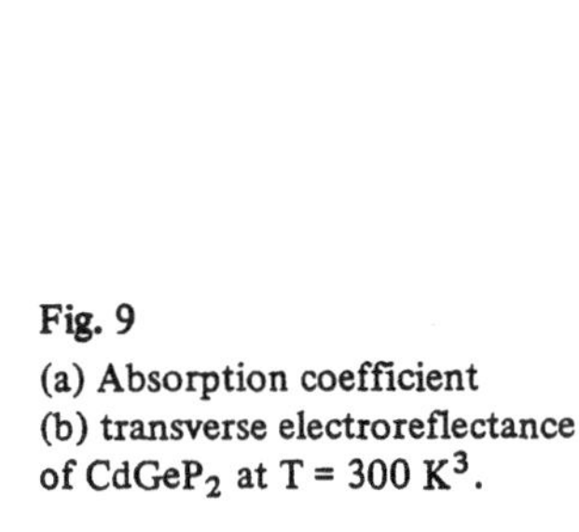

Fig. 9
(a) Absorption coefficient
(b) transverse electroreflectance
of $CdGeP_2$ at T = 300 K[3].

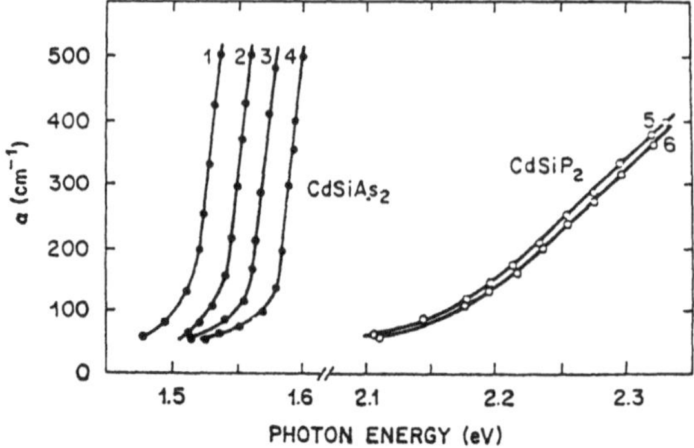

Fig. 10
The absorption adge of $CdSiAs_2$ (curves 1-4) and $CdSiP_2$ (curves 5-6) at various hydrostatic pressures in kg cm^{-1}: (1) 1; (2) 2280; (3) 4360; (4) 6480; (5) 6300); (6) 1^3.

The anisotropy is consistent with a valence band edge where Γ_4, a p_z like state, lies at slightly higher energy than Γ_5, p_x and p_y like states. In chalcopyrites this *crystal field splitting* is mainly caused by the tetragonal compression. In the electroreflectance there are in fact 3 peaks for each polarisation. This is well understood and is related to the relaxation of selection rules in the presence of spin-orbit coupling [9].

In Fig. 10 on the other hand the absorption edges of $CdSiAs_2$ and $CdSiP_2$ are shown under various hydrostatic pressures. They are clearly qualitatively different. $CdSiAs_2$ has a normal direct gap, whereas the absorption edge of $CdSiP_2$ looks like it is due to an indirect gap. It is in fact an example of a pseudodirect gap material. In chalcopyrite the Γ-X matrix elements are zero by symmetry so that the coupling between Γ and X states is only of second order i.e. Γ-W-X. Thus processes due to this interaction are very weak and we see an absorption edge like that of an indirect gap semiconductor.

In contrast to the case of the chalcopyrites the absorption of a typical ordered vacancy compound, $HgIn_2 \square Te_4$ (Fig. 11) shows a much larger anisotropy with opposite sign to that of chalcopyrite. In this case the Γ-X matrix elements are not zero and the crystal field splitting is dominated by the interaction between the states Γ_{15} and X_5 at the top of the valence band of zincblende (Fig. 8a).

In the ternary system these states transform as $\Gamma_4 + \Gamma_5$ and Γ_5 respectively, so that the two Γ_5's, the p_x and p_y like states, repel each other. This results in a crystal field splitting in which Γ_5 lies above Γ_4; the opposite splitting to that of chalcopyrite.

5 The Effect of Ordered Vacancies

A simple example of the effect of vacancies is provided by the elastic properties [10]. The bulk modulus is reduced when the density of vacancies is raised: the material becomes softer (Fig. 12).

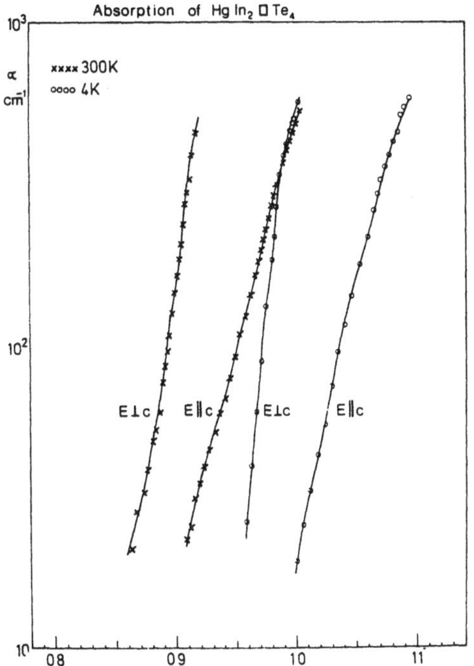

Fig. 11
Absorption coefficient of Hg In_2 □ Te_4 at T = 4 K (0) and T = 300 K (X)[9].

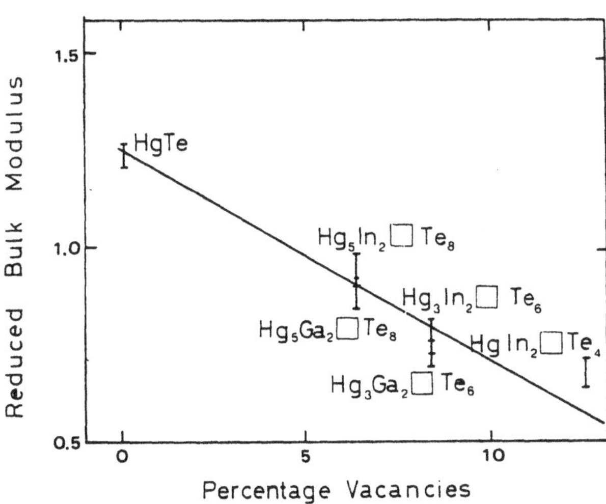

Fig. 12 Reduced Bulk Modulus vs. % Vacancies for various ordered vacancy compounds[10].

Whereas the differences between zincblende and chalcopyrite can be treated as perturbations, it is not reasonable to treat vacancies in this way, especially in considering vibrational properties.

In Fig. 13 we see the infrared reflectivity of $HgIn_2 \square Te_4$ [5] and $Hg_3 In_2 \square Te_6$ [11]. The latter has a cubic unit cell containing 108 primitive zincblende unit cells and 18 vacancies. In the case of $Hg In_2 \square Te_4$ (Fig. 13a) the highest feature in the spectrum lies well under 200 cm^{-1}. In $Hg_3 In_2 \square Te_6$ (Fig. 13b), on the other hand, the spectrum contains features at higher frequencies; up to the limit of the measurement at 500 cm^{-1}, in fact. These features have been interpreted as due to the creation of several phonons [11], a significant non-linear effect.

Raman scattering from the vacancy compounds shows another effect. In Fig. 14 spectra from two such compounds, $HgIn_2 \square Te_4$ (Fig. 14a) [5] and $CdGa_2 \square Se_4$ (Fig. 14b), [12] are shown. In both cases the spectrum is dominated by one very large peak. In the case of $CdGa_2 \square Se_4$ there is a second peak at 277 cm^{-1} which appears to be due to two phonons of the type giving rise to the large peak at 141 cm^{-1}.

It has been shown [5, 6, 11] that the multiphonon effect and the strong Raman scattering are due to the so-called *breathing mode* (Fig. 15). In this vibrational mode the anions surrounding a vacancy move symmetrically towards and away from the vacancy.

The highest valence bands in ordered vacancy compounds consist of states in which the electrons are concentrated around the vacancies [9, 14]. Such states are extremely sensitive to the breathing modes: a condition that leads to strong Raman scattering.

When we consider the atoms as having effective point charges the coulomb potential seen by a breathing mode is very asymmetric, since the vacancy does not have a charge which can attract the neighbouring atoms. This asymmetry is responsible for the multiphonon effects.

The breathing mode is thus responsible for both multiphonon effects and strong Raman scattering in ordered vacancy compounds. Symmetry considerations show that multiphonon effects cannot be seen in infrared absorption when there is only one vacancy in the unit cell [11]. Thus there are no such effects in Fig. 13a.

6 Other Properties

In the physics of semiconductors defects play a very important role. This is just as true for ternaries as for elemental and binary compounds. Indeed the fact that the differences between zincblende and the ternaries can be successfully treated as a perturbation implies that the energies involved are relatively small. It has therefore been suggested by several authors [3, 15] that the energy required to form native defects, especially antisite defects, should be correspondingly small. In the case of the vacancy compounds there is considerable evidence that up to 10 % too many cations can be built in; it is easy to fill up the vacancies with other cations [16].

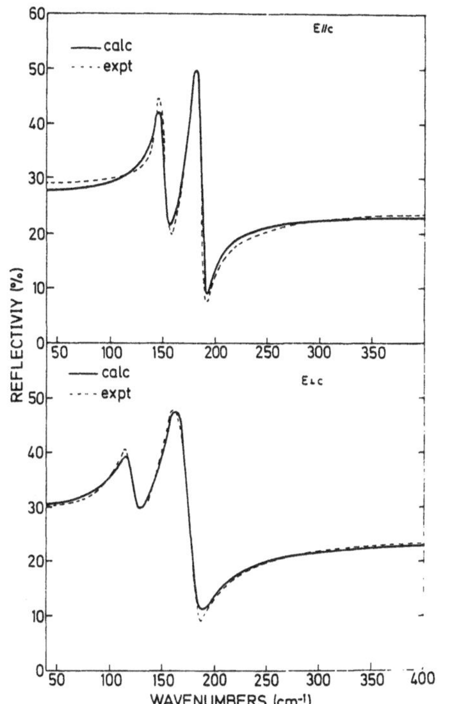

Fig. 13

Infrared Reflectivity of $HgIn_2\square Te_4$ (a)[5] and $Hg_3In_2\square Te_6$ (b)[11].

a)

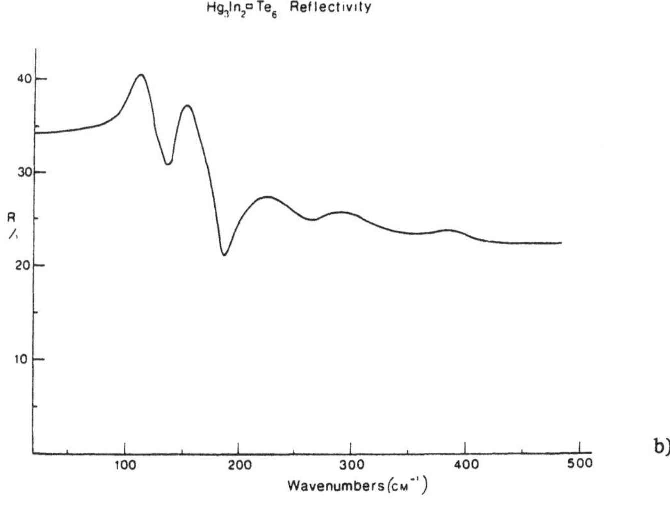

b)

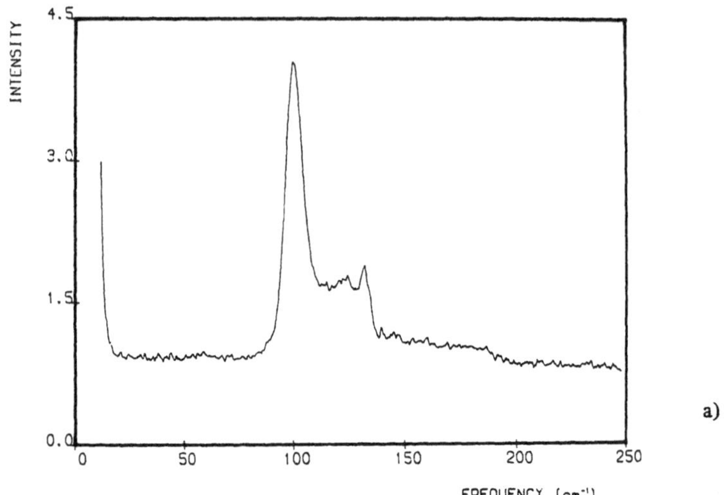

a)

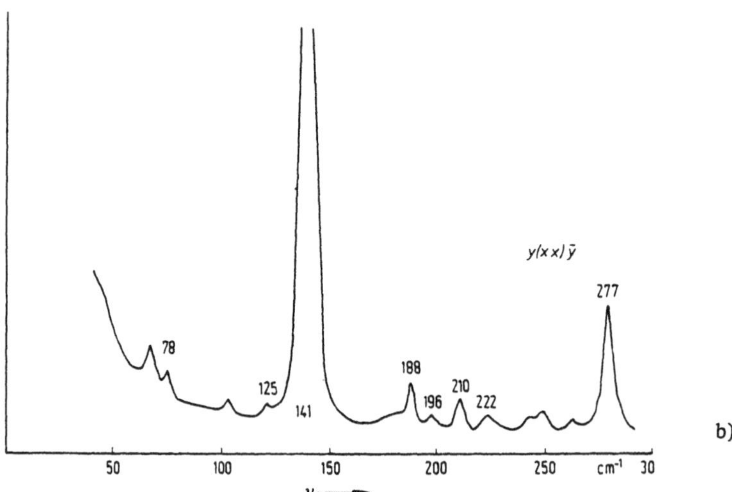

b)

Fig. 14 Raman Spectra of Hg $In_2 \square Re_4$ (a)[5] and Cd $Ga_2 \square Se_4$ (b)[12].

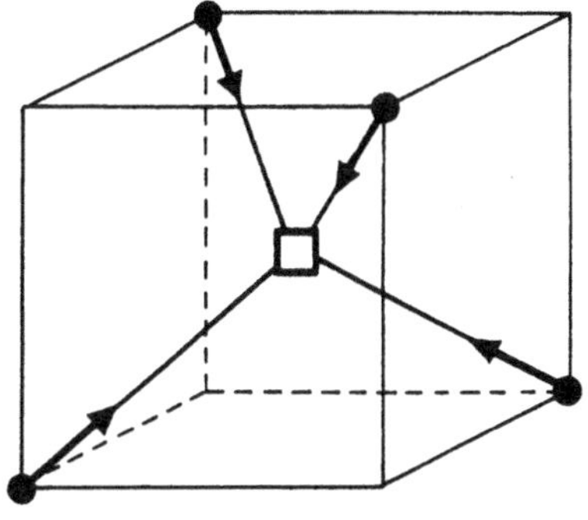

Fig. 15
A breathing mode. □ refers to a vacancy.

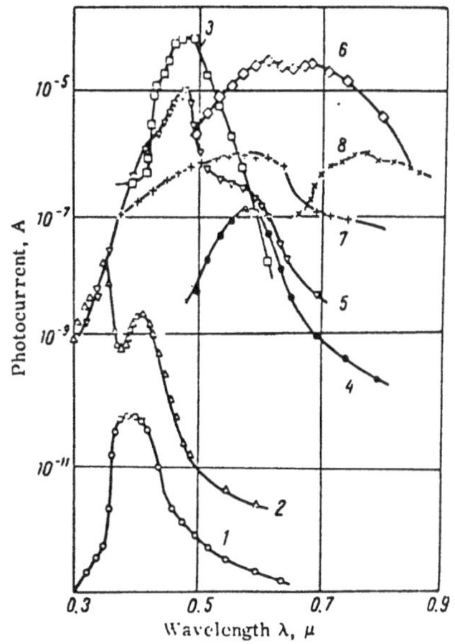

Fig. 16 Photoluminescence spectrum of Cd Si As_2 at T = 1.7 K[3].

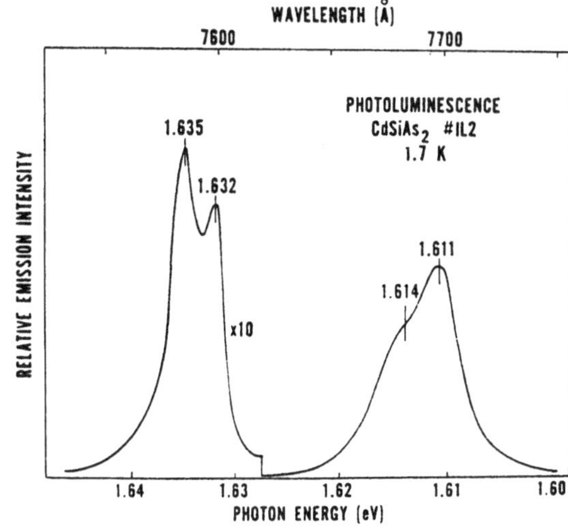

Fig. 17
Photocurrent sprectra of
1) $ZnGa_2\square S_4$; 2) $CdGa_2\square S_4$;
3) $HgGa_2\square S_4$; 4) $ZnGa_2\square S_4$;
5) $CdGa_2\square Se_4$; 6) $HgGa_2\square Se_4$;
7) $ZnIn_2\square Se_4$; 8) $CdIn_2\square Se_4$.

As illustrations of defect controlled processes the photoluminescence of $CdSiAs_2$ (Fig. 16) [3] and the photoconductivity of various vacancy compounds (Fig. 17) will be considered. The band gap of $CdSiAs_2$ is at 1.635 eV at T = 1.7 K so that the four peaks can be interpreted as due to transitions between the valence and conduction band edges, a donor and an acceptor, in this nominally undoped sample. For each compound in Fig. 17 the photoconductivity ranges over several orders of magnitude. In the case of $CdGa_2Se_4$ there are reports of a difference between photoconductivity and dark conductivity of six orders of magnitude [15].

Non-linear optical applications of semiconductors are achieving increasing importance. The majority of elemental and binary semiconductors adopt the cubic zincblende structure, which means that, although their non linear susceptibilities are large, their usefulness is restricted by lack of birefringeance and the consequent inability to *phase match* [3]. The chalcopyrites however have similarly large non-linear susceptibilities and are usually birefringent enough to achieve phase matching. Table 1 lists some non-linear parameters of various semiconductors [17]. It is clear that several chalcopyrites have comparable properties to other conventional non-linear materials.

Table 1 Properties of infrared nonlinear materials.

Material	Transmission range (μm)	Second-order non-linear coefficient $d \times 10^{22}$ (As/V^2)	Birefringence $n_e - n_o$ (10.6 μm)	Index of refraction n	Nonlinear figure of merit d^2/n^3 (relative units)	Pulsed laser-induced surface damage threshold at 10 μm $(J\ cm^{-2})$
Te	4 –25	57	+ 1.45	4.9	405	–
$CdGeAs_2$	2.5 –17.5	26	+ 0.086	3.6	215	5–6
$ZnGeP_2$	0.7 –12	6.6	+ 0.040	3.1	22	–
GaSe	0.65–18	5.0	– 0.376	2.6	21	–
$AgGaSe_2$	0.7 –19	2.9	– 0.033	2.6	7	1.5–3
CdSe	0.7 –25	1.6	+ 0.018	2.5	2.7	–
$AgGaS_2$	0.5 –12.5	1.1	– 0.054	2.4	1.2	3–3.5
Ag_3AsS_3	0.65–13	1.1	– 0.21	2.6	1	1.9
Ag_3SbS_3	0.65–14	1.1	–	2.7	0.9	–

7 Conclusions

Ternary compounds offer the manufacturer of semiconducter devices a great variety of electronic energy gaps and of lattice constants: properties which are of critical importance in solar cells, diodes etc. to name but a few.

In addition to these simple properties we have considered a number of physical phenomena peculiar to the ternary semiconductors, pseudodirect gaps in the chalcopyrites, multiphonon effects etc. in the ordered vacancy compounds.

The birefringeance of the tetragonal systems helps to make the chalcopyrites, and in future probably other ternary and multinary semiconductors, important non-linear materials.

The vacancy compounds show promising photoconductive properties.

On the other hand it is difficult to grow good single crystals of these compounds either because they have a high density of native defects or because they tend to adopt several structures which are difficult to separate.

Ternary semiconductors have a number of possible applications, especially in situations where there are no suitable elemental or binary compounds. With the exception of the chalcopyrites they have been the subject of very little research effort (outside the USSR) and so represent a field for considerable research in future.

Acknowledgements

I would like to thank Prof. D. Weaire and Dr. A. Miller for many useful discussion and considerable encouragement in this work.

References

[1] *H. G. Grimm* and *A. Sommerfeld*, 1926, Z. Phys. **36**, 36.

[2] *E. Parthé*, 1964, "Crystal Chemistry of tetrahedral structures" (Gordon and Breach: New York).

[3] *J. L. Shay* and *J. H. Wernick*, 1974, "Ternary Chalcopyrite Semiconductors –Growth, Electronic Properties and Applications" (Pergamon: New York).

[4] *H. Hahn, G. Frank, W. Klinger, A. D. Störger* and *G. Störger*, 1955, Z. Anorg. Allg. Chem. **279**, 241.

[5] *A. Miller, D. J. Lockwood, A. MacKinnon* and *D. Weaire*, 1976, J. Phys. C: Sol. St. Phys. **9**, 2997.

[6] *A. Miller, A. MacKinnon* and *D. Weaire*, 1981, "Beyond the Binaries – the Chalcopyrites and related semiconducting Compounds", Sol. St. Phys. (in print).

[7] *M. H. Grimsditch* and *G. D. Holah*, 1975, Phys. Rev. **B12**, 4377

[8] *G. D. Holah, A. Miller, W. D. Dunnett* and *G. W. Iseler*, 1977, Sol. St. Commun. **23**, 75.

[9] *A. MacKinnon, A. Miller* and *G. Ross*, 1977, Inst. Phys. Conf. Ser. No. **35**, 171.

[10] *G. A. Saunders* and *T. Seddon*, 1976, J. Phys. Chem. Sol. **37**, 873.

[11] *A. MacKinnon, A. Miller, D. J. Lockwood, G. Ross* and *G. D. Holah*, 1977, Inst. Phys. Conf. Ser. No. **35**, 119.

[12] *R. Bacewicz, P. P. Lottici* and *R. Razzeti*, 1979, J. Phys. C: Sol. St. Phys. **12**, 3603.

[13] *A. MacKinnon*, 1979, J. Phys. C: Sol. St. Phys. **12**, L655.

[14] *A. Baldereschi, F. Meloni, F. Aymerich* and *G. Mula*, 1977, Sol. St. Commun. **21**, 113.

[15] *L. I. Berger* and *V. D. Prochukhan*, 1969, "Ternary, Diamond-like Semiconductors", (Consultants Bureau, New York, London).

[16] *R. N. Sheikh-Zananova* and *V. F. Ponomarev*, 1977, Izv. Akad. Nauk SSSR, Neorg. Mat. **13**, 1308, Inorganic Materials, **13**, 1056.

[17] *G. W. Iseler, H. Kildal* and *N. Menyuk*, 1977, Inst. Phys. Conf. Ser. No. **35**, 73.

Structure and Electronic States in a-Si: H

John D. Joannopoulos, Douglas C. Allan
Department of Physics, Massachusetts Institute of Technology, Cambridge, Mass. 02139, USA

Summary: An overview of recent theoretical accomplishments in the study of the electronic structure of hydrogenated amorphous silicon is presented. Special emphasis is placed on attempts to elucidate the relationships among bonding, structure and electronic states. This includes the search for an understanding of the nature of the properties of a variety of silicon-hydrogen bonding conformations and bonding coordination defects. Particular attention is paid to the interpretation of results in terms of the effects of bulk-like and surface-like environments, the effects of quantitative and topological disorder, and the formation of band-like and gap-like states.

1 Introduction

Since many years ago, it was thought that it was a basic and fundamental property of an amorphous semiconductor that it could not be doped. Mott had explained [1] in a simple and elegant way that the lack of periodic steric constraints could allow the valency of any impurity to be satisfied.

It was therefore a great surprise when in 1975 W. Spear and P. LeComber [2, 3] managed to dope amorphous silicon both n-type and p-type. One very interesting property of this material was that it was prepared by glow discharge silane (SiH_4) decomposition and consequently contained a large amount of hydrogen.

It has been of great importance, therefore, to understand the influence of the hydrogen on the doping process and consequently understand the structural aspects of the material. Is the hydrogen actually allowing the doping of an amorphous network? Do impurities normally occupy substitutional sites in an amorphous network and the hydrogen aids in the doping process by removing gap states? If this is the case, why doesn't the hydrogen also aids in satisfying the valency of the impurities? Could it be that the presence of hydrogen helps to create small microcrystalline regions which can be doped normally? It is still not certain what the answers to these questions are; nevertheless, these questions and the exciting possibilities associated with doping amorphous semiconductors [2–13] and the construction of devices and solar cells [14–30], have inspired a large number of experimental investigations [31–143] of the structural [31–49], electronic [50–68], vibrational [69–82], optical [83–93], luminescent [94–112], magnetic [110–121], and transport and photoconductive [122–144] properties of this material. Recently a moderate number of theoretical studies [145–170] have also appeared, in order to explain the experimental data. In what follows, we review briefly some selected results obtained from these experimental and theoretical investigations. An excellent comprehensive review of the general experimental situation can be found in a recent review article by Fritzsche [144].

A Overview of Experimental and Theoretical Studies

Hydrogenated amorphous silicon (a–Si:H) can be made in a variety of ways. The most popular involve glow discharge silane decomposition [2, 3] thermal decomposition of silane [12], and sputtering of silicon in an inert gas atmosphere containing hydrogen [5]. The properties of the material obtained from these techniques can vary greatly depending on the substrate temperatures and r.f. fields employed.

In describing the properties of hydrogenated amorphous Si it is helpful to divide the experimental information according to whether one is obtaining information about general bonding conformations involving states on a macroscopic scale (i.e. band-like states) or information on a more microscopic scale involving defects and gap-like states.

1 Conformations on a macroscopic scale (band-like states)

The pioneering photoemission measurements by von Roedern et al. [50–51] on the low temperature modifications of a–Si:H were initially interpreted as identifying the presence of large amounts of SiH_3 conformations. This seemed to be in conflict, however, with Raman and IR studies [69–82] which found evidence primarily for SiH_2. Theoretical calculations using tight-binding [149, 156], first principles LCAO [153, 160], and self-consistent $X\alpha$ [161] Hamiltonians, all agreed that the photoemission signatures for SiH_3 and SiH_2 are nearly indistinguishable. Thus all the experiments are actually consistent with the presence of primarily SiH_2 conformations. The gap has also been found to increase with hydrogenation. In particular, it seems that it is the valence band that is shifting downwards. This is reproduced by the aformentioned theoretical calculations [149, 156, 161] and more recently by CPA calculations using realistic tight-binding Hamiltonians [166]. Moreover, it has been argued [156, 164] that fluctuations in the valence band edge with variations in hydrogen content from one region of the sample to the next might account for the much larger mobility of electrons as compared to holes [122, 127, 128].

There is considerable evidence, in fact, that hydrogen is not uniformly distributed throughout the network. Proton NMR studies [38], electron microscopy [34, 36] and small angle neutron scattering [39] indicate that there is a columnar growth morphology in deposited films with most of the hydrogen existing at intergrain boundaries. Only ~3 % of hydrogen is found within the columns of amorphous grains, existing in monohydride (SiH) conformations. Some of this hydrogen is presumably clustered inside the grains at small voids or internal surfaces. These growth characteristics and the seemingly fixed concentration of hydrogen within the grains are consistent with Phillips' topological models and strain-relieving mechanism [150, 158]. The structural inhomogeneities in a-Si:H also seem to be borne out by ball-and-stick structural models [151].

The photoemission results [50–52] for the high substrate-temperature or annealed form of a–Si:H show considerable structure in the valence band density of states (DOS). This structure has been identified as that observed on hydrogenated crystalline Si (111) and (100) surfaces, indicating primarily monohydride conformations. The predominance of SiH in the amorphous material is also substantiated by theoretical DOS calculations [149, 153, 156, 160, 161, 166]. These photoemission results, however, warrant much closer inspection. It is found theoretically [156], for example, that the states associated with monohydride conformations are very sensitive to the topology of the underlying lattice. (This is not true in the case of SiH_2 and SiH_3 configurations [149].) This is very important, for it suggests that photoemission measurements on this material, with varying photon energies, could provide valuable information concerning the nature of the Si structure. It is conceivable, for instance, that one could distinguish between a random-network-like structure and a microcrystalline structure with very small regions of crystalline order. This in turn could be important in understanding the doping and transport properties of the material.

2 Conformations on a microscopic scale (gap-like states)

It is generally agreed that a–Si produced by sputtering or evaporation, without hydrogen, contains a large number ($\sim 10^{20}$ cm^{-3}) of dangling bonds. These dangling bond gap states may explain the insensitivity of a–Si to dopants. When hydrogen is incorporated into the network, it passifies the dangling bonds, removing states from the gap. However, it is typical [8, 31, 117, 118] that considerably more hydrogen is needed than necessary to quench the spin resonance signal. This can be understood by an argument proposed by Phillips [150]. He points out that dangling bonds in amorphous materials are similar to those on surfaces where reconstitution transfers electrons and tends to get rid of most of the spins. Thus the hydrogen is chemisorbing on surfaces of voids where reconstitution is allowing essentially only one spin per void [165].

When a–Si:H is doped, considerably more dopant is necessary than the estimated number of states in the gap that need to be compensated [3, 53]. This can be partially understood on the grounds that most of the dopants have their valency satisfied, as originally suggested by Mott [1]. EXAFS studies [32] have shown, for example, that only a small fraction of arsenic dopants are four-fold coordinated in plasma deposited a–Si:H samples. It may also be the case, however, that dopants introduce more defects and gap states into the material as suggested by decreases in luminescence efficiency and photovoltaic performance [100, 103].

The detailed structure of the gap DOS of a–Si:H remains controversial [53–66]. Typical field effect measurements [63, 54, 58, 59] show structure in the gap at 0.4 and 1.2 eV below the conduction band edge, with the lower energy peak larger by an order of magnitude. Recent deep level transient spectroscopy (DLTS) measurements [61], however, indicate a lower overall density of gap states and *no* sharp structure. Both DLTS and field effect indicate a larger DOS near the valence band edge than the conduction band edge. The differences have been attributed to the sensitivity of field effect to surface and interface states [61, 144]. Nevertheless, photoemission measurements [51, 52] also seem to show structure in the gap about 1.2 eV below the conduction band edge. Various theoretical models have been proposed to account for this possible structure in the gap. These include double dangling bond states [147], three-center-bound [146], and tail-like states localized within a-Si grains because of large fluctuations of the valence band edge [156, 164] and possibly the conduction band edge [164] with the hydrogen inhomogeneities.

Luminescence studies [94–112] generally show a dominant peak near 1.4 eV which has been interpreted in terms of band edge transitions with a moderate Stokes shift (~ 0.2 eV) [97–101]. The idea of a Stokes shift has recently been challenged, however, on the basis of absorption near 1.4 eV [90, 108]. A new peak in the luminescence near 0.9 eV is also observed under conditions of doping or conditions where the defect density is increased in undoped samples [104, 108, 112, 120]. This has been interpreted as a transition between a doubly occupied dangling bond state (~ 0.5 eV below the conduction band with distortion energy ~ 0.1 eV) and a valence-band tail hole [104].

The electron spin resonance (ESR) studies on a-Si:H seem to be fairly well understood [110–120]. There is a characteristic resonance line with g = 2.0055 corresponding to dangling bond states whose concentration can vary between $\lesssim 10^{20}$ cm^{-3}, down to the level of detectability $\lesssim 10^{15}$ cm^{-3}, depending on preparation conditions. This signal decreases and eventually disappears with n- or p-type doping as the dangling bond states fill up or become empty respectively. For undoped specimens, optically induced ESR also decreases this spin signal creating two new resonance lines with g = 2.004 and g = 2.013. This is understood, presumably, in terms of transitions from neutral dangling bond states to tail states in the conduction band (trapped electrons) and transitions from the valence band (trapped holes) forming negatively charged dangling bond states, respectively. For doped materials, the optical excitations increase the dangling bond spin signal.

B Scope of Present Work

The first part of this work is inspired by the photoemission measurements of von Roedern et al. [50–52] on the high temperature substrate or annealed forms of a-Si:H. We will embark on a detailed and careful study of the structure in these spectra with an aim to attempt to unravel the nature of the monohydride conformations and the topological structure of the underlying a-Si lattice. In particular, we will concentrate on identifying how the local environment around atoms affects their electronic states and what the signatures are for various bonding conformations and topologies.

The second part of this work will deal with the effects of defects and quantitative disorder on the electronic structure of a-Si:H. In particular, we will attempt to examine which defects have the lowest formation energies; what are the typical relaxation energies involved when the defects are charged; what are the electronic gap states associated with these defects and what kind of states does quantitative disorder introduce in the gap.

The format of this paper is as follows. In section II we describe the various theoretical methods that we will be using in our studies. In section III, we present results related in a variety of silicon-hydrogen bonding conformations and make comparisons with the photoemission experiments. In section IV we present calculations of ground state and relaxation energies of neutral and charged bonding-coordination defects. In section V we study the effects of these defects, and quantitative disorder, on the electronic states in the the gap. Finally, in section VI we make some concluding remarks.

II Theoretical Methods

A The Cluster-Bethe-Lattice Method

We choose to model the structure of a-Si:H in terms of Si Bethe-lattices (or equivalently Cayley trees) [172]. A Bethe lattice (BL) is an infinite, non-periodic connected system of atoms, where every atom is bonded in the same configuration, but there are no closed rings of bonds. Thus, the system branches out like a tree to infinity as shown schematically in Fig. 1 (c). Since the structure of a-Si:H is highly inhomogeneous (as discussed in I. A.2.) with hydrogen atoms tending to cluster at the boundaries of grains or at internal microvoids, one may think of the structure in terms of hydrogen atoms embedded in a host silicon network. A simple example of this is shown in Fig. 1 (a) where we think of the host as some infinite random network. The electronic states of an infinite random network are generally quite similar to those of a BL. There are three basic reasons for this. First, both are infinite and non-periodic. Second, the local bonding configurations of the atoms in each model are similar. And finally, the topological nature of the random network is such that many rings of bonds of various sizes can exist in the system. These rings impose boundary conditions in the wavefunction which tend to quantize the states and put peaks in the DOS. But a system with many different types of rings will have a DOS with many different peaks which will tend to look smooth and featureless. This, however, is similar to the DOS of the BL that looks smooth and featureless because of the *lack* of rings.

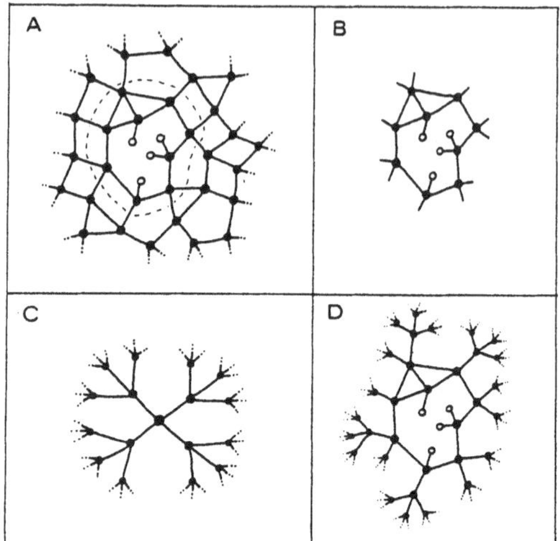

Fig. 1

The cluster-Bethe lattice method involves (a) taking an infinite system of atoms and choosing one or more atoms as a reference point, (b) removing a cluster of atoms surrounding the reference atoms from the system, (c) using a Bethe lattice to model the connected network, and (d) attaching one to each dangling bond of the cluster.

The BL is also very useful from a mathematical point of view because it can be solved exactly for a variety of different Hamiltonians [170, 173]. In this regard the BL is an excellent candidate for an effective medium in which to study the states at local structurally inhomogeneous or disordered sites of an amorphous material. This, in fact, is precisely the basis of the cluster-Bethe-lattice method [174, 175]. The basic idea is shown schematically in Fig. 1. Beginning with an infinite connected network of atoms, a particular set of atoms is chosen as a reference point. A cluster of atoms surrounding and including these atoms is then removed from the system. A BL whose atoms are in the same bonding configuration as in the original system is attached to *each* dangling bond. Thus, the new system in Fig. 1(d) is infinite and non-periodic as the original system. In addition the local environment of the central reference atoms is the same as in the original structure. Moreover, the local DOS of the central atoms can now be solved exactly. Since the density of states of the BL is smooth and essentially featureless, any structure that one obtains in the local DOS of the central atoms can immediately be associated with the local environment of those atoms. It should be remembered that the local DOS of an atom is essentially the DOS of an electron when it is found in the vicinity of that atom.

B Tight-Binding Hamiltonian

To model the valence electrons in a-Si:H we choose the realistic tight-binding Hamiltonian developed by Pandey [176, 177]. The model includes one s-orbital and three p-orbitals per Si atoms, and an s-orbital for each hydrogen. All nearest-neighbor interactions, and second neighbor Si-Si ppπ and ppσ interactions are included. The basis orbitals are assumed to be orthogonal and the interaction parameters are fitted to bulk crystalline Si, gaseous silane and di-silane energy states. With no further empirical adjustments this Hamiltonian gives an excellent description of the monohydrated and trihydrated Si (111) surfaces [176, 177].

C Semi-Empirical Bond Energy Method

It has been popular in the literature to use the sum of the one-electron eigenvalues

$$E_{el} = \int_{-\infty}^{\epsilon_F} \epsilon N(\epsilon)\, d\epsilon, \tag{1}$$

as an estimate of the total energy of a configuration of atoms. Such approximations, however, are in general grossly inaccurate and misleading. The reason is that the real total energy is given by

$$E_{tot} = E_{el} + U, \tag{2}$$

where U represents the energy coming from ion-ion repulsions and corrections to E_{el} for overcounting of electron-electron interactions. The term U is very difficult to calculate exactly but can be estimated empirically by fitting to solid and molecule dissociation energies [178].

Since U is made up of two terms that cancel at large distances, one can approximately write

$$U = \sum_b{}' \Delta U_b, \tag{3}$$

where ΔU_b corresponds to a correction term per bond. For our particular system we need only calculate ΔU(Si-Si) and ΔU(Si-H). This can be accomplished easily by just comparing differences in E_{tot} with experimentally determined dissociation energies D. Thus for bulk Si, the dissociation energy is associated with removing an atom from the system and rebonding the remaining dangling bonds. By modelling bulk Si as a BL we have,

$$D(\text{Si-Si}) = E_{tot}\,(\text{free Si atom}) + E_{tot}\,(\text{SiBL, N-1 atoms}) - E_{tot}\,(\text{SiBL, N atoms}) \tag{4}$$

$$D(\text{Si-Si}) = E_{el}(\text{free Si atom}) - E_{el}(\text{SiBL, 1 atom}) - 2\Delta U(\text{Si-Si}). \tag{5}$$

Using the tight binding Hamiltonian described in II.B we calculate E_{el}(free Si atom) = − 8.03 eV and E_{el}(SiBL, 1 atom) = − 20.69 eV with respect to an arbitrary zero. From the heat of formation of crystalline Si we know $D(\text{Si-Si}) = 4.69$ eV. Therefore equation (5) gives

$$\Delta U(\text{Si-Si}) \simeq 4.0 \text{ eV}. \tag{6}$$

For the SiH bond we need to calculate the energy required to remove a hydrogen atom from silane (SiH_4). Thus

$$D(\text{Si-H}) = E_{tot}(\text{free H atom}) + E_{tot}(SiH_3) - E_{tot}(SiH_4) \tag{7}$$

$$D(\text{Si-H}) = E_{el}(\text{free H atom}) + E_{el}(SiH_3) - E_{el}(SiH_4) - \Delta U(\text{Si-H}). \tag{8}$$

Using $D(\text{Si-H}) = 4.08$ eV we find

$$\Delta U(\text{Si-H}) \simeq 4.4 \text{ eV}. \tag{9}$$

Thus, with $\Delta U(\text{Si-Si})$ and $\Delta U(\text{Si-H})$ known, the total energy of any configuration of atoms in a-Si:H can be calculated by integrating the one-electron energy over the density of states as in (1) and adding correction terms $\pm \Delta U_b$ for each different bond, or broken bond, or extra bond in the structure.

D Quantum Chemical Cluster Calculations

The discussion presented in II.C is primarily valid in the case of neutral configurations of atoms. In order to study the total energies and relaxation process of charged systems care must be taken because of the complexity and long range nature of the interactions. Since we are primarily interested in studying the charged states of localized defects we consider an approach that has been fairly successful in surface studies. This is the Generalized Valence Bond (GVB) approach using small clusters [179–181]. The basic idea is to take a chemical approach whereby a defect in a-Si:H is modelled by a small molecule or complex. One then forms generalized valence bond wavefunctions using one orbital per electron and solves self-consistently in order to obtain optimum orbitals as in Hartree-Fock. One then generally carries out a configuration interaction calculation where all configurations formed by the various occupations of the GVB orbitals are allowed.

In our case, we will be interested in studying the three charged states of a Si dangling bond defect (which we will denote as T_3^+, T_3^0, T_3^-). This defect can be modelled effectively [181] with a Si_4H_9 cluster obtained by selecting a Si atom with a dangling bond, its three Si neighbors and nine "hydrogen" atoms to replace the other bonds to the rest of the Si neighbors. The "hydrogen" atoms are included purely to model the rest of the Si system and are not involved with anything else. A dielectric medium with dielectric constant that of Si is also included half-way around the cluster in order to account for the extra polarization energy introduced by the charged states lying on a surface [181].

The results that are presented in IV.B are based on calculations by Redondo et al. [181] on clusters of this type, with some minor extensions.

III Structure and Electronic States

A Defining the Problem

The valence band photoemission spectra (UPS) of von Roedern et al. [50–51] using $\hbar\omega \sim 40.8$ eV are shown in Fig. 2(a). The results are for a-Si without hydrogen (dashed line), and for a-Si:H (solid line) prepared by sputtering in ambient hydrogen with substrate temperatures $\gtrsim 350$ °C. The valence states between 0 and − 5 eV are mostly Si p-like in character while the states below − 5 eV are mostly Si s-like. There are several new and interesting features that arise with hydrogenation. The peaks near − 5 eV and − 7.5 eV were interpreted [50] as signatures of SiH conformations because of similar structure observed in UPS spectra of monohydrated crystalline Si (111) surface [177]. In Fig. 2(b), for example, we show theoretical calculations of the local DOS of hydrogen atoms [176, 177] for a monolayer of hydrogen on Si (111). (The dashed curve represents the bulk DOS of crystalline Si and is shown for comparison.) The peak near − 10.5 eV in Fig. 2(a) was not discussed in the original experimental measurements, but it too would appear to be essentially consistent with the results in Fig. 2(b).

There are several questions, however, that need to be raised at this point. Does the fair agreement between Figs. 2(a) and (b) imply that a-Si:H consists of microcrystalline regions with exposed monohydrated Si (111) surfaces? Or does the struc-

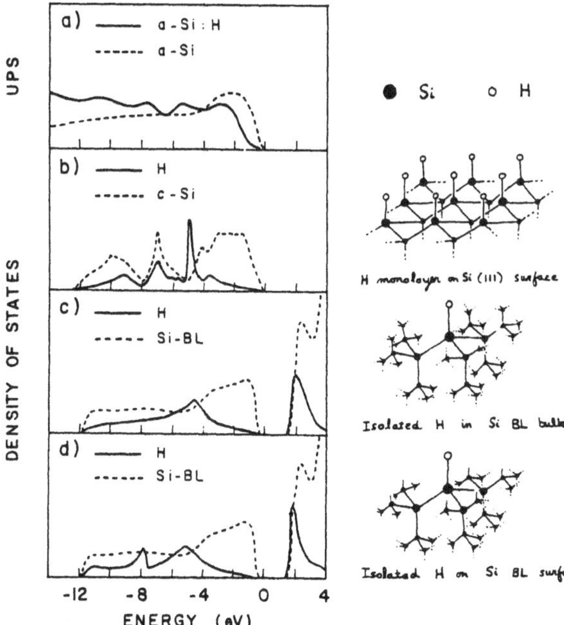

Fig. 2
(a) UPS measurements [50] for a-Si (dashed line) and a-Si:H (solid line).
(b) – (d) Local DOS of H's (solid lines) for atomic configurations shown. The DOS's shown as dashed lines in (b) and (c), (d) are those of crystalline Si and Si-Bethe-lattices respectively. The results in (b) are from [176], and the normalization of the H DOS is 1/4 that of the Si DOS.

ture observed in the DOS arise only from interactions between SiH pairs and is it consequently independent of the nature of the underlying Si lattice? (That is, the same peaks could be observed in a random network system as long as the SiH's are kept at the same distances apart.) Or is it that the structure in the DOS is a completely intrinsic property of the SiH bond, and would be observed even with only one SiH unit embedded in any Si network?

Some of these questions can be answered immediately by considering the configuration of atoms depicted in Fig. 2(c). Here we have removed interactions between SiH units by considering only *one* SiH bond, and have replaced the underlying c-Si lattice (with its six-fold ring topology) with a Si Bethe-lattice (BL) which has a "neutral" topology. The local DOS of the hydrogen atom is shown as a solid curve. (The DOS of the Si BL is shown as a dashed curve and the states at positive energies represent the conduction bands.) It is clear that the peak near − 5 eV remains and *is* an intrinsic property of the SiH bond. The structure below − 6 eV, however, has now disappeared.

It is the purpose of the next two sections to determine how the peaks near − 7.5 and − 10.5 might arise. Before we embark on this investigation, however, it is interesting to compare the differences between a bulk-like and surface-like environment on the electronic states of the SiH bond. This is easily done by considering the configuration of atoms shown in Fig. 2(d). Basically, the upper half of the BL in Fig. 2(c) was removed, exposing a surface of dangling bonds with one chemisorbed hydrogen atom. (This kind of geometry was first studied by Yndurain and Louis [182] as a Bethe-Peierls approximation to a c-Si surface with chemisorbed H atom.) We note two interesting features in Fig. 2(d). First, an s-like surface state near − 8 eV has appeared. This state is very well known in surface studies and is primarily localized on the surface Si atoms. Second, the peak near − 5 eV has shifted to lower energy by ~0.5 eV. This shift is caused by interactions of this Si-H state with the dangling bond orbitals on the surface [182].

It is intriguing that the peaks in Fig. 2(d) seem to match up fairly well with those in the UPS results in Fig. 0(a). Such an identification, however, would imply that there are roughly as many H atoms as dangling bonds in the system. Moreover, no such dangling-bond structure is observed in the non-hydrogenated spectrum of a-Si. Thus, it would be very difficult to explain why more dangling bonds are obtained after hydrogenation. We consequently do not pursue this identification further in this paper.

B Topology and H-Interactions

We wish to determine what structural properties of a-Si:H could give rise to the peaks below − 6 eV observed in Figs. 1(a) and 2(b). We consider two possibilities; the effects of the topology of the underlying lattice and the effects of interactions between SiH units.

The effects of topology on the electronic states in a-Si are fairly well understood [183]. In Fig. 3(a), for example, we show the DOS of c-Si. The states below − 6 eV are very sensitive to the ring topology of the structure. In particular, the peaks near − 7 eV and − 10 eV arise because of the six-fold rings that make up the diamond lattice. This is immediately clear by comparing this c-Si DOS with the local DOS of the Si* atom shown in the same figure. The Si* atom has six six-fold rings of bonds passing through it with BL's connected everywhere outside. The agreement between the two curves is quite impressive. Recall that the DOS of the BL was smooth and featureless in this region (Fig. 2(c)).

If we replace one of the bonds of the Si* atom with a bond to a H atom, we obtain the system shown in Fig. 3(b). The original six rings have been retained. The local DOS of Si* has peaks in the same place as in Fig. 3(a) with the addition of a weak peak near − 5 eV. The latter is, of course, the signature of a SiH conformation, as discussed earlier, and is clearly visible in the local DOS of the H atom. Moreover, the H DOS follows the DOS structure below − 6 eV of the underlying Si atom. This is also consistent with the DOS's in Fig. 2(c). Thus the local DOS of an H atom in a monohydride conformation is very sensitive to the topology of the underlying Si network. (This is not true in the case of dihydride or trihydride conformations because of the strong interactions among the SiH bonds [149, 157].) If the underlying Si network were a random network structure. with many different kinds of rings of bonds, then the average DOS over many "isolated" SiH units would only

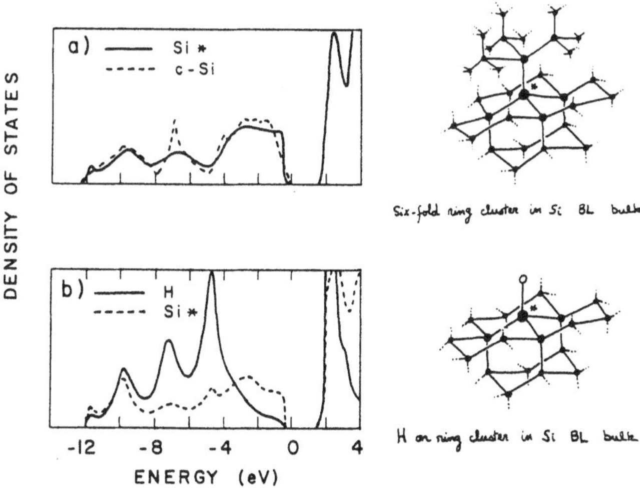

Fig. 3 Local DOS for the specific atoms considered in the configurations shown. The DOS of crystalline Si is reproduced as a dashed line in (a). The normalization of the H DOS is chosen to be the same as that of Si to ease comparisons in this and subsequent figures.

give the peak at − 5 eV as in Fig. 2(c). Only networks with specific kinds of rings of bonds will give well defined structure in the DOS below − 6 eV.

To study the effects of interactions between SiH units, we consider first the system of atoms shown in Fig. 4(a). Here the SiH units are placed at second neighbour distances (with respect to their Si atoms). This configuration is similar to the H monolayer considered in Fig. 2(b) except that the underlying network consists of BL's [182]. We note again the signature of SiH conformations near − 5 eV. In addition, there are two strong peaks below − 6 eV. Clearly, it is the interaction between SiH units that cause this structure since the underlying Si network has the topology of a BL. In Fig. 4(b) we consider a similar configuration of atoms, but now with only seven SiH units. The results clearly show that an entire monolayer is not needed in order to obtain the DOS structure in Fig. 4(a).

Another possible configuration of atoms, where the SiH units are at second neighbour distances (with respect to their Si atoms) but are tilted toward each other, is at a vacancy. This is shown in Fig. 4(c). We note only one peak near − 5 eV. Finally, in Fig. 4(d) we consider a configuration of atoms with SiH units at nearest neighbour distances. This system can be considered as a hydrogenated Si dimer, the kind one would expect to find on a Si(100) surface. Again, we have a peak near − 5 eV but there is also well defined structure below − 6 eV. Again, it is clear that the interactions between SiH units give rise to these peaks.

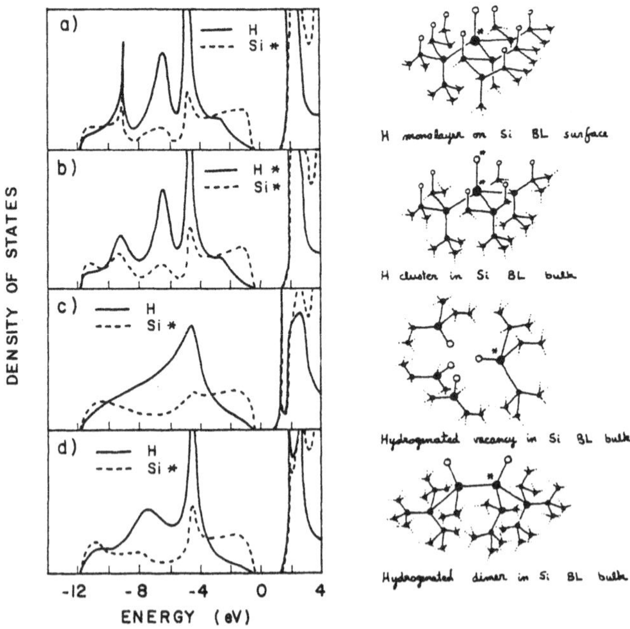

Fig. 4 Local DOS for the specific atoms considered in the configurations shown.

C Comparisons With Experiment

Let us now return to the original photoemission measurements and make comparisons with selected results discussed in the previous section. The UPS measurements of von Roedern et al. [50–51] are reproduced again in Fig. 5(a). In Figs. 5(b) and 5(c) we consider the effects of having SiH units at second neighbour and first neighbour distances respectively. These units are to be thought of as being embedded in a random network structure, modelled by BL's. The DOS's shown correspond to an average over the H atoms and the Si atoms to which they are bonded. The DOS of the Si BL is also shown for comparison. (We note the similarity between the dashed curves in Figs. 5(a) and 5(b) as discussed in II.A.) The results should be interpreted as follows. If the underlying Si network in a-Si:H can be considered as a random network model then Figs. 5(a) – (c) imply that the SiH units are predominantly of the dimer configuration. Ignoring entropy considerations, the energy of formation of a dimer configuration should be very similar to that of the second-neighbor system. (The energy for a non-hydrogenated dimer is in fact one of the lowest energy defects as discussed in IV.A.)

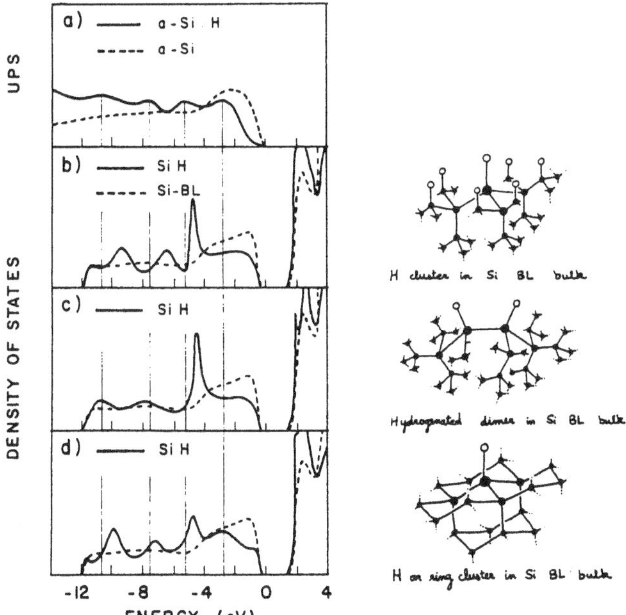

Fig. 5 (a) UPS measurements from [50]. (b)–(d) Average DOS over H and neighboring Si atom. The Si-Bethe lattice DOS is shown as a dashed curve.

In Fig. 5(d) we consider the effect of SiH units embedded in a system whose topology is determined by small regions of crystalline order. Now it is the ring topology of the underlying Si network that plays an important role and it is not very important whether the SiH units are isolated or occur in clusters as discussed above. Since the agreement between the DOS's in Figs. 5(d) and 5(a) is not too bad (given the uncertainties of the Hamiltonian), it is conceivable that the structure of a-Si:H could be viewed as being made up of small regions with medium range crystalline order.

One possible way of distinguishing between these two structural models for a-Si:H is to perform X-ray photoemission measurements on these samples. Since the X-ray scattering cross-section for H atoms is negligible, one would be primarily measuring the DOS of the underlying Si network. If the same kind of peaks are found as in Fig. 5(a), it would suggest that a medium range order model is probably correct. On the other hand, if a smooth spectrum is obtained (like the one for a-Si in Fig. 5(a)) it would suggest that a random network structure for the Si atoms exists and that the H atoms are preferring dimer-like SiH configurations.

IV Ground State Energies and Relaxations

A Tight Binding Studies

In this section we wish to study the ground state energies of a variety of neutral atomic configurations involving Si and H atoms. Our approach is to calculate total energies from the one-electron energies obtained from our tight-binding calculations and correct for the ion-ion interactions and overcounting using the empirical scheme described in II.C.

Our results are summarized in Fig. 6. Both the one-electron E_{el} and total energies E_{tot} are given for comparisons. The energies in the left panel are refered to an isolated H atom and normally coordinated Si atoms in the BL. The energies on the right panel are referred to a normally coordinated Si atom in the BL.

There are three interesting results in the left panel of Fig. 6. First, we note that it seems to be energetically favorable for H atoms to break bonds and form SiH units. This presumably could occur after some barrier is overcome. Secondly, the energy required to remove an H atom from the bulk $\sim$ 3.5 eV. This is $\sim$ 0.5 eV less than in SiH and in good agreement with recent experimental estimates [187]. Thirdly, the monomer, dimer and second-neighbor configurations have essentially the same total energy, if we neglect entropy considerations. Actually the dimer emerges with a slightly lower energy if the values of the calculations are taken seriously to more decimal places.

In the right panel of Fig. 6 we consider some dangling bond defect configurations. We note that the double-dangling bond defect (T_2^0) is not of lower energy than the single dangling bond defect (T_3^0) as earlier speculated [147]. Moreover, an even

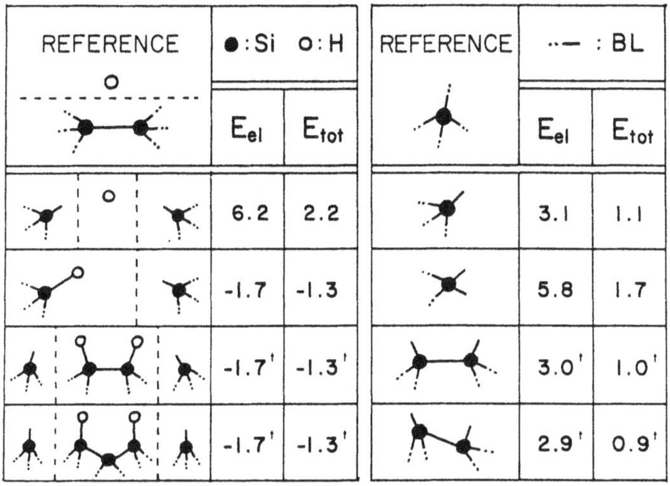

REFERENCE	•:Si o:H		REFERENCE	··— : BL	
	E_{el}	E_{tot}		E_{el}	E_{tot}
	6.2	2.2		3.1	1.1
	-1.7	-1.3		5.8	1.7
	-1.7^{+}	-1.3^{+}		3.0^{+}	1.0^{+}
	-1.7^{+}	-1.3^{+}		2.9^{+}	0.9^{+}

Fig. 6 One electron (E_{el}) and total (E_{tot}) energies of indicated atomic configurations. Energies on right panel are referred to a normally coordinated Si atom in the BL. The energies on the left are referred to an isolated H atom and normally coordinated Si atoms in the BL. The values with (+) refer to energies per defect atom or per SiH conformation.

lower energy defect than T_3^0 is the dimer configuration. The energy gained by relaxation is also presented.

The electronic states associated with these neutral defects are discussed in V.A.

B GVB Studies

Let us now consider the ground state properties and relaxation effects for charged defects. In particular, we consider the three states T_3^+, T_3^0 and T_3^- of the simple Si dangling bond defect, using the GVB results of Redondo et al. [181] discussed in II.D. The configuration energy diagram for these states is shown in Fig. 7. The zero of energy is arbitrarily chosen to be at the equilibrium position of the T_3^-. The configuration coordinate represents the deviation in position of the dangling bond Si atom, from the perfect tetrahedral position, along the (111) direction. The T_3^0, for example, is slightly displaced inward from the ideal position. Note also the two breaks in scale. Finally, we recall (from the discussion in II.D) that it is most appropriate to think of these defects as occurring at the surface of the sample or at large internal voids.

There are two important results that emerge from these calculations. First, the tran sition $2T_3^0 \rightarrow T_3^+ + T_3^-$ is an endothermic reaction. This implies that the Hubbard U is positive for these kinds of defects. This is in contrast to what is observed in the chalcogenide glasses [165]. In particular, the value one obtains from Fig. 7 is

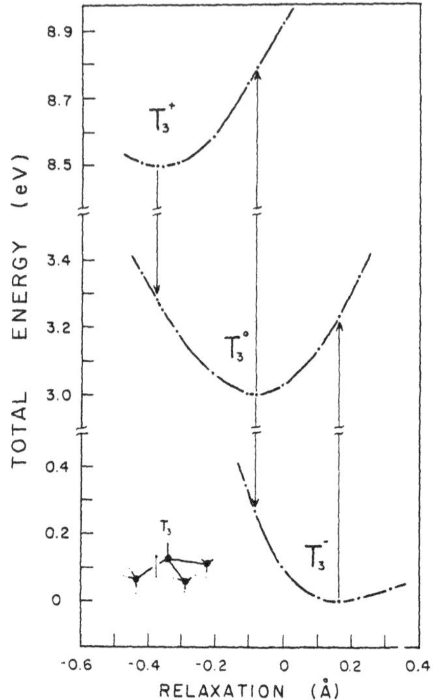

Fig. 7
Configurational coordinate diagram for the three charge states of a dangling bond. The coordinate represents movement of the dangling bond atom along the (111) direction as shown. The results are based on GVB calculations by Redondo et al. [181].

U ~ + 2.5 eV. Secondly, the relaxations associated with charging the dangling bond are by no means negligible. The Stokes shifts for the transitions $T_3^0 \leftrightarrow T_3^+$ and $T_3^- \leftrightarrow T_3^0$ are roughly 0.6 and 0.5 eV respectively.

The implications of these results to the luminescence measurements and states in the gap is discussed in V.B.

V Defects and Electronic States

A Neutral Defect States

The DOS's for a variety of neutral dangling bond defects (using the tight binding Hamiltonian described in II.B) are shown in Fig. 8. The gap state of the T_3^0 lies near 0 eV. Relaxation effects typically move this state to higher energy by ~ 0.1 eV as obtained from Fig. 7. In any case, these states will not be affected by the valence band edge receding with hydrogenation. Defect states of a relaxed T_2^0 are also shown in Fig. 8(a). There are now two states introduced near the gap with the upper one empty. These states can act as hole and electron traps as suggested by Adler [137].

In Fig. 8(b) we show the results for both unrelaxed and relaxed dimer configurations. Again the upper gap states are empty. Moreover, we note that the DOS of the

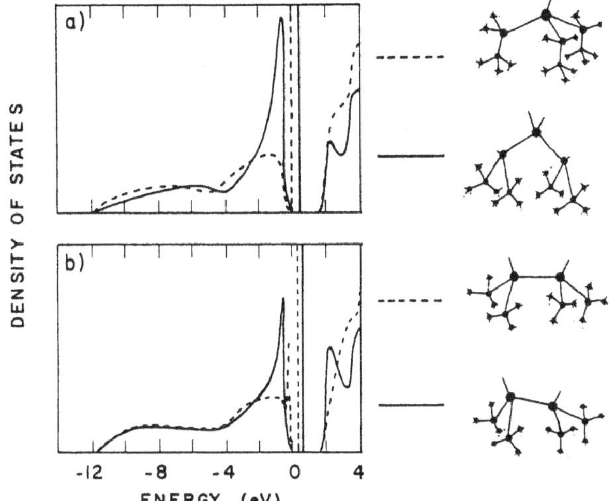

Fig. 8
Local DOS for dangling bond atoms in configurations shown.

relaxed dimer is very similar to that of the relaxed T_2^0. In principle, these states can act as hole and electron traps for undoped samples. Furthermore, these calculations suggest that the Fermi level in undoped samples should tend to lie in the lower half of the gap. This would depend, of course, on how much the valence band edge has receded.

Recent CPA calculations [166] simulating vacancies, di-vacancies, etc., indicate a smear of states in the gap for ~ 5 % vacancies. Calculations on surfaces of crystals [188–190], however, retain well defined peaks in the gap with typical widths of 0.4–0.8 eV. One might expect some structure ought to be retained in the gap.

B Charged Defect States

The energy of the electronic state associated with a charged dangling bond can be estimated from the results in Fig. 7. We note that the ionization energy for the T_3^0 is ~ 5.8 eV. This is in close agreement with the ionization energy of a surface dangling bond. The energy to ionize the T_3^- however, is only ~ 3.2 eV. Thus, the T_3^- state would lie in the conduction band. These results, however, only pertain to surface defects. If the dangling bonds occur inside the bulk, the extra polarization energy would bring the ionization energies of the T_3^- and T_3^0 closer together. Thus, a wide range of levels would seem possible. For example, a rough calculation for a dangling bond in a small void of radius ~ second neighbour distance gives a difference between T_3^0 and T_3^- ionization energies of less than 1 eV. Moreover, if the dangling bond were in the vicinity of other dangling bonds, the interactions would also reduce this energy.

We recall that recent luminescence studies have interpreted a transition near 0.9 eV as occuring via the de-excitation of a T_3^- [104]. If this is correct, it would imply that the optical excitation energy needed to create a T_3^- is ~ 1.4 eV. This is a consequence of the relaxation effects (or Stokes shift) associated with the difference of ~ 0.5 eV in ionization energy for a T_3^- at its equilibrium position and at the equilibrium position of a T_3^0. This is evident in Fig. 7.

C Distortion States

Distortions in bond lengths and bond angles, if they are sufficiently severe, may also produce states in the gap. This is shown in Fig. 9. The upper panel corresponds to a bond being stretched by 20 %. We note that states near the bottom and top of the valence band are shifted to lower binding energies. It is the conduction band that is most sensitive to this distortion, however, and creates a state in the gap near the band edge. These results are consistent with other recent calculations using different models [159, 184, 185].

Distortions in ideal tetrahedral environment can be modelled by "expanding" a disordered tetrahedron into normal-mode-like configurations [186]. These modes are composed of two configurations of symmetry E and three of symmetry F_2. The E mode shown in Fig. 9(b) corresponds to a twist of two bonds with respect to the (100) axis by 40°. The F distortion, shown in Fig. 9(c), involves two bonds bending inward and the other two outward by 40°. Again states are introduced near the edges of the gap. The states near the band edges in Fig. 9(c) are primarily localized on the three atoms that form the bond angle that is increasing. The state at the top of the valence band in Fig. 9(b) is localized primarily on the central atom and its nearest neighbours. As the distortion is made larger it will move further into the gap and be localized primarily around the nearest neighbors. The state at the conduction band edge, however, is very localized on the central atom, and remains so with increasing distortions.

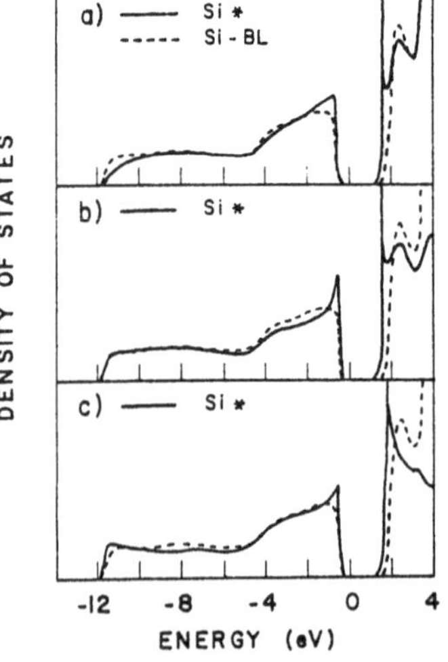

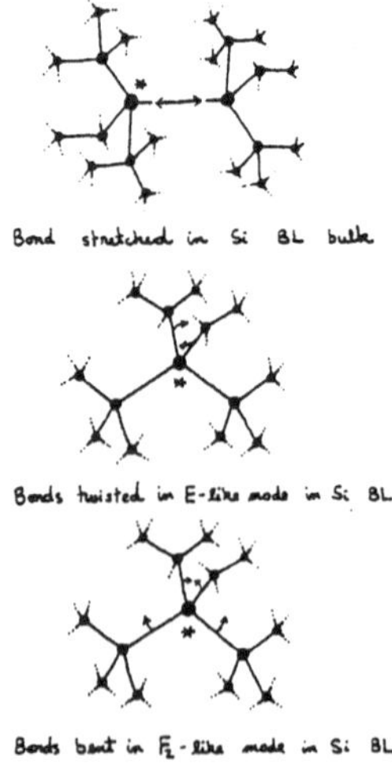

Fig. 9
Local DOS for the atoms designated by an asterisk in the configurations shown.

It is interesting that the effects of all of these distortions on the DOS are quite similar. These results clearly indicate how quantitative disorder can produce tails of states penetrating into the gap.

VI Concluding Remarks

We are still on the frontier of our understanding of the nature of amorphous semiconductors. Many fundamental questions regarding structure, bonding and elementary excitations remain unanswered and are the subjects of on-going research. It was the purpose of this article, however, to emphasize the importance of the intimate relationships between structure and electronic states in these systems and moreover, to show the power and versatility of semi-quantitative (and at times quantitative) theoretical studies of these systems. The approach taken here was that of studying specific kinds of disorder, defects or bonding conformations rather than statistical averages. It is clear that ultimately the latter approach is the most realistic in the final comparison with experiment. It is felt, however, that the former approach is more susceptible to providing insights into the nature of these systems.

References

[1] *N.F. Mott*, Phil. Mag. **19**, 835 (1969).

[2] *W.E. Spear, P.G. LeComber*, Solid State Comm. **17**, 1193 (1975).

[3] *P.G. LeComber, W.E. Spear*, AIP Conf. Proc. **31**, 285 (1976).

[4] *J.C. Knights*, Phil. Mag. **34**, 663 (1976).

[5] *W. Paul, A. Lewis, G. Connell, T. Moustakas*, Solid State Comm. **20**, 969 (1976).

[6] *M. Brodsky*, IBM Tech. Bull. **19**, 4470 (1977).

[7] *M. Brodsky, J. Cuomo*, IBM Tech. Bull. **19**, 4802 (1977).

[8] *W.E. Spear*, Adv. Phys. **26**, 811 (1977).

[9] *A. Madan, S.R. Ovshinsky, E. Benn*, Phil. Mag. **B 40**, 259 (1979).

[10] *A. Madan, S. R. Ovshinsky*, J. of Non-Cryst. Solids **35**, 171 (1980).

[11] *J. Zemek, M. Zavetova, S. Koc*, J. of Non-Cryst. Solids **37**, 15 (1980).

[12] *N. Sol, D. Kaplan, D. Dieumegard, D. Dubreuil*, J. of Non-Cryst. Solids **35**, 291 (1980).

[13] *M. Hirose, M. Taniguchi, T. Nakashita, Y. Osaka, T. Suzuki, S. Hosegawa, T. Shimizu*, J. of Non-Cryst. Solids **35**, 297 (1980).

[14] *D.E. Carlson, C.R. Wronski*, Appl. Phys. Lett. **28**, 671 (1977)

[15] *S.R. Ovshinsky, A. Madan*, Nature **276**, 482 (1978).

[16] *R.A. Gibson, P.G. LeComber, W.E. Spear*, IEE J. S. S. and E. D. **2**, 3 (1978).

[17] *W.E. Spear, P.G. LeComber, A.J. Snell*, Phil. Mag. **B 38**, 303 (1978).

[18] *A. Deneuville, M.H. Brodsky*, Thin Solid Films **55**, 137 (1978).

[19] *R. Williams, R.S. Crandall, RCA Review* **40**, 371 (1979).

[20] *A. Deneuville, M.H. Brodsky*, J. Appl. Phys. **50**, 1414 (1979).

[21] *P.G. LeComber, W.E. Spear, A. Gaith*, Electron. Lett. **15**, 179 (1979).

[22] *M. Konagai, K. Takahashi*, Appl. Phys. Lett. **36**, 599 (1980).

[23] *D.A. Anderson, G. Moddel, W. Paul,* J. of El. Mat. **9**, 141 (1980).

[24] *G. Muller, F. Demond, S. Kalbitzer, H. Damjantschitsch, H. Mannsperger, W. Spear, P. LeComber, R. Gibson,* Phil. Mag. **41**, 571 (1980).

[25] *M. Shur, W. Czubatyj, A. Madan,* J. of Non-Cryst. Solids **35**, 731 (1980).

[26] *R.A. Gibson, W.E. Spear, P.G. LeComber, A. J. Snell,* J. of Non-Cryst. Solids **35**, 725 (1980).

[27] *D.E. Carlson,* J. of Non-Cryst. Solids **35**, 707 (1980).

[28] *T.D. Moustakas, C.R. Wronski, D. Morel,* J. of Non-Cryst. Solids **35**, 719 (1980).

[29] *Y. Hamakawa, H. Ohamoto, Y. Nitta,* J. of Non-Cryst. Solids **35**, 749 (1980).

[30] *L. Vieux-Rochaz, A. Chenevas-Paule,* J. of Non-Cryst. Solids **35**, 737 (1980).

[31] *G.A.N. Connell, J.R. Pawlik,* Phys. Rev. **B 13**, 787 (1976).

[32] *J. Knights, T. Hayes, J. Mikkelsen,* Phys. Rev. Lett. **39**, 712 (1977).

[33] *H. Fritzsche, C.C. Tsai,* Solar Energy Mat. **1**, 471 (1979).

[34] *J. Knights, R. Lujan,* Appl. Phys. Lett. **35**, 244 (1979).

[35] *P.D. Antonio, J.H. Konnert,* Phys. Rev. Lett. **43**, 1161 (1979).

[36] *J. Knights,* J. of Non-Cryst. Sol. **35**, 159 (1980).

[37] *A.J. Leadbetter, A. Rashid, R. Richardson, A. Wright, J. Knights,* Solid State Comm. **33**, 973 (1980).

[38] *J. Reimer, R. Voughn, J. Knights,* Phys. Rev. Lett. **44**, 193 (1980).

[39] *T.A. Postol, C.M. Falco, R. Kampwirth, I. Schuller, W. Yelon,* Phys. Rev. Lett. **45**, 648 (1980).

[40] *E. Schiff, P. Persans, H. Fritzsche, V. Akopyan,* Appl. Phys. Lett. in press.

[41] *A. Triska, D. Dennison, H. Fritzsche,* Bull. APS **20**, 392 (1975).

[42] *M.H. Brodsky, M.A. Frisch, J.F. Ziegler, W.A. Lanford,* Appl. Phys. Lett. **30**, 561 (1977).

[43] *K. Matysik, C. Mogab, B. Bagley,* J. Vac. Sci. Tech. **15**, 302 (1978).

[44] *J. McMillan, E. Peterson, J. Appl.* Phys. **50**, 5238 (1979).

[45] *D. Jones, R. Gibson, P. LeComber, W. Spear,* Solar Energy Mat. **2**, (1979).

[46] *D. Biegelsen, R. Street, C.C. Tsai, J. Knights,* Phys. Rev. **B 20**, 4839 (1979).

[47] *P.S. Peercy, H. Stein, D. Ginley,* Appl. Phys. Lett. **36**, 678 (1980).

[48] *H.J. Stein, P.S. Peercy,* Phys. Rev. **B 22**, 6233 (1980).

[49] *S. Oguz, M.A. Paesler,* Phys. Rev. **B 22**, 6213 (1980).

[50] *B. von Roedern, L. Ley, M. Cardona,* Phys. Rev. Lett. **39**, 1576 (1977).

[51] *B. von Roedern, L. Ley, M. Cardona, F. Smith,* Phil. Mag. **40**, 433 (1979).

[52] *B. von Roedern, G. Moddel,* Solid State Comm. **35**, 467 (1980).

[53] *A. Madan, P.G. LeComber, W. Spear,* J. of Non-Cryst. Solids **20**, 239 (1976).

[54] *W.E. Spear, P.G. LeComber, A.J. Snell,* Phil. Mag. **B 38**, 303 (1978).

[55] *M. Hirose, T. Suzuki, G. Dohler,* Appl. Phys. Lett. **34**, 234 (1979).

[56] *Z. Jan. R. Bube, J. Knights,* J. El. Mat. **8**, 47 (1979).

[57] *M. Shun, W. Czubatyj, A. Madan,* J. of Non-Cryst. Solids **35**, 731 (1980).

[58] *N. Goodmann, H. Fritzsche, H. Ozaki,* J. of Non-Cryst. Solids **35**, 599 (1980).

[59] *N. Goodmann, H. Fritzsche,* Phil. Mag. **B 42**, 149 (1980).

[60] *T. Tiedje, C. Wronski, J. Cebulka,* J. of Non-Cryst. Solids **35**, 743 (1980).

[61] *J.D. Cohen, D.V. Lang, J.P. Harbison,* Phys. Rev. Lett. **45**, 197 (1980).
[62] *P. Persans,* Solid State Comm. **36**, 851 (1980).
[63] *Z. Jan, R. Bube, J. Knights,* J. of Appl. Phys. **51**, 3278 (1980).
[64] *R. Crandall,* J. of El. Mat. **9**, 713 (1980).
[65] *I. Balberg,* J. of El. Mat. **9**, 797 (1980).
[66] *I. Balberg,* Phys. Rev. **B 22**, 3853 (1980).
[67] *B. Goldtein, D.J. Szostak,* Surf. Science **99**, 235 (1980).
[68] *R. Williams, R. Varma, W. Spear, P. LeComber,* J. of Phys. C **12**, L209 (1979).
[69] *M.H. Brodsky, M. Cardona, J.J. Cuomo,* Phys. Rev. **B 16**, 3556 (1977).
[70] *J.C. Knights, G. Lucovsky, R.J. Nemanich,* Phil. Mag. **B 37**, 467 (1978).
[71] *G. Lucovsky,* Solid State Comm. **29**, 571 (1978).
[72] *M. Grimsditch, W. Senn, G. Winterling, M. Brodsky,* Solid State Comm. **26**, 229 (1978).
[73] *M.H. Brodsky, M. Cardona,* J. of Non-Cryst. Solids **31**, 81 (1978).
[74] *E.C. Freeman, W. Paul,* Phys. Rev. **B 20**, 716 (1979).
[75] *C.C. Tsai, H. Fritzsche,* Solar Energy Mat. **1**, 29 (1979).
[76] *J. Knights, G. Lucovsky, R. Nemanich,* J. of Non-Cryst. Solids **32**, 393 (1979).
[77] *G. Lucovsky, R. Nemanich, J. Knights,* Phys. Rev. **B 19**, 2064 (1979).
[78] *C. Fang, L. Ley, H. Shanks, K. Gruntz, M. Cardona,* Phys. Rev. **B 22**, 6140 (1980).
[79] *S. Oguz, D. Anderson, W. Paul, H. Stein,* Phys. Rev. **B 22**, 880 (1980).
[80] *S. Shen, C. Fang, M. Cardona, L. Genzel,* Phys. Rev. **B 22**, 2913 (1980).
[81] *R. Nemanich, J. Knights,* J. of Non-Cryst. Solids **35**, 243 (1980).
[82] *W. Paul,* Solid State Comm. **34**, 283 (1980).
[83] *R.J. Loveland, W. Spear, A. Al-Sharbaty,* J. of Non-Cryst. Sol. **13**, 55 (1974).
[84] *P.J. Zanzucchi, C. Wronski, D. Carlson,* J. of Appl. Phys. **48**, 5227 (1977).
[85] *W.E. Spear,* Fourth Int. Conf. on Phys. of Non-Cryst. Sol. **198** (1977).
[86] *I. Haller, M. Brodsky,* Proc. of 14th Int. Conf. on Phys. of Sem., Edinburgh, 1147 (1978).
[87] *C.C. Tsai,* Phys. Rev. **B 19**, 204 (1970).
[88] *C. Tsang, R. Street,* Phys. Rev. **B 19**, 3027 (1979).
[89] *G. Weiser, D. Ewald, M. Milleville,* J. of Non-Cryst. Sol. **35**, 447 (1980).
[90] *J. Pankove, F. Pollack, C. Schnabolk,* J. of Non-Cryst. Sol. **35**, 459 (1980).
[91] *G. Cody, B. Abeles, C. Wronski, B. Brooks, W. Lanford,* J. Non-Cryst. Sol. **35**, 463 (1980).
[92] *M.H. Brodsky, P. Leary,* J. of Non-Cryst. Sol. **35**, 487 (1980).
[93] *A. Madan, S.R. Ovshinsky, W. Czubatyj, M. Shur,* J. of El. Mat. **9**, 385 (1980).
[94] *D. Engemann, R. Fischer,* Phys. Stat. Sol. **B 79**, 195 (1977).
[95] *T.S. Nashashibi, I.G. Austin, T.M. Searle,* Phil. Mag. **35**, 831 (1977).
[96] *J.I. Pankove, D. Carlson,* Appl. Phys. Lett. **31**, 450 (1977).
[97] *R. Street,* Phys. Rev. **B 17**, 3984 (1978).
[98] *R. Street, J. Knights, D. Biegelsen,* Phys. Rev. **B 18**, 1880 (1978).
[99] *R. Street,* Phil. Mag. **B 37**, 43 (1978).
[100] *C. Tsang, R. Street,* Phil. Mag. **B 37**, 601 (1978).
[101] *C. Tsang, R. Street,* Phys. Rev. **B 19**, 3027 (1979).
[102] *R. Street, N. Johnson, J. Gibbons,* J. of Appl. Phys. **50**, 8201 (1979).

[103] *I. Austin, T. Nashashibi, T. Searle, P. LeComber, W. Spear,* J. of Non-Cryst. Sol. **32**, 373 (1979).

[104] *R. Street,* Phys. Rev. **B 21**, 5775 (1980).

[105] *J. Noolandi, K. Hong, R. Street,* Solid State Comm. **34**, 45 (1980).

[106] *R. Street,* Solid State Comm. **34**, 157 (1980).

[107] *J. Mort, I. Chen, A. Troup, M. Morgan, J. Knights, R. Lujan,* Phys. Rev. Lett. **45**, 1348 (1980).

[108] *R. W. Collins, M. Paesler, G. Moddel, W. Paul,* J. of Non-Cryst. Sol. **35**, 681 (1980).

[109] *B. Weinstein,* Phys. Rev. **B 23**, (1981).

[110] *D. Biegelsen, J. Knights, R. Street, C. Tsang, R. White,* Phil. Mag. **B 37**, 477 (1978).

[111] *R. Street, D. Biegelsen,* Solid State Comm. **33**, 1159 (1980).

[112] *U. Voget-Grote, W. Kummerle, R. Fischer, J. Stuke,* Phil. Mag. **B 41**, 127 (1980).

[113] *S. Hudgens,* Phys. Rev. **B 14**, 1547 (1976).

[114] *J. Stuke,* Proc. 7th Int. Conf. on Amorph. and Liq. Sem., 406 (1977).

[115] *J. Knights, D. Beigelsen, I. Solomon,* Solid State Comm. **22**, 133 (1977).

[116] *H. Fritzsche, C.C. Tsai, P. Persans,* Solid State Tech. **21**, 55 (1978).

[117] *P. Thomas, M. Brodsky, D. Kaplan, D. Lepine,* Phys. Rev. **B 18**, 3059 (1978).

[118] *D. Kaplan, N. Sol, G. Velasco, P. Thomas,* Appl. Phys. Lett. **35**, 440 (1978).

[119] *D. Biegelsen, R. Street, C. Tsai, J. Knights,* Phys. Rev. **B 20**, 4839 (1979).

[120] *R. Street, D. Biegelson, J. Stuke,* Phil. Mag. **B 40**, 451 (1979).

[121] *W. Carlos, P. Taylor,* Phys. Rev. Lett. **45**, 358 (1980).

[122] *P. LeComber, A. Mandan, W. Spear,* J. of Non-Cryst. Sol. **11**, 219 (1972).

[123] *P. LeComber, D. Jones, W. Spear,* Phil. Mag. **B 35**, 1173 (1977).

[124] *D. Jones, P. LeComber, W. Spear,* Phil. Mag. **B 36**, 541 (1977).

[125] *D. Anderson, W. Spear,* Phil. Mag. **B 36**, 695 (1977).

[126] *W. Beyer, H. Mell, H. Overhof,* Proc. 7th Int. Conf. on Amorph. and Liq. Semi., Edinburgh, 328 (1977).

[127] *A.R. Moore,* Appl. Phys. Lett. **31**, 762 (1977).

[128] *D. Allan,* Phil. Mag. **B 38**, 381 (1978).

[129] *D. Ast, M. Brodsky,* Inst. Phys. Conf. Ser. **43**, 1159 (1979).

[130] *W. Beyer, H. Overhof,* Solid State Comm. **31**, 1 (1979).

[131] *P. LeComber, W. Spear, D. Allan,* J. of Non-Cryst. Sol. **32**, 1 (1979).

[132] *D. Jones, W. Spear, P. LeComber, S. Li, R. Martins,* Phil. Mag. **B 39**, 147 (1979).

[133] *D. Ast, M. Brodsky,* Phil. Mag. **B 41**, 273 (1980).

[134] *W. Spear, D. Allan, P. LeComber, A. Ghaith,* Phil. Mag. **B 41**, 419 (1980).

[135] *M. Silver, A. Madan, D. Adler, W. Czubatyj,* 14th IEEE Photo. Sp. Conf. 1062 (1980).

[136] *G. Moddel, D. Anderson, W. Paul,* Phys. Rev. **B 22**, 1918 (1980).

[137] *R.S. Crandall,* Phys. Rev. Lett. **44**, 749 (1980).

[138] *Z. Vardeny, P. O'Connor, S. Ray,* J. Tauc, Phys. Rev. Lett. **44**, 1267 (1980).

[139] *D. Staebler, C. Wronski,* J. of Appl. Phys. **51**, 3262 (1980).

[140] *T. Tiedje, B. Abeles, D. Morel, T. Moustakas, C. Wronski,* Appl. Phys. Lett. **36**, 695 (1980).

[141] *P. LeComber, W. Spear, G. Muller, S. Kalbitzer,* J. of Non-Cryst. Sol. **35**, 327 (1980).

[142] *J. Hvam, M. Brodsky*, Phys. Rev. Lett. **46**, 371 (1981).
[143] *C. Wronski, R. Daniel*, Phys. Rev. **B 23**, (1981).
[144] *H. Fritzsche*, Solar Energy Mat. **3**, 447 (1980).
[145] *S.R. Ovshinsky, D. Adler*, Contemp. Phys. **19**, 109 (1978).
[146] *R. Fisch, D. Licciardello*, Phys. Rev. Lett. **41**, 889 (1978).
[147] *D. Adler*, Phys. Rev. Lett. **41**, 1755 (1978).
[148] *S.R. Elliot*, Phil. Mag. **38**, 325 (1978); Phil. Mag. **39**, 349 (1979).
[149] *J.D. Joannopoulos*, J. Non-Cryst. Sol. **32**, 241 (1979).
[150] *J.C. Phillips*, Phys. Rev. Lett. **42**, 1151 (1979).
[151] *D. Weaire, F. Wooten*, Phys. Stat. Sol. **54**, K41 (1979).
[152] *G. Döhler*, Phys. Rev. **B 19**, 2083 (1979).
[153] *W.Y. Ching, D. Lam, C.C. Lin*, Phys. Rev. Lett. **42**, 805 (1979).
[154] *J. Robertson*, Phil. Mag. **B 40**, 31 (1979).
[155] *J. Marshall, D. Allan*, Phil. Mag. **B 40**, 21 (1979).
[156] *D.C. Allan, J.D. Joannopoulos*, Phys. Rev. Lett. **44**, 43 (1980).
[157] *J.D. Joannopoulos*, J. of Non-Cryst. Sol. **35**, 781 (1980).
[158] *J.C. Phillips*, J. of Non-Cryst. Sol. **35**, 1157 (1980); J. of Non-Cryst. Sol. (1981) in press.
[159] *M.H. Cohen, J. Singh, F. Yonezawa*, J. of Non-Cryst. Sol. **35**, 55 (1980).
[160] *W.Y. Ching, D. Lam, C.C. Lin*, Phys. Rev. **B 21**, 2378 (1980).
[161] *K.H. Johnson, H.J. Kolari, J.P. Deneuville, D. Morel*, -Phys. Rev. **B 21**, 643 (1980).
[162] *W.Y. Ching*, Phys. Rev. **B 22**, 2816 (1980).
[163] *W.Y. Ching*, J. of Non-Cryst. Sol. **35**, 61 (1980).
[164] *M.H. Brodsky*, Solid State Comm. **36**, 55 (1980).
[165] *N.F. Mott*, J. of Phys. **C 13**, 5433 (1980).
[166] *E.N. Economou, D.A. Popaconstantopoulos*, Phys. Rev. (1981) in press.
[167] *D. DiVincenzo, J. Bernholc, M. Brodsky, N. Lipari, S. Pantelides*, Proc. Int. Conf. on Tet. Bonded Amorph. Semi., Carefree, Az (1981).
[168] *L. Guttmann, C. Fong*, Proc. Int. Conf. on Tet. Bonded Amorph. Semi., Carefree, AZ (1981).
[169] *W.Y. Ching*, Proc. Int. Conf. on Tet. Bonded Amorph. Semi., Carefree, AZ (1981).
[170] *D.C. Allan, J.D. Joannopoulos*, Proc. Int. Conf. on Tet. Bonded Amorph. Semi., Carefree, AZ (1981).
[171] *R. Street, D. Biegelsen, J. Knights*, preprint (1981).
[172] *C. Domb*, Adv. Phys. **9**, 145 (1960).
[173] *J.D. Joannopoulos*, Phys. Rev. **B 16**, 2764 (1977).
[174] *F. Yndurain, J. Joannopoulos*, Phys. Rev. **B 11**, 2957 (1975).
[175] *J. Joannopoulos, F. Yndurain*, Phys. Rev. **B 10**, 5164 (1974).
[176] *K.C. Pandey*, Phys. Rev. **B 14**, 1557 (1976).
[177] *K.C. Pandey, T. Sakurai, H. Hagstrum*, Phys. Rev. Lett. **35**, 1728 (1975).
[178] *D. Vanderbilt, J. Joannopoulos*, Phys. Rev. **B 22**, 2927 (1980).
[179] *W.A. Goddard III, R.C. Ladner*, J. Am. Chem. Soc. **93**, 6750 (1972).
[180] *L.B. Harding, W.A. Goddard III*, J. Chem. Phys. **67**, 1777 (1977).

[181] *A. Redondo, W.A. Goddard III, T.C. McGill, G.T. Surratt,* Solid State Comm. **20**, 733 (1976).

[182] *F. Yndurain, E. Louis,* Solid State Comm. **25**, 439 (1978).

[183] *J. Joannopoulos, M.L. Cohen,* Solid State Physics **31**, 71 (1976) and references therein.

[184] *K. Tanaka, R. Tsu,* preprint (1981).

[185] *R. Laughlin, J. Joannopoulos, D. Chadi,* Phys. Rev. **B 21**, 5733 (1980).

[186] *J.D. Joannopoulos,* Phys. Rev. **B 16**, 2764 (1977).

[187] *K. Zellama, P. Germain, S. Squelard, J. Monge, E. Ligeon, J. Non-Cryst.* Solids **35**, 225 (1980).

[188] *M. Schlüter, J. Chelikowsky, S. Louie, M.L. Cohen,* Phys. Rev. **B 12**, 4200 (1975).

[189] *D. Chadi,* Phys. Rev. Lett. **43**, 43 (1979).

[190] *J. Ihm, M.L. Cohen, D. Chadi,* Phys. Rev. **B 21**, 4592 (1980).

Chemical Trends of Deep Impurity Levels in Covalent Semiconductors

Peter Vogl
Institut für Theoretische Physik, Universität Graz, Graz, Austria

Summary: Recent developments in the theory of deep substitutional impurities in semiconductors which focus on chemical trends in the trap level energies are reviewed. These simple but general models provide a conceptual framework for understanding the major chemical trends in deep trap energies, including their dependence on the host energy bands and the impurities' atomic structure. Approximate rules for predicting the ordering of impurity levels are given, which should facilitate the identification and cataloguing of substitutional deep traps in tetrahedral semiconductors.

1 Introduction

Impurities are known to affect and to control the material properties of semiconductors which are relevant to device operation. Shallow impurities such as As or B in Si introduce bound states in the gap with activation energies of a few kT at room temperature and supply carriers for conduction. Other atoms, such as O in GaP, are known as deep impurities and produce states in the midgap region. They act as trapping or recombination centers ("lifetime killers") but are also a main factor limiting device performance.

Experimental techniques for assessing impurity levels in a systematic way have been elaborated only during the last decade. Especially for deep impurities, the capacitance transient techniques [1] and electron spin resonance [2] have provided and still are providing a wealth of new experimental data [3–9]. The need for interpretation of these data has simultaneously led to an unprecedented theoretical activity, by now being documented in many recent excellent books [4, 10, 11] and review articles [12–19]. They cover virtually all theoretical techniques which have been developed during the last 25 years for the calculation of impurity levels. One review article in particular [15] appeared in this series quite recently. It focused on recently developed self-consistent pseudopotential calculations [20–24] which, for the first time, offer a truly satisfactory theoretical assessment of the electronic structure of localized defects in semiconductors. While stille being too laborious to be applied to a great variety of defects, these calculations provide very detailed quantitative information about the electronic and vibronic properties of specific defects.

In this articlę, we will review a recently developed complementary viewpoint. We shall concentrate on very recently developed theories for substitutional deep trapping centers which focus on chemical trends in the impurity level energies

[25–31]. These models suggest an imprecise but global view of deep trap levels, meant to guide future experiments rather than to give quantitative agreement with existing data. Typical questions we are going to deal with in the course of this review are: Which elements in the Periodic Table will most likely produce deep trap states in Si? Why does O produce a deep level in GaP although Te is shallow? Since Te is a shallow donor in GaAs, is it likely to produce a shallow donor level in Ge as well? A major advantage of concentrating on trends rather than on truly quantitative results is that it may permit one to grasp the essential qualitative physics of impurities and to deduce approximate rules [25–31]. Clearly, both – quantitative and chemical-trend-type – approaches should be useful for interpreting data, and none is likely to succeed without the other.

In their recent article, Baraff and Schlüter [15] have given a detailed account of the early theories of deep impurities. We shall therefore only very briefly review some highlights in the development of understanding impurities, meant as a vehicle for presenting those aspects of impurities we want to concentrate on, rather than as a history.

The most successful theory of impurities has been the effective mass theory of shallow impurities [32]. It led to the well-known hydrogenic theory providing both conceptual guidance and – in many cases – quantitative predictions of the shallow binding energies (for a review, including recent extensions of the theory, see Refs. [11, 13, 18, 33]).

Deep traps, on the other hand, are so different from shallow impurities, that extensions of effective mass theory have not proved successful. The crucial factor is that a deep trap can be distinguished from a shallow trap by the characteristic range of its potential. Shallow impurities are controlled by the long-range Coulomb potential while the everpresent short-range potential produces only a slight chemical shift. For deep impurities, the relative importance of the short-range and the long-range components is exactly reversed. Recognizing the importance of the central-cell potential associated with a deep trap, earlier workers simulated it and the surrounding medium with clusters of atoms [34, 35]. An alternative method which exploits both the short-range character of the deep impurity potential as well as the translational symmetry of the undisturbed host crystal has been developed by Koster and Slater [36] and by Callaway and coworkers [37, 38]. All present state-of-the-art calculations of deep impurities use the scattering theoretic Green-function approach basically developed in these early papers. Jaros and coworkers carried out a number of pseudopotential calculations of deep impurities in III-V's [17, 39–41], while the selfconsistent generalizations of this method, developed independently in Refs. [20] and [22], have been applied to the Silicon vacancy [14, 16, 21, 23, 24] and to native defects in GaP [73] so far.

The central idea underlying the simple theories we are going to review is that it may be very hard to predict the energy level of any deep impurity with accuracy, but that it is possible either to predict how a given impurity level will change as the host changes, or to order the energy levels of several impurities in a single host. Most of these recently developed models [25–29] neglect the charge state splittings of these levels and the effects of lattice relaxation – not because they are unimportant [42], but because they represent corrections to the trends (which can be incorporated in principle [26]) and do not lend themselves to the sort of global analysis that the trends do. Furthermore, these effects, in many cases are likely to be monotonic functions of the trap depth, and hence do not disturb the relative ordering of deep trap energies.

Thus we shall concentrate on the understanding and prediction of the major chemical trends in impurity level data [25–27, 53, 57, 58]. The idea of seeking out chemical trends in the electronic structure properties of solids goes back to the pioneering work of Phillips [43] and has recently been substantiated by Harrison and coworkers [44, 45]. A prerequisite for seeking out chemical trends, however, is the availability of *many* experimental data. For both shallow and deep impurities, a greater number of unambigously identified data on defects is becoming accessible only now. It is therefore not too surprising that theoretical models, which can focus on trends, have evolved only recently.

In this article we shall review in greater detail mainly a series of recent papers on the theory of substitutional point defects in covalent semiconductors [25–26, 28] (Sec. 3). Sec. 4 is devoted to an extension of this theory to defect pairs [27, 29]. In Sec. 5, we discuss transition metal impurities [30, 31] which have not been treated in the former papers. Common to all these models is that they do not aim at the quantitative assessment of specific defect levels but to provide mainly a conceptual framework for understanding and interpreting experimental data.

2 Chemical Trends in Impurity Level Data

At the outset, before discussing any calculations, we are going to examine the experimental data of sp^3-bonded (i.e. non-transition-metal) substitutional donor and acceptor binding energies in covalent semiconductors. Plotting the experimental binding energies – irrespective of whether they are shallow or deep – versus the atomic potential strength of the impurity atoms provides convincing evidence for a relationship between trap energies and the impurities' atomic structure, i.e. for chemical trends [25–26].

The impurity potential of a substitutional donor or acceptor consists of a long-range, screened Coulomb tail and a short-range potential which arises from the difference in the atomic potentials of an impurity atom and the replaced host atom. In the shallow impurity levels, this localized potential shows up – strongly attenuated – as the chemical shift or the "central-cell correction"; for deep traps, it is dominating.

To obtain a measure for the short-range part of the impurity potential, we may be guided by plausibility arguments which will be substantiated later (Sec. 3.2). The short-range potential becomes essentially a difference of the atomic orbital energies when one writes the matrix elements of the defect-Hamiltonian in a localized, atomic basis [25–26]. In Table 1 we therefore list the atomic orbital energies of the outermost s- and p-electrons (e.g. 3 s and 3 p for Si), as obtained from Hartree-Fock calculations of free atoms [46]. Actually, the values in Table 1 are scaled by a constant factor, but this is immaterial now. Donor levels are associated with the conduction band, the lower part of which is made up mostly of atomic s-states. Acceptors, on the other hand, are derived from the upper valence bands which are formed by the atomic p-states.

This suggests a two-parameter "electronegativity" scale for impurities: For donors and acceptors, we consider the difference of the impurity and host s-energies ($\epsilon_s^I - \epsilon_s^H$) and the p-energies ($\epsilon_p^I - \epsilon_p^H$) to reflect the central-cell portion of the impurity potential, respectively. A negative energy difference reflects an attractive short-range potential, a positive difference a repulsive short-range potential.

In Figures 1 and 2 we plot the experimental ionization energies of donors and acceptors in Si, Ge, GaP, GaAs, GaSb, and AlSb versus the short-range impurity potential as measured by the atomic orbital energies from Table 1 [25a]. We have been careful to include only well-identified data which have been obtained or

Table 1 Orbital energy parameters. ϵ_s and ϵ_p are proportional to the atomic term values [46, 89], the proportionality factor being 0.8 for the s-energies and 0.6 for the p-energies (see Sec. 3.2.1).

Element	$-\epsilon_s$ (eV)	$-\epsilon_p$ (eV)	Element	$-\epsilon_s$ (eV)	$-\epsilon_p$ (eV)
Be	6.74	3.04	N	20.57	9.26
Mg	5.50	2.26	P	15.15	6.39
Zn	6.37	2.09	As	14.93	6.03
Cd	5.76	1.79	Sb	12.66	5.46
Hg	5.68	0.90	Bi	12.01	5.22
B	10.77	5.06	O	27.08	10.31
Al	8.56	3.43	S	19.14	7.41
Ga	9.24	3.40	Se	18.22	6.58
In	8.11	3.22	Te	15.25	5.87
Tl	7.86	3.14	Po	14.32	5.57
C	15.35	7.07	F	34.22	11.92
Si	11.74	4.85	Cl	23.34	8.26
Ge	12.04	4.69	Br	21.60	7.46
Sn	10.37	4.33	I	17.86	6.581
Pb	9.93	4.17	At	16.66	6.20

confirmed by modern experimental techniques and critically reviewed recently [4–7, 47–50]. As usual, the binding energies are measured from the conduction and valence band edge for donors and acceptors, respectively.

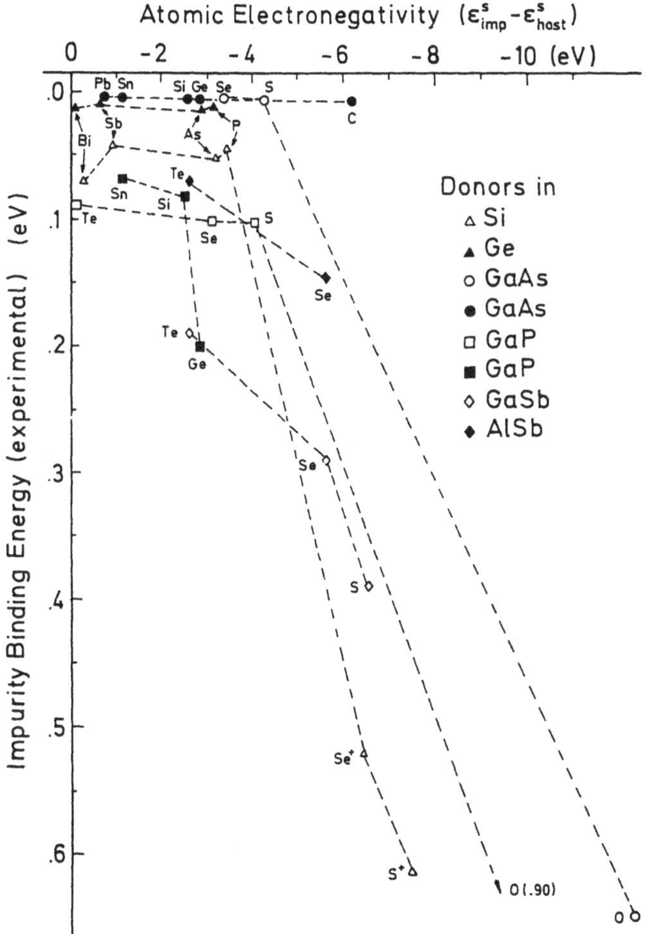

Fig. 1 Experimental ionization energies of donors [4–7, 47–50] in several semiconductors versus the central-cell impurity potential strength. The latter is being measured by the difference between the s-orbital energy of the impurity atom (ϵ^s_{imp}) and the replaced host atom (ϵ^s_{host}). The orbital energies are given in Table 1. The levels which substitute at a particular site in a given host have been interconnected in order to exhibit their chemical trend: The donor binding energies increase with increasing s-electronegativity. Bi and As in Si and Ge do not follow this trend [25a].

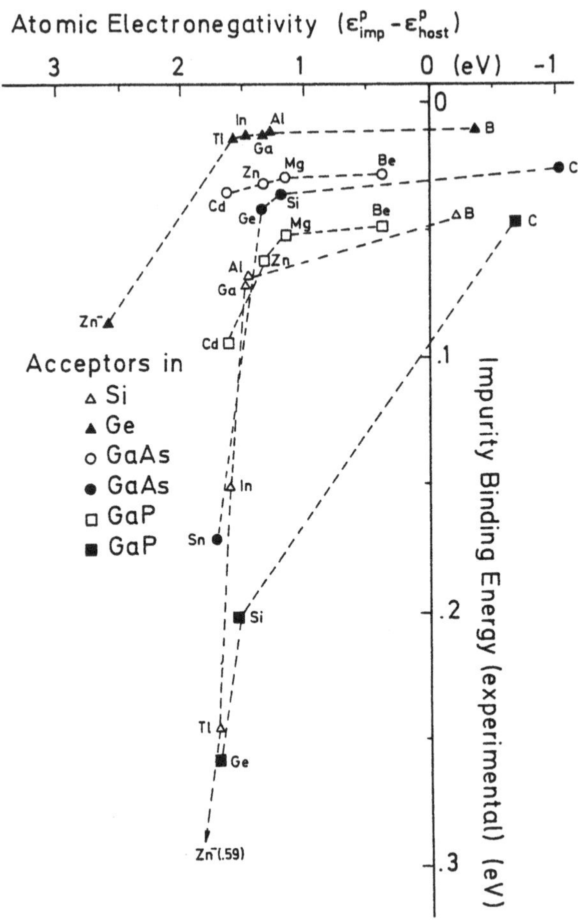

Fig. 2 Experimental ionization energies of acceptors [4–7, 47–50] versus the central-cell impurity potential strength which is measured by the difference between the p-orbital energy of the impurity atom (ϵ^p_{imp}) and the replaced host atom (ϵ^p_{host}). The meaning of the dashed lines is as in Fig. 1 [25a].

In order to make the chemical trends in Figures 1 and 2 apparent, the levels of the impurities which substitute at a particular site in a particular host have been interconnected. There are two striking features in Figures 1 and 2: (i) Upon increasing $(\epsilon^I_s - \epsilon^H_s)$ for donors and $(\epsilon^I_p - \epsilon^H_p)$ for acceptors, the binding energies increase monotonically along the lines interconnecting the impurities of each host and site, with minor exceptions for some Si and Ge donors. (ii) The trend in the interconnecting lines is markedly non-linear.

Qualitatively, these features may be understood in terms of a simple yet illustrative model: Consider a donor electron which is subject to a screened Coulomb potential, $e^2/(\epsilon r)$, and some model-short-range potential. Its precise form is irrelevant for the argument, and we may simply choose a square well. In Fig. 3 we then plot the ground state energy of the electron versus the well depth. This qualitatively illustrates the physical origin of the trends in Figures 1,2: For a small central-cell potential the long-range Coulomb potential controls the bound state. Its energy is determined by the effective Rydberg $Ry^* = (m^*/m\epsilon)$ Ry, which is of the order of several meV as predicted by the effective mass theory. The short-range portion of the potential is then a small perturbation which, however, still determines the *relative* positions of the ground state energies. Upon increasing the atomic potential difference, one reaches a point where the short-range potential alone is able to bind the electron, while the influence of the Coulomb potential decreases. The ionization energy then rapidly increases, following more or less linearly the localized potential.

Thus the tabulated orbital energies of free atoms successfully correlate the relative positions of the substitutional impurity levels associated with a particular host and site. However, no trends in either the host- or the site-dependence of the levels associated with a given impurity are apparent from Figures 1,2. In the subsequent Sections, deep trap data in alloys (Fig. 14) – together with the physical insight provided by the newly developed [25–27] theories – will be seen to uncover further chemical trends.

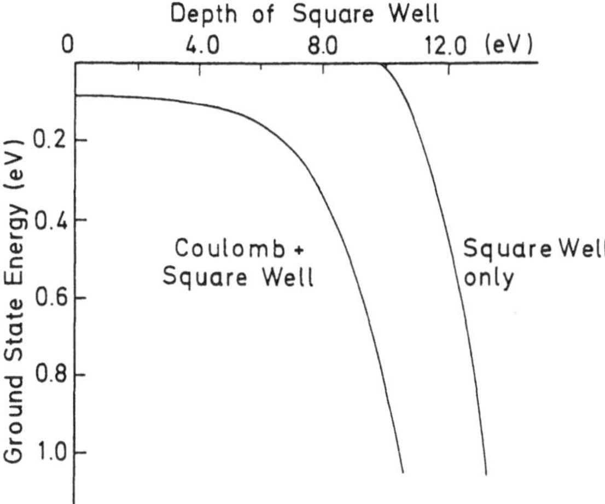

Fig. 3 Schematic illustration of the effect of an increasingly attractive short-range part of the impurity potential on the ionization energy of a donor electron. The potential has been taken to be the sum of a screened Coulomb potential, $e^2/(\epsilon r)$, and an attractive square-well, with $\epsilon = 10$, well-width equal to 5 Å, and effective mass $m^* = 0.1$ m.

In the absence of more detailed calculations, the Figures 1,2 together with Table 1 can provide a crude rule of thumb for predicting the ordering of impurity levels. For example, the donors Te, Se, S, O are likely to have increasingly larger ionization energies in all semiconductors. Among the P-site group IV-acceptors in GaP, the ionization energy of Sn is yet unknown. From Table 1 one infers that it should be the deepest acceptor in the series C, Si, Ge, and Sn. Generally, Fig. 1 shows that for most acceptors one has $(\epsilon_p^I - \epsilon_p^H) > 0$, which reflects a repulsive (electropositive) potential pushing the level deeper into the gap than effective mass theory predicts. Exceptions are C and B and – not shown in Fig. 2 – P in ZnTe, which have weakening central cell potentials. The puzzling fact that P is deep in ZnSe but shallow in ZnTe [4], seems natural in the light of these findings: P is electropositive with respect to Se and consequently deep in ZnSe, but electronegative with respect to Te and therefore shallow in ZnTe.

We finally note that the present atomic energy scale is similar to Phillip's electronegativity [43]. Generally, the latter would work as well, although Phillips concluded in 1970 that no correlations with the data are apparent [51, see also 52]. Yet at this time only the shallow donor levels in Si and Ge were well established, which behave exceptionally also in the present scale. The reason we prefer the atomic two-parameter scale is that we will be able to justify it from a Koster-Slater Green function calculation in Sec. 3.2.

3 Theory of Chemical Trends in Deep Substitutional Trap Levels

We shall now focus on the theoretical understanding of deep impurities, and outline the approach of Ref. [25–27] which exploits the chemical trends found in the data.

Traditonally, deep levels have been defined as those states lying more than 0.1 eV from the nearest band edge. However, we *define* an impurity to be a deep trap theoretically, if its central-cell potential alone, without any long-range Coulombic or elastic potential, is sufficiently strong to bind a state within the band gap of the host. In contrast to a shallow trap, such a deep trap state is not exclusively derived from the nearest bandedge. Thus we experimentally *define* a trap to be deep if it does not follow a nearby band edge when that edge is perturbed by alloying or pressure [53].

With the focus on the localized central-cell portion of the potential of a substitutional defect, the deep trap problem becomes a question of localization: Under what conditions does the central-cell potential bind a state? If it does, the properties of the localized state will primarily depend on the atomic nature of the central-cell potential while the remaining long-range contributions to the potential will have a secondary effect. This viewpoint is quantitatively supported by recent pseudopotential calculations [54] which show that the long-range Coulomb tail affects deep trap energies by typically less than 0.1 eV and may be included *a posteriori* in perturbation theory. In view of these arguments, a multi-band central-site Koster-Slater theory, which omits any long-range impurity potential tail, has been used in Refs. [25–26] to compute the energies of the (non-transition metal) substitutional deep traps; this will be discussed in Sec. 3.2. A crucial finding of these calculations is that the essential *qualitative* physics of deep levels is contained in a simple molecular picture, which captures the essential elements of the sp^3 chemical bonding of the impurity with the host. We present this first.

3.1 The Defect Molecule Picture

As an example, depicted in Fig. 4, we consider N in GaP [25, 26, 28]. N has the same number of valence electrons as P – it is "isoelectronic" to P – and binds an electron through its short-range potential; there is no Coulomb tail [53, 55–59].

We examine N in replacing P and limit our attention to the s-like atomic orbitals on the central atoms, first P and then N, interacting with the s-like atomic orbital residing on the four nearest Ga neighbours.

The Ga atomic level lies above the P s-orbital energy (see Table 1) and is the progenitor of the conduction band in the single s-state model. Bringing Ga and P together

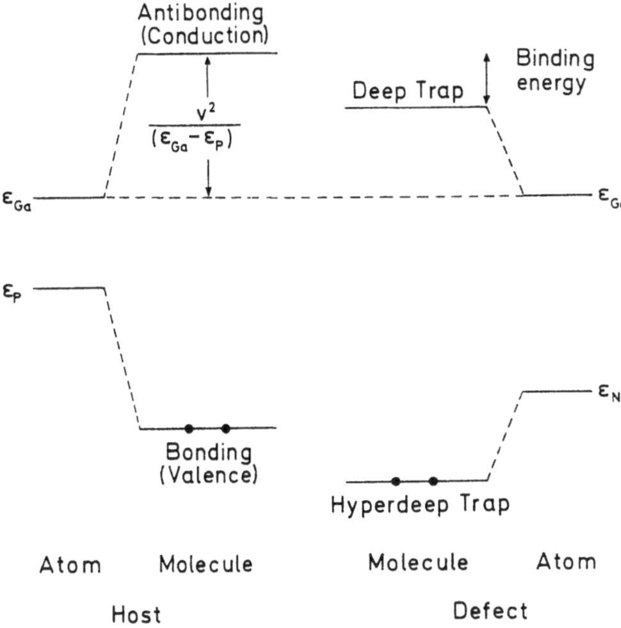

Fig. 4 The two-state model schematically shows the host GaP antibonding and bonding states, derived from atomic energies, compared with the same model with N replacing P. The deep trap level is the GaN antibonding level. It is Ga-like and appears bound relative to the GaP antibonding level. From Ref. [25].

into a molecule leads to a bonding-antibonding splitting, with the antibonding level the parent of the (lower part of the) conduction band in the solid. If one considers a molecule consisting of the defect N and its Ga neighbours, the N s-orbital energy lies ~ 7 eV lower than the P level, causing the Ga-N antibonding state to lie below the Ga-P state and to appear bound relative to it. The antibonding Ga-N state lies lower because the bonding-antibonding splitting is inversely proportional to the energy denominator ($\epsilon_{Ga} - \epsilon_N$), which is ~ 7 eV larger for the defect than for the host. (The splitting is proportional to $v^2/(\epsilon_{Ga} - \epsilon_N)$ where v is the anion-cation transfer matrix element and is known (Sec. 3.2 and Ref. [45]) to depend only on the host bond length. The bond length is the same for the host and the defect, if we neglect the lattice relaxation around the impurity.) The net result is the creation of two impurity states – one is an occupied bonding state within or below the valence band; the other is an unoccupied Ga-like antibonding state slightly below the Ga-P conduction state.

Implicit in this simple picture are many of the features which will be found within the more sophisticated Koster-Slater calculation of Sec. 3.2: (i) Associated with every deep electron trap within the band gap there is a filled, electrically inactive "hyperdeep" level normally below or within the valence band. (ii) While the hyperdeep level is an impurity-like bonding state, the *deep level is a host-like antibonding state.* (iii) When we decrease the impurity's atomic orbital energy to simulate an increasingly electron-attractive substitutional impurity, the Ga-like antibonding trap state becomes asymptotically "pinned" to the Ga-dangling bond energy (let $\epsilon_N \to -\infty$ and see that the antibonding Ga-N level never lies below ϵ_{Ga}). This pinning of the deep level to the host Ga-dangling bond (also representing the unrelaxed, neutral vacancy [34, 60]) is a consequence of orthogonality: The electron-trapping antibonding state is orthogonal to the bonding state which is localized at the impurity site. Hence the trapped electron is to be localized away from this site. An increase of the impurity potential strength only attracts more *bonding* charge on the impurity site without lowering the energy of the trapped electron which resides on the neighbour states.

In Fig. 5 we depict the energies obtained within the two-state defect molecule model for an attractive and repulsive potential. One sees that an attractive potential pulls a state from the conduction "band", and a repulsive potential one from the valence "band". An essential feature of deep traps which is absent in the simple molecular picture is the threshold for binding. If we imagine to turn on the interaction of the defect molecule with the rest of the host crystal, the Ga-P prototype conduction and valence "band" broadens. Then the defect potential ($\epsilon_s^N - \epsilon_s^{Ga}$) must exceed a threshold before a state can be pulled from the conduction band.

3.2 Koster-Slater-LCAO-Scattering Theory

The full exploitation of both the short-range nature of the potential induced by deep impurities and the translational symmetry of the host crystal can optimally be

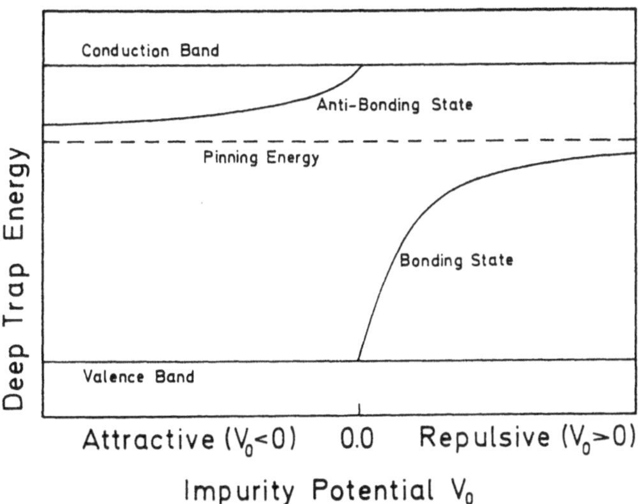

Fig. 5 Schematic plot of the energy levels associated with the two-state model depicted in Fig. 4, versus the impurity potential V_0. The latter is given by the difference between the impurity's and the replaced host-atom's orbital energy. The GaP antibonding and bonding state corresponds to $V_0 = 0$ and represents the conduction and valence "band", respectively. After Ref. [26].

achieved using the scattering theoretic Green function technique. Recently many excellent papers and review appeared where this method and its application to localized perturbations is extensively discussed (unreconstructed vacancies [60], surfaces and interfaces [61], deep impurities [13, 15]). In this approach, originally developed by Koster and Slater [36] and by Callaway [37], one starts out from the perfect crystal described by the Hamiltonian H_0 whose electronic properties can be calculated by using any of the standard techniques. In the second step the defect is introduced, characterized by the defect matrix V. Finally the changes introduced by V in the spectrum of H_0 are calculated by solving exactly the eigenvalue problem of the full Hamiltonian $H = H_0 + V$ in the form

$$|\psi> = (E - H_0)^{-1} V|\psi> = G_0(E) V|\psi>, \qquad (3.1)$$

where $G_0(E)$ is the perfect crystal's Green function. The bound states in the gap of the host crystal (characterized by H_0) are then obtained from the zeros of the determinantal equation

$$\det(1 - G_0(E) V) = 0. \qquad (3.2)$$

The size of this matrix equation is determined by the range of the potential V. Hence it can be solved efficiently in a basis of localized functions to minimize the number of matrix elements needed to define V.

3.2.1 The Perfect Crystal Hamiltonian

The one-electron spectrum of the fully periodic bulk crystal is determined by the Schrödinger equation

$$H_0 |n\vec{k}> = E_{n\vec{k}} |n\vec{k}>, \tag{3.3}$$

n being the band index and $\vec{k}$ the Bloch vector. In terms of these eigenstates, the spectral representation of the Green function $G_0(E)$ is

$$G_0(E) = \sum_{n\vec{k}} \frac{|n\vec{k}><n\vec{k}|}{E - E_{n\vec{k}}}. \tag{3.4}$$

For H_0, the framework of the empirical LCAO method has been used in Ref. [25–26]. In the context of modern solid state theory, this method has been pioneered by Harrison [44]. Since then, this model has been applied to an enormously broad spectrum of solid state properties [45, 61, 62], including vacancies [60]. Motivated by the work of Mele and Joannopoulos [63] and Chadi [64, 65], we have generalized this tight binding model, maintaining its essential features of being simple and preserving and displaying chemical trends, yet making it sophisticated enough so that it reproduces the major features of both the valence and conduction band energies, even in indirect gap semiconductors [66]. In this model, H_0 is represented in an orthonomal set of localized basis functions. We have used five basis states per atom, two s and three p functions; only nearest neighbour interactions have been included. The diagonal or onsite matrix elements of H_0 are ϵ_s^{anion}, ϵ_{s*}^{anion}, ϵ_p^{anion}, ϵ_s^{cation}, ϵ_{s*}^{cation}, ϵ_p^{cation}. In addition, seven nearest-neighbour transfer matrix elements are kept which couple the s, the excited s*, and the p-states on adjacent atoms. Rather than constructing the basis functions explicitly these matrix elements are considered as parameters and are fitted to the known optical gaps of the semiconductors. For the case of Si, the resulting band structure is shown in Fig. 6.

The keypoint of the empirical LCAO method lies in the chemical trends of the Hamiltonian matrix elements [45]: (i) The diagonal matrix elements of H_0 are proportional to the s- and p-atomic energies of the chemical constituents. By fitting band structures, the proportionality constant has been found to be 0.8 for the s-states and 0.6 for the p-states [26, 66]. The atomic orbital energies, scaled by these factors, are given in Table 1. (ii) The nearest-neighbour coupling (off-diagonal) matrix elements are – in a good approximation – universal functions of the nearest-neighbour distance, i.e. they are independent of the chemical nature of the partners, and they scale as the inverse square of the bond length. The physical origin of these trends is discussed in detail in Ref. [45]. Recently, they could also be justified within new electronic structure theories [67–70]. These trends optimally permit the exploration of chemical trends in both perfect and imperfect crystal properties and were recently successfully applied to dielectric and elastic

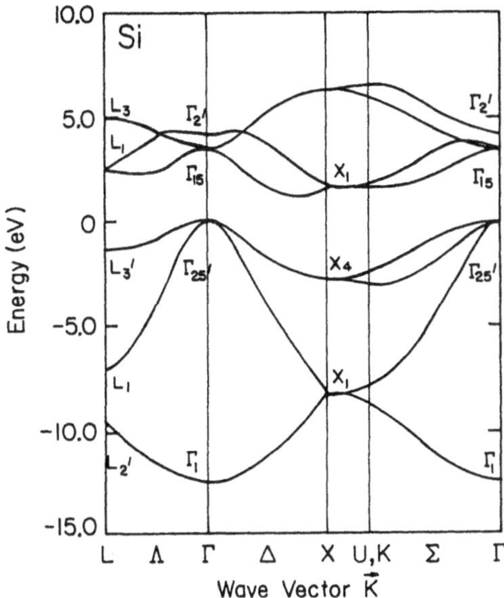

Fig. 6
Energy band structure of Si obtained with the 10-band empirical LCAO-method [66].

bulk crystal properties [45, 65], surfaces and interfaces [61], core excitons [28, 71, 72], as well as to the deep impurity problem presented here [25–29]. They allow one to greatly simplify the defect matrix and to construct it without any adjustable parameters from the tabulated atomic energies only.

3.2.2 The Defect Matrix

The difficult part in any impurity calculations is the defect potential V in Eq. (3.1). Again following Ref. [25–26], we shall sidestep this difficulty by making use of the chemical trends which are found for the perfect crystal Hamiltonian matrix elements and transferring them to the imperfect crystal.

Focussing on the short-range impurity potential, we restrict the impurity potential to just the unit cell containing the defect. Thus we consider V to be non-zero only up to the nearest neighbour sites surrounding the impurity. As discussed above, this has been shown to alter the binding energies only by a small amount on the scale of energies reliably predicted in the deep trap theories [54]. Hence the defect alters at most the diagonal matrix element of the total crystal Hamiltonian H at the impurity site and its off-diagonal elements associated with the impurity-nearest-neighbour coupling. If we limit our attention to impurities in the unrelaxed host, it then follows that the change in the nearest-neighbour coupling matrix elements is zero – by virtue of their dependence solely on the (unaltered) bond length, as discussed in the previous Section. Thus the perturbation due to the impurity, in the empirical tight-binding basis, is diagonal. The diagonality of V then reduces the determinantal equation Eq. (3.2) to a product of scalar equations, one for each basis state associated with the impurity site. The eigenvalue equation Eq. (3.2) for the trap energy E of the trap at site κ then becomes

$$\frac{1}{V_{\ell\kappa}} = \sum_{n\vec{k}} \frac{<\ell\kappa|n\vec{k}><n\vec{k}|\ell\kappa>}{E-E_{n\vec{k}}} = \int_{-\infty}^{\infty} dE' \frac{D_{\ell\kappa}(E')}{E-E'} . \tag{3.5}$$

In this equation, $D_{\ell\kappa}(E)$ is the local spectral density of states of the host, of symmetry ℓ and taken at the site κ which can be either cation or anion. When ℓ is the s-orbital state, Eq. (3.5) corresponds to the bound states of A_1-symmetry, whereas when ℓ is one of the three p-orbitals associated with site κ, Eq. (3.5) corresponds to the bound states of T_2-symmetry. Furthermore, no bound states are found associated with the excited s-state s* which lies high up in the conduction band, and no bound states of symmetry other than A_1 and T_2 are possible in the present model. The A_1-part of the defect matrix is given by the diagonal element

$$V_{A_1} = \epsilon_s^I - \epsilon_s^H , \tag{3.6}$$

in terms of the s-orbital energy of the impurity, ϵ_s^I, and the host, ϵ_s^H. The T_2-part is given by the p-orbital energy difference

$$V_{T_2} = \epsilon_p^I - \epsilon_p^H . \tag{3.7}$$

In Eqs. (3.6), (3.7), we have omitted the site for simplicity. In correspondence with the chemical trends for the LCAO-Hamiltonian matrix elements, discussed in the previous Section, these defect matrix elements can be obtained from the scaled free atomic orbital energies tabulated in Table 1.

Eq. (3.5) shows that the trap energy depends on the atomic properties of the impurity in a straightforward way, while it depends on the host properties mainly through the spectral density of the host band states.

3.3 General Features of the Theory

We are now ready to discuss the results which have been obtained by numerically solving Eq. (3.5), using the 10-band LCAO Hamiltonian from Sec. 3.2.1 and the defect matrix from Sec. 3.2.2 [25-26, 72]. In this Section, we are going to illustrate the general dependence of the deep trap energies on the impurity central-cell potential strength; a detailed comparison between theory and experiment will then be presented in the following Section.

We consider Si first and focus on the A_1-symmetric states and their dependence on the impurity potential. Two ingredients are necessary to solve Eq. (3.5): The localized defect potential, given by the atomic energies, and the A_1 local density of states which is solely a property of the host. Figure 7 displays the A_1 spectral density for Si, as obtained from the tight-binding Hamiltonian, Sec. 3.2.1. It has a form characteristic of all tetrahedral semiconductors. One sees two large peaks, one at

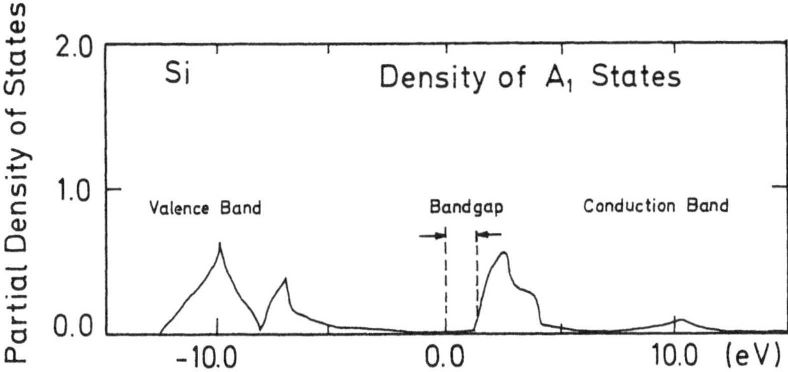

Fig. 7 Projection of the density of states of Si onto the s-orbital state, obtained with the 10-band empirical LCAO-method.

~ -10 eV and the other at ~ 2 eV originating from the s-orbital LCAO-basis states. These peaks correspond to the bottom of the valence band and the conduction band, respectively.

If one approximates this spectral density of states by two weighted delta-functions, we obtain a density of states equivalent to the delta-function density of states of the defect molecule with two A_1-states, which was discussed in Sec. 3.1 [25a]. We see that one can indeed use the defect molecule as a qualitative picture for getting insight about deep traps, much like the hydrogen atom is used in effective mass theory for shallow impurities. With the correspondence between Eq. (3.5) and the defect molecule picture in mind, it will be easy to physically understand the numerical results obtained from Eq. (3.5).

In Fig. 8 we plot $V_{A_1}(E) \equiv G_{A_1}^{-1}(E)$ as a function of energy. The points where $V_{A_1}(E)$ intercepts $V_{A_1} = \epsilon_s^I - \epsilon_s^{Si}$ give the energies of the bound (or resonant) states introduced by the single substitutional defect with central-cell potential V_{A_1}.

The conduction band edge intersects the curve $V_{A_1}(E)$ at a negative and finite potential of -4.1 eV. This threshold is the minimum strength central-cell potential to produce a deep trap state. All impurities with $|V_{A_1}| < 4.1$ eV are therefore predicted to be shallow impurities bound either by the long-range Coulomb tail alone in the case of donors or not at all in the case of isoelectronic impurities (C, Ge, Sn). Experimentally this implies that impurities to the right of $V_{A_1} = -4.1$ eV will follow the band edge upon applying pressure or alloying the host, in contrast to the characteristically deep impurities to the left of the threshold. This threshold behaviour is implicit to all Koster-Slater calculations and has been well recognized earlier [53, 59]. The new and crucial feature of Fig. 8 is the "attenuation" of the central cell potential [25]: In going from C to Cl or to O (which is predicted to produce a resonant state slightly below the valence band edge), the central cell potential

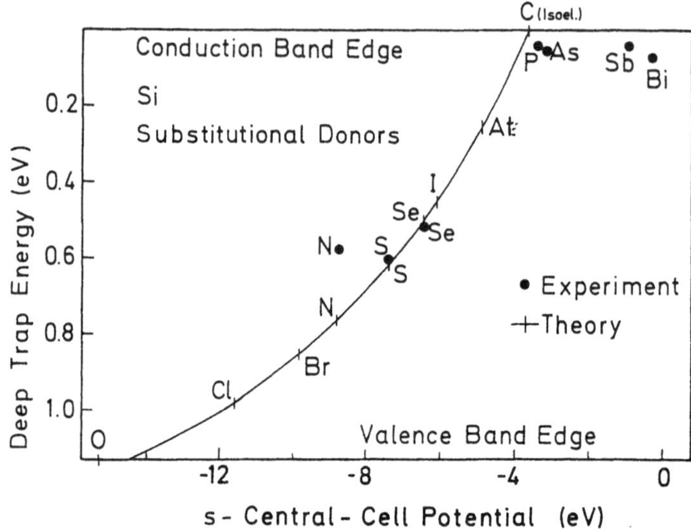

Fig. 8 Predicted energies (relative to the conduction band edge) of A_1-symmetric deep impurity levels in Si [25-26], compared with the experimental data [4–6, 48, 49, 85]. The observed N level may actually be the T_2-state [85]. The theory omits the long-range Coulomb potential; hence, all shallow impurities have zero binding energy. The s-central-cell potential is given in terms of the s-orbital energies.

changes by ~ 12 eV while the binding energy changes only by 1 eV. In other words, $dE/dV \lesssim 0.1$. This attenuation, which is apparent through the nearly hyperbolic shape of the trap-energy function E versus the impurity potential is really the key to understanding the deep trap energies. This concept is solely a consequence of the orthogonality of the crystal's bonding and antibonding states and is independent of our neglect of lattice relaxation or charge transfer [26].

In Fig. 10 the A_1-symmetric anion-site trap energies are plotted versus the s-orbital energies of the impurities for several semiconductors. Focussing on GaP, one finds all P-site impurities that are more electronegative than At to produce a deep donor state. In the very electronegative limit, when $V_{A_1} \to -\infty$, the trap state becomes pinned to the asymptote of the hyperbolic trap-energy function and tends to a pinning energy E, given by

$$0 = \int \frac{D_{A_1,P}(E')}{E' - E} . \tag{3.8}$$

Physically, if one places the impurity's orbital energy at infinity, the impurity no longer couples to the host orbital states. Hence the assumption of an infinite potential is equivalent to removing the impurity without altering the nearby orbital states [60]. The orbital states next to the impurity site therefore become non-bonding, or dangling bonds.

We can see from this that the asymptote of the hyperbola E versus V is – in the case considered – the A_1-type ideal P-vacancy or the Ga-dangling-bond energy. This dangling bond energy is determined solely by the host band structure through Eq. (3.8) and not by any impurity.

In agreement with our previous qualitative conclusions drawn from the defect-molecule picture (Sec. 3.1) we thus find the deep trap wave function to be predominantly hostlike. Eq. (3.5) also yields the impurity-like hyperdeep counterpart of the physically relevant trap state. It lies inside or below the valence band and is orthogonal to the trap state in the gap; it is this hyperdeep state which follows V more or less linearly.

This pinning or, more generally, the attenuation of the E versus V function is the reason why large differences in atomic energies influence trap energies only weakly [25a]. This is particularly seen in GaP, where only very electronegative impurities such as the halogens, are predicted to form deep traps which are also energetically deep in the gap. Other atoms, such as N in GaP, do form deep traps that localize a state, but nonetheless are energetically shallow. In Si, the A_1-pinning energy is embedded in the valence band, yet the attenuation of the central-cell potential is also apparent from Fig. 8. We note that this attenuation also implies that the defect potential needs not to be accurately known. This is apparently one of the reasons for the success of simple modell like the ones presented here.

It is interesting to compare these findings with one-band theories such as the effective mass theory or the earlier deep-trap models of Faulkner [59] and Hsu et al. [53]. The qualitative difference is shown in Fig. 9: The one-band theories exhibit no pin-

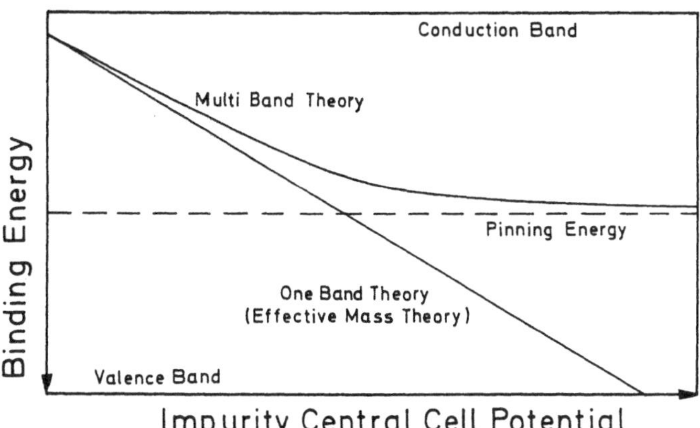

Fig. 9 Schematic illustration of the difference between one-band theories and multi-band theories: The orthogonality of the trap state to the valence states causes the impurity binding energy (full lines) to depend much weaker on the short-range potential than a one-band theory predicts [25a].

ning; consequently the impurity energies respond nearly linearly to the central-cell potential and depend sensitively on it. For small central-cell potential, however, pinning or the central-cell attenuation is not important and a one-band model is more efficient. Also, the long-range potential tail can be more easily dealt with in one-band theories [33].

3.4 Predictions and Comparison with Data

The Figures 10–13 show the predictions for the ionization energies of all sp^3-bonded substitutional impurities, which result from the 10-band-LCAO-Koster-Slater model outlined in the previous Sections [25-26, 72]. Figures 10 and 11 show the s-like A_1-states for the anion site (= P-site in GaP) and the cation-site impurities, respectively. Correspondingly, Figures 12 and 13 display the p-type T_2-states. The calculations have been performed in Ref. [25–26] and [72] for all of the following zincblende hosts: Si, Ge, GaP, GaAs, InP, InAs, InSb, AlP, AlAs, ZnSe, ZnTe. An impurity is predicted to produce a conceptually *deep* trap state in a particular host, site, and symmetry, if the orbital energy of that impurity – given along the

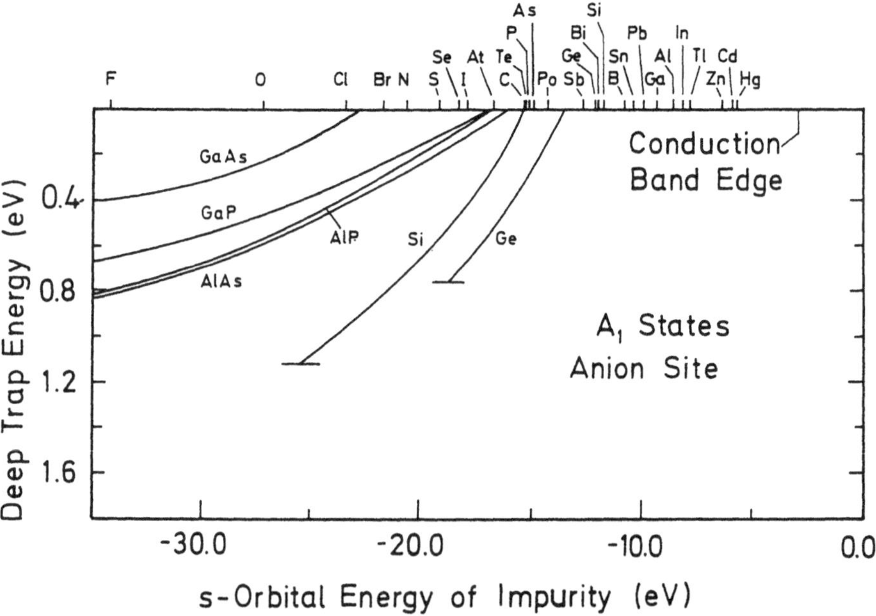

Fig. 10 Predicted ionization energies (relative to the conduction band edge) of the A_1-symmetric, anion-site, deep impurity levels, as functions of impurity orbital energy. The impurites are listed above the figure in the order of decreasing predicted deep-trap binding energy, from F to Hg. Each quasihyperbola is for a different host. The theory omits the long-range Coulomb interaction; hence, all shallow traps have zero binding energy. After Ref. [72].

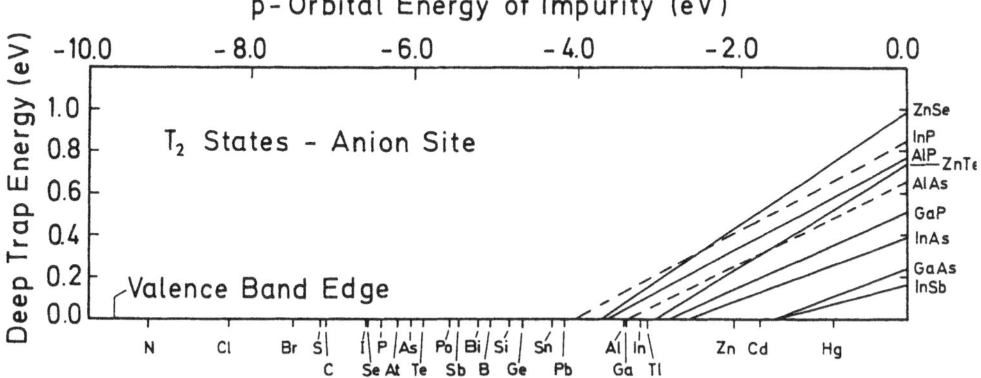

Fig. 11 Predicted ionization energies (relative to the conduction band edge) of the A_1-symmetric, cation-site, deep impurity levels, as functions of the impurity orbital energy. After Ref. [72].

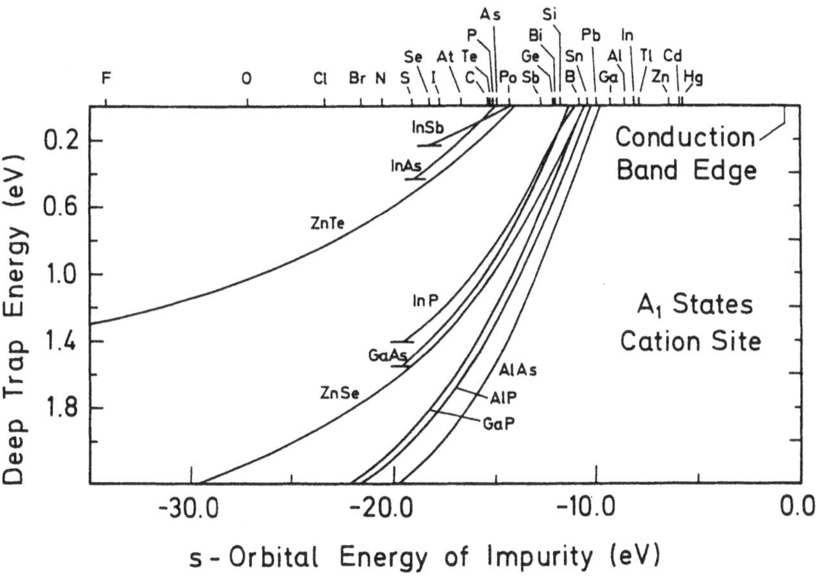

Fig. 12 Predicted ionization energies (relative to the valence band edge) of the T_2-symmetric, anion-site, deep impurity levels, as functions of the impurity orbital energy. After Ref. [72].

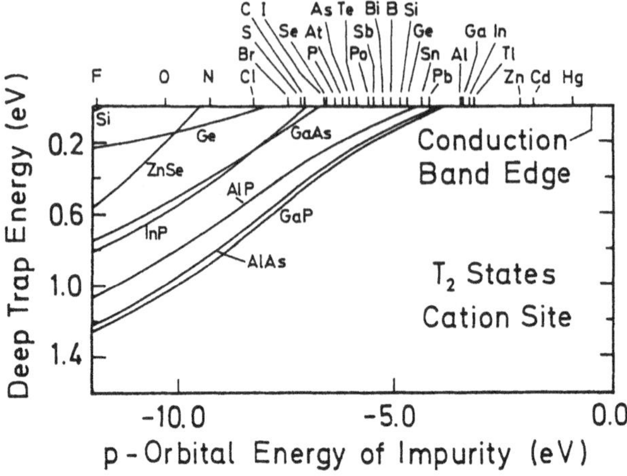

Fig. 13 Predicted ionization energies (relative to the conduction band edge) of the T_2-symmetric, cation-site, deep impurity levels, as functions of impurity orbital energy. After Ref. [72].

abscissas – intersects the quasi-hyperbola associated with this host; the point of intersection gives the impurity's binding energy. If it does not intersect the host-hyperbola, the impurity is predicted to be shallow and effective-mass like. Also, if one of the hosts is missing from a figure, no substitutional (non-transition metal) point defect at that site is expected to produce a deep trap level (e.g. A_1-states, P-site in InP).

The ionization energies which can be deduced from Figures 10–13 represent single-electron (or -hole) states of the electrically neutral crystal. The energy difference between different charge states of impurities (e.g. between S^+ and S^0 in Si) is not predicted by the Koster-Slater model, because the electron-electron repulsion is not explicitely accounted for. However, this energy separation is typically 0.2 eV [73] and does not affect the chemical trends in the impurity levels, as long as one compares either levels in the same charge state or levels with markedly larger energy difference. More elaborate theories are required to systematically treat higher charge states [15, 73]. It has been argued, for example, that strain fields are responsible for localizing two-electron states like O^-in GaP [74].

The S donor in AlAs may serve as an example to illustrate how to make use of the Figures 10–13. S is predicted to produce a deep A_1-donor state on the As site. It is predicted to be deeper in AlAs than in AlP but shallower than in Si, with an ionization energy of approximately 0.2 eV. S produces no T_2 states on the anion site. If S goes "antisite", it produces a deep A_1-trap level rather than a T_2-state which is seen to be energetically less favourable.

Generally, some care is needed in interpreting the cation-site results. A comparison of Figures 11 and 13 shows that the theory predicts deep electronegative impurities to pull in most cases an A_1-state from the conduction band before a localized T_2-state can be formed. The actual situation may be more complex, however, as has been found experimentally for the group-IV donors in GaP [75–79]: The least electronegative donors Sn and Si still form T_2-ground states, while Ge produces an A_1 state. Although the LCAO-theory does not quantitatively reproduce this feature, it suggests that the most electronegative donor C is also most likely to form an A_1-trap state. Recently, this has also been investigated with a generalized effective mass approach [33].

3.4.1 Comparison with Data: $GaAs_xP_{1-x}$

Comparison of the LCAO-Koster-Slater calculation [25–26] with the data for deep traps in $GaAs_xP_{1-x}$ [55, 57, 58, 80] provides convincing testimony to the ability of this theory to predict chemical trends.

In particular, examining N as a function of composition x in the alloy $GaAs_xP_{1-x}$ allows one to explore the dependence of deep trap levels on the host environment. In this alloy, N replaces either As or P but the central-cell potential of N remains nearly constant due to the chemical similarity of As and P (see Table 1).

In Fig. 14 we compare luminescence data for the N [55, 57, 58] and the O [80] deep trap as a function of composition x with the theoretical results; also shown are the O-related substitutional pairs (Zn, O) [80] and the Ga-vacancy – O pair [80] which will be discussed in Sec. 4.

The experimental data [53, 55, 57, 58] show that N produces (i) a resonant state in the As-rich alloy, (ii) an energetically deep state in the mid-composition region, and (iii) a weakly bound state near and in GaP. All of these features are correctly reproduced by the tight-binding model. The most striking feature of both the data and the theory is the linear variation with composition of the energy of all deep impurities [55, 81], with an almost universal slope for all of the trap energies.

The linear variation of the impurity levels, particularly of the N level, implies that they are detached from either the direct or indirect band edge [53, 55, 58, 81]. This detachment is in marked contrast to the behaviour expected of a shallow trap which would appear attached to the bandedge, either direct or indirect, because the effective mass binding energy would be small and only weakly composition-dependent.

The slopes of the deep impurity data points N, O, (Zn, O), and (V_{Ga}, O) are in excellent agreement with the theory of Refs. [25–27, 29] and with the slope of the pinning or dangling-bond energy [82]. The composition dependence of the dangling-bond energy – though not apparent from the defect molecule picture Fig. 4 – can be understood qualitatively as follows. Let us approximate the anion-A_1-spectral density of states (which is qualitatively of the form shown in Fig. 7) by two

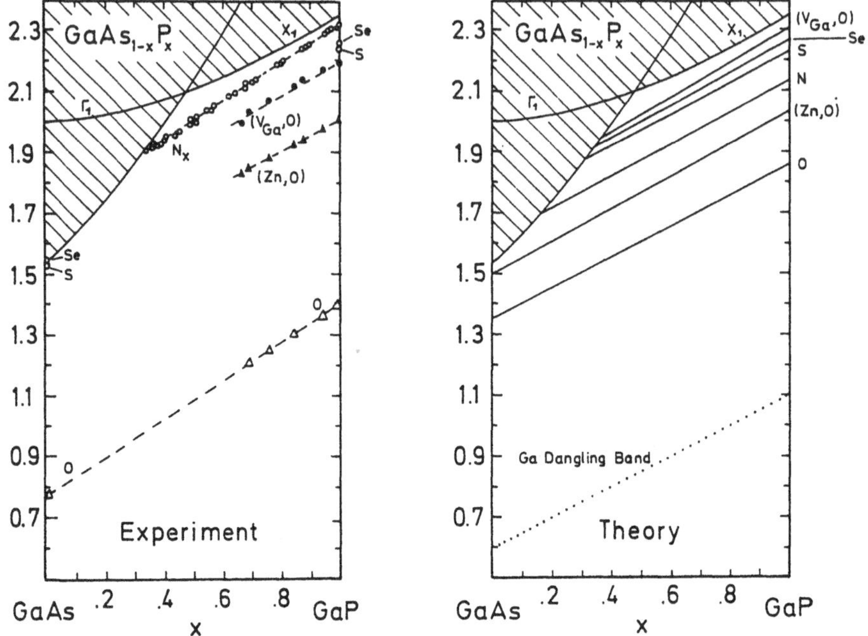

Fig. 14 Trap energies as a function of alloy composition for $GaAs_{1-x}P_x$. Left: Luminescence data for substitutional N, Se, S, O and the substitutional defect pairs (Zn, O) and (Ga-vacancy, O), representing zero-phonons lines. Alloy data for N_x from Ref. [55, 57, 58], with permission. Data for (Zn, O), (V_{Ga}, O), and O from Ref. [80], with permission. Data for Se, S from Refs. [4, 48] O in GaP from Refs. [5, 90], and O in GaAs from Ref. [83] (see also Ref. [82]). Right: The theory predicts the deep traps to follow the compositional dependence of the Ga-dangling bond or pinning energy rather than the Γ_1 or the X_1 point of the band edge. Specially the data for N exhibit this detachment from the bandedge. From Ref. [25, 27], with permission.

delta-functions at $< E_c >$ and $< E_v >$, i.e. at the centers of the A_1-conduction and valence band density of states. Expression (3.8) can then be evaluated immediately and gives the pinning energy $E = (< E_c > - < E_v >)/2$. Thus we see that the host-dependence of the deep trap levels mimics the *average* gap which decreases from GaP to GaAs [43]. Shallow impurities, on the other hand, mimic the *minimum* gap. Also shown in Fig. 14 are the S and Se levels which are known experimentally [4, 5] only in GaP and GaAs. The theory predicts these impurities to be (conceptually) deep in the P-rich alloy but to be shallow in the As-rich alloy, which implies that these levels are expected to change their slope near GaAs.

3.4.2 Comparison with Data: Si

As can be seen from Fig. 8, the group V-donors are expected to be shallow except N, in agreement with the data [4, 85]. The calculated S and Se levels agree (perhaps accidentally) surprisingly well with experiment.

The LCAO-Koster-Slater model predicts no deep T_2-acceptors in Si, which is in disagreement with the data (see Fig. 2). This disagreement is a systematic feature of the T_2-trap level results. Partly, this is thought to originate in the choice of the LCAO-Hamiltonian, and partly, it may have to do with the fact that deep and only partially filled T_2-states are likely to be subject to Jahn-Teller distortions.

3.4.3 Further Predictions

Recently, ESR measurements provided evidence for the existence of antisite defects in III-V semiconductors, where the anions occupy cation sites [2]. Particularly, P_{Ga} in GaP and As_{Ga} in GaAs could be identified [86, 87], although their activation energies are yet unknown. According to Fig. 11, the present theory predicts both P and As to produce deep antisite levels, while antisite-Ga remains shallow. Generally, one deduces from Fig. 11 that almost all anion-site donors appear to give deep states if they go antisite. Further ESR-experiments are needed to clarify this.

4 Chemical Trends in Substitutional Defect Pairs

We have seen that a trap state originating from an isolated non-transition metal deep impurity is largely host-like. From a technological point of view, this fundamentally limits the possibilities to control deep-trap properties. One can overcome this limitation at least in principle by pairing the isolated deep impurity with a "spectator" impurity which replaces one of the host-atoms next to that impurity.

Qualitative insight into the effect of pairing an anion-site impurity like N or O in GaP with a substitutional nearest-neighbour spectator can immediately be gained from the defect molecule picture, Fig. 4. If one replaces the Ga-atom of the defect molecule by an s-electropositive atom, the trap level may be driven into the conduction band. An s-electronegative atom, on the other hand, acts to deepen the isolated trap level.

These ideas have been quantified recently by Sankey et al. [27, 29] who extended the tight-binding-Koster-Slater model discussed in Sec. 3 to the case of extended defect levels. The symmetries of the expected deep pair levels are reduced to "molecular" C_{3v} symmetry. Fig. 15 shows the results of a tight-binding Green-function calculation for the a_1 (C_{3v})-symmetric (spectator, O) pair state, derived chiefly from the isolated A_1-symmetric O level in GaP. The theory shows that pairs with electropositive spectators (e.g. (Mg, O)) produce energy levels which tend to and are "pinned" by the (V_{Ga}, O) level. This pinning is analogous to the one discussed for isolated defects. The theory has also been used to study the trends in the energy levels of paired defects in $GaAs_{1-x}P_x$. The predictions for the pairs (Zn, O) and (V_{Ga}, O) are shown in Fig. 14 and are in excellent agreement with the experimental data. The wealth of information gained by this solid-state chemistry approach becomes most strikingly apparent from Ref. [29] where the deep levels for 841 sp^3-bonded defect-pair combinations in GaAs are predicted.

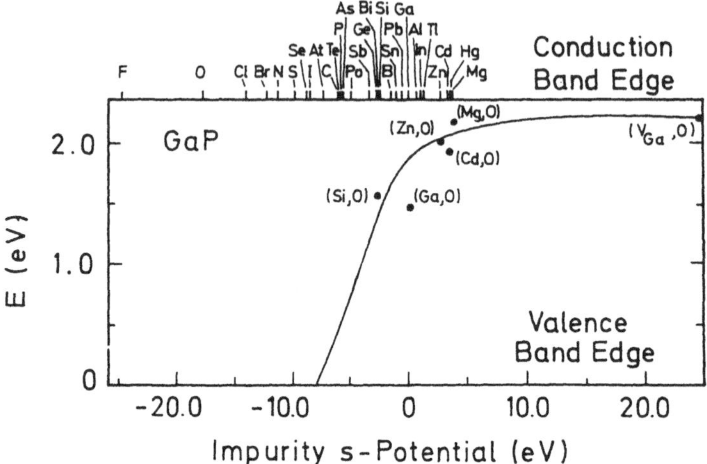

Fig. 15 Plot showing the trends in the data for various Ga-site substitutional impurities complexed with a nearest-neighbour substitutional O impurity on the P-site in GaP. The solid curve is theory, while dots correspond to published data. The spectator impurities appear at the top of the figure, at the appropriate impurity s-potential which may be taken from Table 1. From Ref. [27], with permission.

5 Chemical Trends in Transition-Metal Impurities

We finally turn to the technologically most relevant class of non-sp^3-bonded impurities, the 3d-transition-metal impurities. They generally behave as deep centers and act as most efficient trapping centers. Doping with transition metals gives rise to a remarkable sequence of donor and acceptor levels corresponding to several different charge states of the impurity [2, 4, 10]. The classic picture of a substitutional 3d-transition-metal impurity is the crystal field picture pioneered by Ludwig and Woodbury [88]: It assumes that the defect states are basically atomic d multiplets which have been split by electrostatic interactions with the surrounding host atoms.

The theoretical analysis of transition-metal impurities has been impeded so far by the uncertainty in assigning any observed optical transitions to a particular initial and final state, site-symmetry, and charge state. Haldane and Anderson [30] have developed a simple model which explains qualitatively the physical origin of the many charge states which can appear in the gap. In the free atom, these charge states differ by more than 10 eV. In the covalent solid, on the other hand, the strong hybridization between impurity-d and host-sp^3 valence states allows the *nominal* occupancy of d-like impurity states to vary considerably, while keeping the actual amount of charge in the core-regions of the transition-metal atom almost the same as in the neutral atom. Particularly, for large impurity-ligand interaction, the fraction of the atomic d-character of the levels in the gap becomes small.

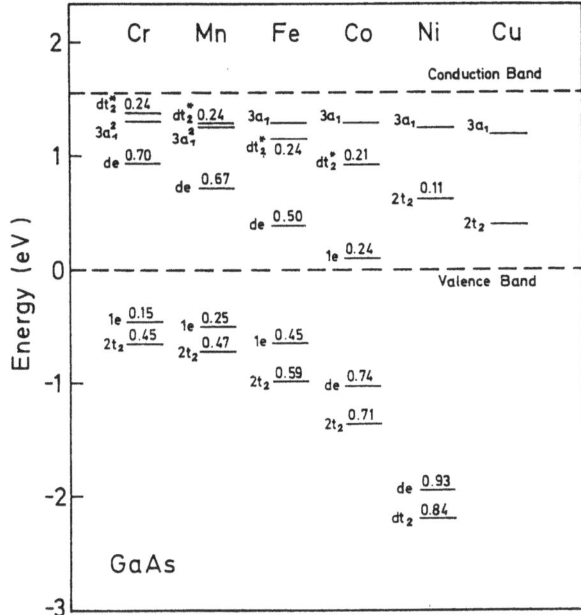

Fig. 16 Impurity-related energy levels calculated for the neutral 3d transition-metal impurities in GaAs. All energies are in eV and are measured from the cluster valence edge. Those states associated with the impurity d states have the fraction of such character shown adjacent to the corresponding level. From Ref. [31], with permission.

This has been substantiated recently by Hemstreet [31] who performed a series of X_α-cluster calculations of the electron states associated with cation-site 3d-transition-metal impurities. This theory predicts the centers Ni, Cu, Zn near the end of the 3d-series to produce host-like T_2-states rather than impurity d-like states. The restults for GaAs are shown in Fig. 16. As one proceeds across the 3d-row from Cr to Cu, the transition-metal-like d-state moves down in energy. In Co, the active trap state is already a host-like valence level, labeled 1e in Fig. 16. In Ni and Cu, the d state is already so deep in the valence band, that the trap states in the gap actually correspond to the A_1- and T_2-As dangling-bond energies.

A direct comparison between theory and experiment seems yet difficult but these qualitative trends are not in contradiction with the data [31] and may prove useful in unraveling the many and varied experimental data currently available for this class of impurities.

6 Conclusions

The central assumption of the presented theories of sp^3-bonded impurities has been that the major chemical trends in deep trap energies are determined by the energy bands of the undisturbed hosts and by the impurities' atomic structure. The response of the host electrons and the lattice to the insertion of an impurity was largely ignored, presuming that it is either insignificant or scales monotonically with the trap depth. By doing so, one arrives at simple but general theories which explain the major chemical trends in deep trap energies surprisingly well and provide a wealth of qualitative insight into the physics of deep impurities. Still, all these simple models are likely, in their present form, to have omitted important factors, which have not shown up clearly in the experimental data yet. By applying the ideas reviewed in this paper to many new experimental data, we hope that it may be possible to extract the essential physical parameters which determine the impurity spectra of semiconductors.

Aknowledgements

We thank Dr. H. P. Hjalmarson for his collaboration on this manuscript and for his permission to use results prior to publication. Useful discussions with Prof. J.D. Dow and Dr. D. Wolford are acknowledged. We are indebted to Drs. S. Modesti, D.J. Wolford, B.G. Streetmann, and P. Williams for permission to use their experimental results prior to publication. This work was supported in part by the Fonds zur Förderung der wissenschaftlichen Forschung in Österreich, Projekt Nr. 4236.

References

[1] *D. V. Lang,* J. Appl. Phys. **45**, 3023 (1974).

[2] *U. Kaufmann* and *J. Schneider,* in Festkörperprobleme XX (ed. by J. Treusch, Vieweg, Braunschweig, 1979), p. 87.

[3] *A. G. Milnes, Deep Impurities in Semiconductors* (Wiley, New York 1973).

[4] *R.K. Watts, Point Defects in Crystals* (Wiley, NY, 1977).

[5] *A.A. Bergh* and *P.J. Dean, Light-emitting Diodes* (Clarendon Press, Oxford, 1976).

[6] *H. G. Grimmeis,* Ann. Rev. Matl. Sci. 7, 341 (1977).

[7] *P.J. Dean, D. Bimberg, U. Kaufmann* and *J. Schneider,* in Landolt-Börnstein (Springer, Berlin), Vol. 17a, (to be published).

[8] *G.L. Miller, D. V. Lang,* and *L.C. Kimerling,* Ann. Rev. Mater. Sci. 7, 377 (1977).

[9] *A. Mircea* and *D. Bois,* Inst. Phys. Conf. Ser. **46**, p. 82 (1979).

[10] *A.M. Stoneham, Theory of Defects in Solids* (Clarendon, Oxford, 1975).

[11] *M. Altarelli* and *F. Bassani, in Handbook of Semiconductors* ed. by W. Paul, (North Holland, NY.), Vol. 1, in print.

[12] *H.J. Queisser,* in Festkörperprobleme XI (ed. by O. Madelung, Vieweg, Braunschweig, 1971), p. 45.

[13] *S. T. Pantelides,* Rev. Mod. Phys. **50**, 797 (1978).

[14] *S.T. Pantelides, J. Bernholc,* and *N.O. Lipari,* J. Phys. Soc. Jap. **49**, Suppl. A, 235 (1980).
[15] *G.A. Baraff* and *M. Schlüter,* Festkörperprobleme XIX (ed. by J. Treusch, Vieweg, Braunschweig, 1979), p. 303.
[16] *G.A. Baraff, E.O. Kane,* and *M. Schlüter,* J. Phys. Soc. Jap. **49**, Suppl. A, 231 (1980).
[17] *M. Jaros,* Adv. Phys. **29**, 409 (1980).
[18] *M. Altarelli, A. Baldereschi,* and *N. O. Lipari,* Solid State Phys. (in press).
[19] *A.M. Stoneham,* Contemp. Phys. **20**, 535 (1979).
[20] *G.A. Baraff* and *M. Schlüter,* Phys. Rev. Lett. **41**, 892 (1978).
[21] *G.A. Baraff* and *M. Schlüter,* Phys. Rev. **B 19**, 4965 (1979), ibid., **B 21**, 5662 (1980).
[22] *J. Bernholc, N.O. Lipari,* and *S. Pantelides,* Phys. Rev. Lett. **41**, 895 (1978).
[23] *J. Bernholc, N.O. Lipari,* and *S.T. Pantelides,* Phys. Rev. **B 21**, 3545 (1980).
[24] *M. Jaros, C.O. Rodriguez,* and *S. Brand,* Phys. Rev. **B 19**, 3137 (1979).
[25] *H.P. Hjalmarson, P. Vogl, D.J. Wolford,* and *J.D. Dow,* Phys. Rev. Lett. **44**, 810 (1980).
[25a] *H. P. Hjalmarson, P. Vogl, D. J. Wolford,* and *J. D. Dow,* to be published.
[26] *H.P. Hjalmarson,* Ph. D. thesis, University of Illinois, 1979, (unpublished).
[27] *O.F. Sankey, H.P. Hjalmarson, J.D. Dow, D.J. Wolford,* and *B. G. Streetman,* Phys. Rev. Lett. **45**, 1656 (1980).
[28] *H.P. Hjalmarson, R.E. Allen, H. Büttner,* and *J.D. Dow,* J. Vac. Sci. Techn. **17**, 993 (1980).
[29] *O.F. Sankey* and *J.D. Dow,* "Deep levels associated with nearest neighbour substitutional defect pairs in GaAs", (to be published).
[30] *F.D.M. Haldane* and *P.W. Anderson,* Phys. Rev. **B 13**, 2553 (1976).
[31] *L.A. Hemstreet,* Phys. Rev. **B 22**, 4590 (1980), ibid., **B 15**, 834 (1977).
[32] *J.M. Luttinger* and *W. Kohn,* Phys. Rev. **97**, 969 (1955).
[33] *M. Altarelli, W.Y. Hsu,* and *A. Baldereschi,* in Physics of Semiconductors 1978 (Institute of Physics, London, 1979) Conf. Ser. **43**, 299 (1978) and to be published.
[34] *C.A. Coulson* and *M.J. Kearsley,* Proc. Roy. Soc. **A 241**, 433 (1957).
[35] *R.P. Messmer* and *G.D. Watkins,* Phys. Rev. **B 7**, 2568 (1973).
[36] *G.F. Koster* and *J.C. Slater,* Phys. Rev. **95**, 1167 (1954) and ibid., **96**, 1208 (1954).
[37] *J. Callaway,* J. Math. Phys. **5**, 793 (1964); Phys. Rev. **154**, 515 (1967).
[38] *J. Callaway* and *A.J. Hughes,* Phys. Rev. **156**, 860 (1967).
[39] *M. Jaros* and *S. Brand,* Phys. Rev. **B 14**, 4494 (1976).
[40] *S. Brand* and *M. Jaros,* J. Phys. **C 12**, 2789 (1979).
[41] *M. Jaros* and *S. Brand,* J. Phys. **C 12**, 525 (1979).
[42] *F.P. Larkins* and *A.M. Stoneham,* J. Phys. **C 4**, 143 (1973); J. Phys. **C 4**, 154 (1971).
[43] *J.C. Phillips, Bands and Bonds in Semiconductors* (Academic, NY, 1973).
[44] *W.A. Harrison,* Phys. Rev. **B 8**, 4487 (1973).
[45] *W.A. Harrison, Electronic Structure and the Properties of Solids* (Freeman, San Francisco, 1980).
[46] *E. Clementi* and *C. Roetti,* Atomic Data and Nucl. Data Tables **14**, 177 (1974).
[47] *W. Scott* and *J.L. Schmit,* Appl. Phys. Lett. **33**, 294 (1978).
[48] *H.G. Grimmeis, E. Janzen,* and *B. Skarstam,* J. Appl. Phys. **51**, 4212 (1980) and ibid., **51**, 3740 (1980).
[49] *J. Schneider,* unpublished.

[50] *R.A. Cooke, R.A. Hoult, R.F. Kirkman,* and *R.A. Stradling,* J. Phys. D **11**, 945 (1978).

[51] *J.C. Phillips,* Phys. Rev. B **1**, 1540 (1970).

[52] *A. Baldereschi,* in Proc. XII. Int. Conf. on the Physics of Semiconductors (Teubner 1974, Stuttgart), p. 345.

[53] *W.Y. Hsu, J.D. Dow, D.J. Wolford,* and *B.G. Streetman,* Phys. Rev. B **16**, 1597 (1977).

[54] *C.O. Rodriquez, S. Brand,* and *M. Jaros,* J. Phys. C **13**, L 337 (1980).

[55] *D.J. Wolford, B.G. Streetman, W.Y. Hsu, J.D. Dow, R.J. Nelson,* and *N. Holonyak, Jr.,* Phys. Rev. Lett. 36, 1400 (1976).

[56] *H. Mariette* and *J. Chevallier,* Sol. State Commun. **29**, 263 (1979).

[57] *D.J. Wolford, W.Y. Hsu, J.D. Dow,* and *B.G. Streetman,* J. Lum. **18/19**, 863 (1979).

[58] *D.J. Wolford,* Ph. D. thesis, University of Illinois, 1979.

[59] *R.A. Faulkner,* Phys. Rev. **175**, 991 (1968).

[60] *J. Bernholc* and *S.T. Pantelides,* Phys. Rev. B **18**, 1780 (1978).

[61] *J. Pollman,* Festkörperprobleme XX (ed. by J. Treusch, Vieweg, Braunschweig, 1980), p. 117.

[62] *K.C. Pandey* and *J.C. Phillips,* Phys. Rev. B **13**, 750 (1976).

[63] *E.J. Mele* and *J.D. Joannopoulos,* Surf. Sci. **66**, 38 (1977).

[64] *D. J. Chadi,* Phys. Stat. Sol. **68**, 405 (1975) and Phys. Rev. B **16**, 790 (1977).

[65] *D.J. Chadi,* Phys. Rev. B **16**, 790 (1977).

[66] *P. Vogl, H.P. Hjalmarson,* and *J.D. Dow,* "A semiempirical tight-binding theory of the electronic structure of semiconductors", (to be published).

[67] *S.G. Louie,* Phys. Rev. B **22**, 1933 (1980).

[68] *D.W. Bullett,* J. Phys. C **8**, 2695 (1975); ibid., C **8**, 2707 (1975); Solid State Physics (ed. by H. Ehrenreich, F. Seitz, D. Turnbull, Academic, N.Y., 1980), Vol. **35**, p. 129; V. Heine, ibid., p. 1.

[69] *E.O. Kane* and *A.B. Kane,* Phys. Rev. B **17**, 2691 (1978).

[70] *C. Tejedor* and *J.A. Verges,* Phys. Rev. B **19**, 2283 (1979).

[71] *R.E. Allen, H.P. Hjalmarson, H. Büttner, P. Vogl, D.J. Wolford, O.F. Sankey,* and *J.D. Dow,* Int. J. Quant. Chem., (in press).

[72] *H.P. Hjalmarson, H. Büttner,* and *J.D. Dow,* "Theory of Core Excitons", (to be published).

[73] *M. Scheffler, S.T. Pantelides, N.O. Lipari,* and *J. Bernholc,* "Identification and Properties of Native Defects in GaP", (to be published).

[74] *T.N. Morgan,* Inst. Phys. Conf. Ser. **43**, 311 (1979).

[75] *F. Mehran, T.N. Morgan, R.S. Title,* and *S.E. Blum,* Phys. Rev. B **6**, 3917 (1972) and Solid St. Commun. **11**, 661 (1972).

[76] *A.C. Carter, P.J. Dean, M.S. Skolnick,* and *R.A. Stradling,* J. Phys. C **10**, 5111 (1977).

[77] *A. Onton,* Phys. Rev. B **4**, 4449 (1971).

[78] *R.A. Cooke,* Ph. D. thesis, University of Oxford, 1979.

[79] *T.N. Morgan,* Phys. Rev. Lett. **21**, 819 (1968).

[80] *S. Modesti, D.J. Wolford, B.G. Streetman,* and *P. Williams,* "Photoluminescence in Oxygen implanted $GaAs_{1-x}P_x$ ($1 \leq x \leq 0.62$)", (to be published).

[81] *D.V. Lang, R.A. Logan,* and *L.C. Kimerling,* Phys. Rev. B **15**, 4874 (1977).

[82] The O-level in GaAs [83] is controversial (see Ref. [84]); however, the recent data [80] for O in $GaAs_{1-x}P_x$ which are plotted in Fig. 14 extrapolate to the controversial point which lends credence to the identification of that level in GaAs.

[83] *B. Deveaud* and *P.N. Favennec*, Proc. 7th Int. Symp. on GaAs and Related Compounds, St. Louis, 1978.

[84] *A.M. Huber, N.T. Link, M. Valladan, J.L. Debrun, G.M. Martin, A. Mittoneau*, and *A. Mircea*, J. Appl. Phys. **50**, 4022 (1979).

[85] *K.L. Brower*, Phys. Rev. Lett. **44**, 1627 (1980).

[86] *U. Kaufmann, J. Schneider*, and *A. Räuber*, Appl. Phys. Lett. **29**, 312 (1976).

[87] *R.J. Wagner, A.M. White, J.J. Krebs*, and *G.H. Strauss*, Inst. of Phys. Conf. Ser. (in press).

[88] *G.W. Ludwig* and *H.H. Woodbury*, Solid State Physics (ed. by F. Seitz, D. Turnball, Academic, N.Y., 1962), Vol. **13**, p. 223.

[89] *C.E. Fischer*, Atomic Data **4**, 301 (1972).

[90] *B. Monemar* and *L. Samuelson*, Phys. Rev. **B 18**, 809 (1978).

Self-Interstitials and Vacancies in Elemental Semiconductors Between Absolute Zero and the Temperature of Melting

Werner Frank

Max-Planck-Institut für Metallforschung, Institut für Physik, and Universität Stuttgart, Institut für Theoretische und Angewandte Physik, Stuttgart, Germany

Dedicated to Professor Dr. Ing. Dr. rer. nat. h.c. Ulrich Dehlinger on the occasion of his 80th birthday on July 6, 1981

Summary: Following an introductory overview of the intrinsic point defects in semiconductors the paper concentrates on Si and Ge, in particular on the exciting recent developments concerning Si self-interstitials. Proceeding from low to high temperatures the discussion covers the following subjects: the low-temperature mobility of point defects induced by particle irradiation or other techniques generating free charge carriers; the geometrical configurations, electric charge states, and thermally activated migration of self-interstitials and vacancies at intermediate temperatures; the predominance of vacancies in Ge and self-interstitials in Si under high-temperature equilibrium conditions. The remainder of the paper deals with the transition of the Si self-interstitials from dumbbell to spread-out configurations between intermediate and high temperatures and – so far as to bring out the relationship to self-interstitials in Si – with concentration-enhanced-diffusion phenomena and with the diffusion of Au in Si.

1 Introduction and Overview

The rapidly proceeding miniaturization of semiconductor devices has led to a considerable increase of the technological importance of intrinsic point defects in semiconductors. Until a few years ago, however, for people concerned with semiconductor technology, dopants and unintended impurities were the only point defects of interest. This is a major reason why fairly little is known on vacancies and self-interstitials in semiconductors: In the case of compound semiconductors there is almost no information on these defects. For Ge a fragmentary picture of the intrinsic point defects begins to arise.

An exception is Si. It has early been chosen as the model substance for basic investigations of intrinsic imperfections in semiconductors. In the sixties, great progress was made in the understanding of vacancy-type defects in this material, and during the past decade we proceeded from the "mystery of the self-interstitials in Si" to a detailed knowledge on these defects.

Intrinsic point defects in Si and Ge play a rôle in the entire regime of temperatures between absolute zero and the melting point. They resemble chameleons, since – depending on temperature and other experimental circumstances – they appear in different geometrical configurations, carry different electric charges, and undergo migration via different mechanisms. They may be present in supersaturations, e.g., after particle irradiation, or in thermal equilibrium, e.g., in self-diffusion experiments. Under non-equilibrium conditions Si and Ge show analogous features, whereas in thermal equilibrium different defect species dominate in these two elemental semiconductors, namely vacancies in Ge and self-interstitials in Si.

The present paper gives a state-of-the-art report on intrinsic point defects in Si, mainly concentrating on recent developments concerning self-interstitials. A comparison to Ge is made as far as admitted by our scarce knowledge on intrinsic point defects in this material.

In order to facilitate the discussion we shall distinguish between the following temperature regimes: (i) the *low-temperature regime* in which self-interstitials become mobile under particle irradiation, illumination with light, or charge-carrier injection (Sect. 2), (ii) the *regime of intermediate temperatures* in which intrinsic point defects present in supersaturations anneal out via thermally activated diffusion (Sect. 3); (iii) the *high-temperature regime* in which intrinsic atomic defects are spread out over several atomic volumes and can be studied under thermal-equilibrium conditions (Sect. 4).

The material presented in Sects. 5 to 7 confirms and completes the picture of self-interstitials in Si developed in Sects. 2 to 4: The gradual transition of these defects from dumbbell configurations at intermediate temperatures to spread-out configurations at high temperatures is discussed in Sect. 5. The phenomena of concentration-enhanced diffusion of substitutional dopants in Si are briefly mentioned in Sect. 6, since they indicate that at high temperatures at least P and B diffuse via an interstitialcy mechanism. Finally, we present the theoretical frame-work for deducing valuable information on Si self-interstitials in thermal equilibrium from studies of the diffusion of Au in Si (Sect. 7).

2 Induced Mobility of Self-Interstitials at Low Temperatures

2.1 Observations

In 1964 Watkins [1, 2] discovered by means of EPR studies on Al-doped p-type Si that during electron irradiation at 4 K equal concentrations of Al^{2+} interstitial atoms (Al_i) and vacancies (V) are produced. He suggested that this observation may be understood if the self-interstitials (I) created by electron irradiation can migrate over long distances already at 4 K: If this is the case the self-interstitials recombine with vacancies,

$$I + V \rightleftharpoons \text{annihilation}, \tag{1}$$

or, alternatively, replace substitutional Al atoms (Al_s) according to

$$I + Al_s \rightleftharpoons Al_i\,. \tag{2}$$

Since electron irradiation introduces vacancies and self-interstitials in pairs, equal concentrations of Al interstitials and vacancies are left behind after the elimination of the self-interstitials via the reactions (1) and (2).

Watkins' observation has been confirmed later by several authors, who demonstrated self-interstitial migration to take place in both n- and p-type Si during electron irradiation down to at least 0.5 K [3–5].

Similar evidences for the migration of self-interstitials at very low temperatures exist for Ge, too [6, 7]. In contrast to Si, in which this low-temperature mobility appears to be a purely athermal effect, the radiation-induced diffusion of Ge self-interstitials is weakly temperature-dependent, corresponding to an activation enthalpy of 4.4 meV [7].

For the sake of completeness we mention that mobility enhancement at low temperatures is not restricted to self-interstitials. E.g., it was also observed for vacancies [2, 8, 9] and Al interstitials [10] in Si. In all cases it can be induced by particle irradiation, illumination, or charge-carrier injection.

2.2 Mechanisms

Various explanations of the induced mobility of defects in semiconductors were suggested in the literature [11–13]. Recently, it was demonstrated [12, 13] that only the so-called *energy-release mechanisms* [11–15] can account for all features of the low-temperature mobility of self-interstitials in Si and Ge.

The *direct energy-release mechanism* works as follows: If conduction electrons and (or) holes produced by irradiation etc. recombine at (are trapped by) defects via non-radiative transitions, the defects may be displaced with the aid of the energy released from the energy reservoir of the free charge carriers. This mechanism is thought to control the low-temperature migration of Si self-interstitials [12, 13].

In the *thermally assisted energy-release mechanism* the released energy is used to populate excited vibrational states localized at the defects. Then by thermal activation the defects can jump from these excited states to neighbouring positions (Fig. 1). The thermally activated step in this mechanism is controlled by an activation barrier H_α^{M*} which is smaller than the migration enthalpy H_α^M of the defect.[1]) We presume that the low-temperature diffusion of Ge self-interstitials occurs via this mechanism and thus interpret the activation enthalpy of 4.4 meV (Sect. 2.1) as H_α^{M*} [12, 13].

---

[1]) The subscript α distinguishes between the various electronic states, e.g., electric charge states, in which defects in non-metals may occur and in which they usually possess different migration enthalpies. As to the charge states of self-interstitials in Si or the corresponding values of H_α^M we refer to Sect. 4.2 or Sect. 3.2 and Table 1, respectively.

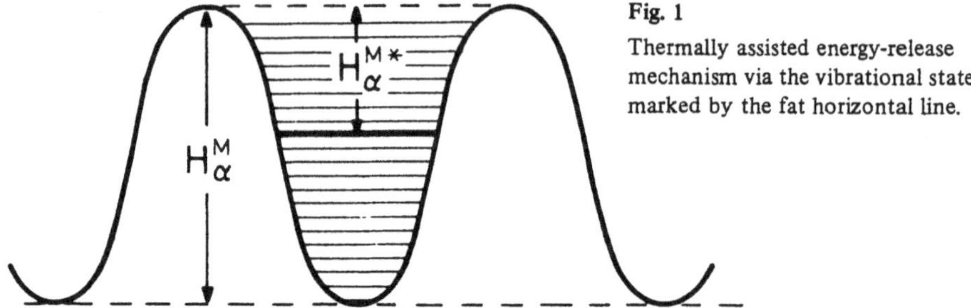

Fig. 1
Thermally assisted energy-release mechanism via the vibrational state marked by the fat horizontal line.

3 Intrinsic Point Defects at Intermediate Temperatures

3.1 Vacancies

Vacancies in Si have been studied extensively during the past two decades. For an overview we refer to review articles by Watkins et al. [2, 9]. Here we only mention some properties that are characteristic of the vacancies in Si up to intermediate temperatures ($\lesssim$ 200 K): (i) The Si vacancy is truly a point defect, i.e., the missing atom is concentrated in about one atomic volume. (ii) It possesses non-cubic symmetry due to a Jahn-Teller distortion. (iii) It may occur in the following electric charge states: double negative, singly negative, electrically neutral, or singly positive.
(iv) The migration enthalpy H_V^M controlling the thermally activated annealing of vacancy supersaturations is strongly charge-dependent. A unique assignment of the H_V^M values observed (0.18 eV, 0.32 eV, 0.45 eV) to the electric charge states is not yet possible.

The only well established information on *vacancies in Ge* at intermediate temperatures is that in n-type Ge vacancies undergo thermally activated migration at about 90 K [16, 17].

3.2 Self-interstitials

In accordance with the theoretical prediction by Blount (tight-binding approximation) [18] and by Watkins et al. (atomic-orbital–molecular-orbital cluster technique) [19] the *self-interstitials in Si* are amphoteric (Fig. 2) with an acceptor level at about $E_C - 0.4$ eV and a donor level at about $E_V + 0.4$ eV [20] (Sect. 4.2). At intermediate (and presumably also at low) temperatures these defects possess dumbbell configurations: From internal friction measurements [21] it was concluded [22] that in the positive (I^+) or neutral (I^0) charge state the dumbbells have a preferential axis

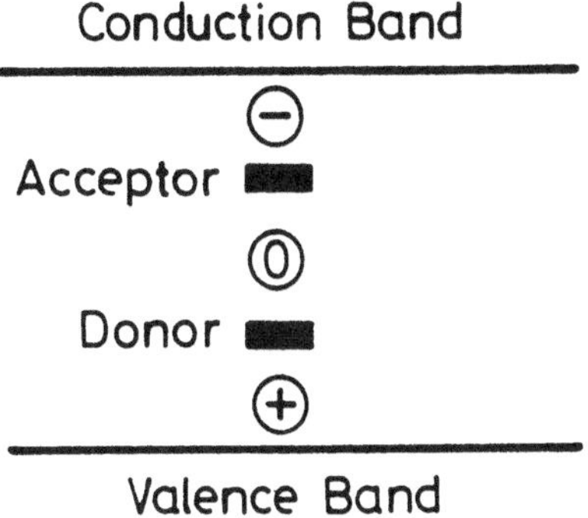

Fig. 2
Energy levels of the self-interstitial in Si (according to Blount [18]).

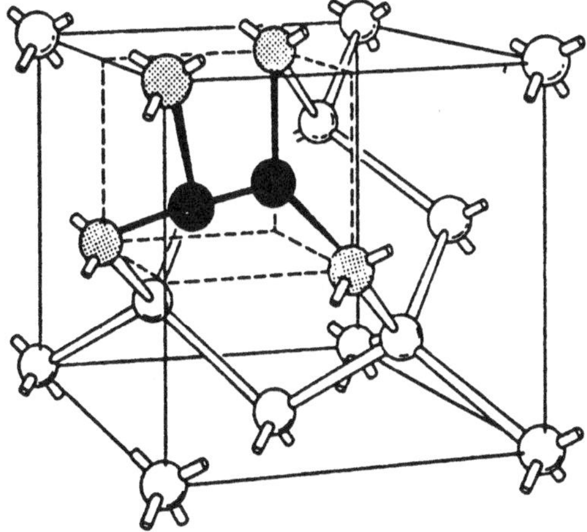

Fig. 3
Singly positively charged self-interstitial in Si (according to Tan et al. [21]).

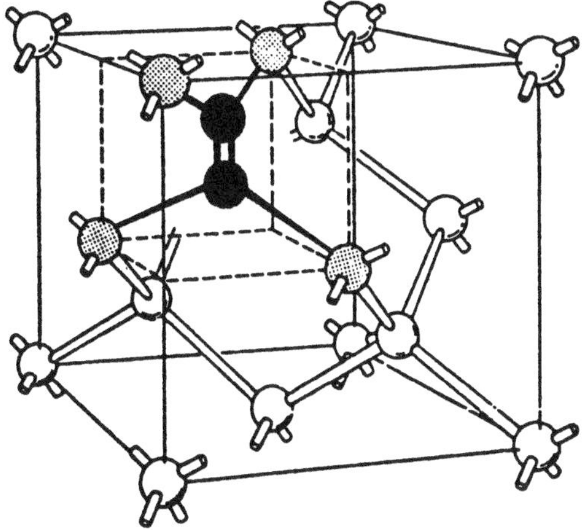

Fig. 4
Electrically neutral self-interstitial in Si (according to Tan et al. [21]).

in ⟨110⟩ (Fig. 3) or ⟨100⟩ (Fig. 4), respectively. In the negative charge state (I^-) the Si self-interstitials give rise to the EPR signal Si-G25. The g tensor has tetragonal symmetry, so I^- presumably is a ⟨100⟩ dumbbell, too [23]. I^-, I^+, or I^0 undergoes thermally activated migration in the temperature regime 140 K to 170 K, 370 K to 420 K, or 540 K to 600 K with a migration enthalpy H_I^M of about 0.35 eV, 0.85 eV, or 1.5 eV, respectively [22–24].

Presumably due to the less frequent technological use of Ge, fewer investigations of intrinsic point defects, in particular radiation experiments, have been performed on Ge than on Si. As a consequence, we know fairly little on *self-interstitials in Ge*. Nevertheless, Seeger and Frank [24, 25] have attempted to develop a tentative picture of these defects: For Ge self-interstitials Blount's model (Fig. 2) has to be generalized by introducing a two-fold acceptor level located between the conduction-band edge and the chemical-donor levels, i.e., in addition to the charge states which may be occupied by Si self-interstitials, Ge self-interstitials may be double negatively charged. The thermally activated migration of I^{2-}, I^-, or I^0 was associated with the recovery stages of irradiated Ge at 27 K, 65 K, or 220 K, respectively. The energy level for the transition I^+/I^0 lies presumably so low that (perhaps except for strongly p-doped Ge) it is crossed by the Fermi level before I^+ undergoes thermally activated migration.

4 Intrinsic Defects under High-Temperature Equilibrium Conditions

4.1 Spread-out vacancies in germanium, spread-out self-interstitials in silicon

The tracer self-diffusion coefficient D^T in a solid may be written as

$$D^T = D_0^T \exp\left(-\frac{H_X^F + H_X^M}{kT}\right) \tag{3}$$

with

$$D_0^T = f_X\, g_X\, a^2\, \nu_{0,X} \exp\left(\frac{S_X^F + S_X^M}{k}\right) \tag{4}$$

provided that self-diffusion is controlled by a single defect species X (H_X^F [S_X^F] or H_X^M [S_X^M] = enthalpy [entropy] of formation or migration of X, $\nu_{0,X}$ = attempt frequency, f_X = correlation factor [26], g_X = numerical factor, a = lattice parameter, k = Boltzmann's constant, T = absolute temperature). In most metals this prerequisite is fulfilled, at least at temperatures sufficiently far below the melting point [27, 28]. There the dominating defect species are monovacancies (X = V). The fact that D_0^T is of the order of magnitude of $10^{-1}\ cm^2 s^{-1}$, corresponding to a self-diffusion entropy $S_V^F + S_V^M$ of about 2 to 3 k, indicates that under high-temperature equilibrium conditions vacancies in metals are truly point defects, i.e., they are concentrated in about one atomic volume.

In contrast to metals, the preexponential factor D_0^T in Ge or Si is about 10^1 cm^2 s^{-1} or 10^3 cm^2 s^{-1}, respectively [29–31]. E.g., Mayer et al. [32] described their ^{31}Si self-diffusion data (full circles in Fig. 5) by

$$D^T = 1\,460 \exp(-5.02\,\text{eV}/kT)\,\text{cm}^2\,\text{s}^{-1}\,. \tag{5}$$

Already in 1968 Seeger and Chik [29, 30] realized that such large values of D_0^T can exclusively originate from large self-diffusion entropies and that these must be due a spreading-out of the intrinsic defects governing self-diffusion in Ge or Si over about 5 or 10 atomic volumes, respectively. Moreover, Seeger and Chik showed that the experimental data on point defects in elemental semiconductors available at that time may be reconciled only under the assumption that the defect species dominating in self-diffusion experiments are monovacancies in Ge but self-interstitials in Si. This picture has been confirmed by later experiments. Some of the experiments which support the unconventional suggestion that extended self-interstitials dominate in Si under high-temperature equilibrium conditions will be discussed in Sects. 4.2, 4.3, and 5.1.

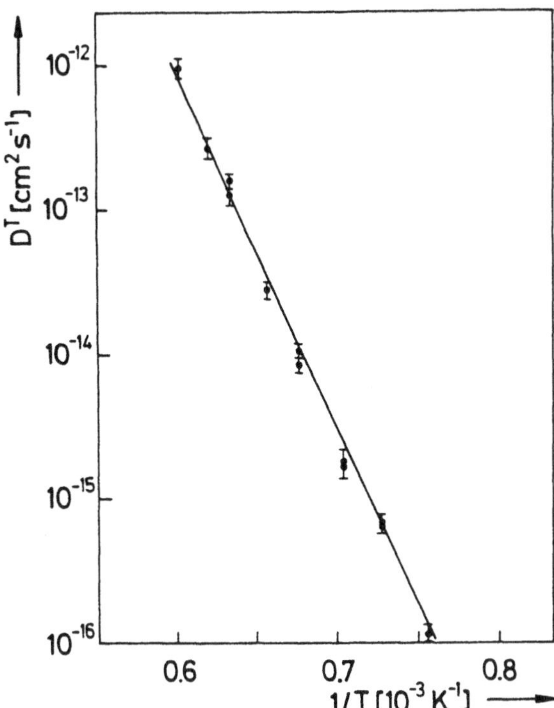

Fig. 5
^{31}Si tracer self-diffusion coefficient measured by Mayer et al. [32] between 1320 K and 1658 K (full circles). The solid curve was adjusted to these data [23] (Sect. 5.1).

4.2 Dependence of the diffusion of germanium in silicon on chemical doping

Among the original arguments presented by Seeger and Chik [29, 30] in favour of an interstitialcy self-diffusion mechanism in Si above about 1200 K are indications in early self-diffusion data that the self-diffusion coefficient is increased by both n and p doping – in agreement with Blount's model, but at variance with what one would expect if self-diffusion were governed by acceptor-type vacancies. The difficulties inherent in these measurements involving the short-lived radioactive isotope ^{31}Si have been avoided in recent experiments by Hettich et al. [33]. Following McVay and DuCharme [34] these authors [33] simulated the self-diffusion in Si by the diffusion of the long-lived tracer ^{71}Ge in Si. Indeed, in these experiments, which extend down to 1130 K, the doping dependence above approximately 1200 K was found to follow the predictions of Blount's model (Fig. 6). We take this as evidence that at high temperatures self-diffusion in Si is governed by self-interstitials present in thermal equilibrium and that the electric charge state of the self-interstitials is negative, neutral, or positive in n-doped, intrinsic, or p-doped Si, respectively.

Various authors [35–37] have attempted to account for the doping dependence of the Si self-diffusion coefficient above about 1200 K in terms of a monovacancy mechanism. These attempts

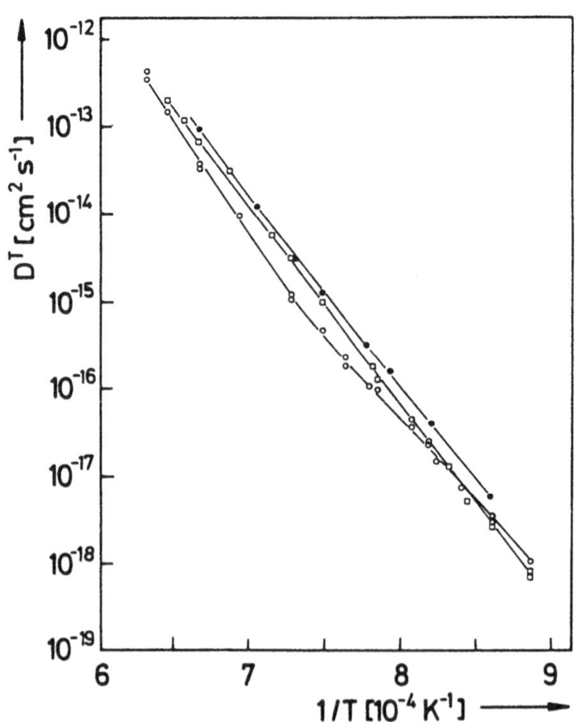

Fig. 6
Temperature dependence of the diffusion coefficient of ^{71}Ge tracers in intrinsic (○), B-doped (□), or As-doped (•) Si, respectively (according to Hettich et al. [33]).

have been shown to be unable to give quantitative agreement [31]. However, we point out that in principle a formal agreement of the vacancy diffusion models with the doping dependence found by experiment can be enforced if, disregarding the position of the donor level established by Watkins et al. [2, 9], the monovacancy is endowed with essentially the same electronic levels that we attribute to the self-intersitial. Hence, experiments giving direct information on the nature of the intrinsic defects present at high temperatures, e.g., those on swirl defects (Sect. 4.3) or the diffusion of Au in Si (Sect. 7) are of great importance.

Fig. 6 shows a change-over of the doping dependence at about 1200 K in the sense that below this temperature diffusion is enhanced or slowed down by doping with As or B, respectively. This doping dependence is compatible with the operation of a vacancy diffusion mechanism below about 1200 K, as suggested originally by Seeger and Chik [29, 30]. We believe that it cannot be decided at present whether under thermal-equilibrium conditions – in spite of the dominant rôle played by self-interstitials – the influence of vacancies must *not* be neglected. This view has been adopted in the combined vacancy-interstitial models of Hu [38, 39] and Sirtl [40].

4.3 Swirl defects

Fairly direct evidence for the interstitial nature of the defects present in Si under high-temperature equilibrium conditions comes from the investigation of the so-called A-type swirl defects or (briefly) A-swirls found in dislocation-free Si single crystals grown by float-zone techniques at suitable rates [41–44]. Using transmission electron microscopy Föll and Kolbesen [43] and Petroff and de Kock [44] have shown independently that A-swirls are interstitial-type dislocation loops. Föll et al. [43, 45] have suggested that A-swirls are formed by clustering of the high-temperature-equilibrium point defects that can neither reach the crystal surface nor disappear at grown-in dislocations during the cooling of the crystals. It then follows that the thermal-equilibrium defects dominating in Si at high temperatures must be self-interstitials. This conclusion is strengthened by the fact that the concentration of self-interstitials 'precipitated" as A-swirls is in good agreement with the self-interstitial concentration at high temperatures estimated with the aid of the values of the migration enthalpy and entropy of Si self-interstitials given in Table 1 (Sect. 5.1) [23, 31].

5 Change of the Configurations of Silicon Self-Interstitials Between Intermediate and High Temperatures

At intermediate temperatures the Si self-interstitials possess dumbbell configurations (Sect. 3.2). At high temperatures the Si self-interstitials are spread-out (Sect. 4.1), at least the electrically neutral ones.[2]) Therefore, proceeding from intermediate to high

---

[2]) At high temperatures the Fermi level is close to the middle of the forbidden gap [46]. Hence, according to Blount's model (Fig. 2) the neutral self-interstitials dominate at high temperatures.

temperatures I^0 (Fig. 4) must gradually lose its dumbbell configuration and become spread-out. This idea is confirmed by an analysis [23] of the ^{31}Si self-diffusion data of Mayer et al. [32] (Sect. 5.1). An atomistic model of this spreading-out process is suggested in Sect. 5.2.

5.1 Experimental evidence

Seeger et al. [23] have shown that, allowing for a linear temperature dependence of the formation and migration enthalpy of I^0 and taking into account the thermodynamic relations

$$(\partial S_{I^0}^{F,M}/\partial T)_p = (1/T)\,(\partial H_{I^0}^{F,M}/\partial T)_p \tag{6}$$

valid for experiments performed at constant pressure p, the self-diffusion data of Mayer et al. [32] may be fitted quite well (curve in Fig. 5) by assuming $S_{I^0}^F = S_{I^0}^M = k$ and $H_{I^0}^M = 1.5\,\mathrm{eV}$ at 570 K. The other parameters given in Table 1 follow from the fitting. These results support the picture that at intermediate temperatures I^0 is a dumbbell (compare $S_{I^0}^F = S_{I^0}^M = k$ at 570 K, which is characteristic of an unextended defect) with a migration enthalpy of 1.5 eV (Sect. 3.2) and that at high temperatures I^0 becomes more and more extended (compare the considerably larger values of $S_{I^0}^F$ and $S_{I^0}^M$ in the temperature range of the self-diffusion measurements [Table 1]).

Table 1 Values of $S_{I^0}^F$, $S_{I^0}^M$, $H_{I^0}^F$, and $H_{I^0}^M$ obtained from fitting the self-diffusion data of Mayer et al. [32] (Fig. 5). $S_{I^0}^F + S_{I^0}^M$ or $H_{I^0}^F + H_{I^0}^M$ is the self-diffusion entropy or enthalpy, respectively.

T [K]	$S_{I^0}^F$ [k]	$S_{I^0}^M$ [k]	$H_{I^0}^F$ [eV]	$H_{I^0}^M$ [eV]
1658	6.11	6.96	3.04	2.02
1320	5.02	5.69	2.90	1.86
570	≈ 1	≈ 1	≈ 2.6	≈ 1.5

5.2 Atomistic model

At low and intermediate temperatures the electrons forming the bonds between the two atoms of an I^0 dumbbell are well localized in the centre of gravity of the defect (Fig. 4). When temperature increases, these electrons occupy excited states and thus become more and more delocalized. As a consequence of this spreading-out of electrons the dumbbell relaxes and spreads out, too. Kimerling [14] proposed that the interstitial may first be extended over a ring of five atoms. Since, independent of such details, the spread-out interstitial may be realized by more microstates and may perform diffusional jumps via more paths than a dumbbell-type interstitial, the spreading-out results in an increase of both $S_{I^0}^F$ and $S_{I^0}^M$ with increasing temperature.

Analogous considerations apply to I^- and I^+, though, according to Blount's model [18], these play a minor rôle under high-temperature equilibrium conditions, when the Fermi level is near the middle of the forbidden gap [46].

Thermodynamics tells us that the increase of the entropies of formation and migration of I^0, I^+, and I^- with increasing temperature is accompanied by an increase of the corresponding enthalpies of formation and migration [see, e.g., the relations (6) in the case of I^0]. An atomistic interpretation of the increase of these enthalpies is that for extended interstitials the formation or migration requires the breaking of a larger number of bonds than for dumbbells. In the case of the migration of the dumbbells Frank and coworkers [22–24] have argued that H_I^M is determined by the strength of the intra-atomic bonds of the dumbbells. This view is supported by the decrease of H_I^M in the order I^0, I^+, I^- provided that the extra electron of I^- is in a fairly extended orbit, antibonding with respect to the dumbbell atoms.

6 Concentration-Enhanced Diffusion of Substitutional Foreign Atoms in Silicon

The diffusion of substitutional foreign atoms (briefly called "dopants") may be described by the diffusion coefficient

$$D^D = \sum_X w_{XD} f_X D_X C_X . \tag{7}$$

Here C_X (D_X) are the concentrations (diffusivities) of the intrinsic defects X used as diffusion vehicles by the dopants. Interactions between X and the dopants are included in w_{XD}.

In the case of *concentration-enhanced diffusion* at least one of the concentrations C_X in (7) is increased over its equilibrium value [12, 13]. In Si this can be achieved by particle irradiation (radiation-enhanced diffusion [47–50]), surface oxidation (oxidation-enhanced diffusion [51–57]), or strong local doping with P or B (so-called anomalous diffusion phenomena [31, 58–61] such as the emitter-push effect, the movement of marked layers, or the kink-and-tail structure of P diffusion profiles). For a detailed discussion of these effects the reader is referred to recent review articles [12, 13, 31, 58, 62, 63]. Here we elaborate on these phenomena only so far as to bring out the relationship to Si self-interstitials.

Concerning *radiation-enhanced diffusion* of dopants in Si, the following facts are of interest: (i) By analyzing their studies of the diffusion of As, B, or P in Si under proton irradiation in terms of the Lomer-Dienes-Damask theory [64–66], Masters and Gorey [47] found 1.5 eV as a lower limit of the migration enthalpy of the less mobile intrinsic point-defect species in the temperature range 870 K to 1070 K. (ii) Since the migration enthalpy of the monovacancies is smaller than 0.5 eV (Sect. 3.1), the less mobile intrinsic point defects were identified as neutral self-interstitials

[31, 67], in accordance with the charge state of the self-interstitials at high-temperatures predicted by Blount's model[2]) and the $H^M_{I^0}$ values given in Table 1. (iii) From items (i) and (ii) it follows that, contrary to the radiation-induced mobility enhancement of Si self-interstitials at low temperatures (Sect. 2.1), in the high-temperature experiments of Masters and Gorey [47] D_{I^0} in (7) was not enhanced over the value determined by thermally activated diffusion. This is expected, since the mobility enhancement via the direct energy-release mechanism (Sec. 2.2) obviously cannot operate if $H^M_{I^0}$ exceeds the width of the forbidden gap [67], as it is the case for I^0 at high temperatures. (iv) The way in which initially homogeneously distributed B or P atoms are redistributed under proton irradiation [48–50] shows that at high temperatures these dopants prefer self-interstitials as diffusion vehicles [12, 13, 58]. This result does not exclude a remarkable contribution to the diffusion of dopants by vacancies at low temperatures [31], as indicated by the formation of E-centres [68].

A common feature of the *oxidation-enhanced diffusion* and of the *anomalous diffusion phenomena* is the enhancement of the diffusion of dopants due to a supersaturation of intrinsic point defects injected into the Si from the surface as a result of surface oxidation or of a high surface concentration of P (or, in some cases, of B), respectively. It is obvious that from these experiments alone one cannot decide whether the enhancement of the diffusion of dopants is due to a supersaturation of vacancies or self-interstitials. However, by studying the climb of dislocations in a transmission electron microscope under experimental conditions giving rise to oxidation-enhanced or anomalous diffusion this decision becomes possible. In all experiments of this type in which the supersaturation of intrinsic point defects was sufficiently high, dislocations were found to climb in directions which are compatible with the emission of vacancies or the absorption of self-interstitials (for details see [62]). Since in the presence of a supersaturation of intrinsic point defects the thermodynamic force driving dislocation climb is such that the point defects are *absorbed*, one is led to the conclusion that the defects inducing oxidation-enhanced or anomalous diffusion are self-interstitials. In agreement with the conclusion drawn from radiation-enhanced-diffusion experiments (see above) this implies that the Group-III and Group-V elements – at least B and P – diffuse in Si at high temperatures predominantly via an interstitialcy mechanism. However, it should be kept in mind that this conclusion is based on experiments in which self-interstitials were present in supersaturations.

7 Diffusion of Gold in Silicon

The diffusion of Au in Si shows extraordinary features, e.g., unusual diffusion profiles in both thick specimens [69, 70] (observation A) and wafers [Fig. 7] (observation B), and an increase of the Au concentration in the middle of Au-diffused

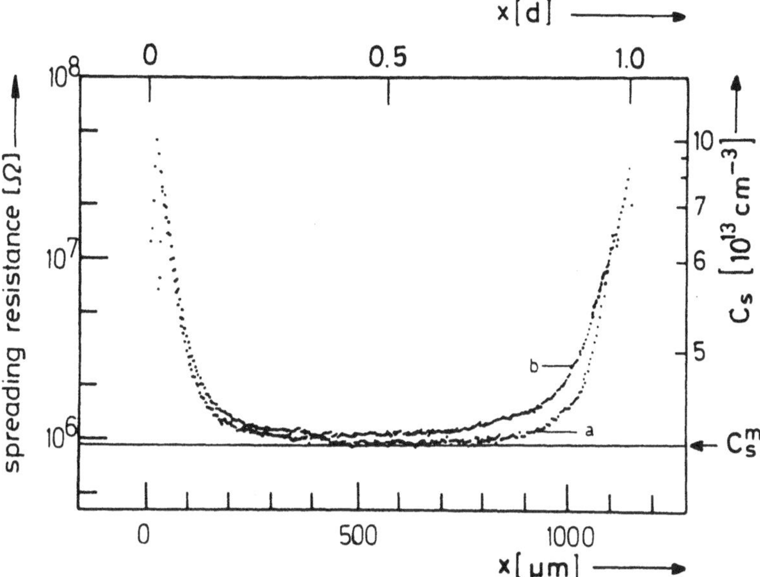

Fig. 7 Distributions of Au_s in Si wafers measured by means of the spreading-resistance technique after 5-hour diffusion annealing treatments at about 1100 K [71]. In the case a, which is of main interest in the present context, the Au was deposited on both surfaces, in the case b on the left-hand surface only.

wafers proportional to the square root of the time of diffusion [Fig. 8] (observation C). In the past it was attempted to explain these findings with the aid of the following hypotheses [69, 74–77]: (a) In Si Au atoms occupy both substitutional (Au_s) and interstitial (Au_i) sites. (b) The solubility of Au_s is much higher than that of Au_i. (c) Au_i diffuses much faster than Au_s. (d) During diffusion the Au atoms interchange between interstitial and substitutional sites. (e) This interchange takes place via the so-called Frank-Turnbull mechanism [78]

$$Au_i + V \rightleftharpoons Au_s \,. \tag{8}$$

The "Frank-Turnbull model" which comprises the items (a) to (e) could indeed account qualitatively for observation B: Due to the high mobility of Au_i its solubility limit is reached extremely fast in the entire wafer. Then the approaching to the higher solubility limit of Au_s takes place via the reaction (8). This happens most rapidly in the vicinity of the wafer surfaces, since these act as efficient sources of V. Thus the U-shaped diffusion profiles (Fig. 7) develop. However, the Frank-Turnbull model fails to explain the observations A and C [79].

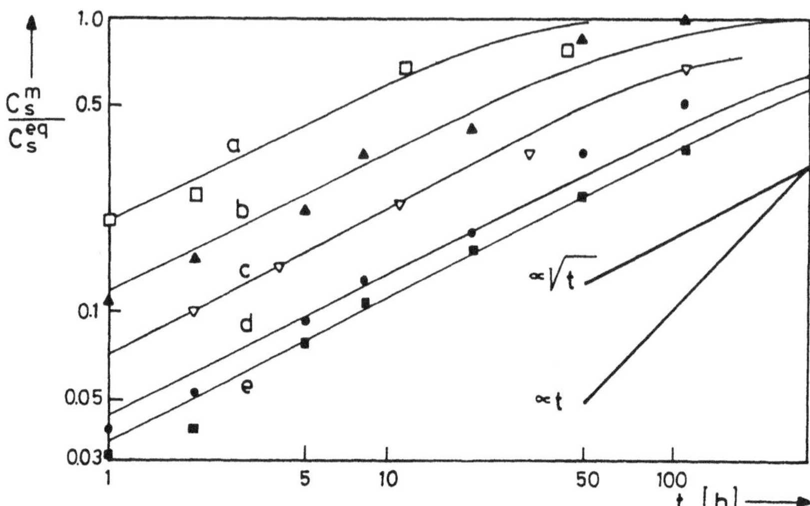

Fig. 8 C_s^m/C_s^{eq} measured as a function of time t on crystals with different dislocation densities ρ or/and at different diffusion temperatures T (curve a [□]: ρ unknown, T ≈ 1375 K [72]; curve b [▲]: $\rho \approx 10^4\ cm^{-2}$, T ≈ 1110 K [73]; curve c [▽]: ρ unknown, T ≈ 1275 K [72]; curve d [●]: $\rho \approx 10^2\ cm^{-2}$, T ≈ 1110 K [73]; curve e [■]: $\rho \approx 0$, T ≈ 1110 K [73]). Except for very long times, all series of measurements follow $\sqrt{t}$ laws.

The facts that the intrinsic point defects dominating in Si under high-temperature equilibrium conditions are self-interstitials (Sect. 4) and that at high temperatures substitutional foreign atoms diffuse via an interstitialcy mechanism (Sect. 6) suggest that the $Au_i - Au_s$ interchange [item (d)] involves self-interstitials according to the "kick-out" mechanism" [80][3])

$$Au_i \rightleftharpoons Au_s + I\,. \tag{9}$$

In the following two points will be demonstrated: (i) The model comprising the features (a) to (d) and the kick-out mechanism (9), briefly called the "kick-out model", accounts for the observations A to C. (ii) An analysis of experimental data on the diffusion of Au in Si in terms of the kick-out model yields valuable quantitative information on Si self-interstitials in thermal equilibrium, which can hardly be obtained in another way.

[3]) From now on the electrically neutral, spread-out high-temperature configuration of the Si self-interstitial is simply called "I", i.e., we drop the superscript "0".

7.1 Basic equations

Diffusion of Au in Si via the kick-out mechanism is described by the following equations:

$$\partial C_I/\partial t = D_I\, \partial^2 C_I/\partial x^2 + \partial C_s/\partial t - A_I\,(C_I/C_I^{eq} - 1), \quad (10)$$

$$\partial C_i/\partial t = D_i\, \partial^2 C_i/\partial x^2 - \partial C_s/\partial t, \quad (11)$$

$$C_i/C_I C_s = C_i^{eq}/C_I^{eq} C_s^{eq}. \quad (12)$$

Here x is the space coordinate in the direction of diffusion, t the diffusion time, C_n the concentration and D_n the diffusivity of the defects of type n, where n = I, i, or s for self-interstitials, Au_i, or Au_s, respectively. The equilibrium value of C_n is denoted by C_n^{eq}. The quantity A_I means the efficiency of dislocations as sinks (sources) for I and is proportional to the dislocation density ρ. By the law of mass action, eq. (12), it is assumed that local equilibrium between Au_i, Au_s, and I is established everywhere at any time via the kick-out mechanism (9).

7.2 Predictions of the kick-out model and comparison with experiments

7.2.1 High dislocation density

In this case, which is characterized by an instantaneous establishing of the self-interstitial equilibrium concentration ($C_I = C_I^{eq}$), (11) and (12) yield for C_s a normal diffusion equation with the effective, C_s-independent diffusion coefficient

$$D_{eff}^{(1)} = C_i^{eq} D_i/(C_i^{eq} + C_s^{eq}). \quad (13)$$

Since the same result is obtained in the Frank-Turnbull model, experiments involving high dislocation densities are not suitable to discriminate between the Frank-Turnbull and the kick-out model.

7.2.2 Dislocation density not high

By definition, the dislocation density is considered *not* to be high if $C_I = C_I^{eq}$ is *not* fulfilled. Then (9) and the establishing of the Au_i equilibrium concentration are the fastest reactions, so that, in addition to the simplified description of (9) by the mass-action law (12), as a further simplification (11) may be replaced by $C_i = C_i^{eq}$. With this assumption (10) and (12) lead to

$$\partial C_s/\partial t = (\partial/\partial x)\,(D_{eff}^{(2)} \partial C_s/\partial x) + A_I(C_s^{eq}/C_s - 1) \quad (14)$$

with the effective, C_s-dependent diffusion coefficient

$$D_{eff}^{(2)} = C_I^{eq} C_s^{eq} D_I/C_s^2 \quad (15)$$

provided that $C_s/C_s^{eq} > (C_I^{eq}/C_s^{eq})^{1/2}$. In practice this restriction is meaningless, since $(C_I^{eq}/C_s^{eq})^{1/2}$ is very small, e.g., of the order of magnitude of 1% at 1100 K

[79]. Therefore, for many purposes it is possible and convenient to define $t > 0$ as the regime of time in which (14)/(15) is valid by choosing the initial condition

$$C_s(0 < x < d, t = 0)/C_s^{eq} \equiv C_s^0/C_s^{eq} = (C_I^{eq}/C_s^{eq})^{1/2} \approx 0. \qquad (16)$$

In the case of a wafer (Sect. 7.2.2.1) d is the wafer thickness; for a semi-infinite solid (Sect. 7.2.2.2) $d = +\infty$.

Under assumptions equivalent to those made in deriving (14)/(15) the Frank-Turnbull model yields a completely different differential equation for C_s: The effective diffusion coefficient is C_s-independent, and the term describing the influence of dislocations as sources (sinks) for vacancies differs from the second term in eq. (14) [79, 81]. Hence, it is not surprising that in all cases in which the dislocation density is not high the Frank-Turnbull model and the kick-out model make different predictions and may thus be distinguished by comparison with experiments.

7.2.2.1 Wafer

In the kick-out model the diffusion of Au from the wafer surfaces into the bulk (observation B) may be described qualitatively as follows: After the establishing of C_i^{eq} throughout the specimen, C_s^{eq} ($> C_i^{eq}$) is reached most rapidly near the surfaces since at these the self-interstitials produced via (9) can disappear most easily. A complete quantitative treatment of this problem requires the solution of the boundary-value problem defined by the partial differential equation (14)/(15), the boundary condition

$$C_s(x = 0, t) = C_s(x = d, t) = C_s^{eq}, \qquad (17)$$

and the initial condition (16). To the author an exact analytical solution of this problem is not known. However, in special cases the time dependence of C_s in the middle of a wafer,

$$C_s^m(t) \equiv C_s\left(x = \frac{d}{2}, t\right), \qquad (18)$$

has been derived and thus may be compared to observation C (Fig. 8).

Wafer with intermediate dislocation density. In this case the first term on the right-hand side of (14) may be neglected, so that (14) degenerates to an ordinary differential equation with the implicit solution [79]

$$\ln[(1 - C_s^m/C_s^{eq})/(1 - C_s^0/C_s^{eq})] + (C_s^m - C_s^0)/C_s^{eq} + A_I t/C_s^{eq} = 0, \qquad (19)$$

which is illustrated in Fig. 9 as curve a. For $C_s^0 \ll C_s^m < C_s^{eq}$ eq. (19) reduces to

$$C_s^m = (2 A_I C_s^{eq} t)^{1/2} \qquad (20)$$

(curve b in Fig. 9).

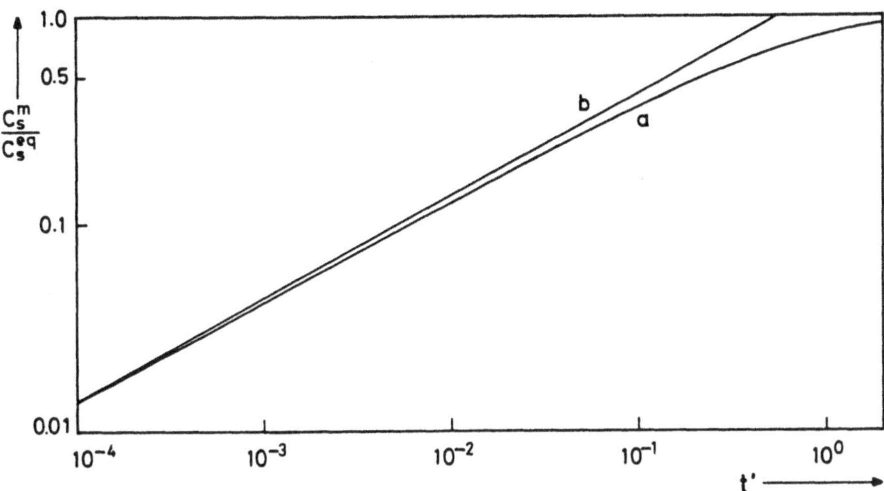

Fig. 9 C_s^m/C_s^{eq} as a function of the normalized time $t' \equiv A_I t/C_s^{eq}$ according to (19) for $C_s^0 = 0$ [curve a] or to (20) [curve b], respectively.

Wafer with low dislocation density. Seeger and Frank [82] have shown that if both terms on the right-hand side of (14) are comparable, for $C_s^0 = 0$ and $C_s^m < C_s^{eq}$

$$C_s^m = (2\,A_I\,C_s^{eq}\,t)^{1/2}\;F\,(A_I\,d^2/C_s^{eq}\,D_I)\,, \tag{21}$$

where the function F does not depend on t.

Dislocation-free wafer. In this case, in which A_I and thus the second term on the right-hand side of (14) are equal to zero, it is helpful to consider the boundary-value problem that is obtained by replacing C_s by C_s^* in (14) to (16) and by replacing (17) by

$$C_s^*\,(x = 0, t) = C_s^*\,(x = d, t) = +\,\infty\,. \tag{22}$$

The solution $C_s^*(x,t)$ of this modified boundary-value problem has been derived by Gösele et al. [79] for $C_s^0 = 0$ with the aid of the ansatz

$$C_s^*\,(x,t) = X(x)\,(D_I\,t)^{1/2} \tag{23}$$

[X(x) = function of x only] and by Meyberg [83] for $C_s^0 \neq 0$. $C_s^*(x,t)$ is a good approximation to the solution $C_s(x,t)$ of the really interesting boundary-value problem (14) to (17) provided $C_s(x,t)$ remains far enough below C_s^{eq}. E.g., this is the case for $x = d/2$ at sufficiently short times. Thus for $C_s^0 = 0$

$$C_s^m(t) \equiv C_s\left(x = \frac{d}{2}, t\right) \approx C_s^*\left(x = \frac{d}{2}, t\right) = X\left(\frac{d}{2}\right)(D_I\,t)^{1/2}$$

$$= (2/d)\,(\pi\,C_s^{eq}\,C_I^{eq}\,D_I\,t)^{1/2} \tag{24}$$

was found [79].

We see that according to the kick-out model the concentration C_s^m must increase proportional to $\sqrt{t}$ except for C_s^m values close to C_s^{eq}, irrespective of whether the dislocation density is intermediate [eq. (20) and Fig. 9], low [eq. (21)], or zero [eq. (24)]. This is in agreement with observation C (Fig. 8). The Frank-Turnbull model predicts other dependences of C_s^m on t [81] and therefore fails to account for observation C. In particular, in the case of intermediate dislocation density the Frank-Turnbull model predicts $C_s^m \propto t$ for $C_s^m < C_s^{eq}$ [79, 81]. The statement [75, 76] that in this case the Frank-Turnbull model would predict a $C_s^m \propto \sqrt{t}$ law as found by experiment (Fig. 8) is not correct [79].

7.2.2.2 Semi-infinite solid (thick specimen)

The diffusion of Au from the surface of a thick Si specimen into the bulk may idealized by the diffusion into a semi-infinite solid. Mathematically this problem is described by the differential equation (14)/(15), the boundary condition

$$C_s(x=0,t) = C_s^{eq}, \tag{25}$$

and the initial condition

$$C_s(x>0,\ t=0) = C_s^0. \tag{26}$$ [4]

Seeger [84] has recently presented an exact analytical solution of this problem for $A_I = 0$ (dislocation-free specimen). He further demonstrated that by this solution the diffusion profiles which were found by Wilcox et al. [69, 70] in a study of the penetration of Au into thick, dislocation-free Si specimens (and which differ considerably from the error-function-type shape predicted by the Frank-Turnbull mechanism [81]) may be fitted satisfactorily well. A corresponding numerical fit by Gösele et al. [85] confirmed Seeger's conclusion.

7.3. Information on silicon self-interstitials in thermal equilibrium from studies of the diffusion of gold in silicon

In order to fully exploit the possibilities to obtain information on self-interstitials in Si, investigations of the diffusion of Au in highly perfect Si crystals should be undertaken. From such measurements $C_I^{eq} D_I$ and C_s^{eq} may be deduced separately, as may be realized from an inspection of the effective diffusion coefficient (15) and the boundary conditions (17) and (25). In particular, the possibility to determine $C_I^{eq} D_I$ in this indirect way is challenging, since direct measurements of the tracer self-diffusion coefficient

$$D^T = f_I C_I^{eq} D_I \tag{27}$$

[4]) We take the opportunity to correct an error in ref. [13]: The eq. (24) or (25) in [13] has to be replaced by eq. (25) or (26) of the present paper, respectively.

with the aid of the short-lived radioactive isotope ^{31}Si are tedious and become impossible at temperatures below 1200 K at which the diffusion times required exceed the half-life of ^{31}Si.

In an attempt to test the possibility to determine $C_I^{eq}D_I$ from the penetration of Au into Si Gösele et al. [63, 85] compared $C_I^{eq}D_I$ values obtained in this way to directly measured values of D^T (Fig. 10). Because of the limited accuracy of the data available at present a distinction between D^T and $C_I^{eq}D_I$ was not attempted. This is justified since for the interstitialcy self-diffusion mechanism in Si (Sect. 4) the correlation factor f_I in (27) is expected to lie in the range $0.5 \lesssim f_I < 1$. The full circles approximately following the dashed straight line 6 are obtained from fitting the kick-out model to the Au diffusion profiles measured by Wilcox et al. [69, 70] on thick, dislocation-free specimens (Sect. 7.2.2.2). The crosses arranged along the dashed straight line 7 are determined from a comparison of eq. (24) with the $C_s^m \propto \sqrt{t}$ law measured by Huntley and Willoughby [76] on dislocation-free Si wafers. The C_s^{eq} values required for the determination of $C_I^{eq}D_I$ were taken from [86] in both cases. The reasonable agreement of the $C_I^{eq}D_I$ values with directly measured values of D^T (represented in Fig. 10 by the solid straight lines 1 to 5) is a convincing confirmation of the kick-out model and an encouragement to perform more precise studies of the diffusion of Au in Si with the objective to determine $C_I^{eq}D_I$ and C_s^{eq}.

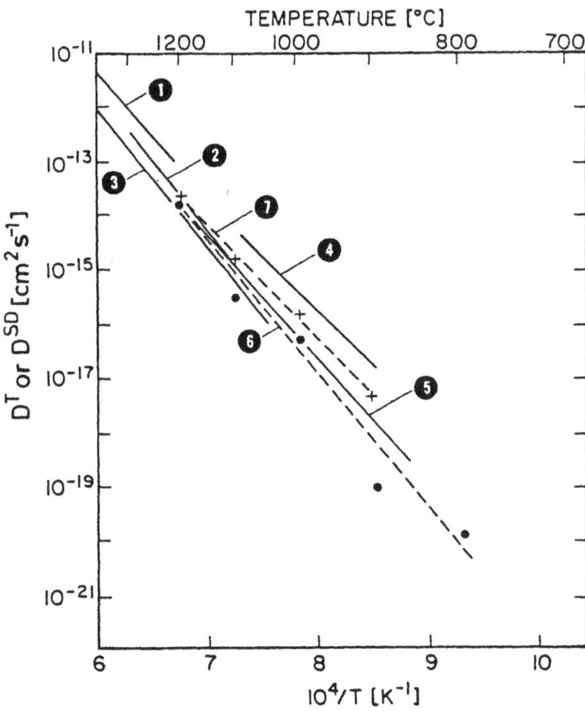

Fig. 10
A comparison of D^T measurements (curves 1 to 5 according to [87], [88], [32], [89], or [90, 91], respectively) with $C_I^{eq}D_I$ values calculated from data on the diffusion of Au in Si (curve 6 from diffusion profiles measured on thick, dislocation-free specimens [69, 70], curve 7 from the $C_s^m \propto \sqrt{t}$ relationship measured on dislocation-free wafers [76]).

Acknowledgement

The author gratefully acknowledges the close cooperation with Dr. U. Gösele and Professor Dr. A. Seeger in the field covered by the present paper.

References

[1] *G. D. Watkins*, in: *Radiation Damage in Semiconductors, P. Baruch* ed. (Dunod, Paris 1964) p. 97.

[2] *G. D. Watkins*, in: *Lattice Defects in Semiconductors 1974, F. A. Huntley* ed. (Institute of Physics, London and Bristol 1975), Inst. Phys. Conf. Ser. No. 23, p. 1.

[3] *K. L. Brower*, Phys. Rev. B **1**, 1908 (1970).

[4] *R. E. McKeighen* and *J. S. Koehler*, Phys. Rev. B **4**, 462 (1971).

[5] *P. S. Gwozdz* and *J. S. Koehler*, Bull. Am. Phys. Soc. **17**, 307 (1972).

[6] *J. Arimura* and *J. W. MacKay*, in: *Radiation Effects in Semiconductors, F. L. Vook* ed. (Plenum Press, New York 1968) p. 204.

[7] *W. D. Hyatt* and *J. S. Koehler*, Phys. Rev. B **4**, 1903 (1971).

[8] *G. D. Watkins*, Phys. Rev. B **12**, 5824 (1975).

[9] *G. D. Watkins, J. R. Troxell*, and *A. P. Chatterjee*, in: *Lattice Defects in Semiconductors 1978, J. H. Albany* ed. (Institute of Physics, Bristol and London 1979), Inst. Phys. Conf. Ser. No. 46, p. 16.

[10] *J. R. Troxell, A. P. Chatterjee, G. D. Watkins*, and *L. C. Kimerling*, Phys. Rev. B **19**, 5336 (1979).

[11] *J. C. Bourgoin* and *J. W. Corbett*, Rad. Effects **36**, 157 (1978).

[12] *W. Frank, U. Gösele*, and *A. Seeger*, in: *Proceedings of the International Conference on Radiation Physics in Semiconductors and Related Materials* (Tbilisi/USSR, September 1979), in the press.

[13] *W. Frank, A. Seeger*, and *U. Gösele*, in: *Defects in Semiconductors, J. Narayan* and *T. Y. Tan* eds. (North-Holland, Amsterdam 1981), in the press.

[14] *L. C. Kimerling*, in: *Defects and Radiation Effects in Semiconductors 1978, J. H. Albany* ed. (Institute of Physics, Bristol and London 1979), Inst. Phys. Conf. Ser. No. 46, p. 56.

[15] *J. D. Weeks, J. C. Tully*, and *L. C. Kimerling*, Phys. Rev. B **12**, 3286 (1975).

[16] *R. E. Whan*, Phys. Rev. A **140**, 690 (1965).

[17] *J. Bourgoin* and *F. Mollot*, phys. stat. sol. (b) **43**, 343 (1971).

[18] *E. I. Blount*, J. Appl. Phys. **30**, 1218 (1959).

[19] *G. D. Watkins, R. P. Messmer, C. Weigel, D. Peak*, and *J. W. Corbett*, Phys. Rev. Lett. **27**, 1573 (1971).

[20] *K. P. Chik*, Rad. Effects **4**, 33 (1970).

[21] *S. I. Tan, B. S. Berry*, and *W. Frank*, in: *Ion Implantation in Semiconductors and Other Materials, B. L. Crowder* ed. (Plenum Press, New York and London 1973) p. 19.

[22] *W. Frank*, Rad. Effects **21**, 119 (1974).

[23] *A. Seeger, H. Föll*, and *W. Frank*, in: *Radiation Effects in Semiconductors 1976, N. B. Urli* and *J. W. Corbett* eds. (Institute of Physics, Bristol and London 1977), Inst. Phys. Conf. Ser. No. 31, p. 12.

[24] *W. Frank*, in: *Lattice Defects in Semiconductors 1974, F. A. Huntley* ed. (Institute of Physics, London and Bristol 1975), Inst. Phys. Conf. Ser. No. 23, p. 23.

[25] *A. Seeger* and *W. Frank*, in: *Radiation Damage and Defects in Semiconductors*, *J. E. Whitehouse* ed. (Institute of Physics, London and Bristol 1972) p. 262.
[26] *H. Mehrer*, J. Nucl. Materials **69/70**, 38 (1978).
[27] *A. Seeger*, J. Phys. F **3**, 248 (1973).
[28] *N. L. Peterson*, J. Nucl. Materials **69/70**, 3 (1978).
[29] *A. Seeger* and *K. P. Chik*, phys. stat. sol. **29**, 455 (1968).
[30] *A. Seeger*, Rad. Effects **9**, 15 (1971).
[31] *A. Seeger*, *W. Frank*, and *U. Gösele*, in: *Defects and Radiation Effects in Semiconductors 1978*, *J. H. Albany* ed. (Institute of Physics, Bristol and London 1979), Inst. Phys. Conf. Ser. No. 46, p. 148.
[32] *H. J. Mayer*, *H. Mehrer*, and *K. Maier*, in: *Radiation Effects in Semiconductors 1976*, *N. B. Urli* and *J. W. Corbett* eds. (Institute of Physics, Bristol and London 1977), Inst. Phys. Conf. Ser. No. 31, p. 186.
[33] *G. Hettich*, *H. Mehrer*, and *K. Maier*, in: *Defects and Radiation Effects in Semiconductors 1978*, *J. H. Albany* ed. (Institute of Physics, Bristol and London 1979), Inst. Phys. Conf. Ser. No. 46, p. 500.
[34] *G. L. McVay* and *A. R. DuCharme*, in: *Lattice Defects in Semiconductors 1974*, *F. A. Huntley* ed. (Institute of Physics, Bristol and London 1975), Inst. Phys. Conf. Ser. No. 23, p. 91.
[35] *D. Shaw*, phys. stat. sol. (b) **72**, 11 (1975).
[36] *J. A. van Vechten* and *C. D. Thurmond*, Phys. Rev. B **14**, 3551 (1976).
[37] *R. B. Fair*, in: *Semiconductor Silicon 1977*, *H. R. Huff* and *E. Sirtl* eds. (Electrochemical Society, Princeton 1977) p. 968.
[38] *S. M. Hu*, J. Appl. Phys. **45**, 1567 (1974).
[39] *S. M. Hu*, J. Vac. Sci. Technol. **14**, 17 (1977).
[40] *E. Sirtl*, in: *Semiconductor Silicon 1977*, *H. R. Huff* and *E. Sirtl* eds. (Electrochemical Society, Princeton 1977) p. 4.
[41] *L. J. Bernewitz* and *K. R. Mayer*, phys. stat. sol. (a) **16**, 579 (1973).
[42] *A. J. R. de Kock*, Phillips Res. Rep., Suppl. **4**, 1 (1973).
[43] *H. Föll* and *B. O. Kolbesen*, Appl. Phys. **8**, 319 (1975).
[44] *P. M. Petroff* and *A. J. R. de Kock*, J. Cryst. Growth **30**, 117 (1975).
[45] *H. Föll*, *B. O. Kolbesen*, and *W. Frank*, phys. stat. sol. (a) **29**, K 83 (1975).
[46] *A. K. Jonscher*, *Principles of Semiconductor Device Operation* (Wiley, New York 1960).
[47] *B. J. Masters* and *E. F. Gorey*, J. Appl. Phys. **49**, 2717 (1978).
[48] *P. Baruch*, *J. Monnier*, *B. Blanchard*, and *C. Castaing*, Appl. Phys. Lett. **26**, 77 (1975)
[49] *P. Baruch*, in: *Radiation Effects in Semiconductors 1976*, *N. B. Urli* and *J. W. Corbett* eds. (Institute of Physics, Bristol and London 1977), Inst. Phys. Conf. Ser. No. 31, p. 126 (1977).
[50] *W. Akutagawa*, *H. L. Dunlop*, *R. Hart*, and *O. J. Marsh*, J. Appl. Phys. **50**, 777 (1979).
[51] *W. G. Allen*, Sol.-State Electr. **16**, 709 (1973).
[52] *D. A. Antoniadis*, *A. G. Gonzales*, and *R. W. Dutton*, J. Electrochem. Soc. **127**, 2243 (1980).
[53] *K. H. Nickolas*, Sol.-State Electr. **9**, 35 (1966).
[54] *G. Masetti*, *S. Solmi*, and *G. Soncini*, Sol.-State Electr. **16**, 1419 (1973).
[55] *G. Masetti*, *S. Solmi*, and *G. Soncini*, Phil. Mag. **33**, 613 (1976).

[56] *R. Francis* and *P. S. Dobson*, J. Appl. Phys. **50**, 280 (1979).
[57] *D. A. Antonidis, A. M. Lin*, and *R. W. Dutton*, Appl. Phys. Lett. **33**, 1030 (1978).
[58] *U. Gösele* and *H. Strunk*, *Appl. Phys.* **20**, 265 (1979).
[59] *S. M. Hu*, in: *Diffusion in Semiconductors, D. Shaw* ed. (Plenum Press, London 1973) p. 217.
[60] *A. F. W. Willoughby*, Rep. Prog. Phys. **41**, 1665 (1978).
[61] *A. F. W. Willoughby*, J. Phys. D **10**, 455 (1977).
[62] *U. Gösele* and *W. Frank*, in: *Defects in Semiconductors, J. Narayan* and *T. Y. Tan* eds. (North-Holland, Amsterdam 1981), in the press.
[63] *U. Gösele, F. Morehead, H. Föll, W. Frank*, and *H. Strunk*, in: *Proceedings of the 4th International Symposium of Silicon Materials Science and Technology* (Minneapolis/USA, May 1981), in the press.
[64] *W. M. Lomer, Harwell Report 1540* (AERE, Harwell 1954).
[65] *J. G. Dienes* and *A. C. Damask*, J. Appl. Phys. **29**, 1713 (1958).
[66] *R. Sizmann*, J. Nuclear Material **69/70**, 386 (1978).
[67] *U. Gösele, W. Frank*, and *A. Seeger*, in: *Defects and Radiation Effects in Semiconductors 1978, J. H. Albany* ed. (Institute of Physics, Bristol and London 1979), Inst. Phys. Conf. Ser. No. 46, p. 538.
[68] *H. Saito* and *M. Hirata*, Japan. J. Appl. Phys. **2**, 678 (1963).
[69] *W. R. Wilcox* and *T. J. LaChapelle*, J. Appl. Phys. **35**, 240 (1964).
[70] *W. R. Wilcox, T. J. LaChapelle*, and *D. H. Forbes*, J. Electrochem. Soc. **111**, 1377 (1964).
[71] *M. J. Hill, M. Lietz, R. Sittig, W. Frank, U. Gösele*, and *A. Seeger*, in: *Proceedings of the International Conference on Radiation Physics in Semiconductors and Related Materials* (Tbilisi/USSR, September 1979), in the press.
[72] *G. J. Sprokel* and *J. M. Fairfield*, J. Electrochem. Soc. **112**, 200 (1965).
[73] *J. L. Lambert*, Wiss. Ber. AEG-Telefunken **45**, 153 (1972).
[74] *W. C. Dash*, J. Appl. Phys. **31**, 2275 (1960).
[75] *F. A. Huntley* and *A. F. W. Willoughby*, Solid-State Electronics **13**, 1231 (1970).
[76] *F. A. Huntley* and *A. F. W. Willoughby*, J. Electrochem. Soc. **120**, 414 (1973).
[77] *F. A. Huntley* and *A. F. W. Willoughby*, Phil. Mag. **28**, 1319 (1973).
[78] *F. C. Frank* and *D. Turnbull*, Phys. Rev. **104**, 617 (1956).
[79] *U. Gösele, W. Frank*, and *A. Seeger*, Appl. Phys. **23**, 361 (1980).
[80] *W. Frank, U. Gösele*, and *A. Seeger*, in: *Defects and Radiation Effects in Semiconductors 1978, J. H. Albany* ed. (Institute of Physics, Bristol and London 1979), Inst. Phys. Conf. Ser. No. 46, p. 514.
[81] *M. D. Sturge*, Proc. Phys. Soc. (Lond.) **73**, 297 (1959).
[82] *A. Seeger* and *W. Frank*, to be published.
[83] *W. Meyberg*, unpublished.
[84] *A. Seeger*, phys. stat. sol. (a) **61**, 521 (1980).
[85] *U. Gösele, F. Morehead, W. Frank*, and *A. Seeger*, Appl. Phys. Lett. **38**, 158 (1981).
[86] *W. M. Bullis*, Solid-State Electronics **9**, 143 (1966).
[87] *R. F. Peart*, phys. stat. sol. **15** , K 119 (1966).
[88] *J. M. Fairfield* and *B. J. Masters*, J. Appl. Phys. **38**, 3148 (1967).
[89] *J. Hirvonen* and *A. Anttila*, Appl. Phys. Lett. **35**, 703 (1979).
[90] *L. Kalinowski* and *R. Seguin*, Appl. Phys. Lett. **35**, 211 (1979).
[91] *L. Kalinowski* and *R. Seguin*, Appl. Phys. Lett. **36**, 171 (1980).

Electronic Structure of Impurities in Transition Metals

Peter H. Dederichs and Rudolf Zeller
Institut für Festkörperforschung, Kernforschungsanlage Jülich, D-5170 Jülich

Summary: We give a review of the recent theoretical developement in our understanding of defects in metals. First we summarize the methods and models used for the theoretical description of point defects. In detail we discuss then the KKR-Green's function method which is suited to describe defects in transition metals as well as in simple metals. Calculations based on this method will be presented for: virtual bound states of 3d-impurities in Cu and Al, magnetic 3d-impurities in Cu, Ag and Al, sp-impurities in Al and Cu and impurities in ferromagnetic Ni. We give results for the local densities of states and for the magnetic moments of the impurities and compare them with the available experimental information.

I Introduction

There is a great need for a theoretical understanding of point defects in metals. For instance, the study of substitutional impurities is of importance for the understanding of the formation and kinetics of alloys, vacancies and interstitials are the basic defects produced by radiation damage in metals, and small interstitials such as H or He in metals are of importance for energy storage requirements, respectively for fast breeder and fission materials. Yet at present we still have no reliable methods to calculate such basic properties. The essential reason for this is that a point defect, when introduced into an ideal crystal, breaks the translational symmetry of the lattice so that band structure methods can no longer be applied.

In the last ten years a dramatic progress has been made in the theoretical understanding of ideal crystals. Due to the developement of density functional theory and linear band structure methods one is now able to reliably calculate ground state properties of metals and ordered alloys [1, 2]. For example, the cohesive energies, lattice constants and bulk moduli of transition metals can be calculated with errors of a few percent only, the magnetic moments of the ferromagnets Fe, Co and Ni are described very well. These calculations do not contain any adjustable parameters. The only input are the nuclear charges and the lattice structure. Admittedly there are some problems as far as excitation energies are concerned, e.g. the d-band of Cu is about 0.5 eV to high [3] or the famous exchange splitting of the d-bands of Ni is only 0.3 eV instead of 0.6–0.8 eV [4, 5]. Nevertheless also excited states are in general astonishingly well described by density functional theory.

Having these successes of density functional theory in mind, it is clear that we should be able to do calculations of similar quality also for inhomogeneous systems such as point defects, surfaces or phonons. Before this, however, can be done methods for inhomogeneous systems have to be developed which are equivalent to the band-structure method for the ideal crystal, naturally a problem of considerably more complexity.

For point defects our task is to calculate the charge and magnetisation densities around the defect, the local densities of states, transport properties, lattice displacements, formation-, binding- and diffusion-energies. A solution of the impurity problem would also be of interest for other fields. For instance the problem of core-holes is directly related to the impurity problem since an ion of valence Z with a core-hole acts chemically as an impurity of valence Z + 1 (without core-hole). The core-hole binding energies are directly related to the formation energies of the Z+1-impurity [6, 7]. Another example are phonons: the harmonic coupling parameters are the forces on the neighbouring atoms due to a unit displacement of a host atom, a special kind of "impurity". The solution of this problem is also necessary for T_c-calculations in superconductivity.

The outline of our paper is as follows: In the next section we review the models used for impurity calculations in the last years and stress the advantages and disadvantages of the different models. We will point out that in our opinion the Green's function method is *the* method for impurity calculations. In section III we outline the general formulas for the KKR-Green's function method which is an appropriate method for defects in metals. It is based on a muffin-tin approximation of the potentials. An important problem is the construction of the ideal lattice Green's function. In the following sections we will discuss applications of this method to special impurity systems. Section IV deals with virtual bound states of 3d-impurities in the noble metals Cu, Ag and Al. Especially we discuss how the virtual bound states of the impurities are influenced by band structure effects. In section V the formation of local moments, i.e. magnetic impurities, in Cu, Ag and Al are considered. Section VI deals with sp-impurities in Cu and Al and finally in section VII we shortly discuss impurities in ferromagnetic Ni.

II Theoretical Models for Impurities

For simple metals most calculations are based on the *jellium model.* For instance to simulate a vacancy, one cuts a spherical hole in an otherwise uniform background of positive charge. For impurities or light interstitials such as H or He one inserts then the corresponding ion potential into the hole. One then calculates the rearrangement of the electronic charge using density functional theory. Due to the spherical symmetry of the problem the calculations are relatively simple; special care is needed to obtain selfconsistency. Such calculations for light interstitials have e.g. be performed by Popovic and Stott [8], Almbladh et al. [9], Zaremba et al. [10], Manninen et al.

[11], Norskov [12], Jena and Singwi [13]. Calculations for vacancies were carried out by Nieminen et al. [14]. These authors obtained reasonable vacancy formation energies for alkali metals, however for polyvalent metals negative values were obtained elucidating the problems and limits of such simple minded models.

The *spherical solid model* of Almbladh and von Barth [15] represents an appreciable improvement of such jellium calculations: The influence of the neighbouring host atoms is described by spherically averaging the ion potentials in the neighbouring shells. Detailed calculations with this model were performed by Manninen and Nieminen [16] for hydrogen and μ^+ in Al, by Nieminen and Puska [17] for 3d impurities in Al (see section IV) and by Kahn, Perrot and Rasolt [18] for hydrogen in Al. The latter authors obtained very encouraging results for the solution and diffusion energy of hydrogen in Al.

Homovalent impurities in simple metals represent weak perturbations and can therefore be treated within *pseudopotential theory*. For heterovalent impurities the perturbation is in general stronger and the application of the usual second order pseudopotential theory is questionable. Higher order contributions should be taken into account. Extensive calculations along this line have e.g. be performed by Solt [19]. For strong scattering centers such as vacancies or H pseudopotential theory is only of rather limited use.

In *cluster calculations* only a finite number of atoms is considered, e.g. the impurity and its nearest neighbours. The multiscattering self-consistent field method of Johnson [20] considers an arrangement of muffin-tin potentials surrounded by a repulsive spherical well potential so that the energy eigenvalues are discrete. Calculations along these lines have been performed by Schwarz and Rösch [21] (C vacancy in NbC), Müller et al. [22] (Fe-Al clusters), Adachi and Imoto [23] (H in transition metals) and Johnson et al. [24] (magnetic impurities in Cu). Harris, Jones and Gunnarsson [25, 26] have further developed this method by using LMTO-trial functions to calculate the ground state properties of diatomic molecules. Special care is taken to take non-muffin-tin contributions into account. When applied to impurity problems the finite size of the cluster is a serious problem. Since in small clusters practically all atoms are at the surface it is very difficult to interpret such calculations. The major advantage of this method is, that contrary to the above models it is not restricted to simple metals.

Due to the development of the recursion method [27] by the Cambridge group the study of very large clusters with thousands of atoms has become possible. These calculations are however restricted to tightbinding hamiltonians with short range interactions. Nevertheless they show that details of the density of states are very sensitive to the size of the cluster.

An interesting variant of cluster calculations is obtained if the outer potential is lowered to the muffin-tin zero thus immersing the *cluster in a free electron sea* (Lloyd and Smith [28]). In this way one can take profit of some of the advantages of the jellium model. Calculations with these model have been performed by Rijsenbrij and Lodder [29], Ries and Winter [30, 31] (nonmetal vacancies in NbC and VN), Lodder and Braspenning [32, 33] (dilute potassium alloys) and Zwicknagel [34] (amorphous metals). Especially Ries and Winter [30] have been able to perform selfconsistent calculations for very large clusters. Yet the density of states of a cluster of 113 Nb atoms is still appreciably different from the one of the bulk. Contrary energy integrated quantities show a much better convergence.

Several attempts have been undertaken to obtain a better *embedding procedure* for the cluster. Two such procedures have been proposed by John and Keller [35] and Lenk [36] and have been studied recently in cluster calculations by van Dijkum [37]. Other attempts of Joannopoulos and Yndurain [38], based on the Bethe approximation, are restricted to tight-binding model hamiltonians.

In *supercell calculations* a certain defect structure is repeated periodically so that a periodic crystal with a rather large unit cell is obtained, which can be handled by standard band structure codes. In this way Gupta and Siegel [39, 40] have performed APW-calculations for a vacancy in Al by using a unit cell of 27 atoms. Calculations for larger unit cells or selfconsistent calculations are limited by the large numerical effort neccessary for such large unit cells. On the other hand the average lattice spacing between the defects is still rather small so that spurious interaction effects cannot be excluded.

These problems are avoided by the *Green's function method* where one really treats a defect in an infinite and otherwise periodic lattice. By introducing the Green's function $\overset{\circ}{G}(\underline{r}, \underline{r}'; E)$ for the ideal periodic lattice, the perturbed Green's function $G(\underline{r}, \underline{r}'; E)$ is determined by Dyson's equation:

$$G(\underline{r}, \underline{r}'; E) = \overset{\circ}{G}(\underline{r}, \underline{r}'; E) + \int d\underline{r}'' \, \overset{\circ}{G}(\underline{r}, \underline{r}''; E) \, \Delta V(\underline{r}'') \, G(\underline{r}'', \underline{r}'; E) \tag{1}$$

where $\Delta V(\underline{r}'')$ is the deviation of the perturbed potential $V(\underline{r}'')$ from the ideal potential $\overset{\circ}{V}(\underline{r}'')$. Thus contrary to cluster calculations or to supercell calculations, only $\Delta V(\underline{r}'')$ or the displaced charge is assumed to be localized, whereas the infinite extent of the wave functions or densities of states is taken into account. Different versions of this method exist depending on the basis used for the expansion of the wave-functions:

a) The method is usually attributed to *Koster and Slater* [41] who used Wannier functions of the host crystal for the expansion. Callaway has put the method into the frame work of scattering theory [42]. However the construction of the Wannier functions proved to be very tedious and difficult [43] thus severely limiting the success of the method.

b) A *hybridized tight-binding-nearly-free-electron* expansion has been proposed by Riedinger [44] and applied by himself and by Cook and Smith [45] to dilute *Cu* Ni, *Cu* Co and *Cu* Fe alloys. Parlebas [46] has studied impurity pairs whereas Sholl and Smith [47] have applied the theory to H in transition metals.

c) In the last three years extensive calculations for defects in semiconductors have been reported by two American groups using an *LCAO-basis.* Baraff and Schlüter [48, 49] and Bernholc et al. [50, 51] have presented selfconsistent calculations for the vacancy in Si. Calculations using *Gaussian orbitals* have been reported by Callaway and Sinhal [52] for the vacancy in Al.

d) A formalism within the framework of the *KKR*-bandstructure *method* has been proposed by Dupree [53], Beeby [54] and Morgan [55]. More recently Holzwarth [56], Lehmann [57], Harris [58] and Hamasaki [59] have worked out the theory. This method seems to be well suited to describe defects in metals. In the last years detailed calculations with this method have been performed by Terakura et al. [60–62]. We have improved this method further by doing selfconsistent calculations within the density functional formalism [63–65] the results of which will be reviewed in this paper. A variant of this method has been recently developed by Léonard and Daniel [66] who used a *LMTO*-expansion set.

The major advantage of the Green's function theory is that it correctly describes the embedding of a disturbed cluster in an ideal lattice. As the KKR-method it is applicable to simple metals as well as to transition metals. The assumption that the potential $\Delta V(\underline{r})$ is localized near the defect is in practice not a severe limitation: Calculations [31, 49, 51] show that the potential or the charge density is indeed strongly localized whereas the wave functions and the densities of states are not. The introduction of the ideal lattice Green's function $\mathring{G}$ also has some practical advantages: e.g. all defects in a given host can be calculated with the same $\mathring{G}$. The selfconsistency problem is divided into two natural steps, first the ideal crystal problem leading to the selfconsistent Green's function $\mathring{G}$ and second the defect problem for which $\mathring{G}$ is unchanged during all selfconsistency cycles.

III The KKR-Green's Function

In this section we give a short summary of the main formulas for the KKR-Green's function. A more detailed description can be found in [56, 57, 59, 65].

We consider a lattice of muffin-tin potentials $v_m(\underline{r}-\underline{R}^m)$. In each cell m, n the Green's function is expanded into eigensolutions of the local muffin-tin potential v_m:

$$\begin{aligned} G(\underline{R}^m+\underline{r}, \underline{R}^{m'}+\underline{r}'; E) &= \\ &= \delta_{mm'} \frac{2m}{\hbar^2} K \sum_L R_l^m(r_<, E)\, Y_L(\underline{r})\, H_l^m(r_>, E)\, Y_L(\underline{r}') + \\ &+ \sum_{LL'} R_l^m(r, E)\, Y_L(\underline{r})\, G_{LL'}^{mm'}(E)\, R_{l'}^{m'}(r', E)\, Y_{L'}(\underline{r}') . \end{aligned} \tag{2}$$

Here $\underline{r}$ and $\underline{r}'$ are restricted to the Wigner-Seitz cell and $r_<$ or $r_>$ are the smaller or larger of $|\underline{r}|$ and $|\underline{r}'|$. Y_L is a real spherical harmonic and $R_l^m(r, E)$ is the regular solution of the radial Schroedinger equation for the potential $v_m(r)$ and the energy E. Further L stands for $L = (l, m)$. $R_l^m(r, E)$ corresponds to the spherical Bessel function, $H_l^m(r, E)$ to the spherical Hankel function. The latter contains both the non-regular solution $N_l^m(r, E)$ and the regular solution $R_l^m(r, E)$

$$H_l^m(r, E) = N_l^m(r, E) - i\, R_l^m(r, E) \,. \tag{3}$$

$R_l^m(r, E)$ and $N_l^m(r, E)$ are real and outside the muffin-tin radius given by combinations of spherical Bessel and Neumann functions $j_l(Kr)$ and $n_l(Kr)$ with $K = \left(\frac{2m}{\hbar^2} E\right)^{1/2}$.

$$\begin{aligned} R_l^m(r, E) &= \cos\delta_l^m(E)\, j_l(Kr) - \sin\delta_l^m(E)\, n_l(Kr) \\ N_l^m(r, E) &= \sin\delta_l^m(E)\, j_l(Kr) + \cos\delta_l^m(E)\, n_l(Kr) \,. \end{aligned} \tag{4}$$

Here $\delta_l^m(E)$ are the phase shifts for the muffin-tin potential in cell m. The first term in Eq. (2) represents the Green's function G_l^m for a single muffin-tin potential $v_m(r)$ embedded in free space. The second term describes the effect of the multiple scattering.

Equation (2) is valid for an arbitrary arrangement of muffin-tin potentials, provided the potentials are non-overlapping and spherically symmetric. The only unknowns are the *structural Green's functions* $G_{LL'}^{mm'}(E)$ which for an infinite lattice with a defect can be related to the structural Green's function $\mathring{G}_{LL'}^{mm'}(E)$ of the ideal crystal by Dyson's equation [65]:

$$\tilde{G}_{LL'}^{mm'}(E) = \tilde{\mathring{G}}_{LL'}^{(m-m')}(E) + \sum_{\substack{n \\ L''}} \tilde{\mathring{G}}_{LL''}^{(m-n)}(E)\, \Delta t_{l''}^{n}(E)\, \tilde{G}_{L''L'}^{n\,m'}(E) \tag{5a}$$

with

$$\tilde{G}_{LL'}^{mm'} = e^{-i\delta_l^m}\, G_{LL'}^{mm'}\, e^{-i\delta_{l'}^{m'}} \,, \qquad \tilde{\mathring{G}}_{LL'}^{mm'} = e^{-i\mathring{\delta}_l}\, \mathring{G}_{LL'}^{mm'}\, e^{-i\mathring{\delta}_{l'}} \tag{5b}$$

The "perturbation" $\Delta t_l^n(E)$ is the difference between the "on-shell" t-matrices in the defect and ideal crystal

$$\Delta t_l^m(E) = t_l^m(E) - \mathring{t}_l(E), \quad t_l^m(E) = -\frac{\hbar^2}{2m}\,\frac{1}{K}\, e^{i\delta_l^m(E)} \sin\delta_l^m(E) \,. \tag{6}$$

On the right hand side of Eq. (5a) only those Green's functions in the perturbed subspace near the defect enter. Therefore Eq. (5a) especially simplifies if only the impurity potential at $n = 0$ is perturbed. Of primary interest is the impurity Green's function $G_{LL'}^{00}(E)$. For cubic symmetry and for $l \leqslant 2$, this Green's function $\tilde{G}_{LL'}^{00}(E) = \tilde{G}_{(L)}(E)\,\delta_{LL'}$ is diagonal and so is the corresponding one for the ideal crystal $(\tilde{\mathring{G}}_{(L)}(E))$ so that the solution of Eq. (5a) trivially becomes

$$\tilde{G}_{(L)} = \frac{1}{1 - \tilde{\mathring{G}}_{(L)}\, \Delta t_l}\, \tilde{\mathring{G}}_{(L)} \tag{7}$$

The major problem of the KKR-Green's function method is the determination of the ideal lattice Green's function $\mathring{G}$, the structural ideal Green's function $\mathring{G}_{LL'}^{mn}(E)$ respectively. Starting with the spectral representation

$$G(\underline{R}^m + \underline{r}, \underline{R}^{m'} + \underline{r}'; E) = \sum_{\nu} \int_{BZ} d\underline{k} \frac{\mathring{\psi}_{\underline{k}\nu}(\underline{R}^m + \underline{r})\, \mathring{\psi}_{\underline{k}\nu}(\underline{R}^{m'} + \underline{r}')}{E + i\epsilon - E_{\underline{k}\nu}} \tag{8}$$

we may insert the usual KKR-ansatz for the wave functions

$$\mathring{\psi}_{\underline{k}\nu}(\underline{R}^m + \underline{r}) = e^{i\underline{k}\underline{R}^m} \sum_{L} i^l \phi_L(\underline{k}\nu)\, \mathring{R}_l(r, E)\, Y_L(\underline{r}) \tag{9}$$

By comparing with the general result (1) for the ideal Green's function $\mathring{G}$ we obtain for the imaginary part of $\mathring{G}_{LL'}^{(m-m')}$

$$\begin{aligned} \operatorname{Im} \mathring{G}_{LL'}^{(m-m')}(E) = {} & \frac{2m}{\hbar^2} K\, \theta(E)\, \delta_{mm'}\, \delta_{LL'} - \\ & - \pi \sum_{\nu} \int_{BZ} d\underline{k}\, \delta(E - E_{\underline{k}\nu})\, \phi_L(\underline{k}\nu)\, \phi_{L'}(\underline{k}\nu)\, i^{l-l'}\, e^{i\underline{k}(\underline{R}^m - \underline{R}^{m'})} \end{aligned} \tag{10}$$

In this form the imaginary part can be directly evaluated by a numerical Brillouin-zone integration. The real part can be obtained from the imaginary one by a Kramers-Kronig integration

$$\alpha_{ll'}(E) \operatorname{Re} \mathring{G}_{LL'}^{(m-m')}(E) = \frac{1}{\pi} \int_{-\infty}^{+\infty} dE' P \frac{1}{E - E'}\, \alpha_{ll'}(E') \operatorname{Im} \mathring{G}_{LL'}^{(m-m')}(E') \tag{11}$$

with

$$\alpha_{ll'}(E) = \int_0^{R_{mt}} r^2 dr\, \mathring{R}_l(r, E)\, \mathring{R}_{l'}(r, E)$$

where for convenience we have integrated out the radial dependence.

In actual calculations the E'-integration in (11) requires the knowledge of the band structure up to infinite energies which is impossible to obtain. For the diagonal part $m = m'$ one can choose a cut-off energy E_c [63]. By comparing with Eq. (1) this means that above E_c one replaces $\mathring{G}(\underline{r} + \underline{R}^m, \underline{r}' + \underline{R}^m; E')$ by the solutions $\mathring{G}_1(\underline{r} + \underline{R}^m, \underline{r}' + \underline{R}^m; E')$ for a single muffin-tin potential. For the non-diagonal terms $m \neq m'$ a simple cut-off in Eq. (11) is not good enough. Instead we replace $\mathring{G}(\underline{r} + \underline{R}^m, \underline{r}' + \underline{R}^{m'}; E')$ for $E' > E_c$ by the solution $\mathring{G}_2(\underline{r} + \underline{R}^m, \underline{r}' + \underline{R}^{m'}; E')$ for two muffin-tin potentials in free space, located at $\underline{R}^m$ and $\underline{R}^{m'}$. Then we obtain from Eq. (11) for $m \neq m'$

$$\alpha_{ll'}(E)\,\mathrm{Re}\,\{\mathring{G}_{LL'}^{(m-m')}(E) - \mathring{G}_{2\,LL'}^{(m-m')}(E)\} = \tag{12}$$

$$= \frac{1}{\pi} \int_{-\infty}^{E_c} dE' P \frac{1}{E-E'}\, \alpha_{ll'}(E')\,\mathrm{Im}\,\{\mathring{G}_{LL'}^{(m-m')}(E') - \mathring{G}_{2\,LL'}^{(m-m')}(E')\}$$

since also $\mathring{G}_2$ obeys a Kramers-Kronig relation.

The imaginary part of the Green's function gives the distribution $n(\underline{r}, E)$ of electrons in space and energy

$$n(\underline{r}, E) = -\frac{2}{\pi}\,\mathrm{Im}\,G(\underline{r}, \underline{r}; E) \tag{13}$$

From this the charge density is calculated by energy integration

$$n(\underline{r}) = \int_{-\infty}^{E_F} dE \frac{-2}{\pi}\,\mathrm{Im}\,G(\underline{r}, \underline{r}; E) \tag{14}$$

whereas the local density of states, e.g. in the impurity cell V_I, follows as

$$n_{loc}(E) = \int_{V_I} d\underline{r} \frac{-2}{\pi}\,\mathrm{Im}\,G(\underline{r}, \underline{r}; E)\,. \tag{15}$$

The total charge in the impurity cell is then

$$N_{loc} = \int_{-\infty}^{E_F} dE\, n_{loc}(E) = \int_{V_I} d\underline{r}\, n(\underline{r})\,. \tag{16}$$

Using Lloyd's formula [67] one can also calculate the change of the total density of states, where also effects due to the polarization of neighbouring host atoms are included. For this we refer to the above articles. Here we want to mention only that Friedel's sum rule, $\Delta N(E_F) = \Delta Z$, connecting the change of the integrated density of states at the Fermi energy E_F with the changed nuclear charge of the impurity, is not exactly satisfied in our calculations, since by e.g. allowing only a single perturbed potential we limit the screening to the impurity muffin-tin sphere. Thus the deviation of $\Delta N(E_F)$ from ΔZ is a measure for the accuracy of our calculations.
In the case of spin polarization we have two different potentials $v_{\pm}^m(r)$ for up spin (v_+) and down spin (v_-). Consequently all Green's functions, phase shifts, etc. obtain an additional index + or − for the spin, e.g. $\delta_{\pm l}^m(E)$, $G_{\pm\,LL'}^{mm'}(E)$, and all

previous equations appear two-fold, one for each spin direction. The local density of states for spin up (+) and down (–) are given by

$$n_{loc}^{\pm}(E) = \int_{V_I} d\underline{r} \, \frac{-1}{\pi} \, \mathrm{Im}\, G_{\pm}(\underline{r}, \underline{r}; E) \tag{17}$$

from which the local moment follows as

$$M_{loc} = \int_{-\infty}^{E_F} dE \, (n_{loc}^{+}(E) - n_{loc}^{-}(E)) = \int_{V_I} d\underline{r} \, (n^{+}(\underline{r}) - n^{-}(\underline{r})) \tag{18}$$

For our calculations we have applied the density functional formalism [68]. For a recent review we refer to the paper of Gunnarsson [95]. Exchange and correlations are treated in the local spin density approximation proposed by von Barth and Hedin [69] with a slight modification of the constants as used by Moruzzi et al. [2]. The host band structures of Cu and Ag were generated from the selfconsistent potentials given by Moruzzi et al. [2]. The ideal Green's functions are then calculated by a Brillouin zone integration. These two time consuming steps of the calculations have to be performed only once for each host metal. This represents a major advantage of a Green's function method since the subsequent calculations for the defect Green's function (Eq. (4) and (1)) and the impurity charge densities (Eq. (13)) are rather fast. Only these fast steps have to be repeated for each impurity, and that many times since the selfconsistent solution for the impurity potential is obtained as usual by an iteration technique. For some problems connected with the iteration process we refer to [65].

IV Virtual Bound States of 3d-Impurities in Cu and Al

By introducing a 3d-impurity into a free electron sea, the localized 3d level is, due to the interaction with the conduction electrons, broadened into a resonance, called virtual bound state after Friedel [70]. For the resonance form one expects a Lorentzian containing 5 electrons for each spin direction.

$$n_{loc}(E) = \frac{5}{\pi} \frac{W}{(E - E_d)^2 + W^2} \quad \text{with} \quad W = \frac{2\pi}{\hbar} |V_{sd}|^2 \, n_{host}(E_d) \tag{19}$$

The halfwidths W is determined by the golden rule and is proportional to the $s-d$ matrix element $|V_{sd}|^2$ and the host density of states at E_d. In this section we will discuss the validity of this model and how it is modified by bandstructure effects.

3d-impurities in Al: Al is usually considered as a nearly-free-electron metal. Indeed the density of states of Al (Fig. 1a) shows the familiar $\sqrt{E}$ – behaviour except for some anomalies caused by band-gaps. Thus it is tempting to describe impurities in Al by an impurity-in-jellium model. Such calculations have been performed by Mrosan et al. [71–73] using a Mattheiss-prescription for the impurity potential.

Recently Nieminen and Puska [74] have performed selfconsistent calculations within density functional theory for 3d-impurities in Al, applying the spherical solid model of Almbladh and von Barth [75]. Fig. 1b shows the resulting local

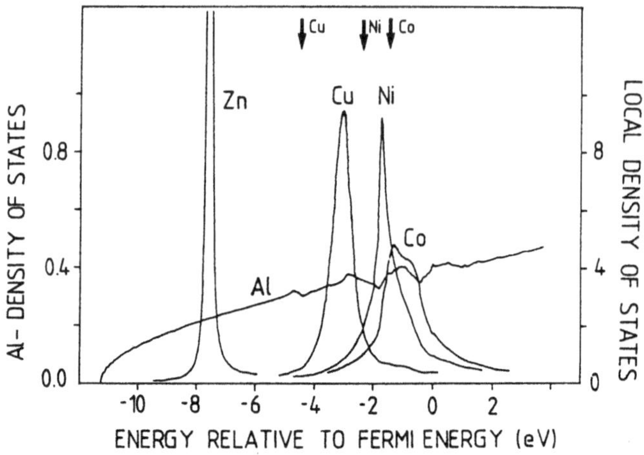

Fig. 1a Local densities of states for Zn, Cu, Ni and Co impurities in Al (right scale) together with the host density of states of Al (left scale). The arrows indicate the position of the virtual bound state peaks obtained by XPS-measurements of Steiner et al. [77].

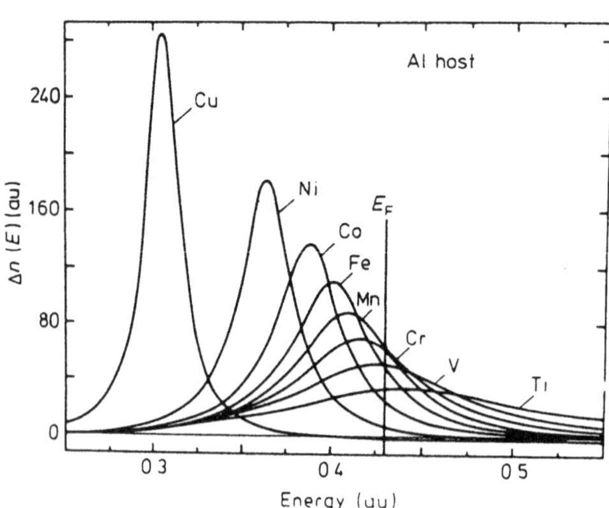

Fig. 1b Local densities of states for 3d-impurities in Al as obtained by the jellium calculations of Nieminen and Puska [74].

densities of states of the 3d-impurities representing Lorentzians which are very broad at the beginning and very narrow at the end of the 3d-series.

Contrary the densities of states calculated with the KKR-Green's function method (Fig. 1a) can be quite different [76]. Whereas the virtual bound states for Zn and Cu are Lorentzian, for Ni we obtain an extremely sharp peak centered at the dip at -1.8 eV in the Al density of states, whereas for Co one gets a double hump structure, centered at about -1 eV. These anomalies are due to critical points in the host band structure and are discussed in detail in [76]. Whereas in jellium calculations essentially only the first term in Eq. (2) is retained describing the scattering at a single muffin-tin potential, i.e. the one of the impurity, the second term in (2) describing the band structure effects of the host can be understood as the effects of "back scattering": The wave evanescing from the impurity is backscattered from the host atoms. Near critical points this effect is especially important due to Bragg reflections at the corresponding lattice planes. E.g. in the case of Ni the critical point at -1.8 eV corresponds to Bragg reflections at 100-planes. The back scattered wave fields then constructively or destructively interfere with the single-muffin-tin solution leading either to a very sharp peak or to a minimum in the local density of states.

The arrows in Fig. 1a indicate the positions of the virtual bound state peaks as obtained by recent XPS-studies of Steiner et al. [77]. Our values are consistently too high which is a typical error of the density functional theory. This will be discussed in more detail in section 6. Part of this discrepancy could however also arise from lattice relaxations, which have been neglected in our calculations.

Virtual bound states in Cu: Fig. 2 shows the density of states of Cu and the local density of states for Ni and Co in Cu (Fe, Mn, Cr and V are magnetic in Cu, see section 5). The virtual bound states of Ni and Co are centered at -1.1 eV and -0.7 eV, which is in good agreement with the observed experimental positions [78]. The curves show an appreciable asymmetry which is induced by the near Cu-d-band. The dotted line gives the integrated local density of states within the impurity cell. One sees that about 3.1 electrons for Ni and 2.5 electrons for Co are broadly distributed over the energy range of the Cu-d-band. The line shape of the virtual bound states has been discussed also by Dagens [79].

Density of States of Nearest Neighbours: We have recently also calculated perturbed charges and densities of states at the neighbouring sites of the impurity. Fig. 3 shows the change $\Delta n^{(110)}(E)$ of the local density of states on a neighbouring Cu site adjacent to a Ni impurity. At about 1 eV one observes a small Lorentzian peak due to the Ni virtual bound state. Below -1.6 eV a strong negativ peak at the onset of the Cu-d-band is seen which can be interpreted as a repulsion between the d-levels of the neighbouring Cu atoms and the d-level of Ni. Within the d-band $\Delta n(E)$ oscillates several times and even more oscillations occur for more distant neighbours. If one would integrate the absolute value of $\Delta n(E)$ up to E_F, one would obtain about

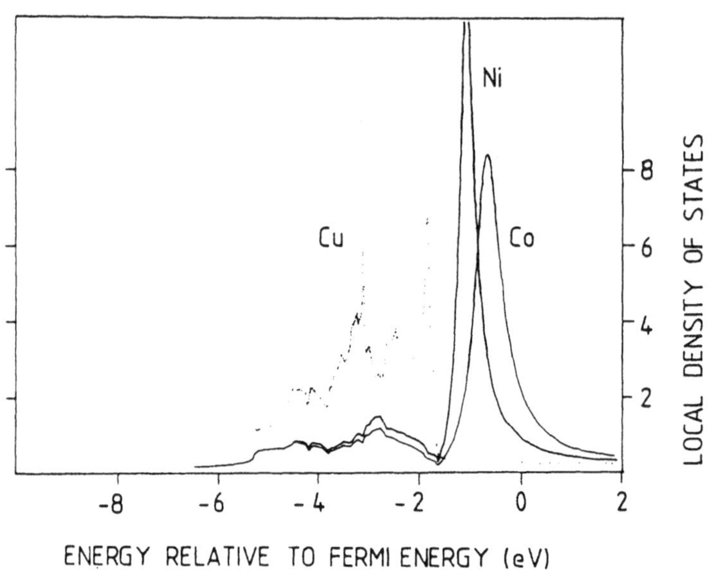

Fig. 2 Local densities of states for Ni and Co impurities in Cu together with the host density of states of Cu (dotted line).

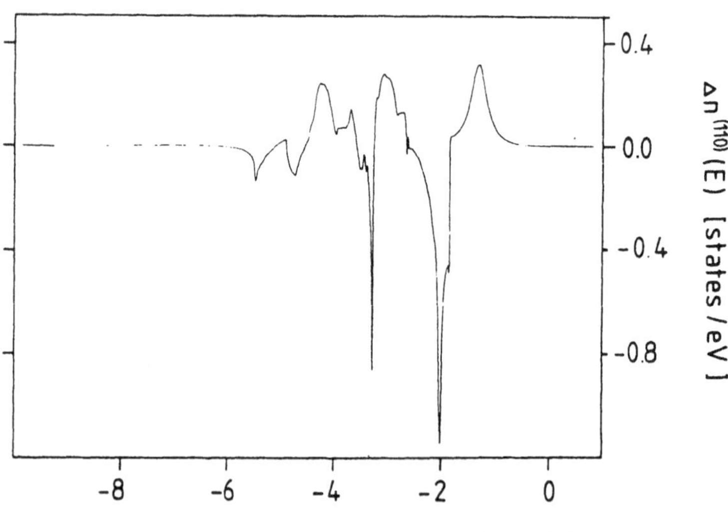

Fig. 3 Change $\Delta n^{(110)}$ (E) of the local density of states of a Cu-atom on a nearest neighbour site (110) adjacent to a Ni impurity.

0.8 electrons on each neighbouring site or about 10 electrons in the first shell. However due to the oscillations there is a nearly perfect cancellation between the positive and negative contributions so that only a very small surplus of 0.011 electrons per neighbouring Cu atom or of 0.13 electrons in the first shell remains.

These observations are in line with the results of other authors [31, 49, 51]: Whereas the perturbed densities of states and wave functions are rather extended, energy integrated quantities like the perturbed charges or the perturbed potentials are strongly localized due to this cancellation effect. This is the reason why the Green's function method and especially our model with only one perturbed potential work so well. The practically 100% cancellation is a speciality of the chosen *Cu* Ni system which does not show an appreciable charge transfer. For other systems the cancellations will be not so perfect, leading to small but important changes of the potentials on the neighbouring sites which have to be taken into account for a reliable calculation of the charge transfer.

V Magnetic Impurities in Cu, Ag and Al

By spin alignment the 3d-electrons can gain exchange energy, resulting in a localized moment of the impurities. As has been first shown by Anderson [80] the strength of the "s-d interaction" between the 3d-electrons of the impurity and the sp-electrons of the host is decisive for the formation of a local moment and a well-defined criterion for the formation can be derived. Here we will give a somewhat simpler derivation in analogy to the Stoner model for ideal crystals. We assume that due to the spin alignment the local density of states $n^{\pm}_{loc}(E)$ for the majority and minority spins are obtained by rigidly shifting the paramagnetic density $n_{loc}(E)$ by $\mp IM_{loc}/2$ (see Fig. 4a, b) where I is the exchange integral and M_{loc} the local moment of the impurity. Since all states up to the Fermi energy E_F are occupied, we get the following selfconsistency condition for the local moment M_{loc}:

$$
\begin{aligned}
M_{loc} &= \int_{-\infty}^{E_F} dE\,(n^{+}_{loc}(E) - n^{-}_{loc}(E)) \\
&\cong \int_{-\infty}^{E_F} dE\,\{n_{loc}(E + IM_{loc}/2) - n_{loc}(E - IM_{loc}/2)\} \equiv F(M_{loc}) \qquad (20)
\end{aligned}
$$

The graphical solution of the condition $M_{loc} = F(M_{loc})$ is illustrated in Fig. 4c. $F(M_{loc})$ is a monotonically increasing function which for $M_{loc} \to \infty$ approaches the saturation magnetization M_s given by $\int_{-\infty}^{\infty} dE\, n_{loc}(E)$. The intersection of

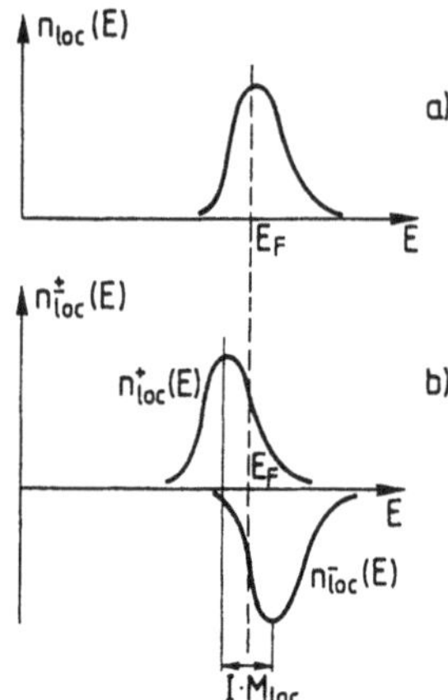

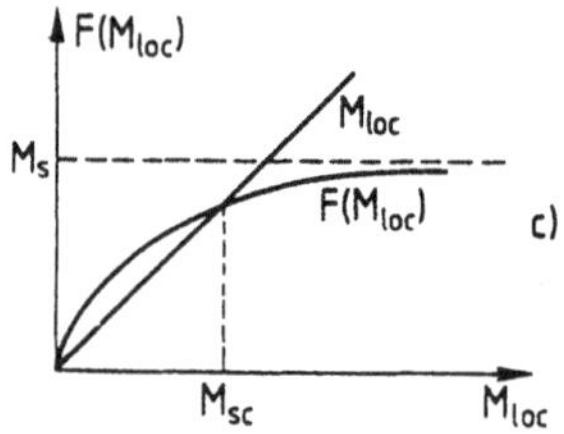

Fig. 4

Stoner model for magnetic impurities

a) local density of state for a paramagnetic impurity

b) splitting of the local density of state for up and down spins

c) graphical solution of the selfconsistency condition for the local moment

$F(M_{loc})$ with the straight line M_{loc} gives the selfconsistent value M_{sc} for the local moment. For symmetry reasons also $-M_{sc}$ is an allowed solution. Such a non-zero solution $\pm M_{sc}$ will exist, if the slope of $F(M_{loc})$ at $M_{loc} = 0$ is larger than one

$$F'(0) = I\, n_{loc}(E_F) > 1 \ . \tag{21}$$

This Stoner criterium for local moments is essentially the same condition as in an ideal crystal, only the density of states of the ideal crystal has been replaced by the local density of states of the impurity. For a jellium model $n_{loc}(E)$ is given by Eq. (19). Thus favourable for magnetism are impurities in the middle of the 3d-series with $E_d \cong E_F$ and with a small width of the virtual bound states, i.e. $|V_{sd}|^2$ and $n_{host}(E_F)$ small. Of course also a large exchange integral is favourable. Due to the dependence on $n_{host}(E)$ one expects local moments to be much stronger in noble metals than in Al.

Contrary to the approximation (20) our selfconsistent calculations [64, 65, 76] do not rely on the assumption of a rigid shifting of the paramagnetic density of states. Furthermore there are no free parameters like the exchange integral. Instead of a constant local moment the full space dependence of the magnetization is taken into account.

The calculated local densities of states $n^{\pm}_{loc}(E)$ for a Mn-impurity in Ag, Cu and Al are plotted in Fig. 5. In all three cases one observes a strong spin splitting, of the order of 3 eV in Ag and 2 eV in Cu and Al. The virtual bound state peaks are narrowest in Ag and broadest in Al. For Al and Cu the majority density of states shows strong backscattering effects. For Mn in Al the majority peak has a van Hove singularity at -1.8 eV due to the backscattering at the 100-planes of Al. The same feature we have already observed for Ni in Al (Fig. 1a). For Mn in Cu the majority virtual bound state is pushed towards the upper edge of the Cu-d-band. As a result an appreciable intensity (2.5 electrons) is distributed over the Cu d-band, ranging from -1.6 eV to -5.3 eV below E_F while only about two electrons are contained in the very narrow majority peak. The majority peak for Mn in Ag is much more like a Lorentzian, since the Ag d-band is much lower in energy.

Experimentally the observation of virtual bound state peaks of magnetic impurities is somewhat controversial. Early photoemission measurements of Norris and Wallden [81] and Wallden [82] at dilute *Ag* Mn show a broad hump in the photoemitted intensity between -1.8 eV and -3.2 eV. On the contrary, for *Cu* Mn no indication of a virtual bound state peak was found, which could be due to the strong overlap of the Mn density of state with the Cu-d-band. Optical measurements lead to similar findings [83, 84]. However in recent XPS studies of Höchst, Steiner and Hüfner [85] no welldefined virtual bound states could be detected for the systems studied (*Ag* Mn, *Au* Fe and *Cu* Fe). Contrary Rechal and Andrews [86] and Andrews and Brown [87] found welldefined virtual bound states for Fe and Cr in Au and for Mn in Ag. In the latter case the majority peak is identified at -2.8 eV which compares favourably with our value of -2.4 eV and also with the older measurements [81, 82].

The calculated local moments M_{loc} are plotted in Fig. 6 for 3d-impurities in Ag, Cu and Al. One sees that the tendency for magnetism is strongest in Ag and weakest in Al. Whereas in Ag only Ni and Sc are nonmagnetic, in Cu in addition Ti and Co become non-magnetic and in Al also V. In agreement with the Stoner model the spin splitting $\Delta E_d = I \cdot M_{loc}$ is proportional to the local moment and the exchange integral I is about 0.7–0.8 eV in Ag and Cu and 0.9 eV in Al. The triangles give the available experimental values for the moments according to different authors. In view of the apparent scattering of these values for the same system the agreement is satisfactory.

Recent calculations [88] show that all 4d-impurities in Ag and Cu are non-magnetic: The virtual bound states are so broad that the Stoner condition (21) cannot be satisfied.

Fig. 7 shows the residual resistivity for the 3d-impurities in Ag and Cu. The resistivity has been calculated by the approximation of Gupta and Benedek [89]. The spin-flip scattering has been neglected. The triangles denote the experimental values. The overall agreement is reasonable.

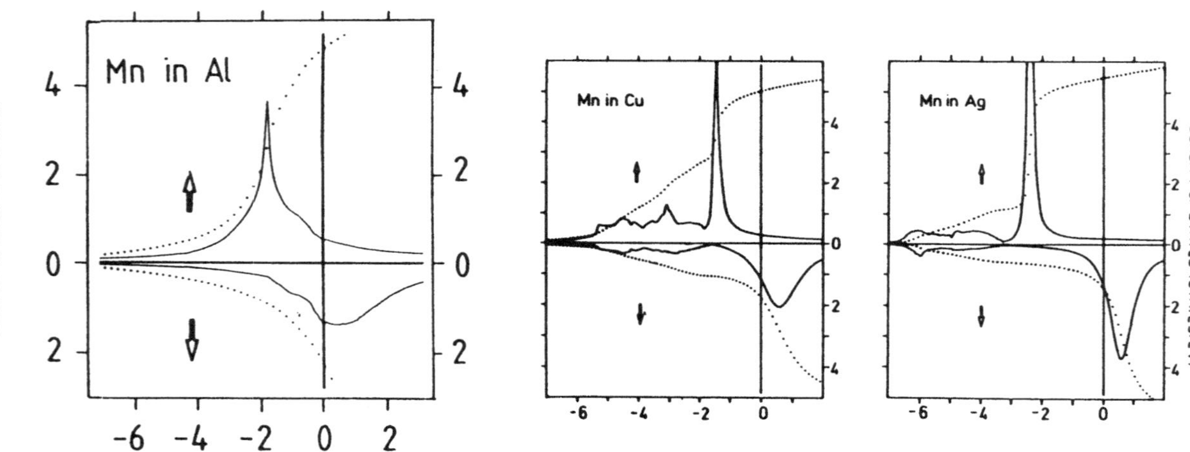

Fig. 5 Local and integrated local density of states for magnetic Mn impurities in Al, Cu and Ag. The local density (full line) refers to the right scale, the integrated local density of states (dotted line) to the left scale. The difference between the integrated local density of states at the Fermi energy gives the local moment of the impurities.

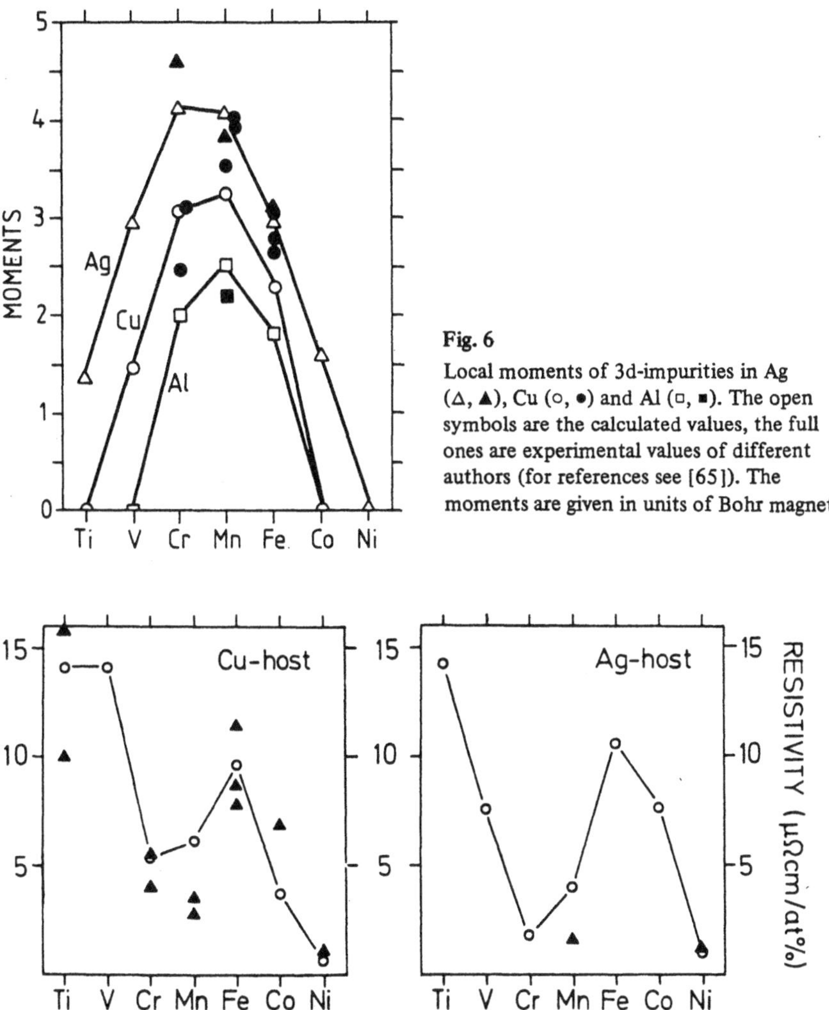

Fig. 6
Local moments of 3d-impurities in Ag (△, ▲), Cu (○, ●) and Al (□, ■). The open symbols are the calculated values, the full ones are experimental values of different authors (for references see [65]). The moments are given in units of Bohr magnetons.

Fig. 7 Residual resistivity for 3d-impurities in Cu and Ag calculated according to the formula of Gupta and Benedek [89]. The triangles are the experimental values of different authors (see [65]).

VI s-p Impurities in Cu and Al

Fig. 8 shows the calculated local densities of states for the impurities Si, P together with the density of states of the *Al* host and the corresponding densities of states for the impurities Ga, Ge and As of the third series of the periodic table [76]. One

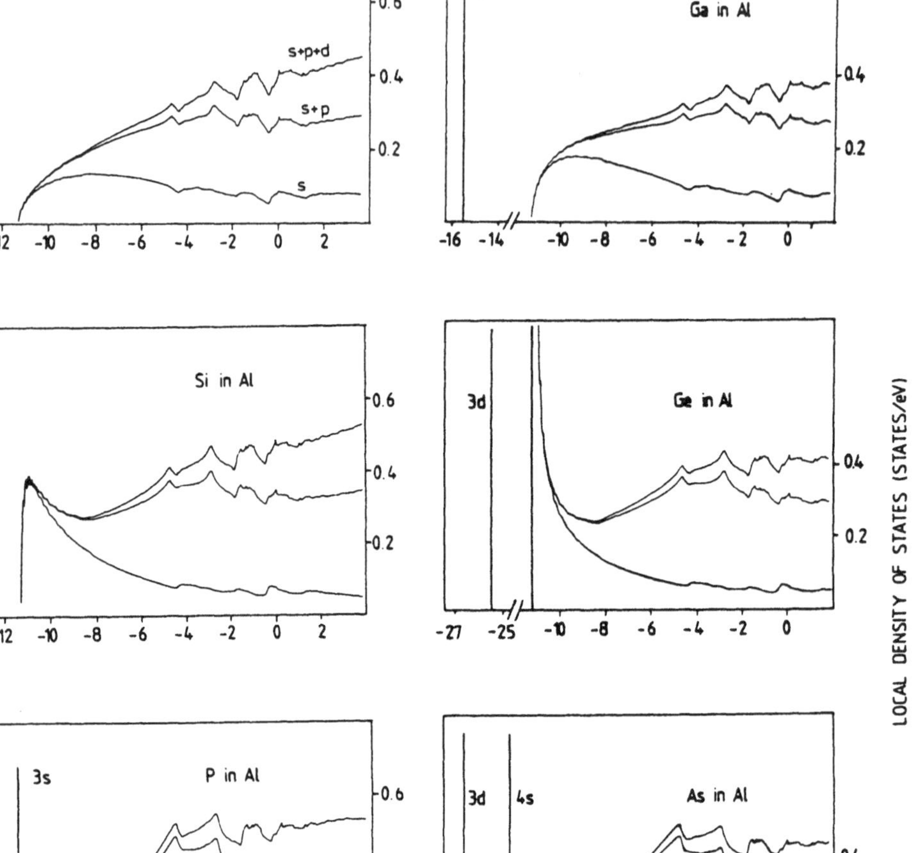

Fig. 8 Host density of Al and local density of states for the s-p impurities S, P, Ga, Ge and As in Al. Within the band the lowest curve gives the s-density of state, the second curve the sum of the s- and p-density of states and the highest curve the total density of states as the sum of the s-, p- and d-contributions.

immediately sees that the chemically equivalent impurities have similar densities of states. For instance the local density of states of As is similar to the one of P, except for the localized 3d-level at −27.2 eV. In the same way Ge shows similar features as Si, and the local density of states of Ga resembles very much the host density of states of Al. The behaviour in one and the same series, e.g. from Ga to As, is characterized by a strong shift of the s-densities of states to lower energies. As and P have a localized 4s-state, Ge and Si are on the verge of getting one. Contrary the p-density of states is filled up, but is unshifted in energy so that the chemistry is mostly determined by the p electrons.

Fig. 9 shows the local density of states for Zn, Ga and Ge in Cu [64]. Some features are similar to the corresponding findings for the Al host: In the series Zn-Ga-Ge the 3d-level moves down in steps of about 10 eV, the s-density of states moves to lower energies leading to a localized 4s-level for Ga already and the p-density of states is filled up. However different to Al is the strong structure observed within the Cu-d-band between −1.6 eV and −5.3 eV. This arises due to a local hybridization of the impurity s and p electrons with the d electrons of the neighbouring Cu-atoms. The local s or p densities of states have minima at −4 eV or at −2 eV respectively. The states below the minimum can be interpreted as bonding, the ones above as anti-bonding orbitals. This effect, already discussed by Terakura [90], should be very important for the formation energies of these impurities. A very simple explanation of the hybridisation minimum has been recently given by Kanamori et al. [91].

The calculated 3d-peaks for Zn, Ga and Ge in Cu are consistently too high: For Zn the virtual bound state appears at −8.2 eV, about 1.5 eV higher than the experimental value found by Norris and Williams [92], for Ga the 3d-peak, calculated at − 16.6 eV, is about 2.0 eV too high [92], for Ge the peak at −26.6 eV is 2.6 eV higher than the experimental value [93]. We have seen a similar discrepancy in Fig. 1 for the virtual bound state of Cu. While a presumably smaller part of this discrepancy could also arise from our approximation, e.g. the neglect of static displacements, such discrepancies also occur in calculations for atoms and ideal crystals and represent a typical error of density functional theory in the local density approximation [95]. For example, the calculated d-band of Cu is about 0.5 eV too high [2], the one for Zn is 1.5 eV too high [94]. The d-level of a Zn atom is even 6 eV too high [95]. Thus the errors are the larger the stronger the localization.

Physically the error made in a ground state calculation is that in the excited final state the relaxation of the electrons due to the hole in the 3d-level is not taken into account. For a localized level this can in principle be described by the transition state concept of Slater [96]. Mårtenson and Johannson [6] and Flynn [7] have recently found an interesting approach to bypass such a calculation. In a good approximation a hole in an inner core shell acts for the valence electrons, as if the nuclear charge would have been increased from Z to Z + 1. For instance, Flynn argues that the energy of the excited state of an Zn-impurity in Cu with a 3d hole is given by the corresponding energy of a free Zn-atom with a 3d hole minus the vapourization

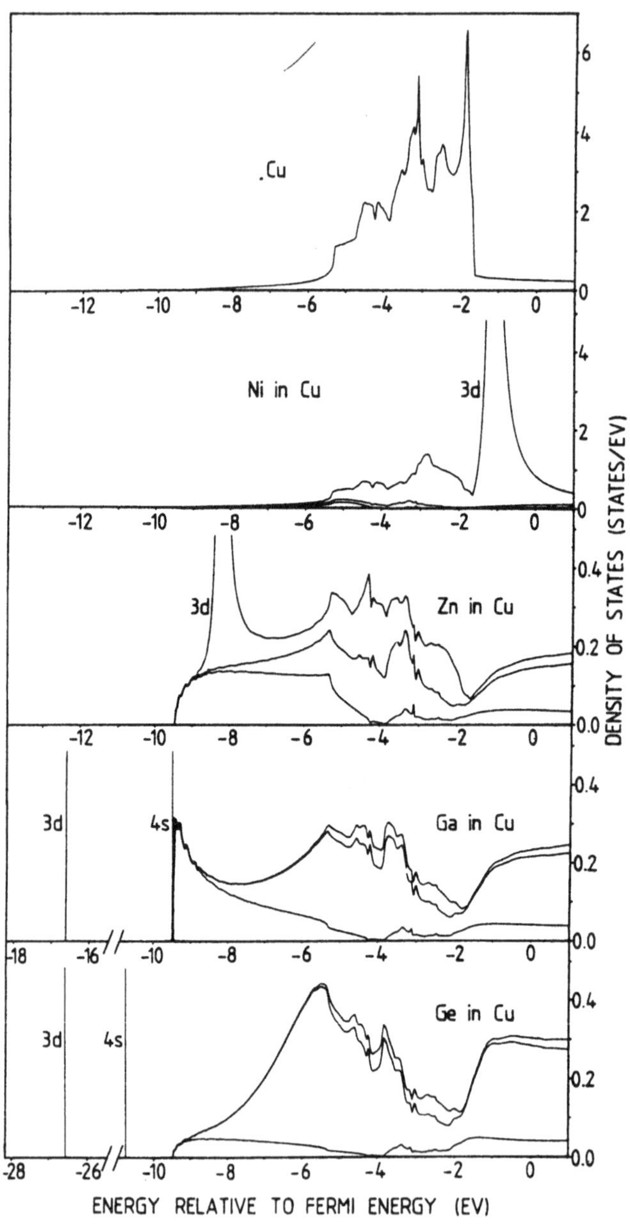

Fig. 9 Local densities of states for Ni, Zn, Ga and Ge in Cu together with the density of states of Cu. Within the band the three curves give the s, s + p and s + p + d contributions.

energy E^V_{Ga} of Ga in Cu, since the excited atom Zn* when brought into the solid, acts like a Ga impurity. Thus the excitation energy for a 3d electron of Zn in Cu is

$$\Delta E^{solid}_{Zn} = \Delta E^{atom}_{Zn} - E^V_{Ga} + E^V_{Zn}$$

since the ground state of Zn in Cu is lowered, compared to the atomic ground state, by $-E^V_{Zn}$. By using the experimental values for the atomic excitation energies and the vapourization energies Flynn could reproduce the experimental excitation energies for the 3d core holes of Zn, Ga and Ge in Cu to within 0.1–0.2 eV. This shows that the (Z + 1)-approximation for the core-hole is quite accurate and that the binding energies of core-levels are directly related to the impurity formation energies. Steiner et al. [103] and Steiner and Hüfner [104] have recently demonstrated for a large number of alloys that the heat of solution can be directly determined from the measured core level shifts.

VII Impurities in Ferromagnets

3d-Impurities in Ni

Whereas in non-magnetic hosts the direction of the local impurity moment is irrelevant, this degeneracy is removed in a ferromagnet: the local moment will either couple ferromagnetically (Co, Fe, Mn in Ni) or antiferromagnetically (Cr, V, Ti in Ni) to the host moment. This additional complication makes the problem especially interesting. Friedel [70] has given an explanation for the antiferromagnetic moments of Cr, V and Ti. Detailed model calculations have been performed by Kanamori [97] and Campbell and Gomez [98]. Recently one of us has performed ab-initio calculations on these systems using the KKR-Green's function method [99]. Here we will only discuss the results for the calculated moments. The resulting local densities of states can be found in ref. [99].

Fig. 10 shows the calculated local moment M_{local} in the impurity cell and the change ΔM of the total moment which includes the perturbation of the neighbouring Ni-moments. One sees that for Co, Fe and Mn the agreement with the experimental values for M_{loc} and ΔM is good. For Cu the local moment is practically zero, whereas the perturbed Ni-neighbours lead to a total change of $\Delta M = -1.59\ \mu_B$ (experiment: $-1.14\ \mu_B$). However for Cr, V and Ti serious problems arise. The calculated local moment for Cr is far too large (exp.: $-0.2\ \mu_B$). More important, the total change ΔM is too small by a factor of two: The experimental ΔM for Cr or V of about $-5\ \mu_B$ indicates a strong perturbation of the neighbouring Ni-moments which is not described correctly by the single site model. Thus perturbations at neighbouring sites have to be taken into account for a more realistic calculation. This shows the limits of our single site model which is only applicable if the perturbation of the neighbouring host atoms is small.

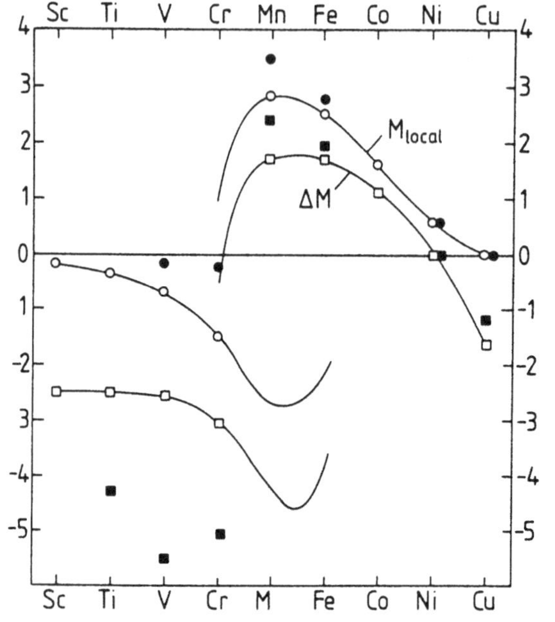

Fig. 10
Local moments M_{local} and changes ΔM of the total moments due to 3d-impurities in ferromagnetic Ni. The open symbols are the calculated values [99], the full ones are experimental values.

The full curves in Fig. 10 have been calculated for hypothetical nuclei with non-integral charges Z. In the range $23.9 \leqslant Z \leqslant 26.2$ one ferromagnetic and one antiferromagnetic solution has been found. Only a total energy calculation could distinguish unambiguously the stable solution from the metastable one. Clearly the ferromagnetic solution for Cr (Z = 24) and the antiferromagnetic solution for Fe (Z = 26) are nearly instable. For Mn we believe the ferromagnetic solution to be more stable from a careful study of the convergence behaviour of the iteration process.

sp-Impurities in Ni and Fe

Very detailed calculations have been performed in the last years by Kanamori, Terakura et al. [61, 62, 100, 101] for sp-impurities in ferromagnetic Ni and Fe. These authors obtained especially nice results for the hyperfine field H_{hf} and the spin lattice relaxation time T_1 of the impurity nuclei. The calculations are also based on the KKR-Green's function method with a single perturbed muffin-tin potential for the impurity which has been determined by the Mattheiss prescription from atomic charge densities (For sp-impurities selfconsistency does not change the potential drastically). Here we will give a short summary of their results. For more details e.g. densities of states etc. we refer to the original articles.

The dominant contribution to the hyperfine field H_{hf} and the nuclear spin-lattice relaxation time T_1 arises from the contact interaction with the conduction electrons. T_1 and H_{hf} are approximately given by [102]

$$(T_1 T)^{-1} = \frac{64\,\pi^3 k}{9\,\hbar} (\gamma_e \gamma_N \hbar^2)^2 \, n_s^+(0, E_F)\, n_s^-(0, E_F)$$

$$H_{hf} = \frac{8\pi}{3} \mu_B \int^{E_F} dE\,(n_s^+(0, E) - n_s^-(0, E))$$

Here γ_e and γ_N are the gyromagnetic ratios of the electron and the nucleus, k is the Boltzmann constant, $n_s^\pm(\underline{r} = 0, E)$ the density of s electrons with energy E at the impurity nucleus. The latter quantity can be directly calculated from the impurity Green's function and is essentially determined by the s-component of the local density of states.

Fig. 11a, b shows the results of these authors for the hyperfine fields of third row elements in Ni and of fourth row elements in Fe. The agreements with the experimental data is excellent. The authors give simple qualitative arguments to explain the main trends of these results and why the results are to an extent insensitive to the atomic potential. However a discussion of this would go beyond the scope of this article.

VIII Summary and Outlook

In the last years good progress has been made in the development of theoretical methods for the electronic structure of point defects. Especially the development of the Green's function method by several groups means a breakthrough for defect

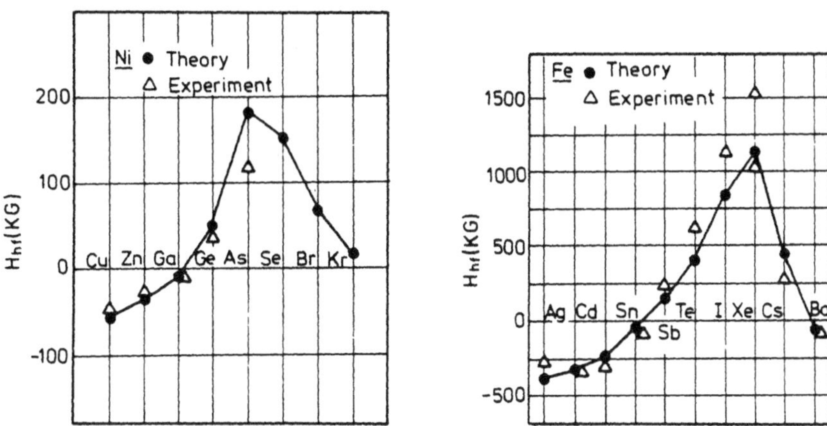

Fig. 11 Calculated hyperfine fields [61, 62, 100, 101] of nuclei of fourth period elements in ferromagnetic Ni and of fifth period elements in ferromagnetic Fe. The triangles are experimental values.

calculations since it allows a correct description of the embedding of a single defect in an infinite lattice which cannot be done by any other method. The calculations show that the perturbed charge densities and the perturbed potentials are strongly localized near the defect, but that the wave functions and densities of state are quite extended. The final aim of calculating defect properties with similar accuracy as the ones of ideal crystals has become feasable.

The calculations with the KKR-Green's function method are based on a muffin-tin approximation. At present only the impurity potential is allowed to be perturbed. Both the host potential and the impurity potential are determined selfconsistently within density functional theory. The problem of determining the host Green's function has been successfully overcome. Several applications have been discussed: Virtual bound states of 3d-impurities in Al and Cu can be strongly influenced by the band structure of the host, i.e. by backscattering. Examples for this are the van Hove singularity in the virtual bound state of Ni and Mn in Al or the anomalies of the majority density of states of Mn in Cu. We have determined the electronic structure of magnetic impurities in Ag, Cu and Al and have shown that the tendency for magnetism is largest in Ag and smallest in Al. The calculated magnetic moments are in satisfactory agreement with the available experimental information. We have discussed the local densities of states of sp-impurities in Al and Cu and the problems related to the positions of the localized 3d-levels of Zn, Ga and Ge in Cu. In the last section impurities in ferromagnetic Ni are considered. The calculated local moments of the 3d-impurities are in good agreement with experiment, however the large perturbations of the neighbouring Ni-moments for Cr, V and Ti cannot be described by a single perturbed potential. The calculations of Kanamori et al. have yielded excellent values for the hyperfine field and nuclear spin lattice relaxation time of sp-impurities in Ni.

Despite of such successes this work should be considered as a beginning. Certainly perturbed potentials also on neighbouring host sites have to be taken into account, even if they will not change the results much in many cases. We are not yet able to calculate total energies such as formation energies or forces on neighbouring host atoms and the resulting static displacements nor do we know how static displacements influence the electronic structure. For many applications the interaction energies of two impurities and its decomposition into electronic, elastic or magnetic contributions is of prime importance. We are optimistic that the Green's function together with density functional theory will allow us to attack and solve many of these problems in the coming years.

Acknowledgements

We would like to thank H. L. Davis and J. F. Cooke, Oak Ridge National Laboratory, for their invaluable help when we started this programm a few years ago. We gratefully acknowledge many discussions with our coworkers and colleges, especially J. Deutz and R. Podloucky.

References

[1] *A. R. Mackintosh* and *O. K. Andersen*, The Electronic Structure of Transition Metals, in: Electrons at the Fermi Surface ed. *M. Springford*, Cambridge University Press Cambridge 1980.

[2] *V. L. Moruzzi, J. F. Janak* and *A. R. Williams*, Calculated Electronic Properties of Metals, Pergamon Press New York 1978 *V. L. Moruzzi, A. R. Williams* and *J. F. Janak* Phys. Rev. **B 15**, 2854 (1977).

[3] *P. Thiry, D. Chandesris, L. Lecante, C. Guillot, R. Pinchaux* and *Y. Pétroff*, Phys. Rev. Lett. **43**, 82 (1979).

[4] *E. Dietz, U. Gerhardt* and *C. J. Maetz*, Phys. Rev. Lett. **40**, 892 (1978).

[5] *D. Eastman, H. J. Himpsel* and *J. A. Knapp*, Phys. Rev. Lett. **40**, 1514 (1978).

[6] *N. Mårtenson* and *B. Johansson*, Solid State Comm. **32**, 791 (1979).

[7] *C. P. Flynn*, J. Phys. F 10, L 315 (1980).

[8] *Z. D. Popovic, M. J. Stott, J. P. Carbotte* and *G. R. Piercy*, Phys. Rev. **B 13**, 590 (1976).

[9] *C. O. Almbladh, U. von Barth, Z. D. Popovic* and *M. J. Stott*, Phys. Rev. **B 14**, 2250 (1976).

[10] *E. Zaremba, L. M. Sanders, H. B. Shore* and *J. H. Rose*, J. Phys. F 7, 1763 (1977).

[11] *M. Manninen, P. Hautojarvi* and *K. Nieminen*, Solid State Comm. 23, 795 (1977).

[12] *J. N. Norskov*, Phys. Rev. **B 20**, 446 (1979).

[13] *P. Jena* and *K. S. Singwi*, Phys. Rev. **B 17**, 1592 (1978). Phys. Rev. **B 17**, 3518 (1978).

[14] *R. Nieminen, M. Manninen, P. Hautojärvi* and *J. Arponen*, Solid State Comm. **16**, 831 (1975).

[15] *C. O. Almbladh* and *V. von Barth*, Phys. Rev. **B 13**, 3307 (1976).

[16] *M. Manninen* and *R. M. Nieminen*, J. Phys. **F 9**, 1333 (1979).

[17] *R. M. Nieminen* and *M. Puska*, J. Phys. **F 10**, L 123 (1980).

[18] *L. M. Kahn, F. Perrot* and *M. Rasolt*, Phys. Rev. **B 21**, 5594 (1980).

[19] *G. Solt*, Phys. Rev. **B** 18, 720 (1978); *G. Solt* and *K. Werner*, Helv. Phys. Acta 1981 (in press).

[20] *K. H. Johnson*, J. Chem. Phys. **45**, 3085 (1966).
K. H. Johnson, in: Advances in Quantum Chemistry Vol. 7, p. 143, ed. *P. O. Löwdin*, New York Academic Press 1977.

[21] *K. Schwarz* and *N. Rösch*, J. Phys. C 9, L 433 (1976).

[22] *Chr. Müller, G. Seifert, G. Lautenschläger, H. Worm, P. Ziesche* and *E. Mrosan*, phys. stat. sol (b) 91, 605 (1979).

[23] *H. Adachi* and *S. Imoto*, J. Phys. Soc. Jap. **46**, 1194 (1979).

[24] *K. H. Johnson, D. D. Vvedensky* and *R. P. Messmer*, Phys. Rev. **B 19**, 1519 (1979).

[25] *J. Harris* and *R. O. Jones*, J. Chem. Phys. **68**, 1190 (1978); 68, 3316 (1978).

[26] *O. Gunnarsson* and *R. O. Jones*, Phys. Scr. **21**, 394 (1980).

[27] *V. Heine, D. Bullet, R. Haydock* and *M. J. Kelly*, Solid State Physics, Vol. **35**, ed. *H. Ehrenreich* et al., Academic Press, New York 1980.

[28] *P. Lloyd* and *P. V. Smith*, Adv. in Phys. **21**, 69 (1972).

[29] *D. B. B. Rijsenbrij* and *A. Lodder*, J. Phys. **F 6**, 1053 (1976).

[30] *G. Ries* and *H. Winter*, J. Phys. F 9, 1589 (1979).

[31] *G. Ries* and *H. Winter*, J. Phys. **F 10**, 1 (1980).

[32] *A. Lodder* and *P. J. Braspenning*, J. Phys. **F 10**, 2259 (1980).

[33] *P. J. Braspenning* and *A. Lodder*, J. Phys. **F 11**, 79, (1981).
[34] *G. Zwicknagel*, Z. Physik **B 40**, 23; **B 40**, 31 (1980/81).
[35] *W. John* and *W. Keller*, J. Phys. **F 7**, L 223 (1977).
[36] *R. Lenk*, phys. stat. sol (b) **86**, 603 (1978).
[37] *C. E. van Dijkum*, Thesis, Amsterdam 1980.
[38] *J. D. Joannopoulos* and *F. Yndurain*, Phys. Rev. **B 10**, 5164 (1974).
[39] *R. P. Gupta* and *R. W. Siegel*, Phys. Rev. Lett. **39**, 1212 (1977).
[40] *R. P. Gupta* and *R. W. Siegel*, Phys. Rev. **B 22**, 4572 (1980).
[41] *G. F. Koster* and *J. C. Slater*, Phys. Rev. **94**, 1392 (1954); **95**, 1165 (1954).
[42] *J. Callaway*, J. Math. Phys. **5**, 783 (1964); Phys. Rev. **154**, 515 (1967).
[43] *J. Callaway* and *A. Hughes*, Phys. Rev. **156**, 860 (1967); **164**, 1043 (1967).
[44] *R. Riedinger*, J. Phys. **F 1**, 392 (1971).
[45] *C. L. Cook* and *P. V. Smith*, J. Phys. **F 4**, 1344 (1974).
[46] *J. C. Parlebas*, J. Phys. **F 4**, 1392 (1974).
[47] *C. A. Sholl* and *P. V. Smith*, J. Phys. **F 10**, 811 (1980).
[48] *G. A. Baraff* and *M. Schlüter*, Phys. Rev. Lett. **41**, 892 (1978).
[49] *G. A. Baraff* and *M. Schlüter*, Phys. Rev. **B 19**, 4965 (1979).
[50] *J. Bernholc, N. O. Lipari* and *S. T. Pantelides*, Phys. Rev. Lett. **41**, 895 (1978).
[51] *J. Bernholc, N. O. Lipari* and *S. T. Pantelides*, Phys. Rev. **B 21**, 3545 (1980).
[52] *S. P. Singhal* and *J. Callaway*, Phys. Rev. **B 19**, 5049 (1979).
[53] *T. H. Dupree*, Ann. Phys. (N.Y.) **15**, 63 (1961).
[54] *J. L. Beeby*, Proc. Roy. Soc. London, Ser. **A 302**, 113 (1967).
[55] *J. Morgan*, Proc. Roy. Soc. London, **16**, 365 (1966).
[56] *N. A. W. Holzwarth*, Phys. Rev. **B 11**, 3718 (1975).
[57] *G. Lehmann*, phys. stat. sol (b) **70**, 735 (1975).
[58] *R. Harris*, J. Phys. **C 3**, 172 (1970).
[59] *M. Hamasaki, S. Asano* and *J. Yamashita*, J. Phys. Soc. Jap. **41**, 378 (1976).
[60] *K. Terakura*, J. Phys. Soc. Jap. **40**, 450 (1976); J. Phys. **F 6**, 1385 (1976); Physica **91 B**, 162 (1977).
[61] *H. Katajama-Yoshida, K. Terakura* and *J. Kanamori*, J. Phys. Soc. Jap. **48**, 1504 (1980).
[62] *H. Katajama-Yoshida, K. Terakura* and *J. Kanamori*, J. Phys. Soc. Jap. **49**, 973 (1980).
[63] *R. Zeller* and *P. H. Dederichs*, Phys. Rev. Lett. **42**, 1713 (1979).
[64] *R. Zeller, R. Podloucky* and *P. H. Dederichs*, Z. Physik **B 38**, 165 (1980).
[65] *R. Podloucky, R. Zeller* and *P. H. Dederichs*, Phys. Rev. **B 22**, 5777 (1980).
[66] *P. Léonard* and *E. Daniel*, paper presented at the conference Physics of Transition Metals, Leeds 1980.
[67] *P. Lloyd*, Proc. Phys. Soc. London **90**, 207 (1967).
[68] *P. C. Hohenberg* and *W. Kohn*, Phys. Rev. **136**, B 864 (1964).
[69] *U. von Barth* and *L. Hedin*, J. Phys. **C 5**, 1629 (1972).
[70] *J. Friedel*, J. Phys. Radium **19**, 573 (1958); Suppl. Nuovo Cimento VII, 287 (1958).
[71] *E. Mrosan* and *G. Lehmann*, phys. stat. sol. (b) **78**, 159 (1976).
[72] *E. Mrosan* and *G. Lehmann*, phys. stat. sol. (b) **87**, K21 (1978).

[73] *G. Lautenschläger* and *E. Mrosan*, phys. stat. sol. (b) **91**, 109 (1979); **96**, 183 (1979).
[74] *R. M. Nieminen* and *M. Puska*, J. Phys. **F 10**, L123 (1980).
[75] *C. O. Almbladh* and *U. von Barth*, Phys. Rev. **B 13**, 3307 (1976).
[76] *J. Deutz, P. H. Dederichs* and *R. Zeller*, J. Phys. **F** (1981).
[77] *P. Steiner, H. Höchst, W. Steffen* and *S. Hüfner*, Z. Physik **38**, 191 (1980).
[78] *P. J. Durham, D. Ghaleb, B. L. Gyorffy, C. F. Hague, J. M. Mariot, G. M. Stocks* and *W. M. Temmermann*, J. Phys. **F 9**, 1719 (1979).
[79] *L. Dagens*, J. Phys. **F 9**, 45 (1979).
[80] *P. W. Anderson*, Phys. Rev. **124**, 41 (1961).
[81] *C. Norris* and *L. Walldén*, Solid State Comm. 7, 99 (1969).
[82] *L. Walldén*, Philos. Mag. **21**, 571 (1970).
[83] *H. P. Myers, L. Walldén* and *A. Karlsson*, Philos. Mag. 18, 725 (1968).
[84] *M. R. Steel* and *D. M. Treherne*, J. Phys. **F 2**, 199 (1972).
[85] *H. Höchst, P. Steiner* and *S. Hüfner*, Z. Physik **B 38**, 201 (1980).
[86] *H. S. Rechal* and *P. T. Andrews*, J. Phys. **F 10**, 1631 (1980).
[87] *P. T. Andrews* and *L. T. Brown*, in: Physics of Transition Metals 1980, p. 141, Conference Series Number 55, The Institute of Physics, Bristol and London.
[88] *P. J. Braspenning* et al., to be published.
[89] *R. P. Gupta* and *R. Benedek*, Phys. Rev. **B 19**, 583 (1979).
[90] *K. Terakura*, Phys. F 7, 1773 (1977).
[91] *J. Kanamori, H. K. Yoshida* and *K. Terakura*, in: Proceedings of International Conference on Hyperfine Interactions V, (Berlin 1980).
[92] *C. Norris* and *G. P. Williams*, Phys. Status solidi (b), **85**, 325 (1978).
[93] *K. Y. Yu, J. N. Miller, P. Chye, W. E. Spicer, N. D. Lang* and *A. R. Williams*, Phys. Rev. **B 14**, 1446 (1976).
[94] *F. J. Himpsel, D. E. Eastman, E. E. Koch* and *A. R. Williams*, Phys. Rev. **B 22**, 4604 (1980).
[95] *O. Gunnarsson*, in: Electrons in Disordered Metals and at Metallic Surfaces, Ed. *P. Phariseau* et al., Plenum N.Y. 1979.
[96] *J. C. Slater*, Quantum Theory of Molecules and Solids, Vol. IV (Mc Graw Hill, New York, 1974).
[97] *J. Kanamori*, J. Appl. Phys. **36**, 929 (1965).
[98] *I. A. Campbell* and *A. A. Gomèz*, Proc. Phys. Soc. 91, 319 (1967).
[99] *R. Zeller*, in: Physics of Transition Metals 1980, p. 265. Conference Series Number 55, The Institute of Physics, Bristol and London.
[100] *H. Katayama, K. Terakura* and *J. Kanamori*, J. Phys. Soc. Jap. **46**, 822 (1979).
[101] *J. Kanamori, H. K. Yoshida* and *K. Terakura*, in: Proceedings of International Conference on Hyperfine Interactions V (Berlin, 1980) **49**, 972 (1980).
[102] *C. P. Slichter*, Principles of Magnetic Resonance (Haper and Row, 1963).
[103] *P. Steiner, S. Hüfner, N. Mårtenson* and *B. Johansson*, Solid State Commun. **37**, 73 (1981).
[104] *P. Steiner* and *S. Hüfner*, Solid State Commun. **37**, 79 (1981).

Intercalated Graphite – A Synthetic Metal
I. Introduction and Electronic Structure

Peter Pfluger * and Hans-Joachim Güntherodt

Institut für Physik, Universität Basel, Basel, Switzerland

Summary: A historical introduction into the field of Graphite Intercalation Compounds (GIC's) is given. It is shown that a great variety of structural variations, including order and disorder phenomena, occur in these layered compounds. After a few short comments on the lattice dynamics of these materials, their electronic structure is discussed in detail. Charges in GIC's are discussed both in reciprocal and real space.

Within the class of *donor* compounds, π-like carriers from the carbon sublattice and s-like carriers from the intercalant sublattice have to be considered. Differences are found between the electronic structures of high- and low-stage compounds („stage" = number of carbon layers between two neighbouring intercalant layers), and between stage 1 compounds of the C_8M (M = K, Rb, Cs) and the C_6Li type. The electronic structure of *acceptor* GIC's can be derived from a system of charged graphite layers alone. In either case, screening of the intercalant layers leads to an inhomogeneous distribution of the counter charges over the carbon sublattice. This means that carbon layers become inequivalent as soon as there are more than two of them between successive intercalant layers.

Finally, a selection of physical properties of GIC's are discussed and related to the electronic structure in these materials.

1 Introduction

1.1 Historical

Swelling of natural graphite after treatment with acid had been observed as early as in 1841 by Schafhäutl [1]. However, Graphite Intercalation Compounds (GIC's) were born in 1926, when Fredenhagen and Cadenbach systematically reacted graphite with potassium metal vapour [2]. It is remarkable that these authors already described in their original paper the two-temperature preparation method in sealed two-bulb glass tubes, which subsequently became and is even today the most widely used method for preparing intercalated graphite. This method is shown in Fig. 1 schematically. The temperature T_2 on one side of the reactor controls the intercalant's

* present address: IBM Research Laboratory,
5600 Cottle Road,
San José, CA 95 193, U.S.A.

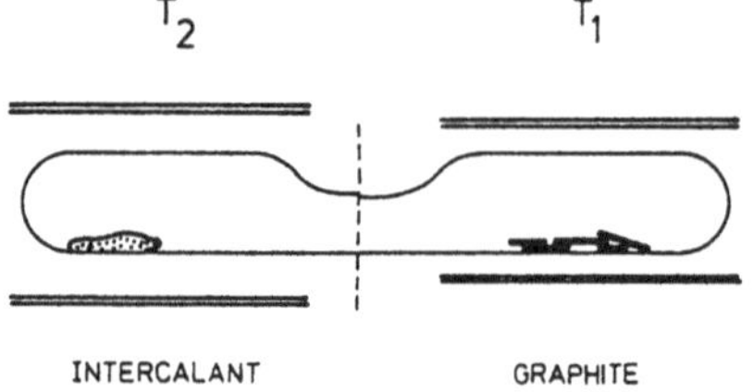

Fig. 1
Method for two-temperature gas-phase preparation of graphite intercalation compounds.

vapour pressure, while the temperature T_1 on the graphite-containing side is used to control the reaction equilibrium. The higher the value of $\Delta T = T_1 - T_2$, the lower the intercalant concentration in the graphite host, and vice-versa. Several other preparation techniques have also been developed since then, among which reactions with the liquid phase intercalant [3, 4], electrochemical preparation methods [5–7] and gas phase reactions in metal containers [8, 9] are the most prominent.

In 1932, Schleede and Wellmann realized that GIC's are lamellar-type compounds, consisting of a succession of Carbon and Intercalant layers (C- and I-layers) [10]. This led directly to the concept of staging: the stage, n, denotes the number of C-layers between two neighbouring I-layers (Fig. 2). In many cases, the stage n can be varied for a given intercalant species by the preparation conditions. The overall intercalant concentration of a GIC is primarily given by the relative amount of I-layers to C-layers, which is represented by the reciprocal stage number $1/n$. However, the intercalant concentration may also be affected by other factors like variation of the in-plane density of the I-layer.

After the early pioneering work a continuous interest prevailed in this class of materials for the next 40 years. It was recognized that charge transfer always accompanies the intercalation process where the intercalant species act either as donors or acceptors. All the synthesized GIC's exhibited metallic character, in contrast to the graphite host material, which is a semimetal in its pristine state. Note

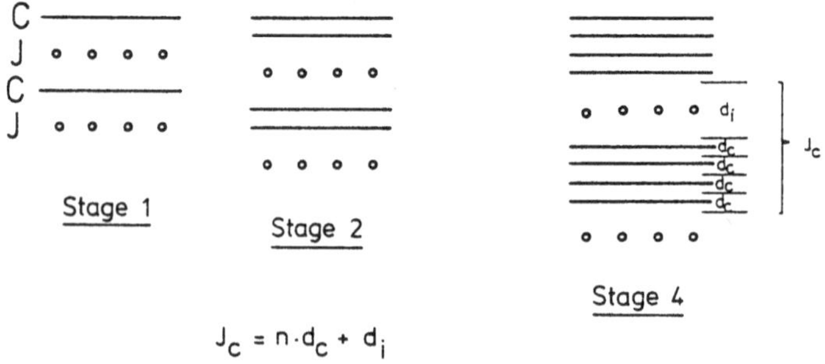

Fig. 2 Stages in intercalated graphite

that there are other types of graphite compounds which are covalent, insulating and in which the graphite layers are buckled rather than planar. Examples are graphite oxide and graphite fluoride, of approximate compositions C_2O and CF, respectively.

Most of the work on intercalation compounds was done by the groups of G. Hennig at Argonne, A. Hérold in Nancy, J. G. Hooley in Vancouver, W. Rüdorff in Tübingen and A. Ubbelohde at Imperial College. Although the search for new GIC's still continues at the present time, some 250 different compounds, covering almost all important species, were known by 1975 [11, 12]. Since then, physicists and materials scientists in particular have taken increased interest in GIC's [13–17]. This spurt in interest was triggered by the work of Vogel and coworkers from the University of Pennsylvania, who claimed that materials surpassing the electrical conductivity of pure copper have been obtained on the basis of graphite intercalated with group V fluorides [18].

1.2 Current Interests

Until now, it has not been possible to obtain in a systematic and reproducible way samples of intercalated graphite with a conductivity clearly larger than that of copper [19]. However, a number of interesting phenomena have been additionally found to occur in these materials, which have motivated the scientific interest of chemists, materials scientists and physicists. Also, applications, other than electrical conduction, have made the GIC's interesting. They are used for the commercial production of graphite foils [20]. They are also used as catalysts in organic chemistry [21, 22] at the laboratory level as well as for separation and purification by selective intercalation. GIC's can be used potentially as electrodes in electrochemical and electrooptical systems [6, 7, 23, 24]. Graphite can conceivably be used as host for (intercalated) material storage and as detector for (intercalating) corrosive or harmful gases. Hydrogen can be either intercalated in graphite to form ternary compounds like $C_8KH_{2/3}$ at room temperature [25] or absorbed in the interstices of not dense-packed I-layers to give compounds like $C_{24}KH_4$ at lower temperatures [26]. Finally, it would obviously be attractive to design the properties of a technologically important material like graphite on a synthetic level by the intercalation processes.

Several review papers have been published in this field in the last few years [11, 12, 14 and several articles in Refs. 13, 15, 16]. The aim of the present article is twofold. *Firstly*, the authors intend to give an introduction for those who are not familiar with intercalated graphite. *Secondly*, the electronic structure of GIC's will be reviewed as important progress has been made in this field since the previous review article (J. E. Fischer in Ref. [15]). Therefore, sections 2 and 3 of the present paper, dealing with structural order/disorder and lattice dynamics in GIC's respectively, have been kept at an introductory level, whereas subsequent sections, dealing with electrons and their properties in intercalated graphite, are described in greater detail and are more based on either the work done in the authors' laboratory or in collaboration with them.

2 Order and Disorder in Intercalated Graphite

2.1 Microscopic Structure

a) Carbon Sublattice

While the carbon sheets are essentially unchanged in the intercalation process, important modifications of the graphite sublattice occur in the *c*-direction, perpendicular to the C-layers. While the interplanar distance is 3.35 Å in pristine graphite, C-layers can be separated by more than 10 Å in intercalation compounds with acceptor molecules. The smallest *c*-axis lattice expansion occurs in lithium-GIC's, where the small size of the Li^+ ions lead to a C—C interplanar distance of 3.71 Å, only 10 % more than in pure graphite [27]. The ... ABAB ... stacking sequence of hexagonal graphite is mostly not preserved in GIC's, the same type of C-layer is usually found on each side of an I-layer, giving for example an ... AIABABIBABAIA... sequence in a stage 4 GIC [27].

Although the honeycomb lattice of graphite is preserved in GIC's, small changes, in the 1 % range, of the in-plane graphite sublattice constant may occur. These changes reflect the change in electronic properties due to charge transfer effects upon intercalation. They have been discussed in detail in the article by Strässler and Pietronero in the present volume (paper II).

b) Intercalant Sublattice

The intercalant sublattice is known to exhibit in-plane order or disorder, depending on the intercalant species, its concentration and the investigated temperature range. For example, Cs layers are liquid-like in Cs-GIC's of stage 2 at 300 K and in Cs-GIC's of stage 1 above 608 K [28, 29]. On the other hand, at room temperature a 2 × 2 Cs-superlattice is formed, epitaxially with the graphite layers, in the stage 1 compound C_8Cs [27]. Figure 3 shows the in-plane structures in first stage compounds of the C_8M type (M = K, Rb, Cs) and of the C_6M type (M = Li, Ba, Ca, Sr, Eu, Sm, Yb) at 300 K [27].

The I-layers can or cannot be regularly stacked in *c*-direction. In the Alkali metal Graphite Intercalation Compounds (AGIC's) of stage 1, ... $\alpha\alpha$ ... , ... $\alpha\beta$ ... , ... $\alpha\beta\gamma$... ,

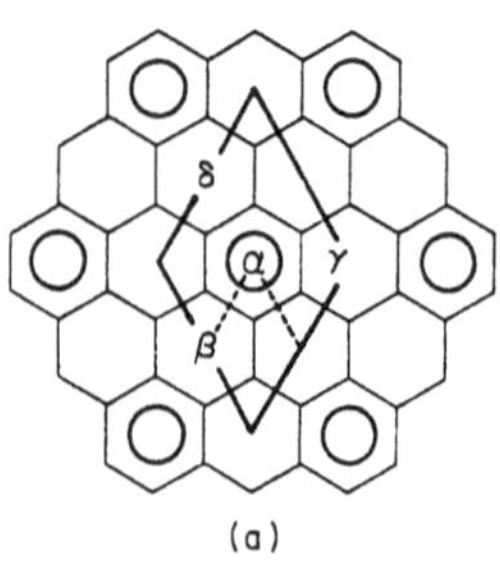

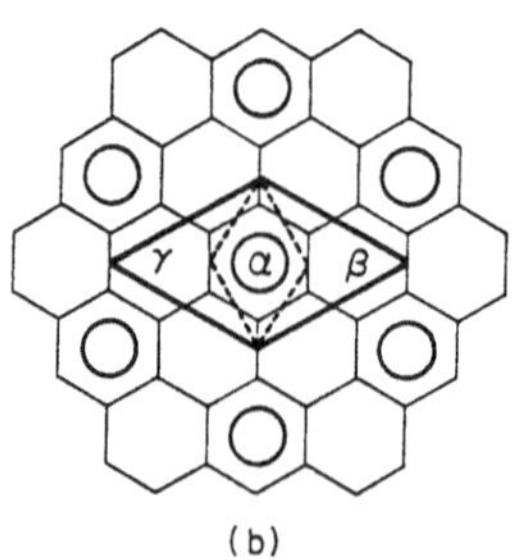

Fig. 3
Basal plane crystal structures of C_8M (a) and C_6M (b) GIC's. The heavy lines show the in-plane unit cells, whereas the dashed lines represent the in-plane unit cell of pristine graphite. The symbols α, β, γ and δ denote equivalent intercalant sites. From Ref. [15], p. 425.

and ... $\alpha\beta\gamma\delta$ ... stacking sequences of the I-layers have been reported [27, 30]. In contrast, in-plane ordered $SbCl_5$-layers do not order three-dimensionally [27].

c) Relative Position of C- and I-Sublattices

After considering the C- and I-layers separately, their organization with respect to each other has to be discussed. In many cases, the sublattices are commensurate and a new common unit cell can be determined. Stage 4 Br_2-GIC's of nominal composition $C_{28}Br$, for example, have a rectangular in-plane unit cell of the dimensions a = 8.5 Å and b = 34.5 Å, containing 8 Br_2-molecules and 112 C-atoms [27]. Incompatible in-plane lattice vectors of the I- and C-sublattice may lead to a incommensurate situation, as in the case of $FeCl_3$-GIC's [27].

d) Structural Phase Transitions

It is clear that these different types of order and disorder in GIC's may give rise to a variety of phase transitions. Some of them have been demonstrated in an X-ray study by Clarke et al. of the stage 2 and 3 Cs-GIC's $C_{24}Cs$ and $C_{36}Cs$ [28]. As a function of increasing temperature up to 300 K, the Cs-layers of $C_{24}Cs$ exhibit four different configurations. A mixed phase in registry with the graphitic lattice is found at low temperatures. Above 50 K, an incommensurate lattice with orientational, but without positional long-range 3-dimensional order is formed. The 3d interlayer correlation is lost above 165 K. Finally, the Cs-layer has a uniform 2d liquid-like structure above 228 K, not in registry with the graphite layers. More work has been done in the past few years on structural phase transitions in AGIC's of stages $\geqslant 2$ [31, 32] including neutron scattering [33] and electron diffraction [34] experiments. Note that discontinuities indicative for such phase transitions in AGIC's have been found in the corresponding curves of electrical resistivity ρ (T) [33], specific heat c_p(T) [35] and magnetic susceptibility χ(T) [36].

e) Staging Phenomenon

One of the most outstanding properties of the GIC's is the existence of stage ordering up to very high stages ($n \sim 10$) in some materials [37–39]. Understanding the nature of the long-range interactions responsible for staging has been the object of recent efforts. It has been found that electrostatic interactions are long-ranged in the GIC's. Total energy calculations show that pure stage configurations are preferred and that the electronic energy plays a significant role in stabilizing the high-stage structures [40, 41].

2.2 Macroscopic Effects

a) The Domain Model of Daumas and Hérold [42]

Obviously, a stage *n* GIC is not perfect over a macroscopic scale. Daumas and Hérold have proposed a pleated layer model (Fig. 4), which offers several advantages over the idealized model shown in detail in Fig. 2. All graphite interspaces are filled equally, but not continuously, thereby forming islands of the intercalant

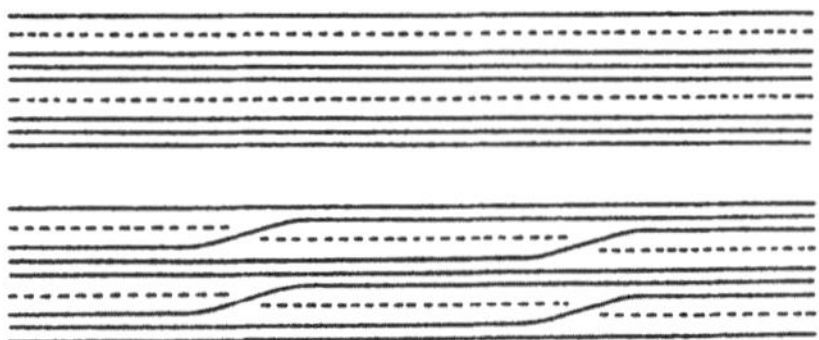

Fig. 4
Idealized and pleated layer model for staging in stage 3 GIC's [42].

element. In particular, stage changes are easier to account for in this model, and X-ray studies of the order-disorder transition in intercalated Cs-layers support it [29].

Using this domain model, the effects of elastic interactions in GIC's have been investigated. While electronic energy terms tend to *stabilize* high stage GIC's, strains have been shown to *drive* a mixed or randomly-staged crystal towards pure stage ordering [43]. The kinetics of the intercalation process should also be determined by elastic interactions.

b) Other Structural Domains in GIC's

More macroscopic structural effects have been found in C_8K, C_8Rb and C_8Cs compounds. Single crystals are built by an assembly of domains characterized by orthorhombic (C_8K and C_8Rb) or hexagonal (C_8Cs) unit cells. These domains are twisted from each other by an angle of 60° around the *c*-axis. This structural model is probably applicable to other GIC's and a similar one has already been proposed for bromine intercalated graphite [27].

Recent electron diffraction and real-space imaging experiments have shown another type of domain formation. Ordered clusters, incommensurate to the graphite basal plane structure, surrounded by disordered intercalant, exist at temperatures between 86 K and 130 K in stage 2 K-GIC's [44].

c) Imperfect Staging

Although the goal of a careful GIC-sample preparation is to get pure staged specimens, mixed stage GIC's may be obtained frequently. They consist of a mixture of domains, each of them being purely staged. Non-pure stages can also occur, particularly in $FeCl_3$-GIC's [37]. There, the regular sequence of C- and I-layers along the *c*-axis is severely disturbed.

However, pure staged samples are not too hard to prepare for an important number of intercalant species. Since such compounds are most widely used for physical measurements, we will restrict ourselves to the discussion of only the pure staged GIC's in this article.

d) Real Crystals

Beyond the scale of several tens to several hundreds of Å, typical for the types of domains discussed above, one should not forget the finite crystallite sizes. Most of the newer experiments were carried out on samples grown from Union Carbide's HOPG (Highly Oriented Pyrolytic Graphite) material, in most cases kindly donated by A. Moore. HOPG is in fact a polycrystalline material, in which the *c*-axes of the crystallites are well-oriented (mosaic spread 0.1–0.4°), while their basal planes are randomly twisted. Typical crystallite sizes for high quality HOPG are 1000 Å in *c*-direction and 3 μm in *a*-direction [45].

Also, single crystals of natural graphite can be used for intercalation. However, their thickness hardly exceeds 0.1 mm and their surface 1 mm^2. While this is sufficient for some experiments (X-rays, electron microscopy, ...) a number of others cannot be performed with this type of samples.

In this section the authors have intended to give some idea of the great variety of structural effects which may be found in GIC's. Readers interested in more details concerning the structure of GIC's are referred to the review article by Hérold on Crystallo-Chemistry of Graphite Intercalation Compounds [27].

3 Lattice Modes in GIC's

The lattice mode structure of GIC's has been discussed in some detail by Dresselhaus and Dresselhaus in Ref. [15], and a summary of Raman- and IR-studies in GIC's has been given by Solin in Ref. [16] about two years ago. A new, up-to-date review of this field is in preparation [46]. We, therefore, restrict ourselves to mentioning some of the recent work and to summarizing the most interesting results already obtained.

In general, the starting point for the interpretation of GIC data is either pure graphite or, less often, the pure intercalant. In particular, the knowledge of the lattice modes of pristine graphite is essential to understand the intercalated material. A satisfactory knowledge of the lattice dynamics of graphite has been obtained only recently from the phonon dispersion curves calculated by Maeda et al. [47].

3.1 Raman and Infrared Reflectivity Studies

Most of the Raman- and IR-studies in GIC's have focused on the graphitic in-plane E_{2g_2} (Raman active)- and E_{1u} (IR active)-derived modes. It has been demonstrated that these experiments are sensitive to two types of graphite layers, namely bounding (C_b)-layers lying next to the intercalant, and interior (C_i)-layers [15, 16]. Figure 5 shows that two Raman peaks associated with the graphite E_{2g_2} mode occur for GIC's of stages $\geqslant 3$. In stage 1 and stage 2 GIC's, only C_b-layers and only one peak are present. The relative intensity of the lower frequency peak increases as the stage number n increases, thus demonstrating its correlation with the C_i-layers.

Studies of the stage-dependent frequency shifts of these graphitic optic modes have been carried out. An upshift (stiffening) for acceptor compounds and a downshift (softening) for donor compounds is detected as the intercalation concentration is increased. The observed linear dependence of $1/n$ might be explained by the an-

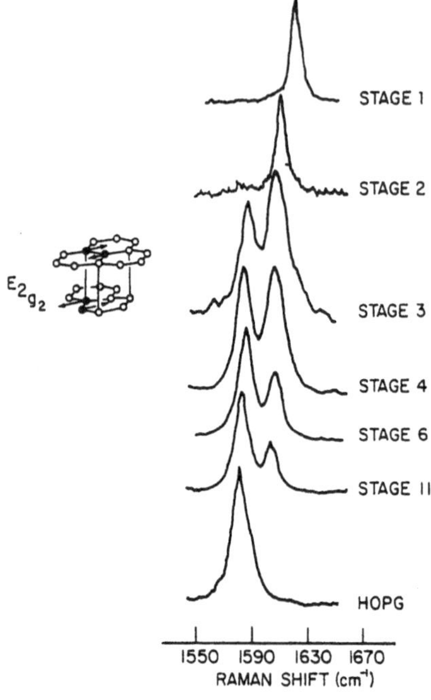

Fig. 5
Raman spectra of the E_{2g_2} mode in pure HOPG and in the $FeCl_3$-GIC's of stages n = 1–4, 6 and 11. The upper frequency component is identified with the bounding layer mode and the lower frequency component with the interior graphitic mode. From C. Underhill et al., Sol. St. Comm. 29, 769 (1979).

harmonicity of the σ bonds upon expansion or contraction of the C–C bonds due to intercalation [48]. A phenomenological model using the concept of zone folding along the k_c-axis has been proposed to account for the stage-dependence of these lattice modes in GIC's [49]. Well-defined and stage-dependent second-order Raman features, closely related to those of pure HOPG, have recently been demonstrated in $SbCl_5$ (acceptor)-intercalated compounds [50].

Little work has been done on intercalant modes in GIC's. Modes of intercalated Br_2 and $FeCl_3$ have been found, but other attempts to detect intercalate modes in donor- and acceptor-compounds have failed [15, 16].

3.2 Neutron Scattering Experiments

In principle, full information on the dispersion relations could be expected from inelastic neutron scattering experiments. Unfortunately, very little of such data is available until now. LA and LO modes propagating along the *c*-axis have been found in C_8Rb [51], and two modes were observed in C_6Li [52]. The *c*-axis sound velocities are increased in both compounds relative to graphite.

Very recently, a more extensive inelastic neutron scattering study of the longitudinal phonon branches along the *c*-axis was carried out for a 50 cm^3 sample of the stage 3 AGIC $C_{36}K$ [53]. An acoustic and two optic branches, separated by frequency gaps, were detected. They were modelled by a simple linear chain along the c-axis, with two different masses representing the I- and C-planes respectively, and with two force constants describing the C–I and C–C interlayer coupling. The measurements and the model have been extended to lower stage AGIC's. In contrast to the third stage, the high-energy optical branch could be detected for $n = 2$ [54]. Measurements on stage 1 compounds are under way [54].

The absence of more detailed experimental data is particularly regrettable. A detailed calculation of the phonon dispersion curves has been made by Horie et al. [55] for stage 1 AGIC's, and a comparison with experiments would be interesting.

4 Electrons in GIC's

4.1 Introduction

After the rebirth of the field around 1975/76, a considerable amount of information concerning the electronic properties of GIC's was collected, among which conductivity, magnetoresistance, Hall coefficient, optical reflectivity, magnetic susceptibility, specific heat, Knight shift, EPR data can be cited [15]. Too little was known on the electronic structure of these compounds to allow a straightforward interpretation of the experimental results.

Attempting to optimize electrical conductivities, reliable values for the density of states (DOS) at the Fermi level, $N(E_f)$, in donor- and acceptor-GIC's were of particular interest. $N(E_f)$ is, of course, coupled to the coefficient of charge transfer through the position of the Fermi level E_f and the shape of the DOS function $N(E)$. In general, it was not clear to what extent GIC's could be approximated by charged graphite. Such a picture would mean that all the states in the vicinity of E_f, contributing to transport, susceptibility, Knight-shift, etc., were of the graphitic π-type.

Graphite itself has 4 valence electrons per C-atom, three of which (2s, $2p_x$, $2p_y$) are used to form the covalent in-plane σ-bonds. The $2p_z$ electrons give rise to the well-known π-state dominating the neighbourhood of the Fermi level. In 2-dimensional graphite, the π-valence band states are just completely occupied, while the antibonding π^*-conduction band states are completely unfilled. In a 3d picture, graphite is semimetal with non-zero DOS at E_f. This is due to interactions between the graphite planes which produce a small overlap (36 meV) of valence and conduction bands [56]. The c-axis dispersion of the energy bands produces Fermi surfaces as shown in Fig. 6, with electron- and hole-like pockets in the corners of the Brillouin zone.

Spain and Nagel have proposed a rigid band model in which the graphitic π and π^* states remain unaffected by intercalation [57]. Donor intercalants raise the Fermi level, while acceptor species lower it (Fig. 7b). For example, in an alkali metal GIC (donor compound) the intercalant's s-band, which may be modified, would either

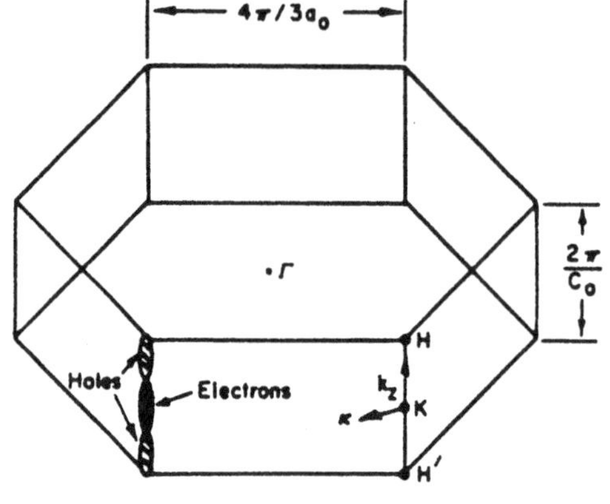

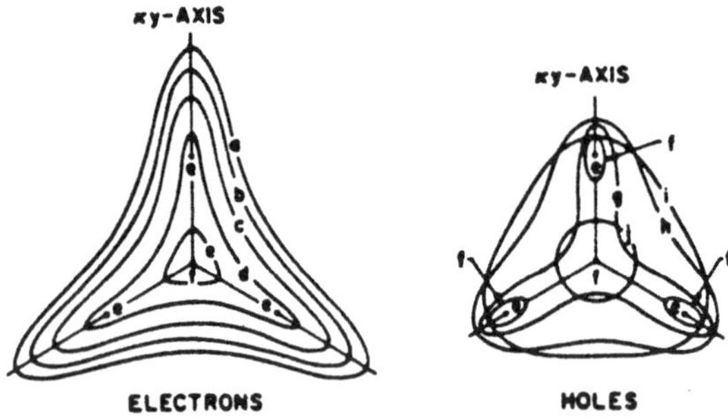

Fig. 6 Brillouin zone and Fermi surfaces in pure 3d graphite [56]

lie above E_f (complete charge transfer, Fig. 7d) or overlap E_f (partial charge transfer, Fig. 7c). Dresselhaus, Dresselhaus and Fischer (DDF) have used the full three-dimensional graphite band structure of Slonczewski, Weiss and McClure [56] to calculate the electronic properties of "dilute" GIC's. In this case of GIC's with low intercalant concentration, no intercalant levels were considered (Fig. 7b). Although the above model was explicitly prescribed for compounds of stages "higher than 4 or 5", surprisingly the DDF model was subsequently employed for other compounds, including those of the first stage. In the DDF model, the charge compensating the I-layers was assumed to be uniformly distributed over the graphite sublattice. This picture has changed subsequently, as described in the following section.

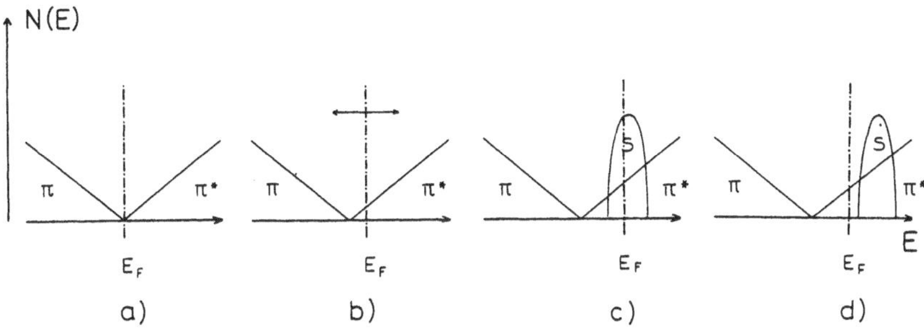

Fig. 7 Schematic densities of states around the Fermi level.
a) Pristine graphite
b) GIC's in the dilute limit
c) Model for a donor-GIC with partial charge transfer
d) Model for a donor-GIC with full charge transfer

4.2 Spatial Charge Distribution in GIC's

a) Charge Distribution in *c*-Direction

In their classical paper of 1978, Pietronero, Strässler et al. calculated the charge distribution $\rho(z)$ along the *c*-axis of acceptor GIC's in a Thomas-Fermi model [59]. The charge distribution was found to be highly nonhomogeneous and to depend nonlinearly on the charge transfer coefficient f per intercalated molecule ($0 \leqslant f \leqslant 1$). For large distances from the intercalant and large f values, $\rho(z)$ varies as z^{-4} (see also Ref. [60]), while for small distances and small f values, a transition to an exponential asymptotic decay occurs. In any case, screening of the I-layers is long-ranged. The results of these authors not only contradict the earlier picture, in which the chemical potential $\mu(z)$ was assumed to be constant throughout the graphite sublattice, but also go beyond the "sandwich model", in which the I-layers were assumed to be screened within one C-layer, i.e. within the bounding C_b-layers. In the latter picture, a GIC consists of C_b-I-C_b sandwiches plus neutral graphitic inner C_i-layers. The sandwich model was experimentally based on the Raman spectra discussed in section 3.1 and shown in Fig. 5, as well as on quantum-magnetothermal oscillations in $SbCl_5$- and Br_2-GIC's [61].

The first experiments probing directly the *c*-axis charge distribution and clearly demonstrating the long-range nature of the I-layer screening were done by Cartier et al. [62]. In these studies, positrons were annihilated in K-GIC's of various stages up to $n = 10$ (Fig. 8). The positively charged β^+ particles are Coulomb repelled from the likewise positively charged K-layers and, thus, sample the center of the *n* graphite layers. For stages > 5, the positron finds pure graphite-like layers in the compound, as evidenced by the angular correlation distribution (ACD) curves identical to those of pristine graphite. While the ACD of the stage 5 compound $C_{60}K$ is still almost graphite-like, a dramatic change to a completely different behaviour occurs on going to the stage 4 compound $C_{48}K$. The foregoing observations imply that the two C-layers neighbouring the I-layer on one side are severely influenced by the intercalant, whereas the third C-layers behaves almost as a pure graphite layer (Fig. 9). Adding even more C-layers by increasing n, a purely graphitic portion is built up in the center of the C-subsystem. Obviously, the K-intercalant is fully screened within a distance of ~ 8.5 Å from the I/C-interface.

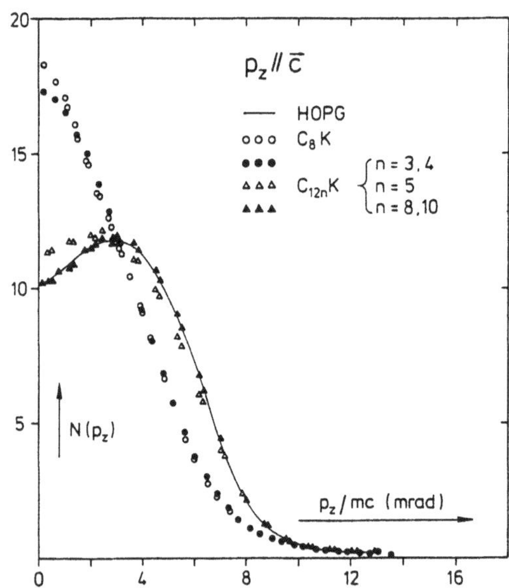

Fig. 8
2γ angular correlation distribution of positron annihilation in pure HOPG and K-GIC's of various stages.

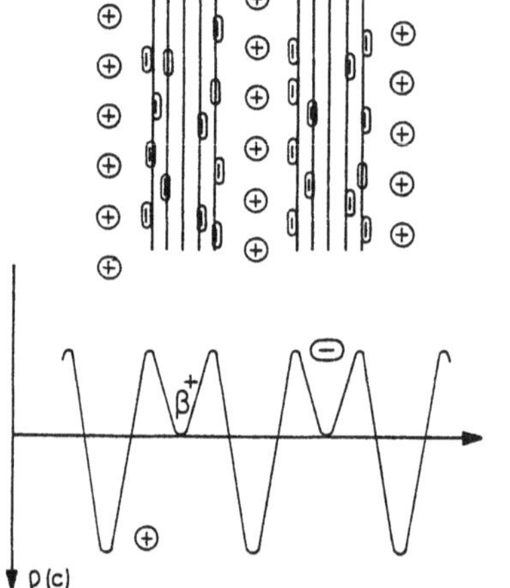

Fig. 9
Schematical c-axis charge distribution in a stage 5 donor-GIC.

Layer potentials and c-axis charge distributions have recently been calculated self-consistently in a full quantum LCAO treatment for films of $n = 3$–8 graphite layers bounded by two, partially ionized I-layers [41].

b) In-Plane Charge Distribution

In donor compounds, the charge distribution within the I-layer is usually assumed to be homogeneous in the sense that all donor atoms equally contribute to the charge transfer towards the C-sublattice. However, this does not imply that f is necessarily 100 %, as will be discussed in the following sections. A modulation of the counter-charge in the a-direction of the C-subsystem, due to the potential of the I-ions, is generally neglected in the discussion of either donor- or acceptor-GIC's.

The distribution of the negative charges in acceptor-type I-layers is a more complicated problem. For a given molecular intercalant, the species existing within the host lattice after intercalation have to be determined. Taking the case of graphite/ SbF_5 for example, SbF_3, SbF_5, SbF_6^- and $Sb_2F_{11}^-$ species have been subject to discussion [63–66]. In particular, one would like to know the location of the accepted negative charges as a function of space and time as well as the relative amount of charged and uncharged species in the I-layers.

In contrast to donor GIC's, the charged species seem to be fully ionized in any stage of most of the acceptor compounds. The overall charge transfer, however, is not 100 %, because the charged molecules are separated from each other by other, neutral species. This happens for example in graphite nitrates, which contain one NO_3^- per 4 neutral HNO_3, resulting in f = 0.2 [67], or in graphite bisulfate, for which a general formula $C_{24n}^+HSO_4^- . 2H_2SO_4$, resulting in f = 0.33, has been determined [68]. It should be noted in this context that the fractional charge f transferred to or from the graphite per intercalant atom or molecule is not necessarily found in the emptied valence or filled conduction bands of GIC's by electronic measurements. As pointed out by Fischer [69], this would only be true if no charge localization by covalent bonding occurs.

N.M.R. investigations of intercalated nuclei eventually supplemented with E.S.R. measurements provide important informations in order to identify the intercalated species [17]. The ^{19}F resonance has been extensively studied [17, 63, 64], but other nuclei like P, B, Al, Li, Rb, Cs are also under investigation.

Spectroscopic techniques have also been used to study intercalated acceptor species and charge locations in the I-layers: XAS for AsF_5-GIC's [70], IR and XPS for SbF_5-GIC's [66]. It was suggested in an XPS study of $FeCl_3$-GIC's [71] that metal ion vacancies might produce the acceptor sites in transition metal halide compounds.

Positron annihilation techniques can also be successfully employed to address these problems. In acceptor-GIC's, the positrons are trapped by the negative charges of the I-layers, giving ACD curves characteristic for the anionic sites. The experiments [62] confirm that the ionized species is the same at any, even very low intercalant

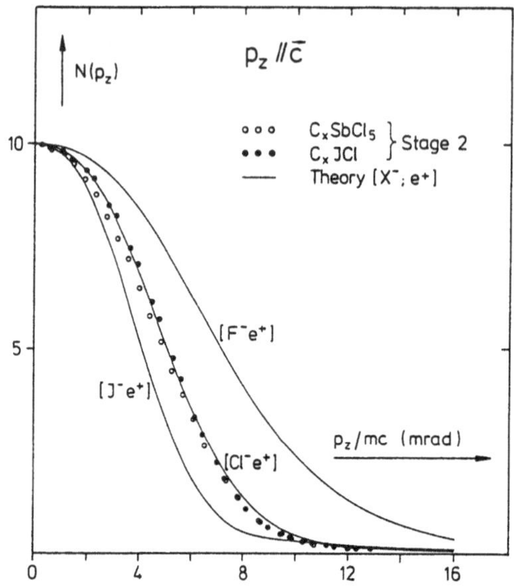

Fig. 10
2γ angular correlation distribution of positron annihilation in acceptor-GIC's. Comparison is made with theoretical distributions calculated by Farazel and Cade. For details see Ref. [62].

concentration. As seen in Fig. 10, the comparison with calculated ACD's of free positronium halides [72] leads to the conclusion that the anion responsible for positron trapping is certainly Cl^- in ICl-GIC's. Similarly, Br-atoms bear the negative charges in IBr-compounds. The method can be tested for other examples, like NO_3^- ions in graphite nitrates, and extended to more complicated systems. The problems concerning the in-plane distribution of the charges in acceptor-GIC's are far from solved. Clearly, more effort is needed to attain a satisfactory knowledge in this area.

4.3 Electronic Structure of Stage 1 Alkali Metal GIC's (AGIC's)

a) Introduction

The saturated compounds C_6Li and C_8M (M = K, Rb, Cs) are among the best defined and most investigated GIC's, both experimentally and theoretically. They are fully ordered at room temperature with a number of atoms per unit cell which still allow full band structure calculations to be made. Furthermore, homogeneous specimens of C_6Li and C_8M are not too difficult to prepare. In contrast, very little data are available on graphite intercalated with alkaline earth or rare earth metals, due to difficulties with their preparation. In interpreting the results of stage 1 compounds, one does not have to worry about electrostatic effects, as only one type of C-layer is present.

Energy band structures were first published by Inoshita et al. [73] from a tight binding calculation of C_8K and by Holzwarth et al. [74] from a KKR muffin-tin calculation for C_6Li (Fig. 11a). The assumptions used in these calculations, concerning the degree of ionization of the alkali atoms, were motivated by the results of early Knight shift measurements [75]. The energy bands of these GIC's in the Fermi region could essentially be described by a superposition of graphitic π- and alkali-like s-bands. However, the relative positions of the s- and π-band systems are not the same in the two cases. For C_6Li, the bottom of the Li 2s band was predicted to lie well above E_f (Fig. 11a), while the lower part of the K 4s band was predicted to lie below the Fermi level in C_8K (Fig. 11b). In C_6Li and C_8M, formation of (different) superlattices on the graphite matrix in real space leads to correspondingly reduced Brillouin zones, into which the graphitic bands are backfolded. In C_6Li, the crossed π-bands from the zone edge of pristine graphite (Fig. 1 of paper II) are folded into the center of the new Brillouin zone. This is not the case in C_8K, because its unit

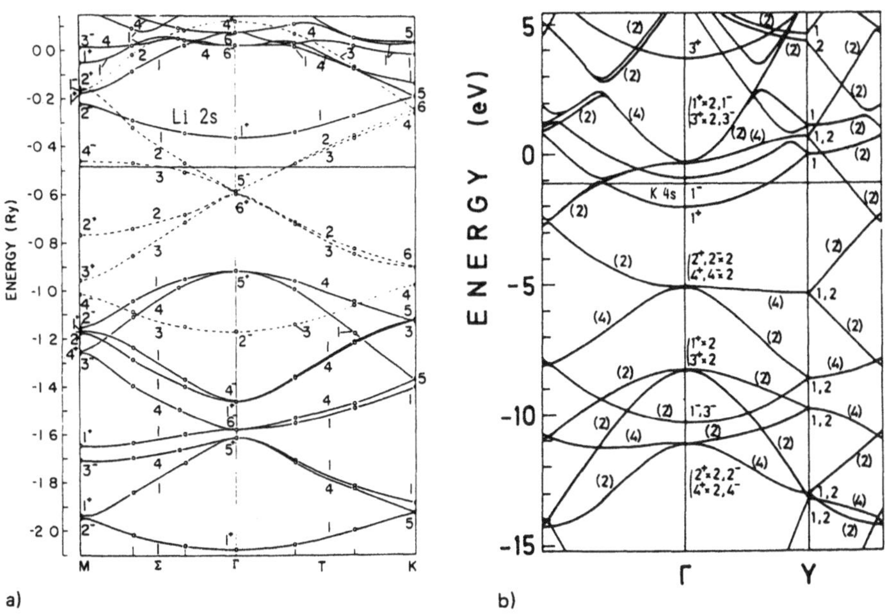

Fig. 11 Energy bands in the first stage GIC's C_6Li (a) and C_8K (b). Brillouin zones and Fermi surfaces are shown in Fig. 19.

a) KKR calculation for C_6Li with non muffin-tin corrections. The carbon π bands are represented by dashed lines, σ bands are represented by full lines [74].

b) Self-consistent calculation for C_8K with use of atomic pseudopotentials and the Madelung-type potential due to the charge transfer [81].

cell is different from that of C_6Li (Fig. 11). C_8K forms a 2×2, but C_6Li forms a $\sqrt{3} \times \sqrt{3}$ superlattice on the graphite substrate (Fig. 3). Coupling between the C- and I-layers produces perturbations of the graphitic bands by hybridization effects with the alkali s-bands. The existence of the above folding effects is generally accepted, while the ideas about the amount and importance of hybridization effects are not definitely settled yet.

Experimentally, the electronic structure of AGIC's was directly investigated in a series of electron spectroscopy experiments (UV-, X-ray photoemission, Auger electron spectroscopy) by Oelhafen et al. [76, 77] on C_8M and higher stage $C_{12n}M$ compounds, while Eberhardt et al. studied the C_6Li compound by angle-resolved photoemission [78].

b) Electronic Structure in C_8M (M = K, Rb, Cs) Compounds

Figure 12 shows the density of occupied states N(E) measured by UPS in annealed C_8K compounds and in pure graphite [79]. The occurrence of a conduction band with a high $N(E_f)$ value, the shift of E_f with respect to structures in the graphitic valence bands, and the small changes in the mainly rigid graphitic DOS structures can be seen clearly. The data obtained from C_8M compounds and from the concentration dependence of photoemission spectra in C_xCs (n = 1–7) GIC's [76, 77] have led to the following interpretation of the conduction band in low-stage heavy alkali metal GIC's [80]: alkali-like s-type states are superimposed on the graphitic π^* antibonding states, giving predominantly an s character to the Fermi level states in the C_8M compounds. The above considerations imply that the transfer of electrons from the K 4s, Rb 5s or Cs 6s into the graphite π^* bands is by no means complete.

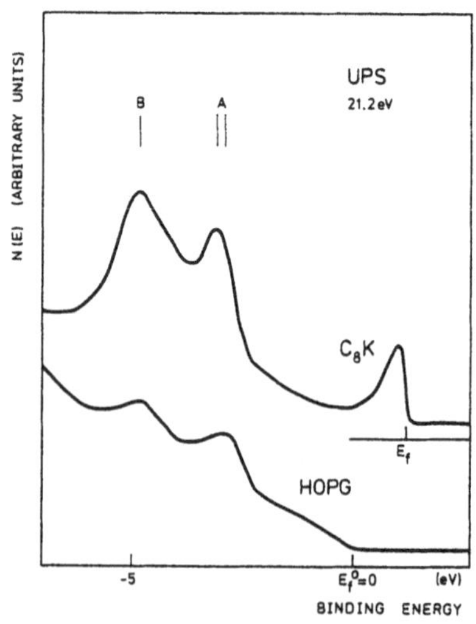

Fig. 12

Valence band spectra of pure graphite and C_8K, obtained from UPS ($\hbar\omega$ = 21.2 eV) measurements [79]. Structure A is due to flat π bands at the Q point, and structure B is mainly due to flat σ bands at the Γ point in the energy band structure of graphite (76; see also Ref. 95, 113 and Fig. 14).

This general picture, compatible with the prediction of Inoshita [73], has been confirmed by several other theoretical and experimental studies [79–86]. In particular, a self-consistent recalculation of the C_8K band structure [81] (Fig. 11b), ^{13}C N.M.R. and E.P.R. studies [82, 83], a detailed UPS investigation of the AGIC conduction bands with varying incident photon energy [79], and positron annihililation studies [85] all reveal an alkali s-like (nearly isotropic, free, parabolic) electron gas. Its existence is now generally accepted, although from a KKR band structure calculation for C_8K using corrected muffin-tin potentials [87], the bottom of the K 4s band is estimated to be 1.5 eV above E_f. It is remarkable that already 20 years ago, a spin resonance study of the conduction carriers in C_8M and $C_{24}M$ compounds gave g-factor shifts and linewidths ΔH indicating alkali metal s-orbitals involved in the carrier motion [88].

c) Electronic Structure of C_6Li

In contrast to the C_8M AGIC's, photoemission by synchrotron [78] and X-ray [89] radiation have indicated a transfer of charge near unity into the graphite π^* bands (Fig. 7d) for the case of C_6Li. The backfolded graphite bands were found by probing the dispersion of the energy bands [78]. However, in C_6Li an enhanced DOS near E_f with respect to the pure graphite π^* conduction band, expected from backfolding effects and also from experimental determinations of $N(E_f)$ [90, 91], was not measured in the synchrotron and X-ray photoemission experiments [78, 89], which is probably due to the unfavourable photoelectric cross-sections at the used excitation energies. On the other hand, the strong DOS peak belonging to the conduction band of C_6Li was found by 21.2 eV UPS (Fig. 13), but unlike in the case of heavy alkali GIC's, an unambiguous assessment concerning the character of the Fermi level states could not be made from a series of electron spectroscopy measurements [92]. As in the case of C_8M, the electronic energy band structure of C_6Li was recalculated self-consistently by Ohno [93]. In contrast to C_8M, the obtained Fermi surface is constructed entirely from folded graphite bands only, the Li 2s states lying well-above E_f, thus confirming earlier results [74, 94]. Since the Li 2s states are slightly mixed with the graphite bands, the charge transfer may be a few percent lower than 100.

d) Fermi Levels and Non-Rigid Band Effects in Stage 1 AGIC's

Non-uniform valence band shifts, indicating notable non-rigid band effects, were already found in C_6Li by photoemission with synchrotron radiation [78]. While photoemission is an excellent tool to study the filled valence bands, information on the unoccupied bands can be obtained from energy loss spectroscopy (ELS) [84] or from secondary electron spectroscopy (SES) [95]. Figure 14 shows that the SES peaks can be correlated with the unfilled bands, and that the combination of UPS and SES measurements can provide detailed information on the rigidity of the graphite bands upon intercalation. If the graphite bands were completely rigid, the sum $E_\sigma + |E_B|$ should be constant, although E_f shifts upon intercalation. It is seen

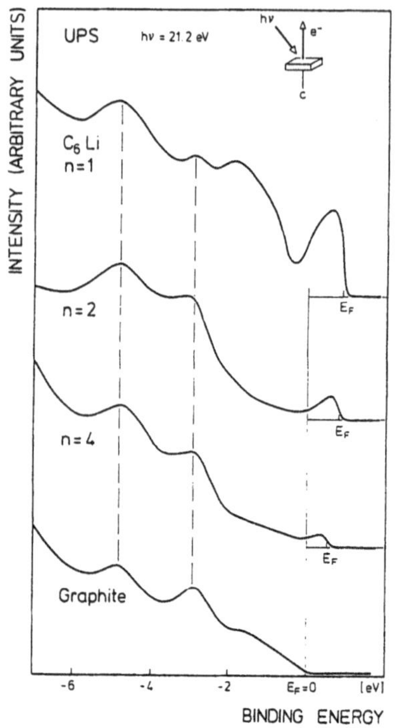

Fig. 13

UPS valence band spectra of pure graphite and lithium-GIC's of stages 1, 2 and 4. Unpublished data, kindly supplied by P. Oelhafen [92].

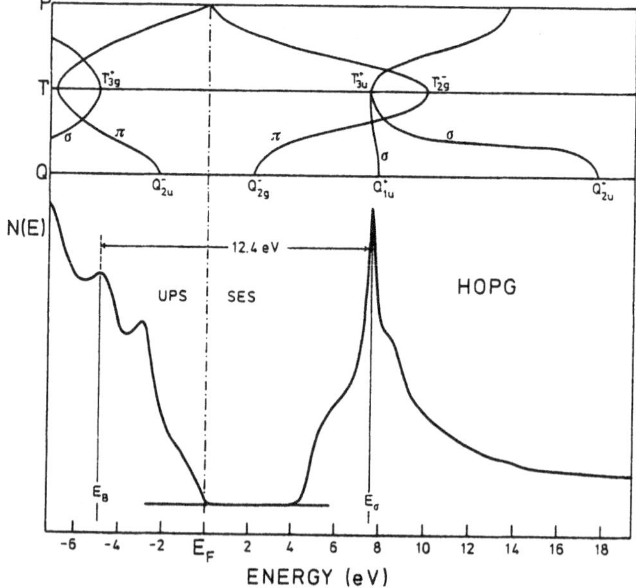

Fig. 14

UPS ($\hbar\omega$ = 21.2 eV) and SES (primary energy E_p = 80 eV) spectra of pure graphite [95]. In the upper part of the figure, a section of the graphite band structure, calculated by Painter and Ellis [113], is shown.

from Table I that non rigid band effects are effective in all AGIC's of stage 1, the most important departure from the 12.4 eV HOPG value occurring in C_6Li [79, 95]. The Fermi energy shift ΔE_f can, in principle, be determined from ΔE_B and ΔE_σ measurements alone. However, since the reference peaks are not completely rigid, the best values for ΔE_f are probably obtained from the width of the conduction band peak itself, as determined from 21.2 eV (Fig. 12) or low-energy UPS. The various data obtained by these procedures for the first stage AGIC's are given in Table I. Table II shows actual values for the DOS $N(E_f)$ in donor-GIC's.

The density of the unfilled states can be measured directly by BIS (Bremsstrahl Isochromat Spectroscopy) or obtained from APS (Appearence Potential Spectroscopy). Such measurements are actually under way and will provide further detailed informations on the electronic structure of GIC's [101, 102].

Table I Fermi energies and non-rigid band effects in stage 1 AGIC's. All data were kindly supplied by J. Krieg and U. Gubler [79, 95]. The symbols are explained in Fig. 14.

	Apparent Fermi energy displacement ΔE_f in eV determined from					σ-bandgap in eV $E_\sigma - E_B$
	ΔE_B (Γ^+_{3g})	ΔE_B (Q^-_{2u})	ΔE_σ (Γ^+_{3u})	Conduction bandwidth β ($h\nu = 21.2$ eV)	Conduction bandwidth β ($h\nu \leq 6.5$ eV)	
C_6Li	1.0		2.0		1.7	11.4
C_8K	1.2	1.2	1.6	1.25	1.4	12.4
C_8Rb	1.2	1.2	1.5	1.2	1.3	12.1
C_8Cs	0.9	0.9	1.4	1.15	1.2	11.9
HOPG						12.4

Table II Fermi level density of states in stage 1 donor GIC's and some of their parent materials obtained from low-temperature specific heat measurements. The given values are not corrected for electron-phonon enhancement factors λ. For these factors, values λ ≅ 0.3 are usually taken for C_8M compounds. Letters a–e refer to Refs. [96–100] respectively, letter f refers to Ref. [90].

$N(E_f)$	C_6Li	C_8K	C_8Rb	C_8Cs	C_4KHg	HOPG	Li	Cs	Cu
states/eV · atom	0.18 [f] 0.21 [e]	0.32 [a] 0.30 [c]	0.49 [a] 0.28 [d]	0.19 [a] 0.27 [c]	0.40 [b]	0.007 [a]	0.69	1.36	0.30
states/eV · molecule	1.3 [f] 1.5 [e]	2.9 [a] 2.7 [c]	4.4 [a] 2.5 [d]	1.7 [a] 2.4 [c]	2.4 [b]				

e) Current Viewpoints

Despite above considerations, which indicate different electronic structures in C_8M and C_6Li, it should be pointed out that some work related to the electronic structure of AGIC's does give similar results for the two types of compounds. For example, calculations of f from C–C bondlength changes give values of ~ 40 % in both cases (paper II). It remains to be established exactly to what extent the simple tight binding approach neglecting non-rigid band effects, used to calculate the bond length changes, is appropriate to treat the C_6Li compound.

As another example, the Knight shift reduction at the alkali nucleus, K_{iso} (GIC)/ K(met), is very similar in ^7Li, ^{87}Rb and ^{133}Cs N.M.R. experiments [103–106], namely 10–20 %. In particular, it can*not* be concluded from the small absolute value of the ^7Li Knight shift in C_6Li that the lithium atom is in its Li^+ state. In general, as C and I atoms can be specifically selected, N.M.R. experiments were expected to give complementary information on the charges lost or gained in the C- and I-subsystems. Unfortunately, the measured N.M.R. resonance line shifts of both, guest and host atoms, are not easy to interpret. Separating the various contributions and finding a proper reference is the task in ^{13}C N.M.R. measurements [82]. The Knight shift of intercalated alkali metal nuclei involves the density of the states at the Fermi level, $N(E_f)$, weighted by their probability density at the site of the nuclei, $\langle |\psi(0)|^2\rangle_{Fs}$. As s- and π-states contribute differently, the character of the Fermi level states should be known in detail at the site of the metal nuclei in real space, as well as the partial DOS $N_s(E_f)$ and $N_\pi(E_f)$ [103]. Such detailed information will hopefully be gained for C_6Li and C_8M compounds from self-consistent LAPW energy band and charge distribution calculations, especially adapted for layered compounds, which are actually under way [107].

The case of the C_6Li compound, where the *c*-axis expansion is only 10 %, and the Li atoms almost completely sink into the hollows of the C-hexagons, demonstrates a basic problem related to the concept of charge transfer. Although 100 % of the Li 2s electrons might have gone into the graphite π^* bands, the electron density at the Li sites can be important, since $|\psi_\pi|^2$ may differ considerably from zero at these locations. In a real space picture then, full charge is not transferred to the carbon subsystem, and it is somewhat arbitrary to decide whether the lithium atoms are in the ionic Li^+ state or not. Therefore, besides the problems of partial covalent bonding and localized charges, one has to distinguish the real and reciprocal space aspects in discussing delocalized transferred charges.

f) Summary

In summary, there is a tendency towards considering C_6Li as a pure π-electron system, although some problems have been pointed out and remain unsolved. Striking differences in the electronic structures of C_6Li and C_8K have recently been detected experimentally and shall briefly be discussed to close this section. Fig. 15 depicts the identical positron annihilation ACD's in pristine graphite and C_6Li,

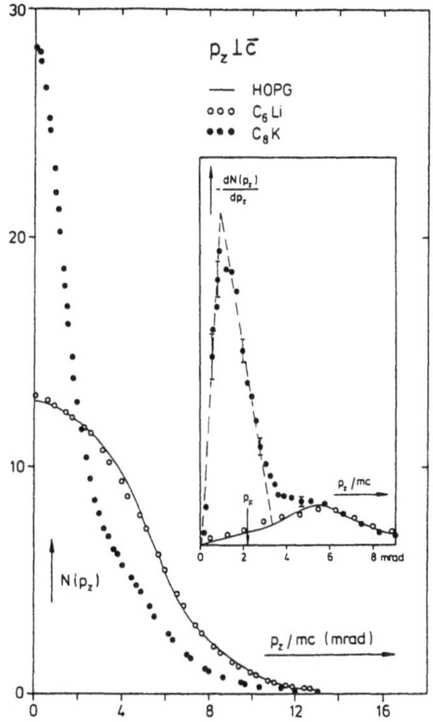

Fig. 15
2γ angular correlation distribution of positron annihilation in pure graphite, C_6Li and C_8K. The γ-emission is parallel to the crystallographic $\underline{c}$-axis. Inset: First derivatives of the angular distributions.

while the analysis of the ACD in C_8K gives a nearly free electron parabola. In the free electron picture, determination of the Fermi momentum p_f leads to $E_f = 1.24$ eV and f = 30 %, in excellent agreement with other experiments. Clearly the β^+ probes do not see a similar alkali-like free electron gas in C_6Li. Information on the conduction band dispersion of AGIC's can be obtained from photoemission by studying the conduction bandwidth as a function of incident UV photon energy [79]. It is known from the pure alkali metals that conduction bands with a parabolic dispersion lead to a likewise parabolic energy dependence of the conduction bandwith. While the expected parabolic free electron-type dependence is found for C_8M compounds, there is no energy-dependence or a weak linear one in low-stage C_xLi compounds. (For details see Ref. [79].)

4.4 Electronic Structure of Higher Stage AGIC's

Much less information has been obtained on the electronic structure in AGIC's of stages $n \geqslant 2$ than on the first stage compounds. The structure of the I-sublattice is more complex and less well-known than in C_6Li and C_8M. At room-temperature, I-layers in $C_{12n}M$ are liquid-like (section 2), while details on the low-temperature structure lack. Due to these complications, full band structures have not been calculated from first principles for these compounds. Even assuming a reasonable, fully periodic crystallographic structure, problems arise from the increasing number of atoms per unit cell in these high-stage GIC's. In the case of Li-GIC's, a calculation for an ideal second stage $C_{12}Li$ model compound has been reported [94]. Unfortunately, contrary to the heavy alkali GIC's where the $C_{12n}M$ formula is well established for $n \geqslant 2$, the pure higher stage Li-GIC's are hard to prepare and still poorly defined with respect to their chemical composition and structure [9].

For $n \geqslant 3$, C-layers start to become inequivalent, as electrostatic screening comes into operation (section 4.2). This aspect produces an additional complication in calculating electronic structures and interpreting experimental results. It seems, therefore, reasonable to approach these materials either from the already discussed first stage limit or from the very dilute limit, in which a GIC is considered as graphite charged by I-impurities [58]. The obvious questions are then: What is the minimum alkali metal concentration necessary for s $(+\ \pi)$-character $(f < 1)$? When does the dilute limit start? When is the charged graphite picture a good approximation? The problem is of particular importance, because band structure models have been elaborated for stage $n \geqslant 2$ GIC's, which consider a GIC precisely as a sum of inhomogeneously charged, but otherwise unperturbed graphite layers [41, 108].

Plotting $N(E_f)$ of various AGIC's, as determined from UPS measurements [76], against the π^* conduction band DOS of graphite [109], a distinct extra-contribution from s-states is seen for stages $\leqslant 3$, while such an extra DOS is not found within experimental error for stages $\geqslant 7$ in $C_{12n}Cs$ GIC's. This procedure, although rather insensitive, already places the loss of s-band behaviour at stages $4 \leqslant n \leqslant 6$.

In his low-energy UPS work, Gubler [79] studied the conduction bandwidth β as a function of photon energy $h\nu$ in stage 1, 2, 4 and 5 K-GIC's (Fig. 16). The stage 1 and 2 compounds C_8K and $C_{24}K$ exhibit a parabolic $\beta(h\nu)$ dependence as expected

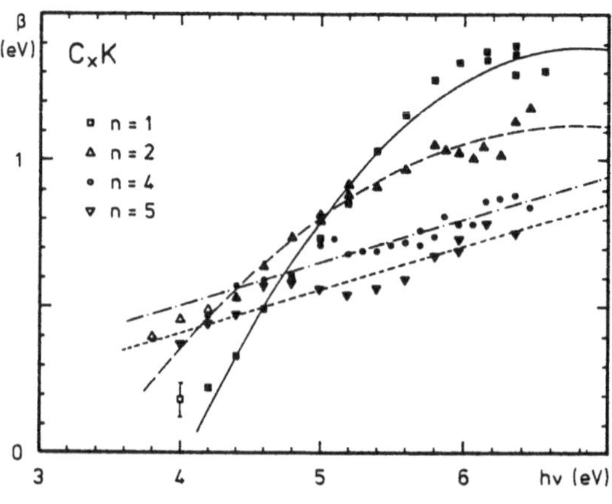

Fig. 16

Conduction bandwidth β, measured by UPS, as a function of incident photon energy in potassium-GIC's of stages 1, 2, 4 and 5 [79].

to be induced by nearly free electrons. In contrast, the stage 4 and 5 compounds $C_{48}K$ and $C_{60}K$ have a weak and linear energy dependence of β, determined by non-s-like bands.

In SES, the observed main peak is split for $n = 4$ and $n = 5$ K-GIC's, while this is not the case for other stages [95] (Fig. 17). This observation also indicates a change of electronic structure at these concentrations which might as well be related to the existence of inequivalent C-layers.

We recall that positron annihilation experiments revealed a drastic change of behaviour between $n = 4$ and $n = 5$ (Fig. 8, section 4.2). For $n > 5$, purely graphitic regions are found by the positrons. Therefore, the K^+ layers must be fully screened and s-wavefunction tails, if any, do not overlap. On the other hand, alkali-like angular correlation distributions are found for $n \leqslant 4$, showing the band character of the s-electrons and incomplete screening of the K^+ layers.

The de Haas-van Alphen effect has been studied in K-GIC's of stages $n = 1-4$ at 1.4 K [110]. The model used to interpret the observed 5 periods in $C_{48}K$ and 2 periods in $C_{36}K$ (rigid graphite bands zone-folded by the *c*-axis superlattices, i.e. homogeneous charge distribution) is not very convincing in view of the actual knowledge on AGIC's. An extension of this model has been proposed to account for Shubnikov-de Haas and magnetoreflection data on high stage K- and Rb-GIC's [108]. In any case, sets of data have been obtained, which would hopefully also be

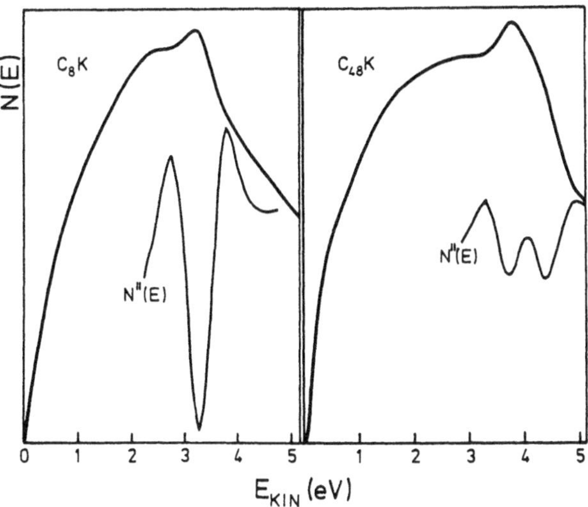

Fig. 17 Secondary electron energy distribution in the potassium-GIC's of stages 1 and 4. The characteristic feature on top of the SE energy distribution is a two-peak structure in the stage 4 compound, as can be seen from the second derivative of the SES curve [95].

explained by the more sophisticated Fermi surface models, as they will become available for higher stage AGIC's.

In summary, several experiments reveal a discontinuity in the electronic structure of $C_{12n}M$ GIC's as a function of stage for $n = 4$–5. Below that limit, s (+ π) character is found, whereas above stage 5, AGIC's can probably be described in a proper way by a system of charged graphite layers. Note that DDF [58] have correctly estimated the onset of perturbation of the graphitic behaviour at stages 4–5 in their dilute limite approach, although the charges are not distributed over the C-sublattice as it was assumed in their original paper.

Similar conclusions cannot be extended to the C_xLi system, because many experiments assign π-metal character already to the stage 1 compound C_6Li. Detailed experimental information on the electronic structure of Li-GIC's as a function of increasing stage number has not been achieved yet.

4.5 Electronic Structure of Acceptor-GIC's

In contrast to the low-stage donor GIC's, there are no intercalant-type states around the Fermi level in acceptor-GIC's. Energy levels of the species constituting the intercalant molecules have been found at ~ 5 eV below E_f and lower in Br_2- and $FeCl_3$-GIC's [111, 112]. In photoemission spectra, these states may mask the σ and π valence states of graphite for some excitation energies, while it can be seen that the graphite valence bands are scarcely affected when radiation of proper wavelength is used [112]. Strong coupling between I- and C-layers, which would lead to warping of the graphite valence bands, does not occur in acceptor-GIC's. The Fermi states must, therefore, simply be obtained by lowering E_f into the graphite π DOS (Fig. 7). As E_f moves away from the band crossing region at the Brillouin zone edge (Fig. 6), Fermi surfaces, nearly cylindrical, grow around these points. This situation is schematically shown in Fig. 1 of paper II within a purely 2-dimensional approximation for the case of the donor GIC's.

The strong DOS peak found at E_f in low-stage AGIC's is absent in acceptor compounds. Indeed, the values $N(E_f) \approx 0.06$ states/eV · atom, experimentally obtained from absolute spin susceptibility measurements in stage 1 and 2 AsF_5-GIC's [109], are an order of magnitude larger than the graphite value, but they are also smaller than the values obtained for low-stage donor-GIC's (Table II), typically by a factor 5.

The N(E) curve can be obtained from one of the numerous band structure models for pure graphite (i.e. [113]). In the vicinity of E_f, N(E) is approximately linear (Fig. 7). The slope of the DOS is largely determined by the tight binding resonance integral γ_0, which describes the nearest-neighbour in plane C–C interaction. By assuming γ_0, an experimentally obtained value for $N(E_f)$ or ΔE_f will give the amount of the transferred charge. However, if $N(E_f)$ *and* ΔE_f are known, the slope of N(E) can be fixed and f can be determined more accurately.

A few years ago, the f values given in the literature for acceptor-GIC's ranged from 1 %–100 % (!), and the indirectly obtained values for ΔE_f were scattered widely. It is reassuring to see that a broad consensus is already being obtained on these points owing to the availability of a large number of experimental data; some recent values for f are tabled in paper II.

In particular, ΔE_f values have been obtained directly by spectroscopic techniques. Carbon 1 s core electron excitation spectra have yielded $\Delta E_f = -0.9$ eV for stage 2 AsF_5-intercalated graphite, as compared with −0.9 and −0.7 eV for $n = 1$ and $n = 2$ $FeCl_3$-GIC's [114]. Stage 2 $FeCl_3$-GIC's were also investigated by photoemission and Auger electron spectroscopy, confirming the result $\Delta E_f = -0.7$ eV [112].

Investigation of the Fermi surface of acceptor GIC's by quantum oscillations has been particularly popular in the recent times. De Hass – van Alphen (magnetization), Shubnikov – de Haas (resistivity) and magnetothermal (temperature) oscillations in GIC's have been studied by 7 research groups [108, 115–120]. The Fermi surfaces, used in the past to discuss the observed periods, were mostly either cylindrical or long cigar-like in shape. Magnetic breakdown effects have also been invoked in order to interpret the experimental data. Folding effects at the new Brillouin zone boundaries due to the *c*-axis periodicity of the I-layers have been increasingly considered. Some of the results are even explained in the 3-dimensional rigid band [58] and in the 2-dimensional metallic sandwich models [61] (see section 4.1). It has been reported that the experimental results are very sensitive to the sample treatment [117, 120]. More careful experiments as well as theoretical work will be necessary to obtain the Fermi surface details in GIC's quantitatively by quantum oscillatory techniques. However, the most recent values for f, E_f or γ_0 obtained from these experiments are converging and agree reasonably well with the corresponding values obtained by other techniques. From these data, typical $N(E_f)$ values of 0.05 states/eV · atom in lowest stage acceptor GIC's can be calculated, thereby confirming the earlier results [109].

As in the donor GIC's, nonhomogeneous screening is effective for $n \geqslant 3$. Blinowski calculated the energy band structure of stage 1–4 acceptor GIC's in a tight binding "model of independent graphite subsystems" [121, 122], including the charge distribution effects for $n = 3, 4$. The use of only two overlap parameters, γ_0 and γ_1 (γ_1 : interlayer interaction between nearest C-atoms), have led already to a reasonably good description of the energy bands in acceptor-GIC's. This situation is in contrast to pure graphite, where up to 7 energy parameters are needed for an accurate description [56]. The reason for this is that graphite has a very singular electronic structure, since E_f lies in a deep minimum of the DOS, and all Fermi surface effects occur in the corners of the Brillouin zone. Many energy parameters of good accuracy are, therefore, necessary to account for all the subtle effects. As one moves out of this critical zone by intercalation, the complexity of the problems is reduced, and many simplifications and approximations in the calculation become admissible.

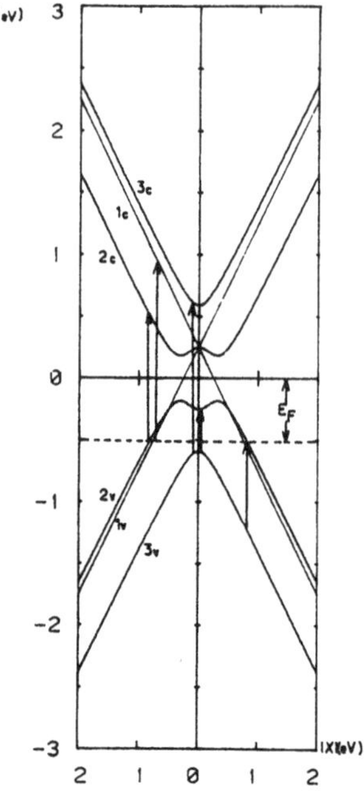

Fig. 18

Structure of the graphite bands near the Brillouin zone edge in an acceptor-GIC of stage 3. The energy bands have been calculated within the model of independent graphite subsystems [121, 122]. The potential energy difference 2δ between the inner and the bounding C-layers has been chosen as $2\delta = 4\gamma_1/3$.

Figure 18 shows what happens to crossed graphite bands at the zone boundary (paper II, Fig. 1) in stage 3 and 4 acceptor GIC's. Due to γ_1 and electrostatic energy terms, degeneracies are lifted and the bands split into strongly warped 3 and 4 graphite band systems which are no more symmetric around zero energy. It is immediately seen that the interband spectrum is very different from stage 1 compounds (no splitting). It is also different from a hypothetical homogeneously charged compound, in which the two-peak structure of the 2v and 2c bands would not occur and the 1v and 1c bands would not be shifted above zero energy.

5 Some Physical Properties of GIC's

5.1 Anisotropy

The anisotropy parameters of several important physical properties in pure graphite are shown in Table III. These anisotropy values are severely affected by intercalation. To summarize, intercalation of donors will weaken the graphite's anisotropy, while intercalation of acceptors will increase it. This is obvious for anybody who ever handled GIC's, as *c*-axis cohesion and cleaving resistance is notable in donor-intercalated HOPG crystals, while acceptor compounds can be cleaved with almost no mechanical resistance.

Table III Anisotropy parameters in pure graphite. All values are given for HOPG graphite, except the electron effective mass ratio which is given for single crystals.

Elastic constants	Thermal conductivity	Electron effective masses	Electrical conductivity	g-factor
$C_{11}/C_{33} = 30$ (45)	$k_{\|}/k_{\perp} = 200$ (45)	$(m^*_{33})_{el}/(m^*_{11})_{el} = 230$ (56)	$\sigma_{\|}/\sigma_{\perp} = 5000$ (45)	$g_{\|} - g_{\perp} = 0.0428$ (83)

Table IV Electrical conductivity anisotropies $\sigma_{\|}/\sigma_{\perp}$ in GIC's. Data for donor GIC's are taken from Ref. [123], while data for acceptor-GIC's are taken from Ref. [75].

Donor compounds		Acceptor compounds	
C_6Li	11	$C_{16}ICl$	$1{,}5 \cdot 10^4$
C_8K	33	$C_{16}Br_2$	$7 \cdot 10^4$
C_8Cs	20	C_6HNO_3	$2 \cdot 10^5$
		$C_{16}AsF_5$	$> 10^6$

There are only few physical properties which have been measured in the *a*- and *c*-directions of donor- and acceptor-GIC's. Among them, electrical conductivity can be cited. Compared to pure HOPG, an anisotropy increase by two orders of magnitude is easily found in acceptor-GIC's, while a decrease of the same amount occurs in donor-GIC's. The most anisotropic compounds measured are the AsF_5-GIC's, while C_6Li is the GIC for which the lowest conductivity anisotropy has been reported (Table IV).

An attempt to obtain effective mass anisotropies in low-stage AGIC's has been made from polarized optical reflectivity measurements [124]. As for the conductivity anisotropy, a drastic reduction is found in the investigated stage 1 AGIC's. Anisotropy factors of 8 and 9 were found for C_8K and C_8Cs. On the other hand, quantum oscillation experiments generally indicate very high effective mass anisotropies in acceptor-GIC's.

The different anisotropies in donor- and acceptor-GIC's are related to the electronic structures. The situation is illustrated by the Fermi surfaces in these compounds. The acceptor-GIC Fermi surfaces are essentially cylinders around the Brillouin zone edges (Fig. 1, paper II). In a donor GIC like C_8K, nearly 2d graphitic portions of the Fermi surface with cylindrical symmetry are found at the corners of the zone, but 3d alkali-like, almost spherical portions occur around the Γ-point and the A-point (Fig. 19b). In C_6Li, the C–C interlayer interactions are much stronger than in any other GIC. C-atoms of neighbouring planes feel each other via their interaction with the lithium ions. As they are not separated far enough geometri-

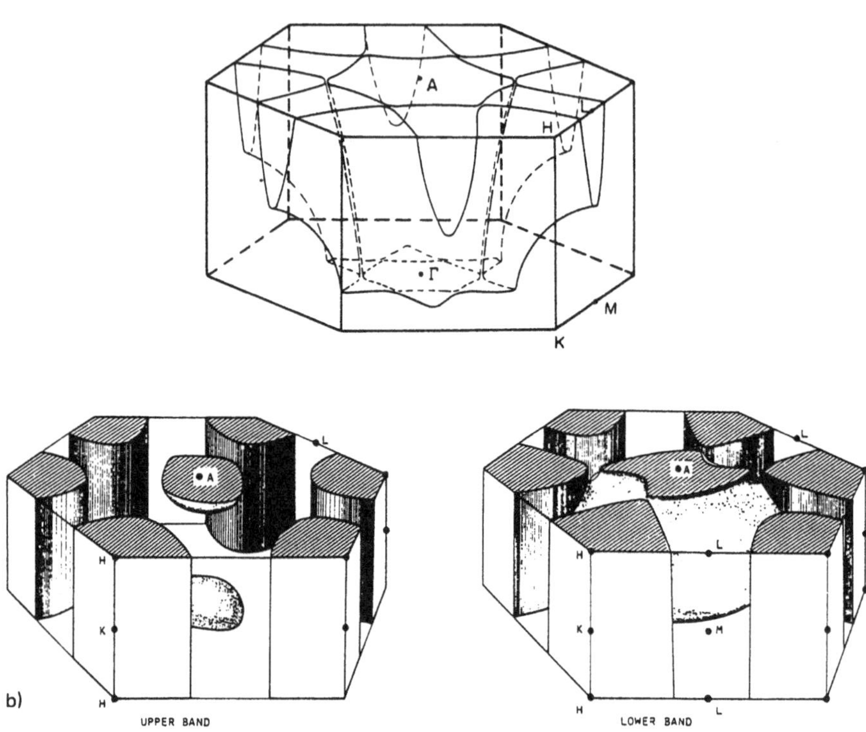

Fig. 19 Fermi surfaces in the first stage GIC's C_6Li (a) and C_8K (b).

a) Fermi surface for C_6Li derived from Fig. 11a [74].

b) Fermi surface for C_8K derived from a tight-binding energy band calculation [73].

cally to reduce this effect, the ideally cylindrical graphitic Fermi surface gets strongly warped (Fig. 19a), leading to a 3d pattern even without the existence of a nearly isotropic s-like portion.

Furthermore, alkali s- and graphitic π-orbitals mix to some extent in AGIC's. Partial covalent bonding can occur in *c*-direction with charges localized between the I- and C-atoms. Acceptor-GIC's are much more salt-like. Charges are bound to the acceptor species, at least as long as the acceptor does not attack the graphite sublattice as a consequence of very high oxidation strength. Normally, there is no covalent bonding and the C-planes are appreciably separated by the intercalant. Some examples of C–C layer separation by the intercalant are given in Table V.

Table V Interplanar distances d_{c-c} of neighbouring carbon layers separated by an intercalant layer. As shown in Fig. 2, d_{c-c} is the sum of the intercalant layer thickness, d_i, and the thickness of a carbon layer, d_c = 3.35 Å. All data were taken from Refs. [27, 66] and obtained from stage 1 compounds, except the values denoted by *, which were obtained from stage 2 compounds.

Graphite	3.35	*Acceptor molecule*	d_{c-c} (Å)
Donor atom	d_{c-c} (Å)	$AuCl_3$	6.80
		Br_2	7.05*
Li	3.71	N_2O_5	7.80*
Sr	4.94	H_2SO_4	7.98
Eu	4.87	SbF_5	8.29–8.46
K	5.35	$MoCl_5$	9.70*
Cs	5.93	$AlBr_3$	10.24

5.2 Electrical Resistivity

a) Concentration Dependence of Electrical Resistivity

It is well-known that in most GIC's, the minimum electrical resistivity is not obtained for maximum intercalant concentration, but rather around 2 atomic percent of I-species, corresponding to $n \approx 4$ or 5 (Fig. 20). This occurs in donor GIC's, like C_xK

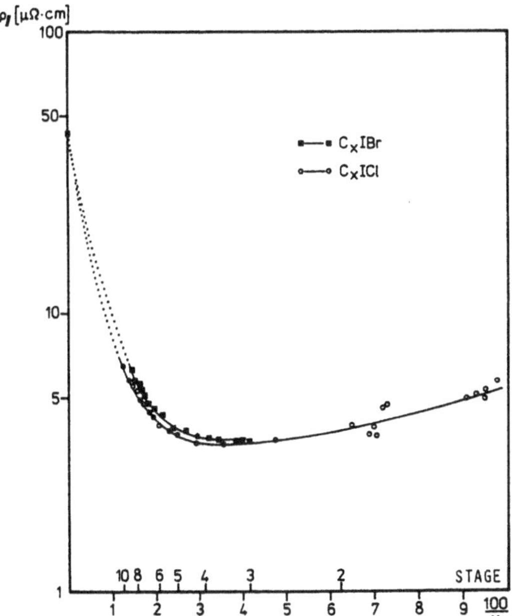

Fig. 20
In-plane resistivity of the inter-halogen-GIC's $C_x(ICl)$ and $C_x(IBr)$ as a function of intercalant concentration. The measurements were done by contactless inductive methods.

[125], and in acceptor GIC's, like $C_x(H_2SO_4)$ [126], and many others. However, in the following we discuss the acceptor-GIC's only. The reasons are that electrical transport is better understood theoretically in acceptor-GIC's due to their simpler electronic structure, and that electrical resistivity is lower than in the donor-compounds, making acceptor-GIC's technologically more interesting.

The concentration (or stage) dependence of electrical resistivity is in fact a trivial geometrical effect. However, one has to keep in mind the real space charge distribution to account for this property. One has to know that the negative charges are rigidly bound on the acceptor molecules and are often separated by neutral species. Therefore, the I-layer will have to be considered as an insulator. One also has to know that the C_b-layers bear most of the delocalized charges and that a decay of the hole density of states occurs towards the inner C_i-layers. This implies that the resistivity will increase from C_b- to C_i-layers. To obtain the overall conductivity, it is sufficient to sum up the conductivity contributions for each layer type and to weight them by their relative frequencies. The in-plane conductivity of an *n*-th stage acceptor GIC is then given by

$$\frac{1}{\rho_{GIC}}(n) = \frac{d_I}{d_I + n \cdot d_c} \cdot \frac{1}{\rho_I} + \frac{2\,d_c}{d_I + n \cdot d_c} \cdot \left(\frac{1}{\rho_{c_b}} + \frac{1}{\rho_{c_i}^{(1)}} + \ldots + \frac{1}{\rho_{c_i}^{(j)}} \right) + \frac{(n - 2(j+1)) \cdot d_c}{d_I + n \cdot d_c} \cdot \frac{1}{\rho_c},$$

with

$$\frac{1}{\rho_I} \ll \frac{1}{\rho_{c_i}^{(j)}} < \ldots < \frac{1}{\rho_{c_i}^{(1)}} < \frac{1}{\rho_{c_b}}.$$

The symbols d_I and ρ_I denote the thickness and the resistivity of the I-layers, d_c is the thickness of a C-layer, $\rho_{c_b}, \rho_{c_i}^{(1)}, \ldots, \rho_{c_i}^{(j)}$ are the resistivities of the bounding C-layer, the first, ... , j-th inner C-layers, and ρ_c is the resistivity of a neutral graphite layer. The resistivity of the single charged C-layers is just what has been calculated by Strässler and Pietronero [127].

It is very easy to see that the electrical resistivity must increase for low stages, because the relative amount of thick insulating layers goes on increasing (Fig. 20). On the other hand, in the dilute limit, the small number of highly conducting C_b layers dominate the resistivity. Then $1/\rho_{GIC}$ goes as $1/n$ + const., which means that the conductivity is a linear function of intercalant concentration.

b) Minimum Absolute Electrical Resistivity

A lot of ink and paper has already been used up on this topic, and there have been many speculations on what should be done to get conductors better than copper out of GIC's. We would like to contribute to this question with two remarks.

The first point concerns the starting graphite which is used to get highly conducting GIC's. Single crystalline and HOPG graphites have basal plane resistivities of 40

$\mu\Omega \cdot$ cm, whereas high-purity annealed copper has a resistivity of 1.7 $\mu\Omega \cdot$ cm. Copper and aluminium used for technological purposes have resistivities which range from 1.9 to 2.3 and from 2.8 to 3.5 $\mu\Omega \cdot$ cm, respectively. This compares with currently obtained values of typically $\rho_{\parallel}$ = 2.5–4 $\mu\Omega \cdot$ cm for acceptor GIC's and $\rho_{\parallel} \geqslant 4.5$ $\mu\Omega \cdot$ cm for donor GIC's [13–17], all made from HOPG or parent materials. It is out of the question to use GIC conductors out of HOPG or single crystalline material. It seems also very difficult to densify and orient graphite powders [18, 19, 66]. One should really try to obtain commercially producable, practically usable graphites of reasonable price with low starting resistivity rather than synthesizing more and more HOPG-GIC's. Several 100 $\mu\Omega \cdot$ cm are typical for the available graphite fibers and non-highly oriented pyrolytic graphites. There is some hope to obtain fibers with $\rho \approx 80$ $\mu\Omega \cdot$ cm along the fiber axis. Reducing its resistivity by a factor of ~ 20 by intercalation would already place such a material at a technologically interesting level, since the resistivity X density product can be very favourable.

A second remark concerns more directly the specific electrical resistivity of GIC's. It has been shown that the mobility of the carriers is already almost at its maximum in the C_b layers. The specific resistivity of these C_b layers is by far better than that of bulk copper. As the macroscopic resistivity has to be related to the whole volume of a GIC including the dead I-volume, the only chance to lower the overall resistivity is to fill a relatively high volume with the highly conducting C_b-layers. This should be the case in a stage 2 acceptor-GIC with I-layers of low thickness. Furthermore, the specific resistivity could be considerably reduced if the I-layers were themselves *2-dimensional conductors*, similar to the graphite C_b-layers, instead of being insulators as in the classical acceptor-GIC's or instead of building up a strongly scattering 3d electronic subsystem as in donor-GIC's.

5.3 Optical Reflectivity

There have been many studies on the optical reflectivity of GIC's since the mid-seventies, covering the energy ranges from the near IR up to 6 eV [75, 128–131 and references therein]. Typical reflectivity curves are shown in Fig. 21. Pure graphite has low IR reflectivity due to the semimetallic early onset of interband transitions. In contrast, IR reflectivity is high for GIC's until a metallic reflectivity drop occurs at energies $\hbar\omega$ typically between 1 and 3 eV. In the investigated energy range, GIC reflectivity spectra exhibit the following features: 1. low energy absorption peaks, 2. reflectivity edges, 3. a broad reflectivity background above a reflectivity minimum $\hbar\omega_{min}$ and 4. a reflectivity peak around 5 eV. These features, known for several years, are now better understood because of the improved knowledge on the electronic structures in AGIC's and acceptor-GIC's.

Within his 2d band structure model of "independent graphite subsystems" [121, 122], Blinowski calculated the frequency-dependent dielectric function including core, intraband and interband contributions in order to interpret the optical properties of stage 1 and 2 acceptor-GIC's. Furthermore, the interband transitions were analyzed within this framework for stages 3 and 4. The experimentally obtained characteristic IR structures of stage $n \approx 2$–4 acceptor-GIC's (Fig. 22) have been shown to be induced by the transitions from valence to valence bands (see Fig. 18) [130].

The reflectivity edges are due to collective excitations of free carriers in the GIC's. Note that the position of this edge shifts within the visible range as a function of intercalant concentration (Fig. 21). In the case of the Cs-GIC's, this gives a lustrous

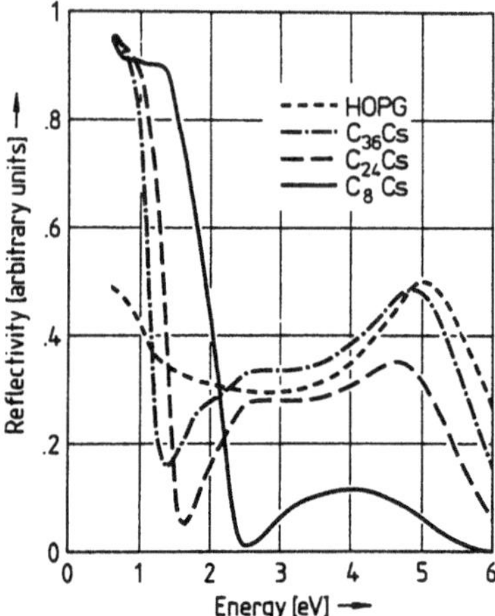

Fig. 21

Optical reflectivity of pristine HOPG and the Cs-GIC's of stages 1, 2, and 3 at near-normal incidence. The curves for the intercalation compounds are normalized to 0.95 at 0.6 eV, whereas the curve for HOPG is drawn on an absolute reflectivity scale.

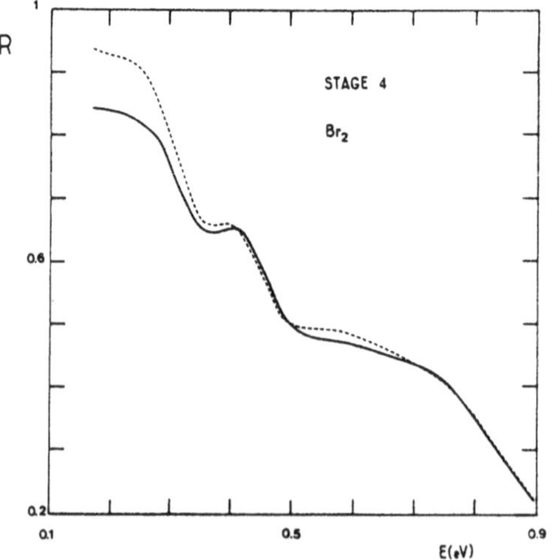

Fig. 22

Infrared reflectance spectra for stage 4 bromine-GIC's. Solid line: experiment. Dashed line: theoretical curve obtained from the model of independent graphite subsystems. From Ref. [130].

metallic golden colour to the stage 1 compound C_8Cs, while the stage 2 compound $C_{24}Cs$ is of shiny dark blue colour. The higher stage compounds look more and more graphite-like. The reflectivity edges of low-stage GIC's have mostly been modelled with a free electron Drude term [75]. This may be appropriate to describe the alkali-like nearly-free electrons which exist in the lower stages of heavy alkali metal GIC's. However, if the π-type free carrier plasmon is excited, the simple Drude model is not expected to work well. In fact, the π-carriers are only free to move essentially in 2 directions. Even in the case of C_6Li which is probably a rather 3d π-electron metal, the conduction carriers are not free electron-like at all, as can be seen from the non-spherical shape of the Fermi surface (Fig. 19a).

Contrary to pure graphite, interband transitions from the graphitic v to c bands have a high threshold energy $\hbar\omega_T$, due to the shift of the Fermi level into the graphite bands (Fig. 18). Such transitions can start only for $\hbar\omega \geqslant \hbar\omega_T = 2\,\Delta E_f$. Through this relation, ΔE_f can be obtained from optical reflectivity curves by the analysis of the v $\rightarrow$ c interband transition onset. This type of transitions produces the broad background in reflectivity above $\hbar\omega_{min}$.

Finally, the reflectivity peak around 5 eV is due to critical point transitions from the π valence to the π^* conduction bands, especially in the Q point (Fig. 14). The shift of this peak, which occurs at 5.05 eV in pristine graphite, as a function of intercalant concentration gives direct information on the rigidity of the graphite π-bands upon intercalation [129].

Phenomenological models for the complex dielectric function $\hat{\epsilon}(\omega)$, supposed to describe the experimental optical reflectivity curves of GIC's, have to include contributions from the above discussed 4 types of features. Such a model has been used successfully to describe optical reflectivity in $SbCl_5$ (acceptor)-GIC's of various stages [131]. It has been extended to analyze the reflectivity curves of C_xLi donor-GIC's of stages $n = 1$–4 [132]. The optical reflectivity data of these compounds represent the first coherent set of data on several stages of the lithium/graphite system (Fig. 23). The employed model is based on the concept of inhomogeneously charged, but otherwise unperturbed graphite layers. Adjustment of the model to the experimental results yields parameters which are completely unreasonable for $n = 1$ and doubtful for $n = 2$. However, the model works well for $n = 3$ and $n = 4$ compounds. This indicates that the charged graphite picture might be a good approximation for Li-GIC's of stages $n \geqslant 3$, while this is only the case for stages above $n = 5$ in heavy alkali metal GIC's (section 4.4).

5.4 Other Electronic Properties

Many other properties dealing with electrons and holes in GIC's could be discussed. However, a choice had to be made and the authors do not wish to go much beyond the present point. They rather wish to end up by mentioning that the magnetic susceptibility of AGIC's as a function of stage shows a peak at high stages on going from strongly diamagnetic pure graphite (stage ∞) to weakly paramagnetic low-stage

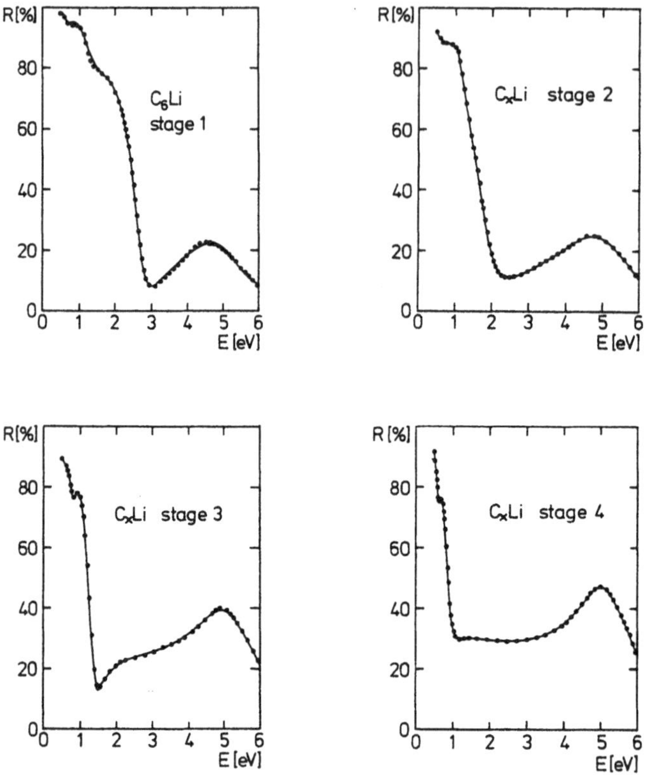

Fig. 23 Absolute optical reflectivity spectra of stage 1–4 lithium-GIC's taken at near-normal incidence.

AGIC's [133]. This behaviour is understood only within the concepts of I-layer screening and non-homogeneous charge distribution [134].

Finally, it should at least be remarked that some GIC's are superconductors (SC). The stage 1 C_8M compounds (M = K, Rb, Cs) have been found to show SC below 1 K, with transition temperatures T_c depending on the samples and on the initial graphite used [135–137]. C_8K shows either type I or type II SC, according to the magnetic field direction with respect to the crystal axes [137]. Very recently, it was reported that the stage 2 heavy metal ternaries C_8KHg and C_8RbHg are SC below T_c = 1.90 K and T_c = 1.44 K, respectively [138].

Conclusion

In the last few years, much progress has been made on the electronic structure of alkali metal and acceptor GIC's. All details are not clear yet, but, in the authors' opinion, an overall satisfactory knowledge seems to be looming. Unfortunately, almost nothing is known on the electronic structure of systems with valence carriers other than the σ and π electrons of graphite and the one s electron of the alkali metals already discussed. It would be interesting to know the influence and behaviour of the $4s^2$, $5s^2$ and $6s^2$ electrons in the alkaline earth metal GIC's C_xCa, C_xSr, C_xBa, of the $4f^6$, $4f^7$, $4f^{13}$ and $4f^{14}$ electrons in the rare earth metal GIC's C_xSm, C_xEu, C_xTm and C_xYb and of the $5d^{10}$ and 6p electrons in the heavy metal ternary GIC's C_xMHg and $C_xMTl_{1.5}$ (M = K, Rb). Certainly, the magnetic properties of the lanthanide GIC's should be studied.

Techniques for mapping GIC Fermi surfaces would be of great interest. Positron annihilation, in particular the new 2d angular correlation distribution technique, would be of great interest. Positron annihilation has proved to be very useful in this respect if the positrons are not trapped in defects [139]. Cartier et al. have shown that this is indeed not the case in GIC's [62, 85]. Quantum oscillation experiments *and* positron annihilation should together reveal the details of the Fermi surfaces in GIC's.

Generally speaking, the electronic transport properties of GIC's are not yet well understood. Only the electrical conductivity problem has been attacked with some emphasis both experimentally and theoretically. The best existing acceptor GIC's already fulfil the requirements for a very good conductor to a large extent, viz. very high mobility due to hard graphite phonons, screening of the intercalate, tight binding type scattering and linear dispersion (paper II). There is not much hope to gain in conductivity unless basically new types of GIC's can be found. Other transport data have been obtained, but detailed interpretations are lacking. Particularly unsatisfactory is the knowledge on the Hall effect in GIC's. The measured data sets of R_H as a function of temperature and stage agree poorly [75, 140–142]. This is perhaps due to problems associated with poor contact, chemical attack and current distribution, as some GIC's are particularly reactive and anisotropic. Also, there have been no attempts to interpret the Hall data from basic principles, i.e. band structures, screening effects and scattering mechanisms.

GIC's offer opportunities to study the intercalant species in quite unusual surroundings. For example, s-band metals like Rb, Cs can be studied in a hexagonal configuration leading to nuclear magnetic resonance shift contributions like anisotropies via direct dipolar interaction, which are normally not obtained from s-electrons. As another example, GIC's offer a possibility to obtain monolayers of magnetic atoms. Such 2d arrangements could be obtained by reducing acceptor GIC's of Fe, Ni, Co, Mn halides in order to keep only the metal between the C-layers, or by synthesizing high stage rare earth metal GIC's. Not very much effort has been spent in these directions until now, although it would clearly have been very valuable to do so.

The study of phase transitions in GIC's has just begun (section 2 and references therein [143, 144]). A great variety of structural ordering phenomena occur, such as in-plane I-ordering, c-axis I-ordering, staging, commensurability/incommensurability of the I- and C-sublattice, etc. Anisotropy introduces additional freedom, since it leads to a distinction between in-plane and out-of-plane interactions. Note that at the present time a great deal of interest exists in the behaviour of absorbates on graphite surfaces. Structures and various order/disorder phenomena of adsorbed atoms and molecules like He, Ar, Kr, Methane, CD_4, N_2, O_2, ... are being studied with enthusiasm (e.g. [145–148]). In fact, a high stage GIC could be considered a model substance, in which atomic or molecular monolayers are repeated periodically on a graphite substrate.

Most of the physical properties in GIC's have been studied as a function of stage or intercalant concentration, temperature, and magnetic field or energy (spectroscopy). It is somewhat surprising to find that there are no systematic investigations as a function of anisotropy. We recall that anisotropy ranges over several orders of magnitude in GIC's, depending on the I-species and the stage number. Anisotropies from 10 to 10^6 have been found in the electrical conductivities of GIC's. The GIC family could provide useful information on the phenomena occurring at the transition from a partly 3d to a practically 2d material.

Several problems in the field of intercalation chemistry still remain unsolved. It is not fully understood yet why some materials intercalate graphite while others do not [149], and how the physical properties of GIC's correlate with the chemical nature and structure of the intercalant elements. Still, much useful information has been obtained on GIC's in the past few years and new compounds with an improved environmental stability are continuously being found. In spite of all the available knowledge, the properties of GIC's are far from predictable to the extent which would allow one to apply them on a large scale as synthetic metals.

Acknowledgements

It is a pleasure for us to acknowledge the fruitful discussions and collaborations we had in this field in the past few years, in particular with E. Cartier and F. Heinrich of the Eidgenössische Technische Hochschule Zürich and with G. Roth and K. Lüders of the Freie Universität Berlin. Additionally, P. Oelhafen, U. Gubler and J. Krieg of our institute kindly supplied us with their data prior to publication. We also would like to thank L. Pietronero, Brown Boveri Research Laboratory, Baden, A. DasGupta, on leave from Oak Ridge National Laboratory, and R. Lapka for critically and carefully reading our manuscript. A. Moore from Union Carbide gratefully supplied us with HOPG graphite, which was carefully reacted to give GIC samples by V. Geiser and S. Stolz. We were financially supported by the Swiss National Science Foundation and by the Schweizerische Volkswirtschaftsstiftung zur Förderung der wissenschaftlichen Forschung.

One of us (P.P.) particularly wishes to thank his wife for her patience during the week-ends and nights when the manuscript was written.

References

[1] *C. Schafhäutl*, J. prakt. Chem. **21**, 129 (1841).

[2] *K. Fredenhagen* and *G. Cadenbach*, Z. allg. anorg. Chem. **158**, 249 (1926).

[3] *D. Billaud* and *A. Hérold*, Carbon **17**, 183 (1979).

[4] *S. Basu, C. Zeller, P. Flanders, C. D. Fuerst, W. D. Johnson* and *J. E. Fischer*, Mater. Sci. Eng. **38**, 275 (1979).

[5] *L. C. F. Blackman, J. F. Mathews* and *A. R. Ubbelohde*, Proc. Roy. Soc. **A256**, 329 (1960).

[6] *P. Pfluger, H.-U. Künzi* and *H.-J. Güntherodt*, Appl. Phys. Lett. **35**, 771 (1979).

[7] *J. O. Besenhard, H. Möhwald* and *J. J. Nickl*, Carbon **18**, 399 (1980).

[8] *D. Guérard* and *A. Hérold*, Carbon **13**, 337 (1975).

[9] *P. Pfluger, V. Geiser, S. Stolz* and *H.-J. Güntherodt*, Synth. Met. **3**, 27 (1981).

[10] *A. Schleede* and *M. Wellmann*, Z. Phys. Chem. **B18**, 1 (1932).

[11] *L. B. Ebert*, Ann. Rev. Mat. Sci. **6**, 181 (1976).

[12] *F. R. Gamble* and *T. H. Geballe*, in "Treatise of Sol. St. Chem.", Vol. III, Ed. N. B. Hannay, Plenum, New York 1976, p. 106.

[13] Proceedings Franco-American Conference on Intercalation Compounds of Graphite, La Napoule, France, Mater. Sci. Eng. **31** (1977).

[14] *J. E. Fischer* and *T. E. Thompson*, Physics Today, July 1978, p. 36.

[15] "Physics and Chemistry of Materials with layered Structures," Vol. 6, "Intercalated layered Materials," Ed. F. A. Lévy, D. Reidel, Dordrecht 1979.

[16] Proceedings of the International Conference on Layered Materials and Intercalates, Nijmegen, The Netherlands, Physica **99B** (1980).

[17] Proceedings Second Conference on Intercalation Compounds of Graphite, Provincetown, Mass. USA, Synth. Met. **2/3** (1980/81).

[18] *F. L. Vogel*, J. Mat. Sci. **12**, 982 (1977).

[19] *S. C. Singhal*, in Ref. [16], p. 536.

[20] *W. H. Smith* and *D. H. Leeds* in "Modern Materials", Vol. 7, Ed. B. W. Gonser, Academic Press, New York and London 1970, p. 139.

[21] *E. Kikuchi, T. Ito* and *Y. Morita*, J. Catal. **57**, 27 (1979).

[22] *J. Golé*, in Ref. [13], p. 309.

[23] *M. Armand* and *P. Touzain*, in Ref. [13], p. 319.

[24] *S. Flandrois, J. M. Masson* and *J. C. Rouillon*, in Ref. [17], **3**, p.

[25] *P. Lagrange* and *A. Hérold*, Carbon **16**, 235 (1978).

[26] *K. Tamaru*, American Scientist **60**, 474 (1972).

[27] *A. Hérold*, in Ref. [15], p. 323.

[28] *R. Clarke, N. Caswell, S. A. Solin* and *P. M. Horn*, Phys. Rev. Lett. **43**, 2018 (1979); Ref. [16], p. 457.

[29] *R. Clarke, N. Caswell* and *S. A. Solin*, Phys. Rev. Lett. **42**, 61 (1979).

[30] *N. Caswell*, Phys. Rev. **B22**, 6308 (1980).

[31] *H. Zabel, S. C. Moss, N. Caswell* and *S. A. Solin*, Phys. Rev. Lett. **43**, 2022 (1979).

[32] *J. B. Hastings, W. D. Ellenson* and *J. E. Fischer*, Phys. Rev. Lett. **42**, 1552 (1979).

[33] *H. Suematsu, M. Suzuki, N. Ikeda* and *Y. Endoh*, in Ref. [17] **2**, p. 133.

[34] *M. S. Dresselhaus, N. Kambe, A. N. Berker* and *G. Dresselhaus*, in Ref. [17] **2**, p. 121; *N. Kambe, G. Dresselhaus* and *M. S. Dresselhaus*, Phys. Rev. **B21**, 3491 (1980).

[35] *D. G. Onn, G. M. T. Foley* and *J. E. Fischer*, Phys. Rev. **B19**, 6474 (1979).

[36] *J. F. Marêché, E. McRae, G. Furdin, D. Billaud* and *A. Hérold*, in Ref. [17] **2**, p. 267.

[37] *W. Metz* and *D. Hohlwein*, Carbon **13**, 87 (1975).

[38] *D. Billaud, J. F. Marêché, E. McRae* and *A. Hérold*, in Ref. [17] **2**, p. 37.

[39] *C. Underhill, T. Krapchev* and *M. S. Dresselhaus*, in Ref. [17] **2**, p. 47.

[40] *S. A. Safran* and *D. R. Hamann*, Phys. Rev. **B22**, 606 (1980).

[41] *S. A. Safran* and *D. R. Hamann*, Phys. Rev. **B23**, 565 (1981).

[42] *N. Daumas* and *A. Hérold*, Compt. Rend. Aca. Sci. Paris **C268**, 373 (1969).

[43] *S. A. Safran* and *D. R. Hamann*, Phys. Rev. Lett. **42**, 1410 (1979).

[44] *A. N. Berker, N. Kambe, G. Dresselhaus* and *M. S. Dresselhaus*, Phys. Rev. Lett. **45**, 1452 (1980).

[45] *A. W. Moore*, in „Chemistry and Physics of Carbon", Vol. 11, Ed. P. L. Walker jr. and P. A. Thrower, Marcel Dekker, New York 1974, p. 69.

[46] *M. S. Dresselhaus* and *G. Dresselhaus*, to be published in "Topics in Current Physics", Eds. M. Cardona and G. Güntherodt: "Light Scattering in Solids", Springer, Heidelberg 1981.

[47] *M. Maeda, Y. Kuramoto* and *C. Horie*, J. Phys. Soc. Jap. **47**, 337 (1979).

[48] *L. Pietronero* and *S. Strässler*, to be published.

[49] *S. Y. Leung, G. Dresselhaus* and *M. S. Dresselhaus*, in Ref. [17] **2**, p. 89.

[50] *P. C. Eklund, D. S. Smith, V. R. K. Murthy* and *S. Y. Leung*, in Ref. [17] **2**, p. 89.

[51] *W. D. Ellensen, D. Semmingsen, D. Guérard, D. G. Onn* and *J. E. Fischer*, in Ref. [13], p. 137.

[52] *J. Rossat-Mignod, D. Fruchart, M. J. Moran, J. W. Milliken* and *J. E. Fischer*, in Ref. [17] **2**, p. 143.

[53] *A. Magerl* and *H. Zabel*, Phys. Rev. Lett. **46**, 444 (1981).

[54] *A. Magerl* et al., to be published.

[55] *C. Horie, M. Maeda* and *Y. Kuramoto*, in Ref. [16], p. 430.

[56] *I. L. Spain*, in "Chemistry and Physics of Carbon", Vol. 8, Ed. P. L. Walker jr. and P. A. Thrower, Marcel Dekker, New York 1973, p. 1.

[57] *I. L. Spain* and *D. J. Nagel*, in Ref. [13], p. 183.

[58] *M. S. Dresselhaus, G. Dresselhaus* and *J. E. Fischer*, Phys. Rev. **B15**, 3180 (1977).

[59] *L. Pietronero, S. Strässler, H. R. Zeller* and *M. J. Rice*, Phys. Rev. Lett. **41**, 763 (1978).

[60] *L. Pietronero, S. Strässler* and *H. R. Zeller*, Sol. St. Comm. **30**, 399 (1979).

[61] *F. Batallan, J. Bok, I. Rosenman* and *J. Melin*, Phys. Rev. Lett. **41**, 330 (1978); *I. Rosenman, F. Batallan* and *G. Furdin*, Phys. Rev. **B20**, 2373 (1979).

[62] *E. Cartier, F. Heinrich, P. Pfluger* and *H.-J. Güntherodt*, Phys. Rev. Lett. **46**, 272 (1981).

[63] *L. B. Ebert* and *H. Selig*, in Ref. [13], p. 177.

[64] *H. A. Resing, F. L. Vogel* and *T. C. Wu*, Mater Sci. Eng. **41**, 113 (1979).

[65] *W. C. Forsman, D. E. Carl* and *T. Birchall*, to be published.

[66] *L. Streifinger, H. P. Boehm, R. Schlögl* and *R. Pentenrieder*, Carbon **17**, 195 (1979).

[67] *S. Loughin, R. Grayeski* and *J. E. Fischer*, J. Chem. Phys. **69**, 3740 (1978).

[68] *A. R. Ubbelohde*, Proc. Roy. Soc. **A309**, 297 (1969).

[69] *J. E. Fischer*, in Ref. [16], p. 383.

[70] *N. Bartlett, R. N. Biagoni, B. W. McQuillan, A. S. Robertson* and *A. C. Thompson*, J.C.S. Chem. Comm. 200 (1978).

[71] *G. K. Wertheim, P. M. Th. M. van Attekum, H. J. Guggenheim* and *K. E. Clements*, Sol. St. Comm. **33**, 809 (1980).

[72] *A. Farazdel* and *P. Cade*, J. Chem. Phys. **66**, 2612 (1977).

[73] *T. Inoshita, K. Nakao* and *H. Kamimura*, J. Phys. Soc. Jap. **43**, 2137 (1977).

[74] *N. A. W. Holzwarth, S. Rabii* and *L. A. Girifalco*, Phys. Rev. **B18**, 5190 and 5208 (1978).

[75] *J. E. Fischer*, in Ref. [15] p. 481.

[76] *P. Oelhafen, P. Pfluger, E. Hauser* and *H.-J. Güntherodt*, Phys. Rev. Lett. **44**, 197 (1980).

[77] *P. Oelhafen, P. Pfluger* and *H.-J. Güntherodt*, Sol. St. Comm. **32**, 885 (1979).

[78] *W. Eberhardt, I. T. McGovern, E. W. Plummer* and *J. E. Fischer*, Phys. Rev. Lett. **44**, 200 (1980).

[79] *U. M. Gubler, Ph. D. Thesis*, University of Basel, Switzerland, 1981; U. M. Gubler et al., to be published.

[80] *P. Pfluger, P. Oelhafen, H.-U. Künzi, R. Jeker, E. Hauser, K. P. Ackermann, M. Müller* and *H.-J. Güntherodt*, in Ref. [16] p. 395.

[81] *T. Ohno, K. Nakao* and *H. Kamimura*, J. Phys. Soc. Jap. **47**, 1125 (1979).

[82] *J. Conard, M. Gutierrez-Le Brun, P. Lauginie* and *H. Estrade*, in Ref. [17] **2**, p. 227.

[83] *P. Lauginie, H. Estrade, J. Conard, D. Guérard, P. Lagrange* and *M. El Makrini*, in Ref. [16] p. 514.

[84] *D. M. Hwang, M. Utlaut* and *S. A. Solin*, in Ref. [17] **3**, p. 81.

[85] *E. Cartier, F. Heinrich, P. Pfluger* and *H.-J. Güntherodt*, Sol. St. Comm., in press.

[86] *G. Volpilhac* and *J. Hoarau*, J. Phys. **C13**, 2281 (1980).

[87] *D. P. DiVincenzo, N. A. W. Holzwarth* and *S. Rabii*, in Ref. [17] **3**, p. 123.

[88] *K. A. Müller* and *R. Kleiner*, Phys. Lett. **1**, 98 (1962).

[89] *G. K. Wertheim, P. M. Th. M. van Attekum* and *S. Basu*, Sol. St. Comm. **33**, 1127 (1980).

[90] *P. Delhaes, J. C. Rouillon, J. P. Manceau, D. Guérard* and *A. Hérold*, J. Physique (Letters) 37, L127 (1976).

[91] *P. Delhaes, J. P. Manceau* and *D. Guérard*, in Ref. [17] **2**, p. 277.

[92] *P. Oelhafen, E. Hauser, F. Greuter, P. Pfluger* and *H.-J. Güntherodt*, unpublished.

[93] *T. Ohno*, Physica B, to be published.

[94] *L. Samuelson* and *I. P. Batra*, J. Phys. **C13**, 5105 (1980).

[95] *J. Krieg, Ph. D. Thesis*, University of Basel, Switzerland, 1981; J. Krieg et al., to be published.

[96] *M. G. Alexander, D. P. Goshorn* and *D. G. Onn*, Phys. Rev. **B22**, 4535 (1980).

[97] *M. G. Alexander, D. P. Goshorn, D. G. Onn, D. Guérard, P. Lagrange* and *M. El Makrini*, in Ref. [17] **2**, p. 203.

[98] *U. Mizutani, T. Kondow* and *T. B. Massalski*, Phys. Rev. **B17**, 3165 (1978).

[99] *M. Suganuma, T. Kondow* and *U. Mizutani*, Phys. Rev. **B23**, 706 (1981).

[100] *C. Ayache, E. Bonjour, R. Lagnier* and *J. E. Fischer*, in Ref. [16] p. 547.

[101] *Y. Baer, P. Cohn, H.-J. Güntherodt* and *P. Pfluger*, to be published.

[102] *V. Dose* et al., to be published

[103] *G. Roth, K. Lüders, P. Pfluger* and *H.-J. Güntherodt*, Sol. St. Comm., to be published.

[104] *J. Conard* and *H. Estrade*, in Ref. [13] p. 173 and private communications.

[105] *Y. Chabre* et al., to be published.

[106] *G. P. Carver*, Phys. Rev. **B2**, 2284 (1970).

[107] *M. Posternak, E. Wimmer* and *A. J. Freeman*, to be published.

[108] *G. Dresselhaus, S. Y. Leung, M. Shayegan* and *T. C. Chieu*, in Ref. [17] **2**, p. 321.

[109] *B. R. Weinberger, J. Kaufer, A. J. Heeger, J. E. Fischer, M. Moran* and *N. A. W. Holzwarth*, Phys. Rev. Lett. **41**, 1417 (1978).

[110] *K. Higuchi, H. Suematsu* and *S. Tanuma*, J. Phys. Soc. Jap. **48**, 1532 (1980).

[111] *B. Bach, E. L. Evans, J. M. Thomas* and *M. Barber*, Chem. Phys. Lett. **10**, 547 (1971); *B. Bach*, Compt. Rend. Aca. Sci. Paris **B273**, 666 (1971).

[112] *P. Oelhafen*, unpublished.

[113] *G. S. Painter* and *D. E. Ellis*, Phys. Rev. **B1**, 4747 (1970).

[114] *J. J. Ritsko* and *E. J. Mele*, in Ref. [17] **3**, p. 73.

[115] *A. S. Bender* and *D. A. Young*, J. Phys. **C5**, 2163 (1972).

[116] *J. A. Woollam, E. J. Haugland, M. B. Dowell, A. Yavrouian, A. G. Lozier* and *G. Matulka*, in Ref. [17] **2**, p. 309.

[117] *R. S. Markiewicz H. R. Hart, Jr., L. W. Interrante* and *J. S. Kasper*, in Ref. [17] **2**, p. 331.

[118] *S. Tanuma, Y. Iye, O. Takahaski* and *Y. Koike*, in Ref. [17] **2**, p. 341.

[119] *F. Batallan, I. Rosenman* and *C. Simon*, in Ref. [17] **2**, p. 353.

[120] *A. Briggs* et al., to be published.

[121] *J. Blinowski, Nguyen Hy Hau, C. Rigaux, J. P. Vieren, R. Le Toullec, G. Furdin, A. Hérold* and *J. Melin*, J. Physique **41**, 47 (1980).

[122] *J. Blinowski* and *C. Rigaux*, J. Physique **41**, 667 (1980).

[123] *C. D. Fuerst, W. D. Johnson* and *J. E. Fischer*, Ext. Abstr. 14th biennal Conf. on Carbon, Penn. State Univ., June 1979, p. 296.

[124] *M. Zanini* and *J. E. Fischer*, in Ref. [13] p. 169.

[125] *E. McRae, D. Billaud, J. F. Marêché* and *A. Hérold*, in Ref. [16] p. 489.

[126] *A. R. Ubbelohde*, Proc. Roy. Soc. **A327**, 289 (1972).

[127] *L. Pietronero* and *S. Strässler*, in Ref. [17] **3**, in paper II and in Phys. Rev., to be published.

[128] *C. C. Shieh, R. L. Schmidt* and *J. E. Fischer*, Phys. Rev. **B20**, 3351 (1979).

[129] *P. Pfluger, K. P. Ackermann, R. Lapka, E. Schüpfer, R. Jeker, H.-J. Güntherodt, E. Cartier* and *F. Heinrich*, in Ref. [17] **2**, p. 285.

[130] *Nguyen Hy Hau, J. Blinowski, C. Rigaux, R. Le Toullec, G. Furdin, A. Hérold* and *R. Vangelisti*, in Ref. [17] **3**, p. 99.

[131] *P. C. Eklund, D. S. Smith* and *V. R. K. Murthy*, in Ref. [17] **3**, p. 111.

[132] *P. Pfluger, E. Schüpfer, R. Lapka* and *H.-J. Güntherodt*, Ext. Abstr. 15th biennal Conf. on Carbon, Philadelphia, June 1981.

[133] *F. J. DiSalvo, S. A. Safran, R. C. Haddon, J. V. Waszczak* and *J. E. Fischer*, Phys. Rev. **B20**, 4883 (1979).

[134] *S. A. Safran* and *F. J. DiSalvo*, Phys. Rev. **B20**, 4889 (1979).

[135] *N. B. Hannay, T. H. Geballe, B. T. Matthias, K. Andres, P. Schmidt* and *D. MacNair*, Phys. Rev. Lett. **14**, 225 (1965).

[136] *Y. Koike, H. Suematsu, K. Higuchi* and *S. Tanuma*, Sol. St. Comm. **27**, 623 (1978).

[137] *Y. Koike, H. Suematsu, K. Higuchi* and *S. Tanuma*, in Ref. [16] p. 503.

[138] *P. Lagrange, D. Guérard, D. G. Onn* et al., unpublished.

[139] *S. Berko*, in "Electrons in disordered metals and at metallic surfaces", NATO Advanced Study Institute Series, Vol. **B42**, Ed. *P. Phariseau, B. L. Györffy* and *L. Scheire*, Plenum, New York and London 1979, p. 239.

[140] *J. J. Murray* and *A. R. Ubbelohde*, Proc. Roy. Soc. **A312**, 371 (1969).

[141] *D. G. Onn, G. M. T. Foley* and *J. E. Fischer*, Phys. Rev. **B19**, 6474 (1979).

[142] *H. Suematsu, K. Higuchi* and *S. Tanuma*, J. Phys. Soc. Jap. **48**, 1541 (1980).

[143] *P. Bak*, Phys. Rev. Lett. **44**, 889 (1980).

[144] *P. Bak* and *E. Domany*, Phys. Rev. **B20**, 2818 (1979) and **B23**, 1321 (1981).

[145] *P. W. Stephens, P. A. Heiney, R. J. Birgeneau* and *P. M. Horn*, Phys. Rev. Lett. **43**, 47 (1979).

[146] *P. Vora, S. K. Sinka* and *R. K. Crawford*, Phys. Rev. Lett. **43**, 704 (1979).

[147] *J. Eckert, W. D. Ellenson, J. B. Hastings* and *L. Passell*, Phys. Rev. Lett. **43**, 1328 (1979).

[148] *P. W. Stephens, P. A. Heiney, R. J. Birgeneau, P. M. Horn, J. Stoltenberg* and *O. E. Vilches*, Phys. Rev. Lett. **45**, 1959 (1980).

[149] *A. Hérold*, in Ref. [13] p. 1.

Festkörperprobleme XXI (1981)

Intercalated Graphite: A Synthetic Metal
II. Theory of Bond Length Change and Conductivity

Sigfrid Strässler and Luciano Pietronero
Brown Boveri Research Center, Baden, Switzerland

Summary: We present a brief review of the present understanding of the electronic properties of graphite intercalation compounds with special attention on bond length changes and electrical conductivity. After a discussion of the band structure and the Fermi surface we show how to extract the values of the charge transfer and information about the electron phonon coupling from the change of the in plane bond length. These values are then used in the variational solution of the Boltzmann equation to compute the resistivity due to the scattering with the in plane phonons and to estimate the maximum ideal conductivity for these compounds. The comparison with experiments is discussed and the main open problems are outlined.

1 Introduction

The vast interest that has fluorished around graphite intercalation compounds is due to both technical and basic reasons [1, 2]. On one hand the flexibility of this class of materials, that can be intercalated with many types of molecules, makes them interesting for potential applications. On the other hand the understanding of their physical properties, in particular transport properties, poses quite challenging questions of basic nature.

These synthetic metals are compounds in which graphite layers alternate with intercalate layers. The intercalate molecules can be of many different types and they can either give (donors) or take (acceptors) charge from the graphite layers. The most striking property is that the conductivity of some of these systems at room temperature is of the order of that of copper while that of pure graphite is about 25 times less. Conduction takes place essentially in the charged graphite layers and the carrier density is less then one tenth that of copper. Since these compounds are very different from simple metals the question whether the reported conductivity is intrinsic or if it could be improved significantly (by improving the sample preparation or the chemistry of the compounds) is of both practical and fundamental importance. In the following we will outline the necessary steps in order to obtain informations about the maximum intrinsic conductivity of these systems. In particular:

In Sect. 2 we briefly discuss the effects of intercalation on band structure, charge distribution and Fermi surface.

In Sect. 3 we show how informations about the electronic structure (charge transfer and electron phonon coupling) can be obtained from an analysis of the bondlength change with intercalation.

In Sect. 4 the resistivity due to scattering with phonons is computed within the variational solution of the Boltzmann equation. Its absolute value and temperature dependence are compared with experiments and discussed.

In Sect. 5 we summarize the results and outline the main open problems.

2 Band Structure and Fermi Surface

Graphite has been one of the most extensive studied systems in solid state physics. The electronic states of a single graphite layer consist of a bonding sp_2 (σ) band plus a bonding p_z (π) band. The empty antibonding p_z (π) band is connected without a gap to the bonding one but the density of states at the Fermi level is zero [3]. This system represents therefore an ideal semimetal or zero gap semiconductor. The band structure in the vicinity of the Fermi level has a linear dispersion

$$\epsilon(k) = \pm (\hbar v_F) |k| = \pm \tfrac{3}{2} J_0 r_0 |k| \tag{2.1}$$

where v_F is the Fermi velocity, k refers to the plane, J_0 is the tight binding matrix element between p^z orbitals and $r_0 = 1.421$ Å is the nearest neighbor carbon carbon distance. In 3–d graphite these layers are weakly coupled, but this small coupling has important effects on the band structure just at the Fermi level. The Fermi surface consists then of electrons and holes pockets along the K–H lines of the Brillouin zone and 3–d coupling is essential for its description [4].

The simplest picture of intercalation consists in assuming a shift of the Fermi level up ($\mu > 0$) or down ($\mu < 0$) for donors and acceptors (Fig. 1b). This is the so called rigid band model. Experimental evidence shows that this approximation is rather appropriate for acceptors (AsF_5, SbF_5, HNO_3, $FeCl_3$, etc.) while for donors (Li, K, Rb, Cs, etc.) the coupling of the π bands to the alkali orbitals can modify the bands

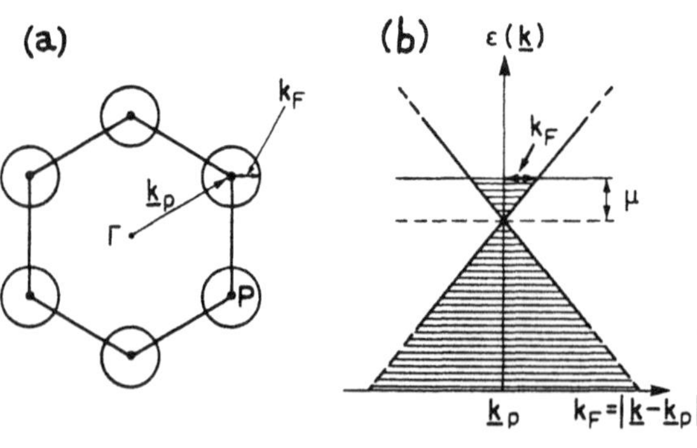

Fig. 1
(a) Brillouin zone for a graphite layer. The circles indicate the Fermi surface for a charged layer.
(b) Band dispersion around a P point. The shift of the Fermi level due to charge transfer is indicated by μ.

in a more complicate way [1, 2]. Typical values for the shift of Fermi level are $|\mu| \sim .5 \div 1.\text{eV}$. These shifts have the effect to move the Fermi surface in a region that is much less affected by 3–d coupling then neutral graphite. As a starting point one can therefore use a model based on independent graphite layers. The band structure is then given by Eq. (2.1) and the Fermi surface consists of circles around the P points of the Brillouin zone (Fig. 1a). For large stages (three or more graphite layers per intercalate layer) the chemical potential (and therefore the charge) can vary within each group of graphite layers. Due to the semimetallic nature of graphite the determination of the charge distribution gives rise to an interesting problem of nonlinear screening. The result is that in general most of the charge is localized in the bounding layers adjacent the intercalate one [4].

Within the independent layer, linear band description, we obtain particularly simple relations between the shift of Fermi level μ and the density of states per carbon atom

$$N_c(\mu) = \frac{2}{\pi\sqrt{3}} \frac{\mu}{J_0^2} \tag{2.2}$$

and the charge transfer per carbon atom

$$f_c = \frac{1}{\pi\sqrt{3}} \frac{\mu^2}{J_0^2} . \tag{2.3}$$

In the following we will always use these relationships even if, for some donors or for properties that are very dependent on the details of the Fermi surface they might be oversimplified. They are certainly appropriate for stage 1 acceptors.

3 Analysis of the Bond Length Changes

The determination of the charge transfer and the electron phonon coupling is the starting point to study the conductivity. Many experiments have been performed especially to determine the charge transfer but the information is in general indirect and a non trivial analysis is needed. This difficulty has given rise to very contrasting values for f and only recently there are signs of convergency between values obtained with different techniques. Here we show how information about the change transfer and electron phonon coupling can be extracted from high precision measurements of the carbon carbon (in plane) bondlength (Fig. 2). We start from the methods that have been used succesfully to explain the bond length changes in carbon based molecules [10] and extend them to intercalated graphite.

The total energy per bond between atoms at positions n and n′ is written as

$$E^T = E^\sigma + E^\pi \tag{3.1}$$

where E_σ and E_π are the energies of the σ and π electrons respectively. Neglecting the terms that do not depend on bondlength we can write

$$E^\pi = 2P\, J^{pp\pi} \tag{3.2}$$

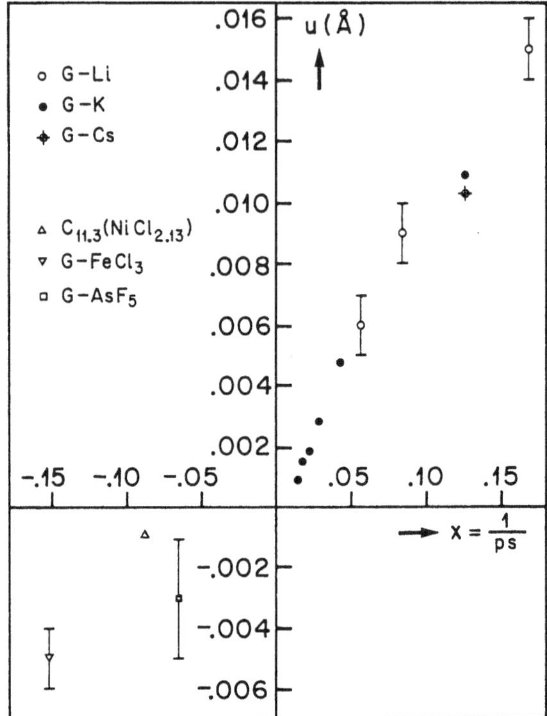

Fig. 2
Experimental values for the bond length change $u = r - r_0$ (r_0 = 1.421 Å) for various intercalation compounds plotted as a function of the ratio between the number of intercalate molecules and that of carbon atoms in a given compound. The nominal formula of a compound is CA_x where $x = 1/ps$, s is the stage and p = 6, 8 etc.. The data are from Ref. (5) (G-Li), Ref. (6) (G-K; G-Cs), Ref. (7) ($C_{11.3}$ $NiCl_{2.13}$), Ref. (8) (G-$FeCl_3$) and Ref. (9) (G-AsF_5).

where $J^{pp\pi} = \langle p_n^z | V_n | p_{n'}^z \rangle$ is the J_0 of the previous section and n and n′ indicate the positions of two nearest carbons. Since we are interested in uniform contractions or expansions we will in general omit these indices.

P is the band order for π electrons defined as

$$P = 2 \sum_{\vec{k}\,(\text{occ.})} c_n^*(\vec{k})\, c_{n'}(\vec{k}) \tag{3.3}$$

where $c_n(\vec{k})$ is the coefficient of the p_n^z orbital corresponding to the state $\vec{k}$. The σ bond is described in terms of bonding states between sp_2 hybrids. This gives (apart from constant terms) [11]

$$E^\sigma = \frac{2}{3} J^{ss\sigma} - \frac{4}{3} J^{pp\sigma} \tag{3.4}$$

where as in Ref. [3] $J^{ss\sigma} = \langle s_n | V_n | s_{n'} \rangle$ and $J^{pp\sigma} = \langle p_n^x | V_n | p_{n'}^x \rangle$.

Let us consider for the moment a single layer charged with f_c electronic charge per carbon atom. This extra charge gives rise to the following three effects:

(a) A modification of the occupancy of the π states that changes the value of the bond order: $P = P_0 + \Delta P(f_c)$ where $P_0 = .525$ is the value for the neutral layer [12].

(b) A modification of all the matrix elements due to the fact that the atomic potential is changed: $J = J_0 + \Delta J(f_c)$.

(c) An additional contribution ΔE_{el} (also defined per bond) to the total energy due to the Coulomb repulsion between all the extra charges.

The total energy is then

$$E^T(f_c;r) = 2\,[P_0 + \Delta P(f_c)]\,[J_0^{pp\pi}(r) + \Delta J^{pp\pi}(f_c;r)] + \\ + \tfrac{2}{3}\,[J_0^{ss\sigma}(r) + \Delta J^{ss\sigma}(f_c;r)] - \tfrac{4}{3}\,[J_0^{pp\sigma}(r) + \Delta J^{pp\sigma}(f_c;r)] + \Delta E_{el}(f_c;r) \tag{3.5}$$

where $r = r_0 + u$ being $r_0 = 1.421\,\text{Å}$ the equilibrium bond length for the neutral layer ($f_c = 0$). We expand then Eq. (3.5) up to quadratic terms in u making use of the condition that r_0 is the equilibrium position for $f_c = 0$. The equilibrium condition for the general case gives now the displacement u as (neglecting the terms $\Delta P \Delta J$, $\Delta J \Delta J$ and $\Delta P \Delta P$) [10]

$$u = -\frac{1}{K}\left[2\Delta P(f_c)\left.\frac{dJ_0^{pp\pi}(r)}{dr}\right|_{r_0} + 2P_0\left.\frac{d\Delta J^{pp\pi}(r)}{dr}\right|_{r_0} + \right. \\ \left. + \left.\frac{d\Delta E^{\sigma}(f_c;r)}{dr}\right|_{r_0} + \left.\frac{d\Delta E_{el}(f_c;r)}{dr}\right|_{r_0}\right] \tag{3.6}$$

where

$$\Delta E^{\sigma}(f_c;r) = \tfrac{2}{3}\,\Delta J^{ss\sigma}(f_c;r) - \tfrac{4}{3}\,\Delta J^{pp\sigma}(f_c;r) \tag{3.7}$$

and

$$K = \left.\frac{d^2 E^T(f_c = 0;r)}{dr^2}\right|_{r_0} \tag{3.8}$$

is the total bond stretching force constant of neutral graphite.

It is interesting to note that only the first term in Eq. (3.5) appears in a similar analysis of the carbon based molecules. The other terms are in fact due to the net charge on the carbon atoms. A discussion of the relations between the two cases will be reported elsewhere.

The total force constant K can be computed from the elastic constant of graphite [13]. It results in $K = 45$ eV/Å. For the evaluation of the matrix elements J and ΔJ we use Slater type orbitals and we optimize the extension of the atomic potential to obtain $J_0^{pp}(r_0) = 2.4$ eV [14]. This gives $dJ^{pp\pi}/dr|_{r_0} = 6.3$ eV/Å [11] that is in agreement with values obtained from the observed gap and Raman modes in polyacetilene [11] but it is about 25 % larger then the value derived from small molecules [10]. For the matrix elements of type ΔJ we consider a ΔV due to the charge transfer from intercalate molecules to spherically averaged p^z orbitals. The charge on the intercalate molecules is replaced by a uniform (compensating) surface charge on a sphere of radius R = 3 Å. The results are not very sensitive to the specific model choosen for the neutralizing charge as the main contribution comes from the extra charge on the carbon atoms. We obtain then [11]

$$\left.\frac{d\Delta J^{pp\pi}(r)}{dr}\right|_{r_0} = -3.34\, f_c\ \mathrm{eV/Å}\,, \qquad \left.\frac{d\Delta J^{ss\sigma}(r)}{dr}\right|_{r_0} = -4.51\, f_c\ \mathrm{eV/Å}\,,$$
$$\left.\frac{d\Delta J^{pp\sigma}(r)}{dr}\right|_{r_0} = .43\, f_c\ \mathrm{eV/Å}\,. \tag{3.9}$$

We use now Eqs. (2.2) and (3.2) to compute the change in bond order within the linear bond approximation described in Sect. 1. We obtain

$$\Delta P(f_c) = \frac{\Delta E^{\pi}(f_c)}{2\, J_0^{pp\pi}(r_0)} = -2\cdot 3^{-7/4}\sqrt{\pi}\,|f_c|^{3/2} = -\,.52\,|f_c|^{3/2}\,. \tag{3.10}$$

To estimate the last term in Eq. (3.6) we consider the electrostatic energy of a uniform charge density within the volume of a graphite layer. This gives

$$\left.\frac{d\Delta E_{el}(f_c\,;r)}{dr}\right|_{r_0} = -\frac{32\pi}{27\sqrt{3}}\,\frac{e^2 c}{\epsilon_\perp r_0^3}\, f_c^2 = -10.60\, f_c^2\ \mathrm{eV/Å} \tag{3.11}$$

where $\epsilon_\perp \simeq 3.4$ is the dielectric constant and c = 3.35 Å is the layer thickness. Since $f_c \leqslant 1/6$, this term is always small.

The final result for stage 1 is then

$$u(1)\,(\mathrm{Å}) = .157\, f_c + .146\,|f_c|^{3/2} + .236\, f_c^2\,. \tag{3.12}$$

For the generalization of this result to higher stages (s) we assume (i) that the charge is localized in the bounding layers [4] (f_c only refers to these) and (ii) that the strain is equally shared by all the layers. This gives

$$u(s) = \frac{2}{s}\, u(1) \qquad (s \geqslant 2)\,. \tag{3.13}$$

In Fig. 2 we have plotted the available data for u as functions of x = 1/ps where p = 6, 8, etc. The roughly linear behavior observed for groups of donors for $s \geqslant 2$ indicates that f_c is about constant.

In view of the availability of several values of u determined with high precision we have used the measured u's in Eqs. (3.12) and (3.13) to determine f_c and then f (charge transfer per intercalate molecule). The results are listed in Table 1 together with values of f determined with the other available methods. The overall excellent consistency of the present analysis provides a new tool for the determination of f and strongly supports the general picture we have been describing and that will be the starting point to study the conductivity.

Table 1: Values of the charge transfer per intercalate molecule obtained from the analysis of the data of Fig. 2 through Eqs. (3.12) and (3.13) compared to the values given from other experiments

	f (experimental)	f (present analysis)
C_6Li	.22 ÷ .42 (a)	.43
	∿ 1 (b)	
$C_{12}Li$		.54
$C_{18}Li$		.55
C_8K	.16 − .45 (a)	.43
	∿ .3 (c)	
$C_{24}K$	∿ 1 (a)	.63
$C_{36}K$	.1 ÷ .4 (d)	.57
$C_{48}K$		.50
C_8Cs	.17 ÷ .44 (a)	.41
	∿ .3 (c)	
	.5 ± .2 (e)	
$C_{16}AsF_5$	−(.26 ÷ .44) (a)	∿ −.35
	− .48 (f)	
	− .41 (g)	
	−(.4 ÷ .5) (h)	
	−(.30 ÷ .51) (i)	
	− .22 (l)	
$C_{6.6}FeCl_3$	− .18 (m)	−.26
$C_{11.3}NiCl_{2.13}$		−.07

(a): Ref. [15]
(b): Ref. [16]
(c): Ref. [17]
(d): Ref. [18]
(e): Ref. [19]
(f): Ref. [20]
(g): Ref. [21]
(h): Ref. [22]
(i): Ref. [14]
(l): Ref. [23]
(m): Ref. [24]

4 Electrical Conductivity

We start from the variational solution of Boltzmann equation [25] that we write in a form that is specially simple for a physical interpretation of the various terms.. The resistivity along a direction x is given by [26]

$$\rho(T) = \left[\frac{e^2}{4\pi^3}\int d\vec{k}\, v_x^2(\vec{k})\left(-\frac{\partial f}{\partial \epsilon}\right)_{\epsilon=\epsilon(\vec{k})}\tau\right]^{-1} \tag{4.1}$$

where $\epsilon(\vec{k})$ is the band energy, v_x is the x component of the electron band velocity, f is the Fermi function and τ is the "transport" relaxation time [25]. The derivative of the Fermi function can be replaced by a delta function to obtain

$$\rho(T) = [e^2 N(\mu) \langle v_x^2(\mu)\rangle \tau(\mu, T)]^{-1} \tag{4.2}$$

where $\langle\ldots\rangle$ is the average over the Fermi surface, μ is the Fermi level and N is the density of states. For the electron-phonon scattering it is convenient to write the relaxation time τ as the product of the asymptotic high temperature expression $\tilde{\tau}\,(\propto T^{-1})$ divided a correction function F (T) that takes care of the fact that the temperature is actually finite and not all phonon states are populated.

$$\tau(\mu, T) = \tilde{\tau} / F(T) \tag{4.3}$$

where

$$\tilde{\tau} = \left[\frac{2\pi}{\hbar} k_B T \langle |g_{\vec{k},\vec{k}'}|^2 (1-\cos\theta)\rangle N(\mu)\right]^{-1} \tag{4.4}$$

can be obtained from the Fermi golden rule plus the "transport" correction $(1-\cos\theta)$ arising from the variational solution. Here $g_{\vec{k}\vec{k}'}$ is the el-ph coupling in the notations of Ref. [27] and θ is scattering angle between $\vec{k}$ and $\vec{k}'$. The correction function F(T) is [26]

$$F(T) = \left(\frac{\hbar}{k_B T}\right)^2 \langle q^2(1-\cos\theta)\, G^{-1}\rangle / \langle q^2 (1-\cos\theta)\,[\omega(\vec{q})]^{-2}\rangle \tag{4.5}$$

$$G = \{\exp[\hbar\omega(\vec{q})/k_B T] - 1\}\cdot\{1-\exp[-\hbar\omega(\vec{q})/k_B T]\} \tag{4.6}$$

where $\vec{q} = \vec{k} - \vec{k}'$ and ω is the phonon frequency. The extension to many phonon branches and many bands is straightforward and we do not include it for simplicity. Of course for $T \to \infty$ we have $F(T) \to 1$.

As discussed in Sect. 1 the Fermi surface of a charged graphite layer consists of circles around the P (or K) points of the hexagonal Brillouin zone and the band dispersion in the vicinity of these points is linear (Eq. (2.1)). We explicitely compute now the resistivity for the specific case of a stage 1 acceptor (C_8AsF_5). Defining an asymptotic $\tilde{\rho}$ in analogy to $\tilde{\tau}$ we have

$$\tilde{\rho} = \rho/F(T) = \frac{4}{9}\,\frac{(d+d')\,\pi k_B T \hbar}{e^2 (J_0 r_0)^2}\,\langle |g_{\vec{k},\vec{k}'}|^2 (1-\cos\theta)\rangle \tag{4.7}$$

where $d = 3.35$ Å and $d' = 4.95$ Å are the thickness of a graphite and an intercalate (AsF_5) layer respectively. The problem is reduced now to the evaluation of the electron-phonon coupling. This will be done within the same tight binding framework that we have been using in the previous sections. The reason for using a tight binding description instead of the rigid ion model [25] is the following: The rigid ion model arises from a description of a metal in terms of a lattice of charged ions plus a gas of quasi free electrons. A phonon produces a distortion of the ionic lattice to which is associated a fluctuation in the charge density. This charge fluctuation, eventually screened by the quasi-free electrons, is the source of scattering in the rigid ion model. The situation is quite different in graphite because the π orbitals corresponding to the conduction band are tightly bound to the carbon atoms [3]. A lattice distortion is therefore expected to produce mainly a change in the overlap between localized states and only to higher order the orbital relaxation gives rise to a charge fluctuation. This naturally leads to a tight binding approach as the correct starting point for the electron-phonon coupling. The explicit calculation has been reported elsewhere in detail [28] and here we only quote the results. For shifts of the Fermi level that are not too large ($|\mu| \lesssim 1$ eV) one obtains for the scattering within the same pocket

$$\left| g^{\ell(t)}_{\vec{k},\vec{k}'} \right|^2 = \frac{9}{32\,\rho_c} \left(\frac{r_0\,\alpha}{c} \right)^2 [1 \overset{(+)}{-} \cos 3\,(\phi + \phi')] \tag{4.8}$$

for longitudinal (ℓ) and transverse (t) phonon respectively. Here ρ_c is the 2-d mass density of graphite ($\rho_c = m_c \times 3.82 \times 10^{15}$ cm^{-2}) and $\phi(\phi')$ is the angle corresponding to $\vec{k}(\vec{k}')$. The term c is an average speed of sound $c = 2.1 \times 10^6$ cm/s and $\alpha = dJ_0/dr \simeq 6.3$ eV/Å as computed in the previous section. We have then

$$\left\langle \left| g^{\ell}_{\vec{k},\vec{k}'} \right|^2 (1-\cos\theta) \right\rangle = \left\langle \left| g^{t}_{\vec{k},\vec{k}'} \right|^2 (1-\cos\theta) \right\rangle = \frac{9}{32\,\rho_c} \left(\frac{r_0\,\alpha}{c} \right)^2 . \tag{4.9}$$

Note that Eqs. (4.8) and (4.9) do not depend on f_c and therefore on the charge transfer. For larger shift of the chemical potential ($|\mu| \gtrsim 1$; in C_8AsF_5 we have $\mu \sim 1.3$) it is necessary to include other terms that are dependent on the charge transfer. The extra contribution to the coupling with ℓ phonons gives [26]

$$\left\langle \left| g'^{\ell}_{\vec{k},\vec{k}'} \right|^2 (1-\cos\theta) \right\rangle = \frac{2.22\,(r_0 k_F)^2}{\rho_c} \left(\frac{r_0\,\alpha}{c} \right)^2 \tag{4.10}$$

where k_F is the Fermi momentum. The extra contribution to the coupling with t phonons is negligible. For C_8AsF_5 the charge transfer per intercalate molecule (consistent with $J_0 = 2.4$ eV) is $f \simeq .44$ [14]. This corresponds to $k_F = 2.56 \times 10^7$ cm^{-1} so that the term given by Eq. (4.10) is as large as each of those given by Eq. (4.9). The resistivity ρ_0 (T) due the graphite phonons is shown by the continuous line in Fig. 3a while the dashed line gives the corresponding asymptotic behavior. Fig. 3b

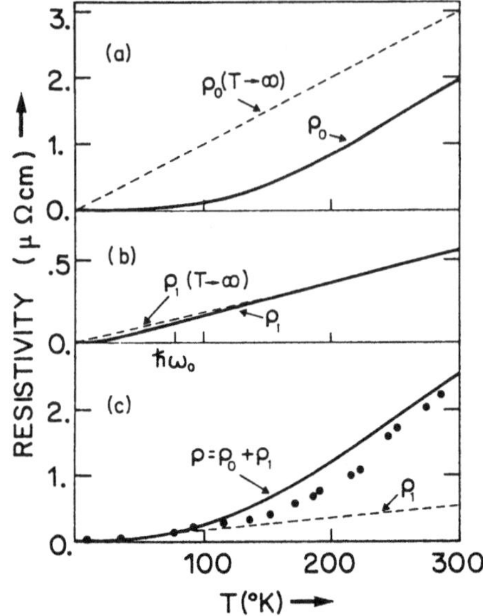

Fig. 3

(a) $\rho_0(T)$ represents the computed contribution to resistivity due to graphite phonons for C_8AsF_5 with f = .44. The dashed line shows the asymptotic linear behavior ($T \to \infty$).

(b) $\rho_1(T)$ is the computed resistivity due to the intercalate phonons and ω_0 is the pinning frequency of the intercalate molecules.

(c) $\rho(T)$ is the total computed resistivity due to el-ph interaction for C_8AsF_5 (without any ajustable parameter). The dots are the experimental values (29) once the residual ($T \to 0$) resistivity has been detracted.

shows the contribution to resistivity due to the local modes (ω_0) of the intercalate molecules that we have not discussed here and it is in general quite small. In Fig. 2c the two contributions are added and compared to the experimental values for C_8AsF_5 [29] once the residual ($T \to 0$) resistivity has been detracted. Note that no adjustable parameter has been used.

The good agreement between theory and experiments supports our general picture from which we estimate that the measured values for the conductivity are already close to the theoretical maximum for this class of compounds.

5 Summary and Open Problems

We have described how using a tight binding approach for the orbitals of graphite a variety of properties of intercalation compounds can be properly described. In particular we have shown how the concepts used for carbon based molecules can be extended to graphite to analyse the change of bond length with intercalation and extract the value of the charge transfer f and informations about the electron phonon coupling. The same general picture is then used to compute the electrical resistivity due to the scattering with phonons and to estimate the maximum ideal conductivity for this class of systems that results to be close to the presently observed values.

Finally we would like to point out two important problems that should receive consideration in the future. The first is that the value of $\alpha = dJ_0/dr$ appearing in the electron phonon coupling can be obtained with different methods: The analysis of the optical properties and Raman modes of polyacetilene gives a value of about 6 also in agreement with our simple calculation with Slater orbitals. On the other hand from a study of various properties of carbon based molecules it results in $\alpha \sim 4.5$. This discrepancy seems to be beyond the intrinsic indetermination of the two analysis and might have to do with the electronic correlation.

Another problem arises from the fact that the charge distribution within a plane of graphite is in general (and also in the present paper) assumed to be uniform. In some cases instead it is known that the intercalate molecules undergo chemical reactions whose result is that only a few of the molecules are actually charged. This creates strong local fields that could make the charge very inhomogeneous within a graphite layer [30]. The consequences of this effect on transport properties have not yet been analyzed.

References

[1] Proceedings of the Provincetown Conference on Intercalation Compounds of Graphite, Ed. by *F. L. Vogel*, Synthetic Metals v. 3/4 (1981).

[2] *J. E. Fischer* and *T. E. Thompson*, Physics Today, July 1978, p. 36.

[3] *F. Bassani* and *G. Pastori Parravicini*, Nuovo Cim. **B 50**, 4259 (1967).

[4] *L. Pietronero, S. Strässler, H. R. Zeller* and *M. J. Rice*, Phys. Rev. Lett. **41**, 763 (1978). *L. Pietronero, S. Strässler* and *H. R. Zeller*, Solid State Comm. **30**, 399 (1979).

[5] *D. Guérard, C. Zeller* and *A. Hérold*, C. R. Acad. Sci. Paris, C **283**, 437 (1976).

[6] *D. E. Nixon* and *G. S. Parry*, J. Phys. C **2**, 1732 (1969).

[7] *S. Flandrois, J. M. Masson, J. C. Rouillon, J. Gaultier* and *C. Hauw*, Synthetic Metals, in print.

[8] *A. Hérold*, private communication.

[9] *R. S. Markiewicz, J. C. Kasper* and *L. V. Interrante*, preprint.

[10] *T. Kakitani*, Progr. Theor. Phys. **51**, 656 (1974).

[11] *L. Pietronero* and *S. Strässler*, to be published.

[12] *L. A. Coulson* and *R. Taylor*, Proc. Phys. Soc. **A 65**, 815 (1952).

[13] *A. P. P. Nicholson* and *D. J. Bacon*, J. Phys. C **10**, 2295 (1977).

[14] *J. Blinowski, Nguyen Hy Hau, C. Rigaux, J. P. Vieren, R. Le Toullec, G. Furdin, A. Herold* and *J. Melin*, J. Physique **41**, 47 (1980).

[15] *J. Conard, M. Gutierrez-Le Brun, P. Lauginie, H. Estrade-Szwarckopf* and *G. Herman*, preprint. A preliminary analysis that gives larger values for f can be found in: *J. Conard, H. Estrade, P. Langinie, H. Fuzellier, G. Furdin* and *R. Vasse*, Physica **99 B**, 521 (1980).

[16] *W. Eberhardt, I. T. McGovern, E. W. Plummer* and *J. E. Fischer*, Phys. Rev. Lett., **44**, 200 (1980).

[17] *P. Oelhafen, P. Pfluger, E. Hauser* and *H. J. Güntherodt*, Phys. Rev. Lett. **44**, 197 (1980).

[18] *H. Suematsu, K. Higuchi* and *S. Tanuma*, J. Phys. Soc. Japan, **48**, 1541 (1980).

[19] *L. E. Campbell, G. L. Monet* and *G. J. Perlow*, Phys. Rev. **B 15**, 3318 (1977).

[20] *B. R. Weinberger, J. Kaufer, A. J. Heeger, J. E. Fischer, M. Moran* and *N. A. W. Holzwarth*, Phys. Rev. Lett. **41**, 1417 (1978).

[21] *R. S. Markiewicz, H. R. Hart* Jr., *L. V. Interrante* and *J. S. Kasper*, preprint.

[22] *J. J. Ritsko* and *E. J. Mele*, preprint.
[23] *Y. Iye, O. Takahashi* and *S. Tanuma*, Solid State Comm., **33**, 1071 (1980).
[24] *E. J. Mele* and *J. J. Ritsko*, preprint.
[25] *J. M. Ziman*, "Electrons and Phonons" Oxford Univ. Press, London (1960).
[26] *L. Pietronero* and *S. Strässler*, Synthetic Metals (Ref. 1) **3**, 209 (1981).
[27] *A. A. Abrikosov, L. P. Gorkov* and *I. I. Dzyaloshinski*, "Methods of Quantum Field Theory in Statistical Physics, Prentice Hall (1963).
[28] *L. Pietronero, S. Strässler, H. R. Zeller* and *M. J. Rice*, Phys. Rev. **B 22**, 904 (1980).
[29] *C. Zeller, L. A. Pendrys* and *F. L. Vogel*, J. of Mat. Sci. **14**, 2241 (1979).
[30] *E. Tosatti*, private communication.

Dynamics of Fast Ion Conductors

Wolfgang Dieterich
Fakultät für Physik, Universität Konstanz, Konstanz, Germany

Summary: Current theoretical concepts are outlined, which are useful for the interpretation of the dynamic properties of fast ion conductors at low frequencies. The discussion emphasises the hydrodynamic theory and models for many-particle diffusion.

1 Introduction

The electrical conductivity of ionic solids is usually due to the migration of a small number of vacancies or interstitials. Both the concentration and the mobility of those defects is thermally activated with activation energies of the order of several electron volts. Consequently the ionic conductivity at room temperature is normally very low, e.g. $\sigma_{ion} \sim 10^{-15}$ $(\Omega\,cm)^{-1}$ for NaCl.

In contrast the so-called "fast ion conductors" or "superionic conductors" are characterized by a large number of highly mobile defects. Correspondingly, the ionic conductivity goes up to $\sigma_{ion} \sim 1\,(\Omega\,cm)^{-1}$, an order of magnitude which is characteristic for liquid electrolytes like the molten salts.

There is a variety of materials showing fast ion transport [1, 2]. Most experiments have so far been done with α-AgI-type compounds which have a relatively simple structure. From the more complicated structures we mention only the β-alumina and compounds like K-hollandite [3], where the ionic motion is confined to planes or to linear tunnels, respectively.

The physics as well as the practical use of fast ion conductors has been reviewed several times under different points of view [1–8]. This paper gives an outline of current theoretical concepts which should help to understand the experimentally observed dynamic properties. Emphasis is put on the many-particle aspect of ionic diffusion.

2 Some basic properties

In order to elucidate the nature of the highly conducting state we briefly discuss a few basic properties known from thermodynamic and structural investigations or from computer simulation. The discussion refers mainly to AgI-typ compounds.

Specific heat

By increasing the temperature, the AgI-type compounds undergo a first-order phase-transition at a critical temperature T_{crit}, before reaching the highly conducting state. From specific heat measurements one can deduce the entropy changes at T_{crit} and at the melting point T_m, which are both about half the entropy change at the melting point of an ionic insulator [4]. This indicates a large degree of disorder in the temperature range $T_{crit} < T < T_m$, suggesting that the crystal is melting "in two steps".

Structural properties

The highly conducting state is investigated further by means of X-ray or neutron diffraction. Fig. 1 shows the measured Ag^+-density distribution ρ_{Ag} in the bcc-lattice of I^--ions in α-AgI [9]. The results closely agree with recent molecular-dynamics studies [10]. ρ_{Ag} assumes maxima at the tetrahedral sites and minima at the octahedral sites. By going from a tetrahedral site to an adjacent one via the intermediate saddlepoint, ρ_{Ag} varies only by a factor of about two. Therefore a network of pathways exists along which the silver-density does not vary drastically.

Recent studies of the Ag-pair-correlation function $g_{Ag\text{-}Ag}$ (r), shown in Fig. 2, provide further support for a liquid-like behaviour of the subsystem of Ag^+-ions [11]. An interesting point is that the correlation hole due to the Coulomb-repulsion between the Ag^+-ions extends to the second-nearest tetrahedral site, showing that we are dealing with a strongly correlated system.

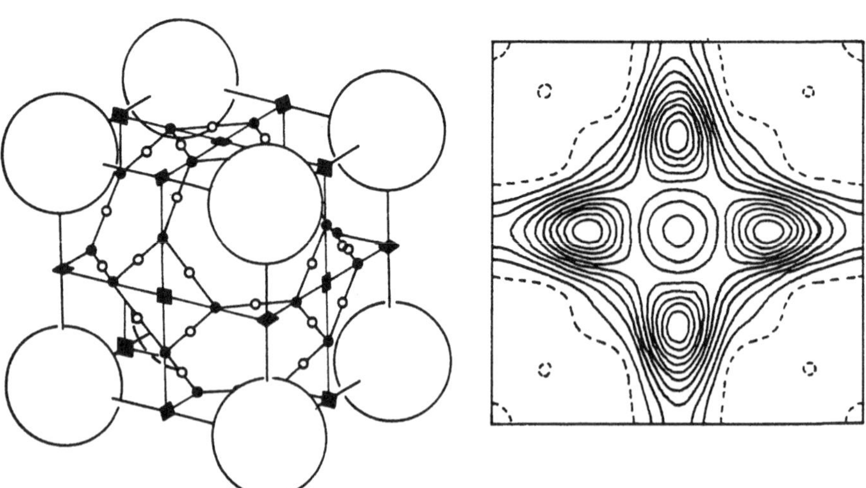

Fig. 1 Structure and silver density-profiles of α-AgI [9].

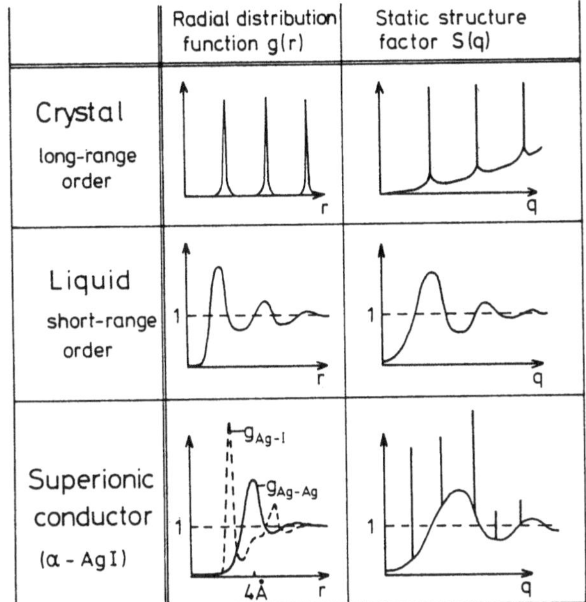

Fig. 2
Comparison of the radial distribution function g(r) and the static structure factor S (q) for a crystalline solid, a liquid and a fast ion conductor. The curves $g_{Ag\text{-}Ag}$ (r) and $g_{Ag\text{-}I}$ (r) for α-AgI are taken from Ref. [11]. The other curves are only schematic.

Single-particle motion

The single-particle motion is conveniently characterized by the mean-square displacement $\langle \Delta r^2 (t) \rangle$. So far most information about its time-dependence comes from molecular dynamics studies. As $t \to \infty$ one finds for the conducting ions $\langle \Delta r^2 (t) \rangle \sim Dt$ where D is the single-particle diffusion constant. On the other hand, for the lattice ions $\langle \Delta r^2 (t) \rangle$ approaches a constant which corresponds to the mean-square amplitude of the thermal vibration [12]. A typical behaviour of the mean-square displacement is shown in Fig. 3 [13].

These remarks show that in fast ion conductors features characteristic of both liquids and crystalline solids coexist on a microscopic scale. This leads to interesting dynamic properties which will be discussed in the subsequent sections.

3 Classical diffusion versus quantum diffusion

Before constructing a model for diffusion in an ionic conductor one has to consider the elementary jump process, by which a particle is transferred from site to site. Depending on the particle mass, the potential barriers and the temperature, several processes are possible [14], which are summarized schematically in Fig. 4.

The first regime (I) refers to very light particles at low enough temperatures. Here essentially coherent tunneling takes place, known from the conventional electronic conductivity in metals.

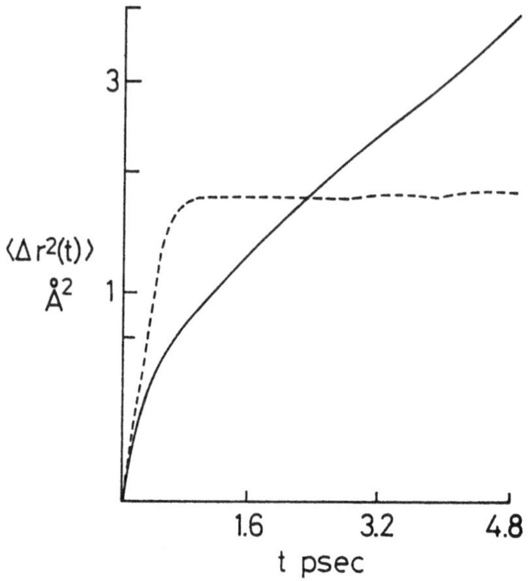

Fig. 3
Mean-square displacement of diffusing ions (full line) and lattice ions (dashed line). The actual curves refer to non-stochiometric and stochiometric Na-β-alumina, respectively (De Leeuw and Perram [13]).

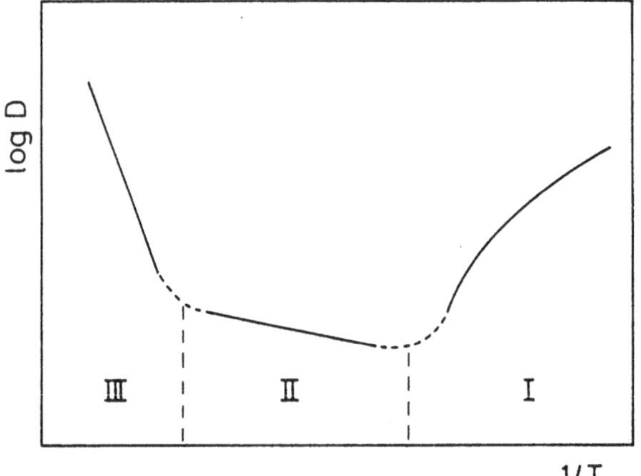

Fig. 4
Schematic illustration of different regimes for diffusion in solids. The relevance of the regimes I (coherent motion), II (small-polaron hopping) and III (over-barrier hopping) for a given particle depends on its mass and on the crystalline potential.

In the second regime (II) the coupling of the particle to the lattice leads to a localized state. Tunneling to a neighbouring site is associated with large rearrangement processes in the lattice. This regime is covered by the small polaron hopping theory [15] which leads to an Arrhenius law for the diffusion constant,

$$D \sim \omega_0 \exp(-E_A/k_B T). \tag{3.1}$$

In this case the activation energy E_A is of the order of a typical lattice energy $\hbar\omega_D$. The prefactor ω_0 is proportional to the square of the tunneling integral, usually satisfying $\omega_0 \ll \omega_D$.

As the temperature increases, over-barrier processes become dominant (III). Quantum effects may still play a role through the discreteness of the energy levels of the localized state in a potential well. The diffusion constant is again thermally activated, see Eq. (3.1). However, E_A now corresponds to the potential barrier height, which usually satisfies $E_A \gg \hbar\omega_D$, whereaus $\omega_0 \sim \omega_D$.

The diffusion of hydrogen in bcc-transition metals seems to cover the regimes III and II [26], while recent muon-diffusion experiments indicate a transition between the regimes II and I [17]. For ionic conductors it is important to note that the masses of the particles are larger by about two orders of magnitude than the hydrogen mass. Therefore, the energy levels of a particle inside a potential well may be regarded as a continuum, and the motion becomes completely classical if the temperature is not too low. Diffusion in ionic conductors is therefore a problem of classical statistical mechanics.

4 Diffusion of independent particles

Our aim is now to discuss some theoretical methods for treating various types of classical diffusion processes relevant to ionic conduction. From the experiments described in section 2 we know that the many-particle aspect is very important. Nevertheless, in order to introduce some basic concepts we shall start with independent-particle models. Their possible experimental relevance is limited to single-particle properties, like self-motion.

4.1 Single-particle hopping

The simplest model we are considering is a single particle hopping between well-defined sites of a crystalline lattice. For illustration we begin with a linear chain of equivalent sites with spacing a. The occupation number n_ℓ of site ℓ changes because of jumps to neighbouring sites. The rate of change of the occupation numbers is given by the set of equations

$$\frac{dn_\ell}{dt} = \frac{1}{2\tau}(-2n_\ell + n_{\ell+1} + n_{\ell-1}) \tag{4.1}$$

τ is the mean residence time which determines the overall time scale. From Eq. (4.1) the correlation function of the Fourier-transformed quantities n (q, t) is readily found to be

$$\langle n(q,t)\, n(-q,0)\rangle = \exp(-\gamma(q)\,|t|) \tag{4.2}$$

with the decay rate $\gamma(q) = \frac{1}{\tau}(1 - \cos aq)$. In the limit $q \to 0$ we obtain $\gamma(q) \simeq Dq^2$ with the diffusion constant $D = (1/2)\, a^2/\tau$. Fourier-transforming Eq. (4.1) with respect to time yields the "quasielastic" spectrum

$$S_{inc}(q, \omega) = \frac{1}{\pi} \frac{\gamma(q)}{\omega^2 + \gamma^2(q)} \tag{4.4}$$

of Lorentzian shape. Thus the diffusion constant D can be obtained directly from the quasielastic line-width observed in a scattering experiment.

The extension to jump diffusion on a non-primitive lattice is straightforward. If the elementary cell contains s sites, s relaxation branches $\gamma_\lambda(q)$ ($\lambda = 0, 1, \ldots, s-1$) are obtained. The lowest branch $\lambda = 0$ with the property $\gamma_0(q \to 0) \simeq Dq^2$ describes the relaxation in density among different elementary cells or macroscopic diffusion as $q \to 0$. The remaining $s - 1$ branches with finite limiting values $\gamma_\lambda(q \to 0)$ describe relaxation within an elementary cell. (Note the analogy to harmonic lattice dynamics and to the dispersion relations for acoustic and optical phonons!) The simple form Eq. (4.4) for the dynamical structure factor is replaced by a superposition of s Lorentzians with certain weighting factors,

$$S_{inc}(q, \omega) = \sum_{\lambda=0}^{s-1} |M_\lambda(q)|^2 \frac{1}{\pi} \frac{\gamma_\lambda(q)}{\omega^2 + \gamma_\lambda^2(q)}. \tag{4.5}$$

For example, in the case of nearest-neighbour jumps on tetrahedral sites in a bcc-lattice, Eq. (4,5) has been evaluated in connection with hydrogen-diffusion in bcc-transition metals [16]. As mentioned in section (2.2), the same lattice geometry also applies to α-AgI. The corresponding "dispersion relations" $\gamma_\lambda(q)$ for the relaxational modes are shown in Fig. 7a.

4.2 Continuous diffusion

Hopping models are adequate for ionic conductors with activation energies much larger than thermal energies, so that the hopping particles remain well localized near a lattice site over most of the time. This applies for example to ionic conductors with the fluorite structure [18]. The liquid-like features observed in AgI-type compounds, however, strongly suggest a description in terms of continuous diffusion [19]. Replacing the action of the vibrating lattice on the diffusing particle by frictional and random forces, one is lead to the picture of a Brownian particle moving in a periodic potential V(x). The corresponding Langevin-equation reads

$$m\ddot{x} + m\gamma\dot{x} + \frac{\partial V}{\partial x} = f(t). \tag{4.6}$$

The random force f(t) and the friction constant γ are related by

$$\langle f(t)\, f(0) \rangle = 2m\gamma\, k_B T\, \delta(t). \tag{4.7}$$

The dynamic mobility

$$\mu(\omega) = \mathrm{Re} \int_0^\infty dt\, e^{i\omega t} \langle \dot{x}(t)\, \dot{x}(0) \rangle \tag{4.8}$$

and the dynamic structure factor

$$S_{inc}(q, \omega) = \frac{1}{\pi} \mathrm{Re} \int_0^\infty dt\, e^{i\omega t} \langle e^{iq(x(t)-x(0))} \rangle \tag{4.9}$$

have been calculated by different methods [20, 21]. Fig. 5 shows their high- and low-temperature limits, well-known from free diffusion and from damped harmonic oscillations, respectively. An actual spectrum at intermediate temperatures is shown in Fig. 6, in remarkable qualitative agreement with incoherent neutron scattering spectra observed in $RbAg_4I_5$ [22].

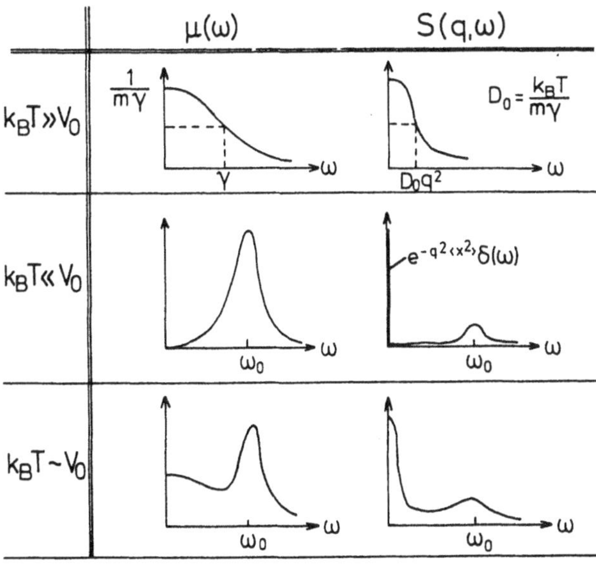

Fig. 5 Dynamic mobility $\mu(\omega)$ and dynamic structure factor $S(q, \omega)$ for a Brownian particle moving in a periodic potential (schematic). V_0 denotes the potential barrier height, ω_0 is the natural vibration frequency.

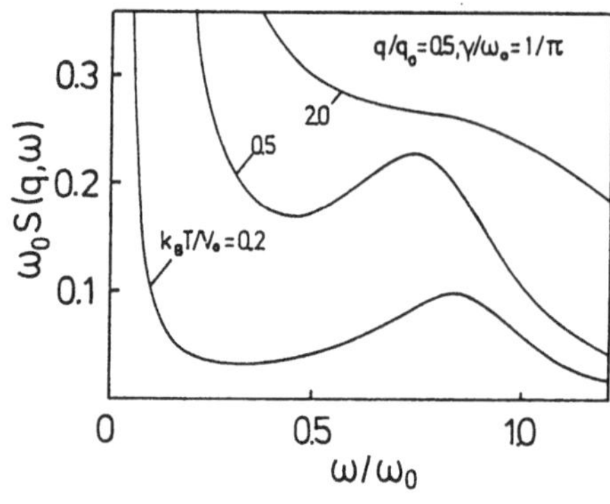

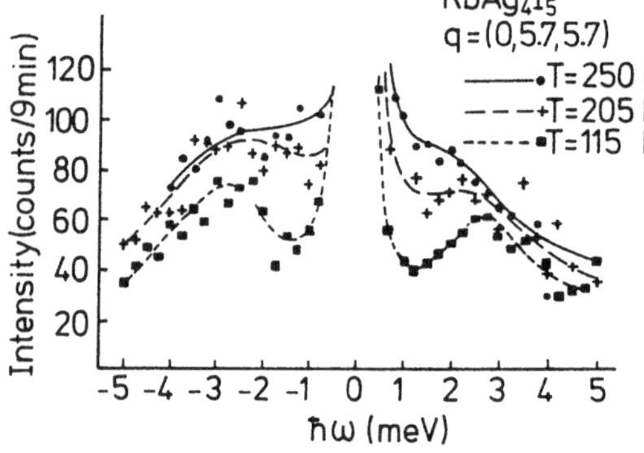

Fig. 6

a) Dynamic structure factor for Brownian motion in a cosine potential at low friction $\gamma/\omega_0 = 1/\pi$ [20].

b) Results on inelastic neutron scattering from $RbAg_4I_5$ [22].

5 Hydrodynamics

In section 2 we have seen that the diffusing ions in a fast ion conductor represent a strongly correlated system. Collective quantities like the conductivity or the spectrum of density fluctuations are therefore strongly affected by interactions between the diffusing particles. Quite generally, such a many-particle problem simplifies in the limit of low frequencies $\omega\tau \ll 1$ and long wave-lengths, where τ means a relaxation time for establishing local thermodynamic equilibrium. In this hydrodynamic limit the conservation laws and the symmetry of the system allow one to obtain the form of the dynamic correlation functions for the conserved quantities without considering microscopic details. In the complex frequency-plane those correlation functions assume a simple pole-structure, each pole corresponding to a well-defined hydrodynamic mode, whose frequency goes to zero as the wave-vector q goes to zero. For uncharged systems the number of hydrodynamic modes is equal to the number of conserved quantities plus the number of broken continuous symmetries [23]. We will not discuss this in detail but rather concentrate on the question how this rule is modified for charged systems.

Let us consider the concentration fluctuations in a binary fluid. In the neutral case their spectrum is determined by a pole $\omega = -i\,Dq^2$, where D denotes the interdiffusion constant. For a charged binary system, the concentration fluctuations are identical to charge-density fluctuations. Hydrodynamic considerations lead to a pole $\omega = -i\,4\pi\sigma$, where σ denotes the conductivity. The corresponding charge-fluctuation spectrum takes the form [24, 25]

$$S_{cc}(q, \omega) = S_{cc}(q) \frac{4\sigma}{\omega^2 + (4\pi\sigma)^2} \tag{5.1}$$

with

$$S_{cc}(q) \simeq q^2/q_D^2 \,. \tag{5.2}$$

Here $q_D = (4\pi e^2 \rho/k_B T)^{1/2}$ is the inverse Debye-Hückel-screening length. The q^2-dependence of $S_{cc}(q)$ expresses the perfect screening condition, which guarantees that the static dielectric constant $\epsilon(q) = (1 - 4\pi e^2 \rho S_{cc}(q)/q^2 k_B T)^{-1}$ diverges as $q \to 0$. It is clear now that in the presence of Coulomb-forces the concentration fluctuations are not really hydrodynamic modes, because their characteristic frequency remains finite as $q \to 0$. As a consequence, Eq. (5.1) is not always valid. Nevertheless, the hydrodynamic concept remains useful, if the collisional effects are sufficiently strong. The question, however, then arises, whether there are additional slow non-hydrodynamic modes with relaxation-rates comparable to or lower than the relaxation-rate $4\pi\sigma$ for long wave-length charge-density fluctuations. Such modes are in fact expected in view of Fig. 7a which shows the dispersion of modes describing relaxation within a unit cell of a non-primitive lattice. The inclusion of such modes ('quasihydrodynamic theory') is of course important for the low-frequency dynamics [26]. This point will be discussed further in the next section.

6 Models for many-particle diffusion

The hydrodynamic treatment needs generalization in various directions. First, as discussed above, relaxational processes within a unit cell must be taken into account. This suggests the construction of hopping models including Coulomb-forces between the hopping particles. Second, one would like to study the dependence of the conductivity on microscopic quantities. In particular, the observed high ionic conductivities pose the question under which circumstances σ will be enhanced or suppressed by the pair-interactions. Furthermore, it is interesting to study the frequency-dependence $\sigma(\omega)$. Since these problems are far from beeing solved, we shall discuss only a few aspects.

6.1 Many-particle hopping models

The dynamics of a many-particle hopping model will be determined by a master-equation, which generalizes the simple rate-equations discussed in section 4. Restricting ourselves to nearest-neighbour jumps only, the rate of change of the probability $p(\mathbf{n}, t)$ to find a configuration $\mathbf{n} = \{n_\ell\}$ of occupation numbers $\mathbf{n}_\ell$ takes the form

$$\frac{dp(\mathbf{n}, t)}{dt} = \frac{1}{2} \sum_{\ell,\delta} \left[- w_{\ell,\ell+\delta}(\mathbf{n})\, p(\mathbf{n}, t) + w_{\ell,\ell+\delta}(\mathbf{n}^-)\, p(\mathbf{n}^-, t) \right]. \tag{6.1}$$

Here $w_{\ell,\ell+\delta}(\mathbf{n})$ is the rate for a jump from site ℓ to a nearest-neighbour site $\ell + \delta$. This quantity, which contains the interaction, must satisfy detailed balance. This guarantees that any initial configuration finally goes over into the equilibrium configuration $p_0 \propto \exp[-\beta H(\mathbf{n})]$, $H(\mathbf{n})$ representing a lattice gas Hamiltonian. $\mathbf{n}^-$ denotes the configuration which results from $\mathbf{n}$ by interchanging the occupation of sites ℓ and $\ell + \delta$. Much work has been devoted to the problem of nearest-neighbour interactions (for a review see Ref. [7]). Here we give a few comments on a lattice gas in the presence of Coulomb-forces. The RPA-approximation leads to a modification of the "dispersion relations" $\gamma_\lambda(q)$ obtained in the independent-particle model (section 4). For the case of tetrahedral sites in a bcc-lattice the result is shown in Fig. 7 [26]. While the Coulomb-force have no drastic influence on the branches $\lambda \neq 0$, the dispersion of the $\lambda = 0$-branch is changed qualitatively. As $q \to 0$, it describes the relaxation of long wave-length density fluctuations of the mobile ions relative to the lattice ions. In the non-interacting case we have $\gamma_0(q \to 0) \simeq Dq^2$, whereas in the presence of Coulomb-forces $\gamma_0(q \to 0) = 4\pi\sigma$, in accord with the hydrodynamic considerations.

Let us estimate the ratio $(\gamma_0/\gamma_{\lambda \neq 0})_{q \to 0}$. First we note that $\gamma_{\lambda \neq 0}$ is of the order of the bare jump rate $1/\tau$. Putting $\sigma \sim \rho (Ze)^2\, D_0/k_B T$ with $D_0 \sim \ell^2/\tau$, where ℓ is a typical jump distance and writing $\rho = 3/4\, \pi\, r_0^3$ we find

$$(\gamma_0/\gamma_{\lambda \neq 0})_{q \to 0} \sim (r_0/\ell)^2 \cdot \Gamma. \tag{6.2}$$

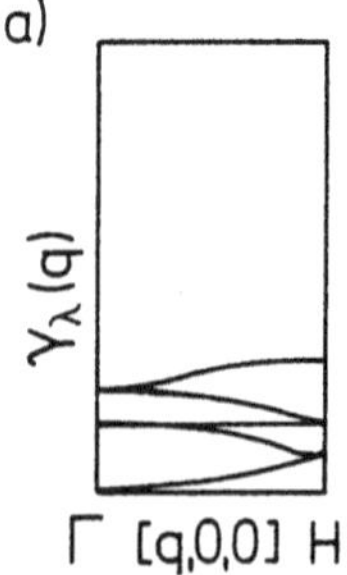

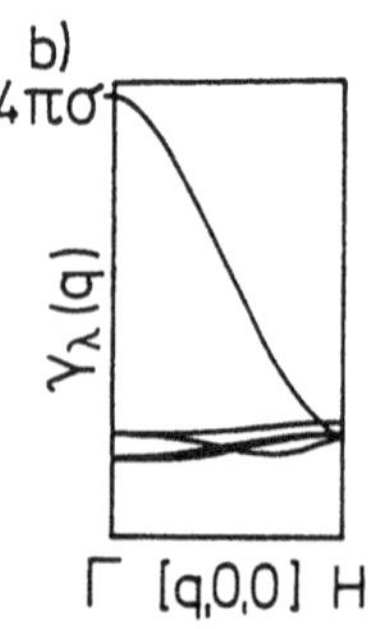

Fig. 7
a) Dispersion of relaxational modes for jump-diffusion of independent particles on tetrahedral sites of a bcc-lattice.
b) Dispersion of relaxational modes when Coulomb-interactions between the hopping particles are taken into account [26].

Here

$$\Gamma = \frac{(Ze)^2}{r_0 k_B T} \tag{6.3}$$

denotes the plasma parameter, which measures the average Coulomb interaction energy in relation to $k_B T$. For AgI-type ionic conductors we have $r_0 \sim \ell$ and $\Gamma \gg 1$ [27], giving $(\gamma_0/\gamma_\lambda)_{q\to 0} \gg 1$. However, in other materials with much lower density of charge carriers (mobile ions or vacancies), one may find a situation where $(\gamma_0/\gamma_\lambda)_{q\to 0} < 1$.

Fig. 7 allows one to discuss various scattering experiments on α-AgI. For small momentum transfer coherent neutron scattering should give a spectrum containing two distinct quasi-elastic contributions. One expects a broad component with a half-width $\sim 4\pi\sigma/\epsilon_\infty$, resulting from the $\lambda = 0$-mode, and a narrow component with a half-width $\sim 1/\tau$ resulting form the other branches $\lambda \neq 0$. Incoherent scattering should also contribute to the narrow component. Interestingly, the observed spectra [28, 8] do show such features, see Fig. 8. The half-width of the broad component is nearly independent of q, while its weight increases as q increases, in accord with Eqs. (5.1) and (5.2). The experiments done so far, however, are limited to $q \sim 1$ Å^{-1} and therefore do not really refer to the hydrodynamic regime. For testing the above predictions, experiments below $q \sim 0.6$ Å^{-1}, which corresponds roughly to a point midway between Γ and H, would be very interesting. A further test could be achieved by measuring the temperature-dependence of the spectra. At larger q, where S(q) is close to unity, the intensity of the narrow component was found to be anisotropic, in agreement with single-particle jump models [8].

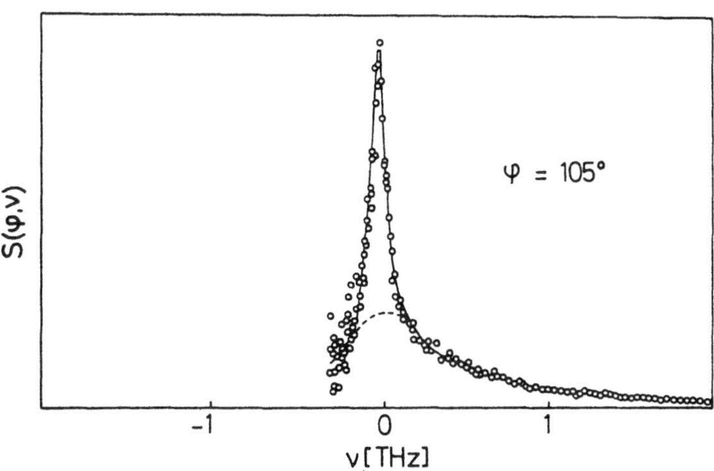

Fig. 8 Quasielastic neutron scattering on α-AgI [28].

Concerning light-scattering, where $q \sim 0$, the $\lambda = 0$-mode does not contribute, because its weight vanishes as $q \to 0$. A symmetry analysis shows, however, that polarizability-fluctuations arise from the mode with Γ_3-symmetry [20]. This leads to a quasielastic spectrum with a half-width of the order of the bare jump-rate $1/\tau$, in agreement with the experimental findings [29].

6.2 Liquid-like models

The drawbacks of a hopping description of AgI-type ionic conductors have been pointed out in section 4. Let us therefore turn now to the treatment of continuous diffusion of interacting particles. Concerning the static properties, our system can be regarded as a liquid embedded in a periodic medium. The total potential takes the form

$$V(\mathbf{r}_1, \dots \mathbf{r}_N) = \sum_k V^{(1)}(\mathbf{r}_k) + \frac{1}{2} \sum_{k \neq \ell} V^{(2)}(\mathbf{r}_k - \mathbf{r}_\ell). \tag{6.4}$$

$V^{(1)}$ is a periodic single-particle term and $V^{(2)}$ represents the pair-potential. Generalizing Eq. (4.6), the equations of motion are taken to be coupled Langevin-equations [7],

$$m\ddot{\mathbf{r}}_i + m\gamma\,\dot{\mathbf{r}}_i + \frac{\partial}{\partial \mathbf{r}_i} V(\mathbf{r}_1, \dots \mathbf{r}_N) = f_i(t) \tag{6.5}$$

Let us classify a system according to the relative strength of $V^{(1)}$ and $V^{(2)}$. Three different regimes have to be distinguished. First, in the weak-coupling case the average interaction energy $\overline{V^{(2)}}$ per particle is much smaller than the amplitude V_0 of the periodic potential $V^{(1)}$, i.e. $\overline{V^{(2)}} \ll V_0$. We shall not discuss this dilute-gas-like regime any further. Second, in the opposite limit of strong coupling, $\overline{V^{(2)}} \gg V_0 \sim k_B T$, the pair-potential induces an almost regular arrangement of the particles. Then the problem can considerably be simplified by an expansion of $V^{(2)}$ up to quadratic terms about the equidistant particle positions. This maps the system on the Frenkel-Kontorova model, pertaining for example to submonolayer films adsorbed on the surface of a crystalline substrate [30] or to the Hg-chains in $Hg_{3-\delta}AsF_6$ which form a quasi-one-dimensional liquid [31]. In one dimension the potential energy takes the form

$$V = \frac{V_0}{2} \sum_k \cos \frac{2\pi x_k}{a} + \frac{A}{2} \sum_k (x_{k+1} - x_k - b)^2 \tag{6.6}$$

where a means the lattice constant and b the mean separation of particles. The static properties at finite temperature can be calculated exactly by means of the transfer-integral method [32]. Finally, the regime of intermediate coupling $\overline{V^{(2)}} \gtrsim V_0$ corresponds to a dense fluid with large fluctuations of the nearest-neighbour distances. Superionic conductors of the AgI-type fall into this category.

Our aim is to compute the conductivity from Eqs. (6.4) and (6.5). Thereby the following particular aspect of our model turns out to be important. As $V^{(2)}$ increases there is increasing competition between two intrinsic lengths, namely the lattice constant a and a second length which characterizes the short-range order induced by the pair-forces. In the Frenkel-Kontorova model this second length is simply b. Generally, one may identify it with the position of the main peak in the radial distribution function.

To illustrate the dependence of the conductivity on the 'commensurability' of those two lengths we consider first the case of a weak periodic potential, such that $k_B T > V_0$. One can show that for large friction and to second order in $V_0/k_B T$ the conductivity is approximately given by

$$\sigma(0) = \sigma_0 \left[1 + \frac{1}{2}\left(\frac{\beta V_0}{2}\right)^2 S_0^2(q_0)\right]^{-1} \tag{6.7}$$

where $q_0 = 2\pi/a$ denotes a reciprocal lattice vector and $\sigma_0 = \rho(Ze)^2 D_0/k_B T$. $S_0(q)$ denotes the static structure factor in the absence of $V^{(1)}$. This shows that the conductivity depends sensitively on the relation between q_0 and the peak-positions of $S_0(q)$. For example, $\sigma(0)$ is suppressed as compared to the non-interacting case if q_0 coincides with the main peak of $S_0(q)$, so that $S_0(q_0) > 1$. In this 'commensurate' situation the interaction tends to lock the particles in the wells of the periodic potential. On the other hand, an incommensurate situation ($S_0(q_0) < 1$) corresponds to particle delocalization and to an enhanceement of $\sigma(0)$. Clearly, these effects are most pronounced in the strong-coupling limit [19], where the approximation of $V^{(2)}$ in terms of harmonic forces (see Eq. (6.6)) leads to

$$S_0(q) = \frac{\sinh \frac{1}{2} q^2 \sigma^2}{\cosh \frac{1}{2} q^2 \sigma^2 - \cos qb}. \tag{6.8}$$

Here $\sigma^2 = k_B T/A$ denotes the mean-square fluctuation of the nearest-neighbour distance, which must satisfy $\sqrt{\sigma^2} \ll b$. Substituting Eq. (6.8) into Eq. (6.7), one obtains pronounced oscillation of σ as a function of the mean particle density [19].

As mentioned before, for fast ion conductors the intermediate coupling regime $V^{(2)} \sim V_0 \sim k_B T$ is most relevant. It can be treated with the aid of the approximate relation

$$\sigma = \sigma_0 \left[\overline{\rho(x)}\,\overline{\rho^{-1}(x)}\right]^{-1} \tag{6.9}$$

between the conductivity and the density $\rho(x)$, which holds for large friction [33]. The bar in Eq. (6.9) means an average over a unit cell. The inhomogeneous version of the Percus-Yevick theory has been used to evaluate Eq. (6.9) for the case of a repulsive pair-potential of the form $V^{(2)}(x) = V_0 (b/x)^2$ [34]. The dependence of

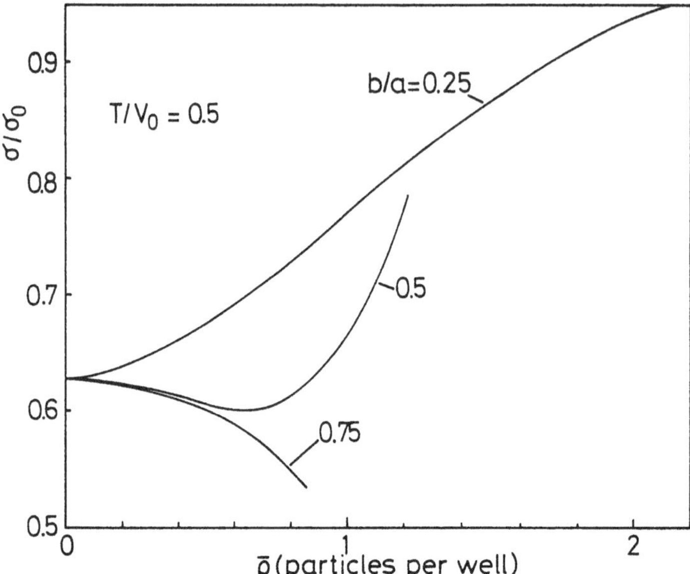

Fig. 9 Conductivity of strongly damped Brownian particles in a periodic potential for various strengths of the pair-potential [34].

the conductivity on the average density is shown in Fig. 9. For the case b/a = 0.75 one can show that the radial distribution function g_0 (x) of the homogeneous system develops a peak at $x \approx a$, as ρ increases. In this 'commmensurate' case the conductivity decreases. On the other hand, in the 'incommensurate' case b/a = 0.25 the pair-forces bring the system to an almost homogeneous state, i.e. the barrier height of the effective single-particle potential $U(x) = -k_B T \log \rho(x)$ decreases and the conductivity increases. Such considerations may help to understand the low activiation energies observed in ionic conductors with a liquid-like distribution of the mobile component.

References

[1] *S. Geller* (ed.), Solid Electrolytes, Springer-Verlag, Berlin, Heidelberg, New York (1977).

[2] *P. Hagenmuller, W. van Gool,* Solid Electrolytes, General Principles, Characterization, Materials, Applications, Academic Press, New York (1978).

[3] *B. U. Beyeler,* in: Physics in One Dimension, p. 323, J. Bernasconi, T. Schneider (eds.), Springer-Verlag, Berlin, Heidelberg, New York (1981),

[4] *A. Rabenau,* in: Festkörperprobleme (Advances in Solid State Physics), Vol. XVIII, p. 77, J. Treusch (ed.), Vieweg, Braunschweig (1978).

[5] *J. B. Boyce*, B. A. Huberman, Phys. Rep. **51**, 189 (1979).
[6] *M. B. Salamon* (ed.), Physics of Superionic Conductors, Springer-Verlag, Berlin, Heidelberg, New York (1979).
[7] *W. Dieterich, P. Fulde, I. Peschel*, Adv. in Physics **29**, 527 (1980).
[8] *K. Funke*, in: Festkörperprobleme (Advances in Solid State Physics), Vol. XX, p. 1, J. Treusch (ed.), Vieweg, Braunschweig (1980).
[9] *R. J. Cava, F. Reidinger, B. J. Wuensch*, Solid St. Commun. **24**, 411 (1977).
[10] *P. Vashishta, A. Rahman*, Phys. Rev. Lett. **40**, 1337 (1978).
[11] *Y. Tsuchiya, S. Tamaki, Y. Waseda*, J. Phys. C: Solid St. Phys. **12**, 5361 (1979).
[12] *A. Rahman*, J. chem. Phys. **65**, 4845 (1976).
[13] *S. W. De Leeuw, J. W. Perram*, preprint (1979).
[14] *S. Fujii*, J. Phys. Soc. Japan, **46**, 1833 (1979).
[15] *T. Holstein*, Ann. Phys. (N. Y.) **8**, 343 (1959); *C. P. Flynn, A. M. Stoneham*, Phys. Rev. **B 10**, 3966 (1970).
[16] For a review see e.g. *K. Sköld* in: Hydrogen in Metals I, p. 267, *G. Alefeld, J. Völkl* (eds.), Springer-Verlag, Berlin, Heidelberg, New York (1978).
[17] *H. Graf, G. Balzer, A. Weidinger*, Phys.. Rev. Lett. **20**, 1333 (1980).
[18] *M. Dixon, M. J. Gillan*, Journal de Physique **41**, C 6-24 (1980).
[19] *T. Geisel*. in: Physics of Superionic Conductors, p. 201, *M. B. Salamon* (ed.), Springer-Verlag, Berlin, Heidelberg, New York (1979).
[20] *W. Dieterich, T. Geisel, I. Peschel*, Z. Phys. B. **29**, 5 (1978).
[21] *H. Risken, H. D. Vollmer*, Z. Phys. B **31**, 209 (1978).
[22] *S. M. Shapiro*, in: Physics of Superionic Conductors, p. 45,, *M. B. Salamon* (ed.), Springer-Verlag, Berlin, Heidelberg, New York (1979).
[23] *P. C. Martin, O. Parodi, P. S. Pershan*, Phys. Rev. A **6**, 167 (1972).
[24] *J. Jäckle*, Z. Phys. B**30**, 255 (1978).
[25] *R. Zeyher*, Z. Phys. B **31**, 127 (1978).
[26] *R. Zeyher*, in: Fast Ion Transport in Solids, Electrodes and Electrolytes, p. 509, *P. Vashishta, J. N. Mundy, G. K. Sehnoy* (ed.), North-Holland, New York, Amsterdam, Oxford (1979).
[27] Including a dielectric constant $\epsilon_\infty \simeq 5$ for the lattice, one finds $\Gamma = (Ze)^2/\epsilon_\infty r_0 k_B T \simeq 20$ for α – AgI at T ≈ 500 K.
[28] *G. Eckold, K. Funke, J. Kalus, R. E. Lechner*, J. Phys. Chem. Solids, **37**, 1097 (1976).
[29] *G. Winterling, W. Senn, M. Grimsditch, R. Katiyar*, Proc. Int. Conf. on Lattice Dynamics, p. 553, *M. Balkanski*, (ed.), Flammarion, Paris (1977).
[30] See e.g. *M. Nielsen, J. P. McTague, W. Ellenson*, Journal de Physique, **38**, C4-10 (1977).
[31] *J. D. Axe*, in: Solitons and Condensed Matter Physics, p. 234, *A. R. Bishop, T. Schneider* (eds.), Springer-Verlag, Berlin, Heidelberg, New York (1978).
[32] *R. A. Guyer, M. D. Miller*, Phys. Rev. Lett. **42**, 718 (1979).
[33] *R. A. Guyer, M. D. Miller*, Phys. Rev. **A17**, 1774 (1978).
[34] *A. Bunde, W. Dieterich*, Solid State Commun., in print.

Thermoelectric Effects in Superconductors

Gerd Schön
Institut für Theorie der Kondensierten Materie, Universität Karlsruhe, Karlsruhe, Germany

Summary: A summary of recent theoretical and experimental work on thermoelectric and related effects in superconductors is given. Whereas this field for a long time was characterized by large discrepancies between theory and experiment, more recent experiments confirm the theoretical concepts. Experiments which investigate the difference between the electrochemical potential of excitations and Cooper pairs (charge imbalance) play a prominent role. They allow one to determine the thermoelectric transport coefficient. Also a new type of thermoelectric effect is discussed: a charge imbalance is created if a supercurrent is flowing in the presence of a temperature gradient.

I Introduction

Thermoelectric effects in superconductors have been studied as early as 1927, when Meissner found evidence that no thermoelectric voltage exists in a steady state superconductor. From the absence of the familiar thermoelectric effects (Seebeck effect, Peltier effect, or Thompson heat), at some time the conclusion was drawn that thermoelectric effects would vanish completely in the superconducting state [1]. However, in 1944 Ginzburg [2] pointed out that there still exist thermoelectric effects in superconductors, although they manifest themselves in a different way. Using a two fluid description where the total current is $\vec{j} = \vec{j}_s + \vec{j}_n$, he argued that a temperature gradient gives rise to the flow of a normal current $\vec{j}_n$, which however, is locally compensated by the supercurrent $\vec{j}_s$. Observation of this effect requires to distinguish between these two types of currents.

Later Galperin, Gurevich and Kozub [3] showed that indeed a temperature gradient gives rise to an electric normal current $\vec{j}_n$ carried by the excitations in the superconductor, and they calculated the thermoelectric transport coefficient in the superconducting state. They found that the coefficient develops in a continuous fashion from the normal state value as the temperature is reduced below T_c, and the difference is small (of relative order $(1 - T/T_c)^{3/2}$). Also they, and independently Garland and van Harlingen [4] proposed a method which allows the experimental observation of thermoelectric effects in superconductors. It relies on the fact that in a superconducting bimetallic ring the phase of the order parameter is subject to the constraint of fluxoid quantization. As a consequence the cancellation of supercurrent and thermoelectric normal current in general cannot be complete, which gives rise to a (nonquantized) magnetic flux through the ring. Several experiments utilizing this principle have been conducted [5–8]. However, gross disagreement (by as much as a factor 10^5) between theoretical prediction and experimentally determined transport coefficient was obtained. So far the reason for the discrepancy in the latest of these experiments (Ref. [8]) remains unresolved. Also several other experiments remained inconclusive [9].

Only very recently, in a different type of experiment, van Harlingen [10] found values of the thermoelectric transport coefficient in superconductors which agree reasonably well with those of the normal state. The experiment was first proposed by Artemenko and Volkov [11]. It makes use of the fact that the counterflow of the normal thermoelectric current and the supercurrent in the bulk of the superconductor requires conversion between these two types of currents to take place near two opposite ends. In these conversion regions an imbalance between normal charge (in the system of excitations) and superfluid charge (in the condensate of Cooper pairs) i.e. a "charge imbalance" builds up [12, 13]. It is accompanied by a difference in the electrochemical potential of the normal and superfluid component, which can be detected,and from which the strength of the bulk thermoelectric currents can be deduced.

Also,recently a different effect was predicted by Pethick and Smith [14] and observed experimentally by Clarke, Fjordbøge, and Lindelof [15]. A temperature gradient in the presence of an imposed supercurrent gives rise to a charge imbalance in the bulk of the superconductor proportional to $(\vec{j}_s \cdot \vec{\nabla} T)$. This effect has in common with the other thermoelectric effects that some kind of internal symmetry has to be distorted. In the usual thermoelectric effects,a temperature gradient may give rise to an electric current because of an asymmetry between the properties of particle and hole like excitations. Similarly, a temperature gradient can give rise to a charge imbalance because the supercurrent $\vec{j}_s = en_s \vec{v}_s$ changes the energy of the excitations by $\vec{p} \cdot \vec{v}_s$, i.e. depending on the direction of the momentum. The experimental result and the original prediction differed by a large factor since the dominant relaxation channel, namely impurity scattering, was overlooked in Ref. [14]. Soon thereafter, an adequate theoretical description was given [16] and a very good agreement with experiments was found. The outstanding role of impurity scattering, which in other charge imbalance phenomena only plays a trivial part, is the subject of further recent theoretical work [17–20]. Also,recently it was shown that a charge imbalance proportional to $\vec{j}_s \cdot \vec{\nabla} T$ is to be expected from a systematic analysis of dissipative processes in superconductors [21], which leads to a scheme of transport equations. Actually further effects are implied by this scheme which, however, have not yet been detected experimentally.

The plan of the paper is as follows. We first will describe the experiments which use the flux through a bimetallic ring as a probe for thermoelectric effects. In Chapter III the physical ideas and necessary theoretical framework describing the charge imbalance phenomena are developed. Then van Harlingen's recent experiment [10] is presented. Finally in Chapter IV we turn to the $(\vec{j}_s \cdot \vec{\nabla} T)$ effect. We discuss in some detail the dynamics of the involved quasiparticles, and compare different limiting theoretical results. The phenomenological scheme of transport equations will serve as a guideline throughout this paper and expressions for the various transport coefficients will be derived.

II Thermoelectric Currents in Superconductors

A Theory

Thermoelectric effects in normal metals have been widely studied. They may be summarized by the transport equations

$$\vec{j} = \sigma \vec{E}' - L \vec{\nabla} T , \qquad \vec{q} = LT\vec{E}' - \kappa \vec{\nabla} T . \tag{1}$$

Since a gradient in the chemical potential μ leads to a diffusion current, in addition to the current driven by the electric field $\vec{E}$, the total driving force $\vec{E}' = \vec{E} - \frac{1}{e} \vec{\nabla} \mu$ appears. In stationary situations, we can write $\vec{E}' = -\frac{1}{e} \vec{\nabla} \varphi$ where $\varphi = eV + \mu$ is the electrochemical potential. The electric conductivity is denoted by σ. A temperature gradient produces a heat current $\vec{q}$, which at low temperatures is mainly carried by the electrons. Then the electric and thermal conductivity κ are related (in simple models) by Wiedeman-Franz's law $\kappa/T\sigma = \pi^2 k_B^2/3e^2$. The off-diagonal terms in Eq. (1) describe the thermoelectric effects. Onsager's relations require the same thermoelectric coefficient L to appear in both equations. Since thermoelectric effects only arise because of small asymmetries in the properties of particles and holes near the Fermi surface, they are weak and the coefficient L is small. In a simple model one finds

$$L = \frac{\pi^2}{3e} k_B^2 T \frac{d\sigma(\epsilon_F)}{d\epsilon_F} . \tag{2}$$

The conductivity,for a given value of the Fermi energy ϵ_F depends on the density of states, the Fermi velocity, and the scattering rate at this energy as $\sigma(\epsilon_F) = \frac{2e^2}{3} N_0(\epsilon_F) v_F^2 \tau_{imp}(\epsilon_F)$. Here we take into account that at low temperatures impurity scattering is the dominant scattering mechanism and ignore effects due to phonons.

A typical experiment in the field of thermoelectricity makes use of the Seebeck effect. If a temperature gradient is imposed on an open-circuited metal, an electric field (more precisely E') will be established,such that it balances the thermally induced current. The ratio of $\vec{E}'$ and $\vec{\nabla} T$ yields the thermopower $S = L/\sigma$, from which L can be obtained. The agreement between theoretical expectations and experimental results is often poor. One reason may be the uncertainty in details of the impurity scattering processes. In fact, experiments often show a strong dependence of L on the impurity concentration. But the fact that L is sometimes found to vary strongly with temperature is certainly beyond the simple model refered to above. The effect of phonons on the electronic transport properties (e.g. phonon drag) may well be important.

Also,in a superconductor an electric field or a temperature gradient lead to a flow of excitations,and hence to an electric normal current and to a heat current. The transport relations given by Eq. (1) are still valid where the electric current is now the normal current $\vec{j}_n$. It is clear, however, that in a superconductor in equilibrium (for extension see below) no stationary electric field can exist since it is shorted out by the supercurrent. Therefore, the usual Seebeck effect is not found. Rather in an open-circuited superconductor the flow of normal electrons $\vec{j}_n$ driven by the temperature gradient is locally cancelled in the bulk by a counterflow of superconducting electrons $\vec{j}_s = -\vec{j}_n$. The existence of this supercurrent in the bulk, that means deeper than the penetration depth $\lambda(T)$, gives rise to unusual effects which we will discuss below.

The transport coefficients $\sigma^{(S)}$, $\kappa^{(S)}$, and $L^{(S)}$ in superconductors differ from those in the normal state, but theoretical calculations reveal that they approach the normal state values in a continuous fashion as the temperature is raised to T_c. The conductivity for the normal current at low frequencies,and wave vectors close to T_c, is [21] $\sigma^{(S)}/\sigma^{(N)} = 1 + \frac{\Delta}{2k_BT} \ln \frac{\Delta}{\Gamma}$ where Γ is a pairbreaking energy. At low temperatures it vanishes proportional to $\exp(-\Delta/k_BT)$. The thermal conductivity in superconductors is [22]

$$\kappa^{(S)}/\kappa^{(N)} = G(\Delta/k_BT)\,. \tag{3}$$

The function G is given by

$$G(\Delta/k_BT) = \int_{\Delta}^{\infty} dE\, E^2 \frac{\partial n^{th}(E)}{\partial E} \Bigg/ \int_{0}^{\infty} dE\, E^2 \frac{\partial n^{th}(E)}{\partial E} \tag{4}$$

and Δ is the energy gap in the superconductor and $n^{th}(E)$ the Fermi function. Limiting values are

$$G(\Delta/k_BT) = \begin{cases} 1 - \dfrac{1}{2\pi^2}\left(\dfrac{\Delta}{k_BT}\right)^3 & \text{for } T \to T_c \\ \\ \dfrac{6}{\pi^2}\left(\dfrac{\Delta}{k_BT}\right)^2 e^{-\Delta/k_BT} & \text{for } T \to 0\,. \end{cases} \tag{5}$$

The thermoelectric coefficient L in superconductors has been calculated by Galperin, Gurevich, and Kozub [3]. Again,impurity scattering is the dominant scattering mechanism and they find the ratio of superconducting and normal state value to be

$$L^{(S)}/L^{(N)} = G(\Delta/k_B T)\,. \tag{6}$$

There may exist a significent temperature dependence of the normal state value $L^{(N)}$ (on a scale of order 1 K), which may be accompanied by a considerable uncertainty. However, there is much more confidence that the ratio $L^{(S)}/L^{(N)}$ is given by Eq. (6), at least in a narrow range near T_c (on a scale of order $10^{-3}\ldots10^{-2}$ K).

B Thermoelectric Flux

An experimental arrangement to detect thermoelectric effects in superconductors was proposed by Galperin et al. [3] and by Garland and van Harlingen [4]. It exploits the fact that upon application of a temperature gradient to a superconducting bimetallic ring, the normal thermoelectric current cannot be cancelled completely by the supercurrent. As a result, a magnetic field and flux through the ring is established. This, in turn, can be detected experimentally.

One can easily convince oneself that a magnetic flux has to exist: A complete local cancellation of normal thermoelectric current and supercurrent, and the requirement of fluxoid quantization cannot be met simultaneously. For a quantitative analysis we denote the complex order parameter by $\underset{\sim}{\Delta} = \Delta e^{i\theta}$. The supercurrent is $\vec{j}_s = en_s\vec{v}_s$ where the superfluid velocity is $\vec{v}_s = \frac{\hbar}{2m}(\vec{\nabla}\theta - \frac{2e}{\hbar c}\vec{A})$, and $\vec{A}$ is the vector potential. We consider a bimetallic ring of thickness much larger than the penetration depth $\lambda(T)$, where one end is kept at temperature T_1, the other at T_2 (see Fig. 1a). Deep in the interior of the superconducting materials (A or B), the supercurrent balances the normal current, e.g. $\vec{j}_s^A = -\vec{j}_n^A = L^A\vec{\nabla}T$. By integrating $\vec{\nabla}\theta$ along a path in the interior of the superconductors around the closed bimetallic ring, we find

$$n\Phi_0 = \Phi_m + \Phi_T \tag{7}$$

where n is an integer, Φ_0 the flux quantum, and Φ_m the magnetic flux determined as usually by $\vec{A}$. In addition, a thermoelectric flux associated with the nonvanishing supercurrent along the contour of integration appears, which can be written as

$$\Phi_T = \frac{mc}{e^2}\int_{T_1}^{T_2} dT\left(\frac{L^A}{n_s^A} - \frac{L^B}{n_s^B}\right). \tag{8}$$

Actually, there is a small extra contribution to Φ_T resulting from the regions near the interfaces of material A and B, where normal and supercurrents are converted (see below). However, if the dimension of the ring is much larger than the quasiparticle diffusion length, it can be ignored [11]. The result Eq. (7) is an interesting example of the fact that the fluxoid and not the magnetic flux is quantized. This is familiar from thin rings (of thickness smaller than the penetration depth), but here it is found in a thick ring. The origin is the nonvanishing supercurrent in the bulk of the superconductor in the presence of a temperature gradient. On the other hand, the cancellation of normal and supercurrent is not complete everywhere as we can conclude from the existance of the magnetic flux through the ring, $\Phi_m \neq 0$. This nonbalanced current is flowing within a penetration depth $\lambda(T)$ near the surface of the inner walls of the ring. It is much smaller (for small trapped flux) than the total thermoelectric normal current in the ring, by a factor of order $\lambda^2(T)/Rd$ where 2R is the diameter of the ring, and d is the thickness from inner wall to outer wall of the ring. (For general trapped flux this inequality is still found as far as changes of the currents are concerned.)

Usually, experiments are conducted at a temperature close to the critical temperature of one material (say A) but well below that of the other material. Therefore, $n_s^A \ll n_s^B$ and from the temperature dependence of Φ_T the temperature dependence of L^A/n_s^A can be deduced. Since n_s vanishes as $(1 - T/T_c)$ near T_c, Φ_T is expected

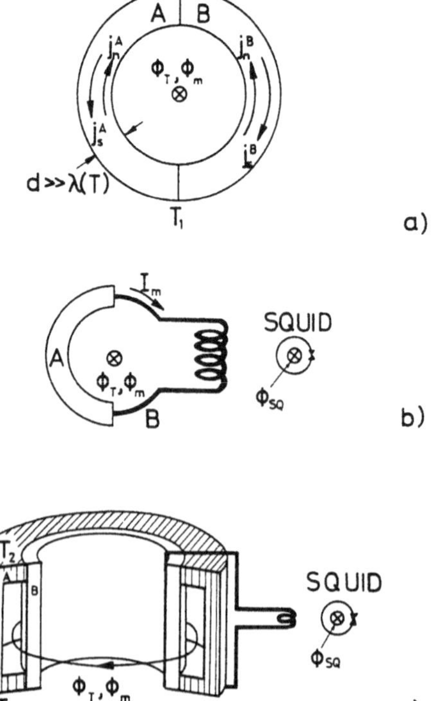

Fig. 1

a) A bimetallic ring is formed by two superconductors A and B and the two junctions are kept at different temperatures. The normal thermoelectric currents are cancelled in the bulk by a counterflow of supercurrents. Fluxoid quantization requires the existance of a magnetic flux Φ_m, which itself is not quantized, produced by surface currents.

b) Superconductor A is the sample, superconductor B forms the signal coil of a SQUID galvanometer. The non-balanced current I_m producing the magnetic flux Φ_m is measured.

c) A temperature difference between bottom and top of the bimetallic toroid produces a circulating magnetic flux. This is coupled into a SQUID magnetometer via a superconducting flux transformer.

to diverge logarithmically near T_c. The temperature dependence of Φ_T can be found by measuring the magnetic flux Φ_m. This, in turn, can be done in two ways. Either the nonbalanced current which flows near the surface and produces the magnetic flux can be measured, or the magnetic flux is detected directly. The former method was used by Zavaritskii [5], Pegrum et al. [6] and Falco [7]. In these experiments, the superconducting signal coil of a SQUID galvanometer forms one side of a bimetallic loop (see Fig. 1b), while the material to be studied forms the other side. The experiments yielded a divergent flux at T_c. However, the magnitude was much larger than predicted by theory. Later, Pegrum and Guénault [23] suggested that all of these observations could have been caused by temperature dependent changes of the penetration depth $\lambda(T)$. Also, these experiments were very sensitive to external magnetic fields.

Later, van Harlingen, Heidel, and Garland [8] performed experiments where they measured the flux directly. They extended the bimetallic ring into a toroid formed by two superconducting materials, and measured the magnetic flux by coupling it into a SQUID magnetometer via a superconducting flux transformer (see Fig. 1c). It is interesting to note, that even though the magnetic field of interest is completely shielded by the sample, and is therefore confined to the interior of the toroid, the flux transformer is able to detect it since the vector potential is not shielded. The toroidal geometry has significant advantages over the simple loop geometry used before. For a detailed discussion see Ref. [8]. In summary, the improvements are 1) increased sensitivity, 2) less perturbation by external fields, 3) less perturbation by changes of the penetration depth.

In the experiments which were performed on indium-lead samples, the lower temperature T_1 was kept constant at a value below T_c, and T_2 was increased up to T_c. The changes of $\Phi_T(T_2)$ were observed from which the ratio L/n_s of indium was obtained. If the usual result for temperature dependence and magnitude of n_s is used, $L(T)$ can be extracted. Also, the normal state value of $L^{(N)}(T_c)$ was measured. The experimental results for seven samples in a temperature range $1\,\mathrm{mK} < T_c - T_2 \lesssim 10\,\mathrm{mK}$ can be summarized by

$$L^{(S)}(T) = \frac{2.0 \times 10^3}{\sqrt{1 - T/T_c}} L^{(N)}(T_c)\,. \tag{9}$$

Clearly, there is gross disagreement between this result and theoretical predictions Eq. (6). For the temperature closest to $T_c \approx 3.4\,\mathrm{K}$, $T_c - T_2 \approx 1\,\mathrm{mK}$, the ratio $L^{(S)}/L^{(N)}$ is 10^5. To date no satisfactory explanation for this large discrepancy has been given, neither as far as the experiment nor as far as the theory is concerned. We want to mention, however, one feature. If the sample were normal in the experiments just described, the signal would be larger by many orders of magnitude, namely proportional to the total thermoelectric current, rather than to the unbalanced current.

III Charge Imbalance Effects

A Theoretical concepts

In the following, we want to summarize some concepts of the theory of non-equilibrium superconductivity. Simple pictures can be developed for clean superconductors where the impurity scattering rate is small[1] $\frac{1}{\tau_{imp}} \ll \Delta$ under certain restrictions: The space (wave vector q) and time dependence (frequency ω) of the various physical quantities has to be sufficiently weak (qv_F, $\omega \ll \Delta$), such that there exists a local Bogliubov transformation leading to almost stationary excitations. Furthermore, in order that the energies of the excitations are sufficiently well defined, inelastic scattering events (inelastic electron phonon collision rate $1/\tau_E$) and other pairbreaking effects (pair breaking energy Γ) have to be weak too, ($1/\tau_E$, $\Gamma \ll \Delta$). In this case the energy of the excitations is $\hat{E}_{\vec{p}} = E_p + \vec{p}\vec{v}_s$, where $E_p = \sqrt{\xi_p^2 + \Delta^2}$ and ξ_p is the energy of electrons in the normal state measured relative to the chemical potential of the Cooper pairs $\xi_p = \epsilon_p - \mu_s$. In equilibrium μ_s is simply the Fermi energy ϵ_F, but in a general situation, the Cooper pairs form the reference level for the excitations [24]. We include in our discussion the effect of a superfluid velocity $\vec{v}_s$, which will be important in the next chapter, but for the moment we assume that $\vec{v}_s$ is small. The excitations in the superconductor are a linear combination of electrons with amplitude u_p and holes with amplitude v_p, as described by the Bogoliubov transformation. The amplitudes are given by $|u_p|^2 = \frac{1}{2}(1 + \xi_p/E_p)$ and $|v_p|^2 = \frac{1}{2}(1 - \xi_p/E_p)$. As a consequence, the charge of an excitation is $e\xi_p/E_p$. Far from the Fermi surface, the excitations are either electrons (for $\xi_p \gg \Delta$) or holes (for $\xi_p \ll -\Delta$) but at small energies they change their character gradually.

In thermal equilibrium the occupation of the levels on the two branches of the excitation spectrum ($\xi_p > 0$ and $\xi_p < 0$) is equal. Most perturbations, e.g. external radiation preserve this symmetry. An excess population of excitations in this case relaxes by recombination of a more electron- and a more hole-like excitation, under emission of a phonon. On the other hand, in a superconductor there exists the possibility that the two branches are occupied differently [12] (see Fig. 2). As a result, the net charge in the system of excitations is nonzero. In a normal metal this degree of freedom would strongly be restricted by the requirement of charge neutrality. However, in a superconductor charge neutrality can be maintained by a corresponding change of the charge in the system of the Cooper pairs. The relaxation of such a charge imbalance takes place by inelastic electron phonon scattering process. This is visualized in Fig. 2; the charge of the excitation in the initial and in the final state differ. Again, the condensate of Cooper pairs adjusts in a way such

---

1) In the following we use units where $\hbar = k_B = 1$.

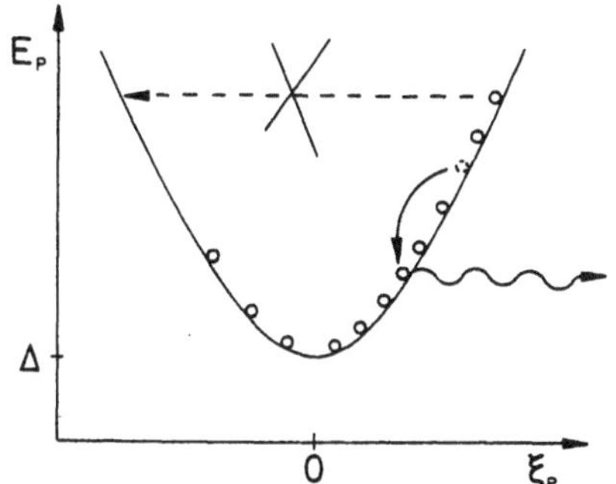

Fig. 2
A charge imbalance is obtained if the two branches of the excitation spectrum are occupied differently. Inelastic electron phonon collisions relax the charge imbalance. Elastic scattering at nonmagnetic impurities where initial and final state are on different branches do not take place.

that the total charge is conserved. In contrast, elastic scattering events at nonmagnetic impurities,although they are usually much more frequent than the inelastic scattering processes, cannot (in isotropic systems as long as $\vec{v}_s \approx 0$) relax a charge imbalance,since the branch crossing scattering is forbidden. (This is clearly so in a normal metal, and remains true in an isotropic superconductor.) Nonequilibrium phenomena can be classified according to the parity of the nonequilibrium distribution. In the former example,where the distribution function is even with ξ_p the total number of excitation and their energetic distribution is modified. These phenomena carry the label "energy mode" [25]. It is linearly independent [13] (for $\vec{v}_s \approx 0$) of the "charge mode" discussed above as the second example. Quantitatively, the dynamics of the quasiparticles is described by the Boltzmann equation which in the clean limit takes the intuitive form [24]

$$\frac{\partial n_{\vec{p}}}{\partial t} + \frac{\partial \hat{E}_{\vec{p}}}{\partial \vec{p}} \frac{\partial n_{\vec{p}}}{\partial \vec{r}} - \frac{\partial \hat{E}_{\vec{p}}}{\partial \vec{r}} \frac{\partial n_{\vec{p}}}{\partial \vec{p}} = I_{imp} + I_{ep} . \tag{10}$$

The collision integrals on the right hand side can be derived by simple Golden Rule type arguments [22]. The impurity collision operator becomes (for isotropic scattering)

$$I_{imp} = -\frac{1}{\tau_{imp}} \int d\xi_{p'} \int d\Omega_{\vec{p}'} \, |u_p u_{p'} - v_p v_{p'}|^2 \, (n_{\vec{p}} - n_{\vec{p}'}) \, \delta(\hat{E}_{\vec{p}} - \hat{E}_{\vec{p}'}) . \tag{11}$$

(It is obvious that the branch crossing scattering processes for $\vec{v}_s = 0$ are forbidden by the coherence factors.) We do not write down the electron-phonon collision integral, but we want to stress again that it does not conserve the charge in the system of excitations.

In principle, e.g. for a given external perturbation, one has to solve the Boltzmann equation (10) to find the distribution function. For a qualitative discussion of the charge imbalance phenomena, it is sufficient to proceed on a more macroscopic level, using a two fluid type of description developed in this context by Pethick and Smith [25]. It requires to introduce the concepts normal charge and superfluid charge. The total charge is

$$Q_{tot} = 2e \sum_{\vec{p}} |u_p|^2 n_{\vec{p}} + |v_p|^2 (1 - n_{-\vec{p}}) . \tag{12}$$

It may change for two reasons. Either the distribution of the excitation changes. The corresponding change of charge is denoted as normal charge

$$Q_n = 2e \sum_{\vec{p}} \frac{\xi_p}{E_p} \delta n_{\vec{p}} . \tag{13}$$

Or it may change because the coherence factors change. These, in turn, depend on the chemical potential of the Cooper pairs. (Changes of the magnitude of the order parameter refer to the independent energy mode!) The corresponding change of the charge is denoted as superfluid charge. It can easily be calculated that

$$Q_s = \chi_s \ (\mu_s - \epsilon_F) \tag{14}$$

where near T_c one finds $\chi_s = 2eN_0 \cdot \frac{\pi\Delta}{4T}$.

In many situations, it turns out that the distribution function of the excitations has a thermal form, however, corresponding to a chemical potential μ_n which differs from the equilibrium value and from the chemical potential of the Cooper pairs [26]. Then

$$Q_n = \chi_n \ (\mu_n - \epsilon_F) \tag{15}$$

where $\chi_n + \chi_s = 2eN_0$, i.e. near T_c

$$\chi_n = 2eN_0 \left(1 - \frac{\pi\Delta}{4T}\right) . \tag{16}$$

By integrating the Boltzmann equation one can obtain approximately a continuity equation for the normal charge

$$\dot{Q}_n + \operatorname{div} \vec{j}_n = \left(\frac{dQ_n}{dt}\right)_{ep} . \tag{17}$$

(Strictly, a current differing slightly from the normal current $\vec{j}_n = \frac{2e}{m} \sum_{\vec{p}} \vec{p}\, \delta n_{\vec{p}}$ appears in this equation (see Ref. [25]). However, close to T_c the difference can be neglected.) The right hand side reflects the fact that normal charge can be converted into supercharge by electron phonon collisions. A reasonable ansatz for this conversion term, as long as Q_n is of the form (15), is

$$\left(\frac{dQ_n}{dt}\right)_{ep} = -\frac{1}{\tau_{Q^*}} \chi_n (\mu_n - \mu_s) \tag{18}$$

Here it is taken into account that the part of the normal charge $Q^* = \chi_n (\mu_n - \mu_s)$ which is out of a local equilibrium, defined by the Cooper pairs, is converted into supercharge with a certain rate $1/\tau_{Q^*}$. This ansatz is confirmed by detailed calculations [13]. These also yield the rate $\frac{1}{\tau_{Q^*}}$, which for T near T_c is

$$\frac{1}{\tau_{Q^*}} = \frac{\pi\Delta}{4T} \frac{1}{\tau_E} . \tag{19}$$

Here $1/\tau_E$ is the total inelastic electron phonon collision rate for low energy excitations at temperatures near T_c. The prefactor Δ/T results from the fact that only those inelastic collision processes which involve excitations at low energies (E_p of order Δ) change the charge. These are only a fraction Δ/T of the total number since the levels up to E_p of order T are occupied.

Conservation of total charge yields a similar continuity equation for the superfluid charge. Hence,

$$\dot{Q}_s + \operatorname{div} \vec{j}_s = \frac{1}{\tau_{Q^*}} \chi_n (\mu_n - \mu_s) . \tag{20}$$

This last relation has to be added to the set of transport equation discussed before. In stationary situations we have

$$\begin{aligned} \vec{j}_n &= \sigma \vec{E}' - L \vec{\nabla} T , \qquad \vec{q} = LT \vec{E}' - \kappa \vec{\nabla} T , \\ \varphi_n - \varphi_s &= + \frac{\tau_{Q^*}}{\chi_n} (\vec{\nabla} \cdot \vec{j}_s) . \end{aligned} \tag{21}$$

Furthermore, in the steady state,the electrochemical potential of the Cooper pairs is constant $\varphi_s = \mu_s + eV = \text{const}$,since $m\dot{\vec{v}}_s = -\vec{\nabla}\varphi_s = 0$. Also, it is clear that $\operatorname{div} \vec{j}_s = -\operatorname{div} \vec{j}_n$. Experimentally, the difference between the electrochemical potentials $\varphi_n - \varphi_s$ is obtained by coupling a normal and a superconducting contact to the sample of interest and measuring the voltage difference.

As an application of these concepts,let us consider the situation where an electric current is driven through a superconductor – normal metal interface. Then the normal current has to be converted into a supercurrent. Since this takes place with a finite rate,the normal current penetrates somewhat into the superconductor. It has to be driven by a gradient of the normal electrochemical potential (see Fig. 3). As a consequence,there is also a potential drop in the superconductor which can be expressed by an extra resistance.

For the geometry shown in Fig. 3,and the boundary condition that in the normal metal the current is j_0,we find from the set of equations (21) (with $\vec{\nabla}T = 0$) the normal current in the superconductor

$$j_n(x) = j_0 \cdot e^{-x/\lambda_{Q^*}} . \tag{22}$$

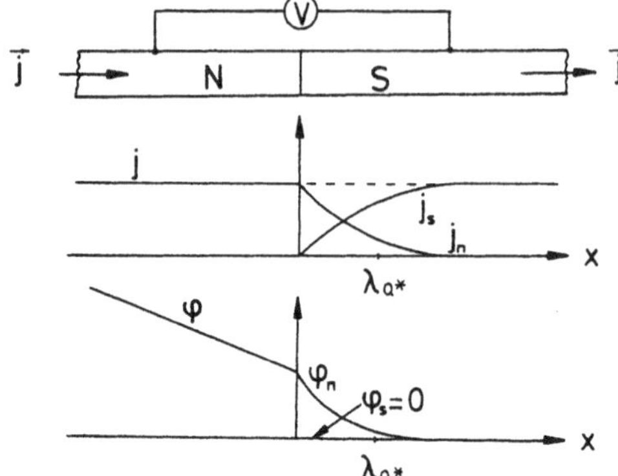

Fig. 3
If a current is flowing through a normal metal superconducting interface, the normal current penetrates to depths of order λ_{Q^*}. This is accompanied by a gradient of the normal electrochemical potential φ_n.

The diffusion length λ_{Q^*} to lowest order in Δ/T is

$$\lambda_{Q^*}(T) = \sqrt{D\,\tau_{Q^*}} \tag{23}$$

and $D = \sigma/2e^2 N_0$ is the diffusion constant in the superconductor. The potential drop in the superconductor therefore is $\varphi_n(0) - \varphi_s = j_0\, e\lambda_{Q^*}/\sigma$. The weak temperature dependence of λ_{Q^*} proportional to $(1 - T/T_c)^{-1/4}$ allows to separate this contribution from the voltage drop in the normal metal. Experiments [27] confirm these results quantitatively well.

B Thermoelectric Generation of a Voltage at a Superconductor-Normal Metal Interface

If a uniform heat current is flowing through a SN interface,it leads to temperature gradients which are uniform,though in general different in S and N. In the normal metal under open-circuited conditions, a uniform electrical driving force $\vec{E}'^{(N)} = -\vec{\nabla}\varphi^{(N)}/e = L^{(N)}/\sigma^{(N)} \cdot (\vec{\nabla}T)^{(N)}$ is established,preventing electric current flow. The currents and fields in the superconductor (see Fig. 4) are obtained from the set of equation (21), subject to the boundary conditions, that at the interface $j_n(x=0) = j_s(x=0) = 0$, whereas deep inside the superconductor $j_n(x=\infty) = -j_s(x=\infty) = -L^{(S)} \cdot (\frac{\partial T}{\partial x})^{(S)}$. Hence, in the superconductor

$$j_n(x) = -L^{(S)} \cdot \left(\frac{\partial T}{\partial x}\right)^{(S)} (1 - e^{-x/\lambda_{Q^*}}) \tag{24}$$

and

$$\varphi_n(0) - \varphi_s = \frac{e\lambda_{Q^*}^{(T)}}{\sigma^{(S)}} \cdot L^{(S)} \cdot \left(\frac{\partial T}{\partial s}\right)^{(S)} . \tag{25}$$

Again, the drop $\varphi_n(0) - \varphi_s$ of the normal electrochemical potential in the superconductor can be separated from that in the normal metal by its temperature dependence.

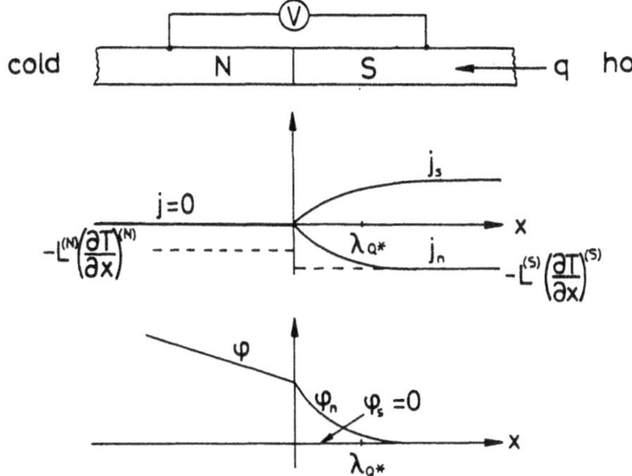

Fig. 4
A heat current is flowing through on N S interface. The space dependence of currents and electrochemical potentials is shown. The signs of $\vec{\nabla}\varphi$ in N and S depends on the signs of $L^{(N)}$ and $L^{(S)}$.

The described effect was first discussed by Artemenko and Volkov [11], and recently investigated experimentally by van Harlingen [10]. In the experiments, first the resistance to an imposed current was measured in order to determine material parameters in the product $\lambda_{Q*}/\sigma^{(S)}$. Then the potential drop in the presence of an imposed temperature gradient was measured. Near T_c it diverges as λ_{Q*}, in very good agreement with Eq. (25) and the theoretical result that $L^{(S)}$ and also $\kappa^{(S)}$ depends only very weakly on $T-T_c$. From the prefactor of the divergent contribution, the value of the thermoelectric transport coefficient can be derived. The experiment yields a value of the thermopower of the lead sample $S^{(S)}=L^{(S)}/\sigma^{(S)}=-7.8\ 10^{-7}$ V/K,which agrees reasonably well with the normal state value of pure lead at T_c: $S^{(N)}=-2.2\ 10^{-7}$ V/K. The small difference can be accounted for by impurities, which may change,even in small concentrations, the value of the thermopower by as much as an order of magnitude. Similar agreement is found for indium samples. These results confirm the theoretical concepts as far as the transport properties are concerned. It leaves the discrepancy in the experiments measuring the magnetic flux unresolved.

In the above analysis the temperature gradient was uniform. On the other hand in situations where it is nonuniform we find

$$(\varphi_n-\varphi_s)-\lambda_{Q*}^2\nabla^2\varphi_n=\frac{\tau_{Q*}}{\chi_n}\vec{\nabla}(L\vec{\nabla}T)\,. \tag{26}$$

Hence, in regions where $\nabla^2 T$ is nonzero one also expects a difference between normal and superfluid electrochemical potentials. It has been proposed by Tinkham [28] that this effect has been observed in a recent experiment by Falco [29].

IV $\vec{v}_s$-Dependent Phenomena

A Phenomenological Scheme

A systematic account of the transport coefficients and relaxation times in superconductors can be given on the basis of the thermodynamic theory of irreversible processes. In the case of a superfluid also the normal fluid may move without dissipation, and the corresponding scheme has been discussed by Khalatnikov [30]. In contrast, in a superconductor the flow of a normal current always leads to dissipation. The required modifications have been discussed by Putterman [31]. Recently Schmid [21] has shown that in the spirit of this scheme the normal current, the heat current, and $\varphi_n - \varphi_s$ have to be interpreted as generalized dissipative currents, whereas $\vec{E}'$, $\vec{\nabla}T$, and $\operatorname{div}\vec{j}_s$ have to be interpreted as generalized forces. If, as usual, a linear connection between currents and forces is assumed, the requirement of nonnegative entropy production leads to certain restrictions and relations between the various coefficients. Also Onsager's relations connect pairs of nondiagonal coefficients. The transport equations (21) are exactly this linear relation between currents and forces. In this context we learn that $e\tau_{Q^*}/\chi_n$ has to be interpreted as coefficient of second viscosity ζ. So far we have left some blanks. In an isotropic superconductor as long as $\vec{v}_s = 0$, the coefficients can only be scalars and therefore e.g. the scalar force $\operatorname{div}\vec{j}_s$ cannot be connected with the vector current $\vec{j}_n$. However, this is possible if an equilibrium supercurrent is flowing. In this case, the complete scheme becomes

$$\begin{aligned} \vec{j}_n &= \sigma\vec{E}' - L\vec{\nabla}T + \gamma\vec{v}_s(\vec{\nabla}\cdot\vec{j}_s) \\ \vec{q} &= LT\vec{E}' - \kappa\vec{\nabla}T + \beta(T/e)\,\vec{v}_s(\vec{\nabla}\cdot\vec{j}_s) \\ \varphi_n - \varphi_s &= \gamma e\vec{v}_s\cdot\vec{E}' - \beta\vec{v}_s\cdot\vec{\nabla}T + (\zeta e)(\vec{\nabla}\cdot\vec{j}_s)\,. \end{aligned} \tag{27}$$

The extension suggest a series of new nonequilibrium phenomena. One of them,a charge imbalance induced by the simultaneous presence of a supercurrent and a temperature gradient,has been observed experimentally [15] and the coefficient β was calculated by a number of authors [14, 16–20] in different ranges of the parameters.

B Supercurrent Induced Generation of a Charge Imbalance in the Presence of a Temperature Gradient

In the presence of a temperature gradient,the two branches of the excitation spectrum are occupied differently, in the superconductor just as well as in a normal system. This is shown in Fig. 5a. However, no net charge imbalance results, since opposite directions in momentum space have an opposite imbalance. In a superconductor, however, if a supercurrent is imposed,the excitation energy is shifted to $\hat{E}_p = E_p + \vec{p}\vec{v}_s$. As a result of this asymmetry a net charge imbalance may be obtained (see Fig. 5b). This statement can be substantiated by considering the

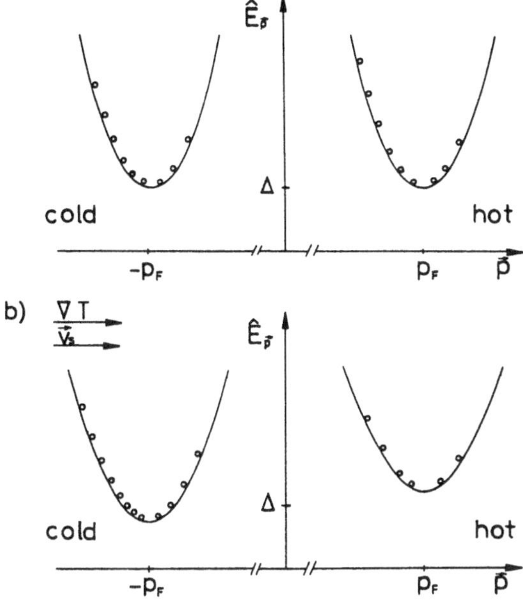

Fig. 5

The excitation spectrum and the occupation of the levels for two opposite directions in momentum space is visualized.

a) A temperature gradient is applied.

b) A temperature gradient and a supercurrent are applied.

Boltzmann equation. In zero order the spatial variation of the temperature leads to a distribution function which is thermal, however, depending on the local value of the temperature $n^0(\vec{r},\vec{p}) = n^{th}(T(\vec{r}), \hat{E}_{\vec{p}})$. This enters into the convective terms of the Boltzmann equation, which in stationary situations, including the $\vec{v}_s$ dependence [32] becomes

$$-\left(\frac{\xi_p}{E_p}\frac{\vec{p}}{m} + \vec{v}_s\right) \cdot \frac{\vec{\nabla} T}{T}\, \hat{E}_{\vec{p}}\, \frac{\partial n^{th}(\hat{E}_{\vec{p}})}{\partial \hat{E}_{\vec{p}}} = I_{imp} + I_{ep} \,. \tag{28}$$

In practically all cases, the elastic scattering at impurities is by far the most frequent process. And it is important to notice that the actual energy $\hat{E}_{\vec{p}}$ is conserved during these scattering events. The collision operator is given by Eq. (11) and it is $\hat{E}_{\vec{p}}$ which appears in the energy conserving δ-function. On the other hand, the coherence factor $|u_p u_{p'} - v_p v_{p'}|^2$ depends on E_p only. Hence, for $\vec{v}_s \neq 0$ it needs no longer to vanish if $\vec{p}$ and $\vec{p}'$ refer to different branches. An interesting consequence is that a branch imbalance may also relax by impurity scattering if the superfluid velocity is finite $\vec{v}_s \neq 0$ (or if the superconductor is anisotropic for other reasons). Because of the dominant role of impurity scattering,it is clear that the magnitude of $\vec{p}$, which is not conserved, is an unsuitable variable. Rather we introduce the actual conserved energy $E = \hat{E}_{\vec{p}}$, and the direction of the momentum $\hat{p} = \vec{p}/|\vec{p}|$ as the new independent variables. Furthermore we consider the charge at a given energy E and direction of $\hat{p}$ which is

$$\psi(E, \hat{p}) = \int_{-\infty}^{\infty} d\xi_p \, (\xi_p/E_p) \, \delta n_{\vec{p}} \, \delta(E - \hat{E}_{\vec{p}}) \,. \tag{29}$$

Here, $\delta n_{\vec{p}}$ is the deviation of the distribution function from a local equilibrium solution $n^{th}(\hat{E}_{\vec{p}})$. Only in this quantity the collision integrals can be linearized. The total charge imbalance Q^* then is

$$\frac{Q^*}{2eN_0} = \int_0^{\infty} dE \, \langle \psi(E, \hat{p}) \rangle_{\hat{p}} \,. \tag{30}$$

By corresponding integration of the Boltzmann equation, we find the equation for the charge

$$\begin{aligned} &\frac{1}{\tau_{ep}} \left\{ N_1(E, \hat{p}) \, \psi(E, \hat{p}) + 2 \, \frac{\partial n^{th}(E)}{\partial E} \int_0^{\infty} dE' \, \frac{1}{2} \left[\psi(E', \hat{p}) + \psi(E', -\hat{p}) \right] \right\} + \\ &+ \frac{1}{\tau_{imp}} \left\{ \langle M(\hat{p}, \hat{p}') \rangle_{\hat{p}'} \, \psi(E, \hat{p}) - \langle \psi(E, \hat{p}') \rangle_{\hat{p}'} \right\} = \\ &= 2 \, \frac{(\vec{p} \cdot \vec{\nabla} T)}{mT} \, \frac{\partial n^{th}(E)}{\partial E} \, E \,. \end{aligned} \tag{31}$$

Here, $N_1(E, \hat{p})$ is a shifted BCS reduced density of states, i.e. $N_1(E, \hat{p}) = (E - \vec{p}\vec{v}_s)/\sqrt{(E - \vec{p}\vec{v}_s)^2 - \Delta^2}$ for positive arguments of the square root and zero otherwise. Furthermore, $M(\hat{p}, \hat{p}') = N_1(E, \hat{p}) \, N_1(E, \hat{p}') \, [1 - \Delta^2/(E - \vec{p}\vec{v}_s)(E - \vec{p}'\vec{v}_s)]$, and the angular average $\langle \ldots \rangle_{\hat{p}}$ extends over those $\hat{p}'$ only where $E - \vec{p}'\vec{v}_s \geqslant \Delta$.

In Eq. (31) we have written the "scattering in term" i.e. the integral in the electron-phonon collision operator in "reduced form" [33]. This approximation preserves the most important properties of the exact operator. The "scattering out term" $\frac{1}{\tau_{ep}} N_1(E, \hat{p}) \, \psi(E, \hat{p})$ and the scattering in term do not balance completely. This reflects the fact that the charge in the system of excitations decays by electron-phonon scattering. Also, in the impurity scattering operator the "scattering out" and the "scattering in" term (i.e. the first and second contribution, respectively) do not completely balance. If $\vec{v}_s = 0$ one has $M(\hat{p}, \hat{p}') = 1$ (for $E > \Delta$). For finite $\vec{v}_s$ however, $M(\hat{p}, \hat{p}') \neq 1$, which again is an expression for the fact that a net angular averaged charge may relax by impurity scattering.

The inhomogeneity in Eq. (31) vanishes upon angular averaging. Nevertheless, even at energies $E > \Delta + p_F v_s$ a net angular averaged charge $\langle \psi(E, \hat{p}) \rangle$ may result, because of the asymmetry of the scattering probabilities $\langle M(\hat{p}, \hat{p}') \rangle / \tau_{imp}$ and $N_1(E, \hat{p})/\tau_E$. We mention at this point that the inhomogeneity in Eq. (28) in the variable $\vec{p}$ seems to create directly a net charge (it contains a part $\propto \xi_p (\vec{p}\vec{\nabla}T)(\vec{p}\vec{v}_s)$). However, it would be difficult to draw quantitative conclusions from this term considering the properties of impurity scattering.

Solutions of Eq. (31) have been obtained in various limits [16, 19]. It is important to keep in mind that it is valid only in the clean limit where $1/\tau_{imp}$ is small. It is instructive to consider first the limit where, furthermore, electron-phonon scattering is completely neglected $1/\tau_E = 0$. (The precise restriction on $1/\tau_{imp}$ and $1/\tau_E$ will be discussed below.) Under these restrictions an exact solution of Eq. (31) was given by Beyer Nielsen, Ono, Pethick and Smith [19]. The net charge at energy E can be written as

$$\langle \psi(E, \hat{p}) \rangle_{\hat{p}} = A_E \, \tilde{\psi}(E) \tag{32}$$

where $A_E = -\dfrac{(\vec{v}_s \cdot \vec{\nabla} T)}{v_S T} \, \ell_{imp} \, \dfrac{\partial n^{th}(E)}{\partial E} \, \Delta$, and

$$\tilde{\psi}(E) = -\frac{E}{\Delta} \, \frac{\langle \cos\theta / \langle M(\hat{p}, \hat{p}') \rangle_{\hat{p}'} \rangle_{\hat{p}}}{1 - \langle 1/\langle M(\hat{p}, \hat{p}') \rangle_{\hat{p}'} \rangle_{\hat{p}}} \, . \tag{33}$$

Furthermore, $\ell_{imp} = v_F \tau_{imp}$ and $\cos\theta = (\vec{p} \cdot \vec{v}_s)/p v_s$. The solution shows again that at energies $E > \Delta + p_F v_s$ a net charge is only obtained due to asymmetries in the impurity scattering. Otherwise the numerator in Eq. (33) vanishes. On the other hand, it also relaxes only because of these asymmetries, which makes the denominator nonzero. At energies larger than $\Delta + p_F v_s$, for small values of $p_F v_s/\Delta$ the solution can be expanded to yield

$$\tilde{\psi}(E > \Delta + p_F v_s) \approx \frac{4}{15} \, \frac{p_F v_s}{\Delta} \, \frac{E^2}{E^2 - \Delta^2} \, .$$

It is small, of order $p_F v_s/\Delta$ only. In contrast in the energy range $\Delta - p_F v_s \leqslant E \leqslant \Delta + p_F v_s$ the solution $\tilde{\psi}$ is of order 1; a "pocket" appears. The main reason is that states are available and contribute in the angular averaging only for the directions of the momenta $\vec{p}$ for which $E \geqslant \Delta + \vec{p} \cdot \vec{v}_s$. Therefore, the approximate cancellation of the contribution from opposite directions is reduced and the numerator in Eq. (33) is large. The solution is plotted as curve a in Fig. 6. Because of the large difference in magnitude of the solutions in the pocket range and outside, the pocket yields a noticeable contribution to Q^*, though being restricted to a narrow energy range. The result is approximately [19]

$$\frac{Q^*}{2eN_0} = p_F \, \ell_{imp} \, \frac{\vec{v}_s \vec{\nabla} T}{T} \left\{ 1.93 \, \frac{\Delta/T}{\mathrm{ch}^2 \, \Delta/2T} + \frac{8}{15} \, n^{th}(E = \Delta) \right\} \tag{34}$$

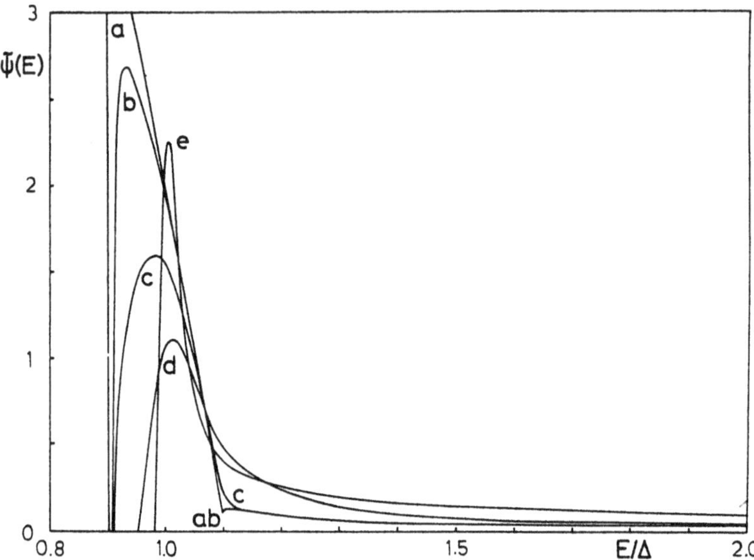

Fig. 6 The net charge imbalance (in reduced units) $\widetilde{\psi}$ (E) at energy E is shown for the case where electron phonon scattering can be ignored. The superfluid velocity has a value such that $p_F v_s/\Delta = 0.1$. Curve a is the solution of Ref. [19] which shows a pocket in the range $\Delta - p_F v_s \leqslant E \leqslant \Delta + p_F v_s$. It is the solution in the limit $1/\tau_{imp} \to 0$. The other curves are obtained for the parameters $1/\Delta\tau_{imp} = 0.01$ (b), 0.1 (c), 1 (d), 4 (e). The last curve agrees already very well with the dirty limit result.

where the first term comes from the pocket. The range of parameters where the solution (34) is valid is very restricted. We will see that the condition

i) $\dfrac{1}{\Delta \tau_{imp}} \left(\dfrac{\Delta}{p_F v_s}\right)^{1/2} \ll 1$ and ii) $\dfrac{1}{\tau_E} \ll \dfrac{1}{\tau_{imp}} \left(\dfrac{p_F v_s}{T}\right)^2$ have to be satisfied.

The first restriction i) results from the impurity scattering. This restriction has rarely been noticed in other problems of nonequilibrium superconductivity. The reason is that in most other problems the superconductor could be assumed to be isotropic and impurity scattering, after it made the distribution function (at least nearly) isotropic, has no (or only simple) further effects. Therefore, calculations based on the simple clean limit Boltzmann equation turned out to cover the whole range of parameters. As an example we mention that the same charge imbalance relaxation rate $1/\tau_{Q*}$ is found in the dirty limit [13] and the clean limit [26]. Some other transport coefficient simply scale with τ_{imp}.

In contrast, in the present problem, the superconductor is anisotropic (due to $\vec{v}_s$), and the perturbation creates a highly anisotropic distribution. In this case, effects of impurity scattering carefully have to be taken into account. The requirement

$1/\tau_{imp} \ll \Delta$ necessary for Eq. (10) or Eq. (31) to be valid is rarely satisfied in an experimental situation. We therefore, derived from the Green's functions technique the kinetic equations valid for arbitrary impurity scattering [20]. An interesting feature is that with increasing strength of $\frac{1}{\tau_{imp}}$ the $\vec{p}\vec{v}_s$ shifted densities of states $N_1(E, \hat{p})$ (given after Eq. (31) for the clean limit) are first smeared out, most pronounced in the vicinity of the gap [34]. Also,the shift is reduced. The minimum energy gap becomes larger than $\Delta - p_F v_s$. The amount of the smearing is characterized by the parameter $\frac{1}{\Delta\tau_{imp}} \left(\frac{\Delta}{p_F v_s}\right)^{1/2}$ for small $1/(\Delta\tau_{imp})$. When $1/(\Delta\tau_{imp})$ becomes larger than one, the transition to the familiar dirty limit result takes place. In this case the density of states is isotropic. However, it differs from the BCS form due to the pairbreaking effect of the supercurrent [35] $\Gamma_J = \frac{2}{3} (p_F v_s)^2 \tau_{imp}$.
The development of the densities of states with increasing $1/\tau_{imp}$ is strongly reflected in the solution $\widetilde{\psi}(E)$. In Fig. 6 we plot it for different values of the parameter $\frac{1}{\Delta\tau_{imp}}$. It shows the transition from the clean limit result to the dirty limit result. An energies larger than $\Delta + p_F v_s$ the condition $\frac{1}{\tau_{imp}} \ll \Delta$ is sufficient in order for the clean limit solution to be valid. However, if we are interested in details very close to the gap, as for example the pocket, stronger restrictions are required. The solution close to the gap is given by Eq. (33) only if the strong condition i) listed after Eq. (34) is satisfied. For completeness we list the result in the dirty limit, when no phonons are present. Then one obtains approximately

$$\frac{Q^*}{2eN_0} = p_F \ell_{imp} \frac{\vec{v}_s \vec{\nabla} T}{T} \frac{2}{3} \left(1 + \frac{\Delta}{4T} \ln \frac{4}{\Gamma_J}\right). \tag{35}$$

The second restriction to the solutions (34) refers to the action of phonons. The electron phonon scattering rate $1/\tau_E$ is much smaller than the impurity scattering rate, but electron phonon scattering processes relaxes a charge imbalance more effectively. Outside a region close to the gap, i.e. for $E > \Delta + p_F v_s$ impurity scattering relaxes the charge in the clean limit at an average rate of order

$$\frac{1}{\tau_{imp}} (\langle\langle M(\vec{p},\vec{p}')\rangle\rangle_{\hat{p},\hat{p}'} - 1) = \frac{1}{\tau_{imp}} \frac{(p_F v_s)^2 \Delta^2}{3 (E^2 - \Delta^2)^2}.$$

In contrast,phonons do so at a rate $\frac{1}{\tau_E} \cdot \frac{E\Delta^2}{(E^2 - \Delta^2)^{3/2}}$. Therefore, in order that phonons can be ignored at all energies the requirement ii) listed above has to be satisfied.
On the other hand, from this analysis we also learn that if $\frac{1}{\tau_E} \gg \frac{1}{\tau_{imp}} \left(\frac{p_F v_s}{\Delta}\right)^2$ we may set M equal to 1 at energies above the gap. If furthermore, $p_F v_S \ll 1/\tau_E$ the level broadening effect of electron phonon scattering will prevent any pronounced

effects due to $p_F v_s$ in the vicinity of the gap edge. In this case we may expand Eq. (31) in $p_F v_s$ and obtain near T_c [16]

$$\frac{Q^*}{2eN_0} = p_F \ell_{imp} \frac{\vec{v}_s \vec{\nabla} T}{T} \frac{1}{3\pi} \ln(8\Delta\tau_E) . \tag{36}$$

A net charge is now found due to asymmetries in the electron-phonon scattering (in the variables E, $\hat{p}$). In a dirty material superconductor,we had found [16] a similar result. No restrictions on $p_F v_s$ had to be made in this case. If furthermore, the pairbreaking effect of the supercurrent is weak,such that $\Gamma_J \ll 1/\tau_E$ then the dirty limit result differs from Eq. (36) only in so far as it is larger by a factor 2.

Summarizing, we may say that in a large variety of different limits (except in very clean superconductors, when in addition electron phonon scattering can be ignored), the results for Q^* only differ by factors of order one. They all agree in the linear dependence on $\vec{v}_s \cdot \vec{\nabla} T$ and in the temperature dependence in leading order in Δ/T. The experiments [15] were conducted on tin samples in a temperature range $0.5 \leqslant T/T_c \leqslant 1$. The superfluid velocity was varied between $0 \leqslant p_F v_s \lesssim 0.01\,\Delta$. The electron phonon scattering in tin is $\tau_E \approx 2.5 \;\; 10^{-10}$ sec. Most of the samples were of dirty material with $\tau_{imp} \approx 8.8 \;\; 10^{-14}$ sec, one sample was cleaner with $\tau_{imp} = 6.6 \;\; 10^{-13}$ sec. (Here we used a value [17] of $v_F = 0.65$ cm/sec) The potential difference between normal and superfluid component was measured. It is related to Q^* by [36] $V = Q^*/2e^2 N_0\, g_{NS}$, where g_{NS} is the normalized conductance between normal contact and the superconductor. For a comparison with experiment we have to use the dirty limit result referred to after Eq. (36). Using the free electron value $p_F/\hbar = 1.6 \;\; 10^8\,\mathrm{cm}^{-1}$, we find

$$\frac{V g_{NS} T}{(\vec{j}_s \cdot \vec{\nabla} T)} = 0.29 \ln(8\Delta\tau_E) \frac{1}{(1 - T/T_c)} \cdot 10^{-16}\, \Omega\, \mathrm{cm}^3 . \tag{37}$$

The value of the logarithm varies between 5.5 and 6.5 in the interesting temperature range. For clean tin samples nearly the same result is found. If we express the superfluid density by the penetration depth, for which the value $\lambda(0) = 5 \times 10^{-6}$ cm is quoted in Ref. [17], we find that the prefactor is 0.25 instead 0.29. The temperature dependence of (37) is in very good agreement with the experimental results in a temperature range $0.8 \lesssim T/T_c \leqslant 1$. At lower temperatures,$Q^*$ is expected to vanish exponentially, which indeed is found in the experiments. Also the absolute magnitude of the result (37) is in good agreement with experiment. Numerical factors varying between 0.8 and 3.8, which have to be compared with $0.29 \ln(8\Delta\tau_E)$ are reported in Ref. [15]. Considering this variation and the uncertainty in different parameters,the agreement has to be considered good.

V Conclusion

The experiments of van Harlingen, Heidel, and Garland [8] raised the suspicion that the theory of thermoelectricity in superconductors is incomplete. On the other hand, the more recent experiments of van Harlingen [10] suggest that the basic theoretic concepts are sufficient. More experimental work would by very desirable in order to decide this question definitely. As far as the other transport coefficients in the scheme Eq. (27) are concerned,good agreement between theory and experiments is found. This is certainly the case as far as the diagonal coefficients normal conductivity (in particular at nonzero frequencies), thermal conductivity, and second viscosity i.e. the charge imbalance relaxation rate, are concerned. The charge imbalance proportional to $\beta(\vec{v}_s \cdot \vec{\nabla} T)$ has been observed in experiments,and the theoretical concepts are verified.

The scheme of transport equations suggests further effects. There is little hope to detect in an experiment a heat current proportional to $\vec{v}_s \operatorname{div} \vec{j}_s$, which according to Onsager's relation carries the same transport coefficient β (up to dimensional factors) as the $\vec{j}_s \vec{\nabla} T$ effect. It should be noted that these effects are very small: The differences in the electrochemical potentials in the $\vec{j}_s \vec{\nabla} T$ effect lead to signals of order 10 pV only. Also,a charge imbalance proportional to $\gamma \vec{v}_s \cdot \vec{E}'$ is difficult to detect, since it requires to create first a (large) gradient in $\varphi_n - \varphi_s$,and then to detect small changes in $\varphi_n - \varphi_s$ due to $\vec{v}_s$. However, it is conçeivable that this can be done [37]. Theoretically, in clean samples one expects a coefficient $\gamma = \tau_{Q*}$.

Acknowledgement

The author is pleased to acknowledge many pleasant and stimulating discussions concerning these topics with J. Clarke, D. van Harlingen, C. Pethick, A. Schmid, H. Smith and M. Tinkham.

References

[1] For a recent review and further references see *V. L. Ginzburg* and *G. F. Zharkov*, Usp. Fiz. Nauk **125**, 19 (1978) [Sov. Phys. Usp **21**, 381 (1978)].

[2] *V. L. Ginzburg*, Zh. Eksp. Teor. Fiz. **14**, 177 (1944) [Sov. Phys. JETP **8**, 148 (1944)].

[3] *Yu. M. Galperin, V. L. Gurevich*, and *V. I. Kozub*, Zh. Eksp. Teor. Fiz. **66**, 1387 (1974) [Sov. Phys. JETP **39**, 680 (1975)].

[4] *J. C. Garland* and *D. J. van Harlingen*, Phys. Lett. A **47**, 423 (1974).

[5] *N. V. Zavaritskii*, Pisma Zh. Eksp. Teor. Fiz. **19**, 205 (1974) [JETP Lett. **19**, 126 (1974)].

[6] *C. M. Pegrum, A. M. Guénault*, and *G. R. Pickett*, in Low Temperature Physics – LT 14, edited by *M. Krusius* and *M. Vuorio* (North Holland, Amsterdam, 1975), Vol. II, p. 513.

[7] *C. M. Falco*, Solid State Commun. **19**, 623 (1976).

[8] *D. J. van Harlingen*, and *J. Garland*, Solid State Commun. **25**, 119 (1978); *D. J. van Harlingen, D. F. Heidel* and *J. Garland*, Phys. Rev. B **21**, 1842 (1980).

[9] For example, *J. Clarke* and *S. M. Freake*, Phys. Rev. Lett. **29**, 588 (1972); *A. A. J. Matsinger, R. de Bruyn Ouboter*, and *H. van Beelen*, Physica (Utrecht) **93 B**, 63 (1978).

[10] *D. J. van Harlingen*, preprint.

[11] *S. N. Artemenko* and *A. F. Volkov*, Zh. Eksp. Teor. Fiz. 70, 1051 (1976) [Sov. Phys. JETP **43**, 548 (1976)].

[12] *M. Tinkham*, Phys. Rev. B **6**, 1747 (1972); *J. Clarke*, Phys. Rev. Lett. **28**, 1363 (1972).

[13] *A. Schmid* and *G. Schön*, J. Low Temp. Phys. **20**, 207 (1975).

[14] *C. J. Pethick* and *H. Smith*, Phys. Rev. Lett. **43**, 640 (1979).

[15] *J. Clarke, B. R. Fjordbøge*, and *P. E. Lindelof*, Phys. Rev. Lett. **43**, 642 (1979).

[16] *A. Schmid* and *G. Schön*, Phys. Rev. Lett. **43**, 793 (1979).

[17] *J. Clarke* and *M. Tinkham*, Phys. Rev. Lett. **44**, 106 (1980).

[18] *P. E. Lindelof, B. R. Fjordbøge* and *J. Bindslev Hansen*, preprint (1979).

[19] *J. Beyer Nielsen, Y. A. Ono, C. J. Pethick* and *H. Smith*, Solid State Commun. **33**, 925 (1980).

[20] *G. Schön* (to be published).

[21] *A. Schmid*, J. Low Temp. Phys. **41**, 37 (1980).

[22] *J. Bardeen, G. Rickayzen*, and *L. Tewordt*, Phys. Rev. **113**, 982 (1958).

[23] *C. M. Pegrum* and *A. M. Guénault*, Phys. Lett. A **59**, 393 (1976).

[24] *A. G. Aronov* and *V. L. Gurevich*, Fiz. Tverd. Tela **16**, 2656 (1974) [Sov. Phys. Solid State **16**, 1722 (1975)].

[25] *C. J. Pethick* and *H. Smith*, Ann. Phys. **119**, 133 (1979).

[26] This is found also from the numerical calculations, see e.g. *C. C. Chi* and *J. Clarke*, Phys. Rev. B **19**, 4495 (1979).

[27] *G. J. Dolan* and *L. D. Jackel*, Phys. Rev. Lett. **39**, 1628 (1977). *T. Y. Hsiang* and *J. Clarke*, Phys. Rev. B **21**, 945 (1980).

[28] *M. Tinkham*, Phys. Rev. B **22**, 2594 (1980).

[29] *C. Falco*, Phys. Rev. Lett. **39**, 660 (1977).

[30] *I. M. Khalatnikov*, An Introduction to the Theory of Superfluidity (Benjamin, New York, 1965).

[31] *S. J. Putterman*, Superfluid Hydrodynamics (North Holland, Amsterdam, 1974) Chapter IX.

[32] *A. G. Aronov*, Zh. Eksp. Teor. Fiz. **67**, 178 (1974) [Sov. Phys. JETP **40**, 90 (1975)].

[33] *U. Eckern* and *G. Schön*, J. Low Temp. Phys. **32**, 821 (1978).

[34] In the context of gap anisotropy this effect was discussed by *P. Hohenberg*. Zh. Eksp. Teor. Fiz. **45**, 1208 (1963) [Sov. Phys. JETP **18**, 834 (1964)].

[35] *K. Maki*, in Superconductivity, Vol. 2, *R. D. Parks*, ed. (Marcel Dekker, New York, 1969), Chapter 18.

[36] *J. Clarke, U. Eckern, A. Schmid, G. Schön*, and *M. Tinkham*, Phys. Rev. B **20**, 3933 (1979).

[37] *D. J. van Harlingen* (private communication).

Aspects of Nonlinear Spectroscopy

Dietmar Fröhlich
Institut für Physik, Universität Dortmund, Dortmund, Germany

Summary: Nonlinear spectroscopy allows to study optical processes where the induced polarization depends in higher order on the external photon fields. After some general remarks on the expected magnitude of nonlinear optical effects and their inherent additional degrees of freedom specific examples of $\chi^{(2)}$ and $\chi^{(3)}$ spectroscopy in solids are discussed.

1 Introduction

Nonlinear optics has established itself as a broad field of research in modern optics. It has a great impact both in fundamental physics as well as in applied physics. There is indeed a very fruitful mutual stimulation between both fields. With the development of high power lasers which allow to bring easily electric fields into a solid of the order of the field within an atom, nonlinear effects are expected to be detectable. Starting with the observation of second harmonic generation by Franken et al. [1] in 1961 the field of nonlinear optics grew rapidly. Nonlinear optical effects as second harmonic generation, parametric amplification and oscillation were soon used to create new laser lines and even tunable laser devices. These in turn stimulated nonlinear *spectroscopic* investigations in solids, liquids and gases. The number of possible experiments is drastically increased as compared to linear optical spectroscopy, since starting with two or more photon fields one has many more degrees of freedom. It is then the task of the spectroscopist to choose the right combination of laser frequencies, the appropriate polarization for the different photon beams, and even certain K combinations might be more favorable than others, to study a specific transition.

There are indeed very different motivations to do nonlinear spectroscopy. One might be interested in studying optical transitions which are forbidden for linear spectroscopy (one-photon absorption OPA). This point was already discussed 50 years ago by Goeppert-Mayer [2], who derived the two-photon transition probability from 2nd order perturbation theory.

If one is interested in experiments where incoming photon beams are mixed to create photon beams of other photon energies and other propagation vectors, the macroscopic approach is appropriate. Wave equations for the macroscopic electric fields, which contain nonlinear source terms have then to be solved to account for propagation effects. In addition the macroscopic approach allows a natural expansion of the induced polarisation into terms of increasing order in the applied photon fields.

The expansion parameters are the nonlinear susceptibilities $\chi^{(n)}$. In the following we will first discuss rather generally the nonlinear polarisation in order to get a feeling for the magnitude of the expected nonlinear effects. We will then focus on some selected coherent $\chi^{(2)}$ and $\chi^{(3)}$ effects in detail. Finally we will consider two-photon absorption (TPA) with special emphasis on forbidden transitions in semiconductors.

This paper is not meant as a review on nonlinear spectroscopy. We would rather like to discuss only some aspects of this field of modern optical research. According to the author's own interest we will limit ourselves to applications in solid state physics.

2 General discussion of nonlinear optical effects

Starting with the Fourier transform $P(\omega, K)$ of the nonlinear polarisation we will discuss effects up to 3rd order in the external fields:

$$\frac{1}{\epsilon_0}\vec{P}(\omega) = \chi^{(1)}(\omega)\cdot\vec{E}(\omega) + \chi^{(2)}(-\omega, \omega_1, \omega_2) : \vec{E}(\omega_1)\vec{E}(\omega_2) + \chi^{(3)}(-\omega, \omega_1, \omega_2, \omega_3) \vdots \vec{E}(\omega_1)\vec{E}(\omega_2)\vec{E}(\omega_3) + \ldots \quad (1)$$

ω_i stands for $\omega_i K_i$. The K-dependence in the fields is trivial, the K-dependence in the nonlinear susceptibilities is caused by spatial dispersion, which will not be considered here. $\chi^{(n)}$ are (n + 1)-rank tensors with 3^{n+1} components. The number of independent components is drastically cut down by symmetry considerations. For a discussion of this point we refer to a recent review article by Chemla [3] and literature cited therein. One important point can be deduced immediately by inspection of equ. 1.: for a system with inversion symmetry all components $\chi^{(2)}_{ijk} = 0$, since $\vec{E}$ and $\vec{P}$ transform as polar vectors and $\chi^{(2)}$ should be in such a system invariant with respect to inversion. By nonlinear spectroscopy experiments one gets information on $\chi^{(2)}$ and $\chi^{(3)}$ and higher terms. One naturally expects the higher order terms to be smaller than lower ones. Using an anharmonic oscillator model Bloemberger [4] showed that for the nonresonant case

$$P^{(n+1)}/P^{(n)} = \frac{E}{E_{at}} \quad (2)$$

The electric field strength E of the laser, which is easily derived from its intensity I by

$$E\,(\mathrm{V/cm}) = \left(\frac{2I}{c\,n_0\,\epsilon_0}\right)^{1/2} = 27\left(\frac{I\,(\mathrm{W/cm^2})}{n_0}\right)^{1/2} \quad (3)$$

has to be compared with a typical field inside an atom, which can be estimated as $E_{at} \approx 10^8$ V/cm. Even from a laser intensity of 10^{10} W/cm^2 we get a field of $2 \cdot 10^6$ V/cm which is well below E_{at}, we thus expect generally higher order effects to be much smaller than lower order effects. In addition the application of high intensities is limited by dielectric breakdown. As shown by Smith et al. [5] the dielectric breakdown threshold E_b (rms field strength $E_b = \frac{1}{\sqrt{2}} E$) is for alkalihalides and other transparent media in the order of $10^6 - 10^7$ V/cm for picosecond pulses at 1.06 μm. Though there is no problem to get high fields by focussing pulsed laser beams, limitations are set by laser damage even in transparent regions of the spectrum. Research on laser damage is of great importance because of its technical implications [6].

Akhmanov [7] discusses nonlinear susceptibilities up to $\chi^{(5)}$. He gets for SiO_2 $\chi^{(3)}/\chi^{(2)} = 0.9 \cdot 10^{-5}$ (esu) which corresponds to $3 \cdot 10^{-8}$ (cm/V). A factor of 10^{-8} between nonlinear susceptibilities of successiv order follows from the simple anharmonic oscillator model $\chi^{(n+1)}/\chi^{(n)} = \frac{1}{E_{at}}$ with an atomic field of 10^8 V/cm. The numbers quoted apply only to the nonresonant case. Considering again the anharmonic oscillator model as done by Bloemberger [4], one gets a good physical insight in the many possibilities of *resonance* nonlinear spectroscopy. For a quantitative discussion one has to derive the susceptibilities as macroscopic parameters from the microscopic properties of the substance under consideration. The various resonance denominators come then in naturally through the application of higher order perturbation theory. As discussed in detail recently by Flytzanis [8] one can give expressions for the susceptibilities in terms of matrixelements. A typical term of $\chi^{(3)}$ is given by:

$$\chi^{(3)}(-\omega_4, \omega_1, \omega_2, \omega_3) = A \cdot \sum_{ijk} \frac{x_{gi} \cdot x_{ij} \cdot x_{jk} \cdot x_{kg}}{(\omega_{ig} - \omega_1 - i\Gamma_{ig})} \times \frac{1}{(\omega_{jg} - \omega_1 - \omega_2 - i\Gamma_{jg})(\omega_{kg} - \omega_4 - i\Gamma_{kg})} + \ldots \quad (4)$$

The nonlinear polarization $P_i(\omega_4 K_4)$ leads to a source term in the wave equation, where all combinations of frequencies $\omega_4 = \pm\omega_3 \pm\omega_2 \pm\omega_1$ are possible. Apart from propagation effects which manifest themselves in the K-dependence, we expect many "resonances" in $\chi^{(3)}$ as a result of the different combinations of ω_1 to ω_4. Considerable enhancement can be achieved if one takes advantage of these resonance denominators. Besides the additional degrees of freedom introduced by the possibility to select different combinations of photon energies one can choose different polarization combinations. As known from linear spectroscopy, symmetry considerations lead to selection rules for these many photon transitions. Necessary conditions on the incoming polarizations directions $\vec{e}_n$ to induce a certain polarization P_i can be directly derived from the cartesian tensor components of the nonlinear susceptibilities $\chi^{(n)}$. For measurements with circular polarized light it might be more useful to introduce spherical tensors. The polarization selection rules are a very important tool to descriminate between various transitions. This point will be discussed in more detail for the specific example of two-photon absorption in section 6.

Besides energy conservation and polarization selection rules, one has to consider K-conservation. This additional degree of freedom to combine incoming K-vectors to excite a propagating wave in the medium is of great relevance in nonlinear spectroscopy as will be discussed in the next sections.

The many possibilities of incoming and outgoing photon beams lead to a large number of nonlinear optical experiments, which either probe $\chi^{(2)}$ or $\chi^{(3)}$ or both of them. Many experiments have been done up to now. We list in the following some of them by their characteristic abreviations.

$\chi^{(2)}$ *methods*

SHG: Second Harmonic Generation
SFG: Sum Frequency Generation
DFG: Difference Frequency Generation
OPA: Optical Parametric Amplification
OPO: Optical Parametric Oscillation

$\chi^{(3)}$ *and higher order methods*

THG: Third Harmonic Generation
TPA: Two Photon Absorption
SRS: Stimulated Raman Scattering
SSRS: Stimulated Spinflip Raman Scattering
SRRS: Stimulated Resonance Raman Scattering
CARS: Coherent Antistokes Raman Scattering
CSRS: Coherent Stokes Raman Scattering
ARCS: Angular Resolved Coherent (Raman) Scattering
RIKES: Raman Induced Kerr Effect Studies
HORSES: Higher Order Raman Spectral Excitation Studies
ASCS: Active Spectroscopy of Combinatorial (Raman) Scattering
ANLS: Active Non Linear Spectroscopy
HRS: Hyper Raman Scattering
SHRS: Stimulated Hyper Raman Scattering

One can easily think of other experiments by just playing with these abreviations. Replacing e.g. Raman by Brillouin, one gets coherent stokes or antistokes Brillouin scattering (CSBS or CABS).

3 $\chi^{(2)}$ effects

In the following we discuss briefly $\chi^{(2)}$ effects, where the necessity to consider K-conservation becomes obvious. Though we stay within the dipol approximation and do not consider spatial dispersion (K does not appear in $\chi^{(n)}$ explicitly), the K-vectors or finally the phase-mismatch ΔK enters into the propagation equations. This was immediately recognized by Maker et al. [9] after the first experiments of second harmonic generation (SHG) by Franken [1]. In the so-called parametric approximation, which assumes that bleaching of the fundamental intensity I_ω can be disregarded, one gets for the second harmonic intensity [3] $I_{2\omega}(l)$

$$I_{2\omega}(l) = \frac{2\,\omega^2\,(\chi^{(2)})^2 \cdot l^2}{\epsilon_0\, n_{2\omega}\, n_\omega^2\, c^3} \cdot I_\omega^2 \cdot \left(\frac{\sin(\Delta k l/2)}{\Delta k l/2}\right)^2 \qquad (5)$$

The 2nd harmonic intensity depends strongly on the phase-mismatch $\Delta K = K_{2\omega} - 2K_\omega = \frac{\pi}{l_c}$ which is expressed by the so called coherence length l_c. For efficient SHG one takes advantage of birefringence to assure phase matching ($\Delta K = 0$). Phase matching results in a coherent effect throughout the crystal. Though $\chi^{(2)}$ is small as discussed before, one might get high conversion efficiency. 50 % transfer from the fundamental I_ω to the second harmonic $I_{2\omega}$ can be achieved. In this case the depletion of the pump has to be taken into account [3]. Sum frequency (SFG) and difference frequency generation (DFG) are a direct application of $\chi^{(2)}$ spectroscopy. As mentioned before these techniques can only be applied in systems without inversion symmetry. GaP and CuCl are the mostly studied materials. One of the motivations to apply these techniques of nonlinear spectroscopy, where one has the additional degree of freedom to choose K_{tot} and ω_{tot} in certain kinematic limits independently of each other, was to study the polarition despersion $\omega(K)$. Faust and Henry [10] were the first to study a phonon resonance in the nonlinear susceptibility by DFG. As an example we will briefly report experiments by De Martini [11] on DFG in GaP. As scetched in Fig. 1 two coherent beams of frequencies ω_1 and ω_2 which are both derived from a ruby laser by stimulated Raman effect are directed under a certain angle on a GaP prism. The difference frequency $\omega_1 - \omega_2$ is monitored. These measurements allow to determine the lower polariton branch as well as linear absorption data, which are difficult to get directly.

For SFG we refer to experiments by Haueisen and Mahr [12] in CuCl and ZnO. Two photons of a pulsed dye laser which is pumped by a ruby laser, are tuned through the exciton resonance. The SHG signal clearly reveals the nonlinear resonance, which can nicely be fitted by an anharmonic oscillator model as seen in Fig. 2.

4 $\chi^{(3)}$ effects

The degrees of freedom in ω and K can be further increased by turning to the 3rd order susceptibility $\chi^{(3)}$. Since one has now the possibility to choose independently three incoming photon beams $\omega_1, \omega_2, \omega_3$ to create a polarisation P at ω_4. The additional degrees of freedom in the choice of K-vectors opens up many possibilities for coherent spectroscopy. In the following we will briefly discuss coherent anti-stokes Raman scattering (CARS). For reviews we refer to a paper by Tolles et al. [13] and Levenson and Song[14]. The first experiments were done by Maker and Terhune [15], who were able to create in a three-wave mixing experiment antistokes light at $\omega + \Delta\omega$ by shining in light at ω and $\omega - \Delta\omega$. The authors studied the propagation

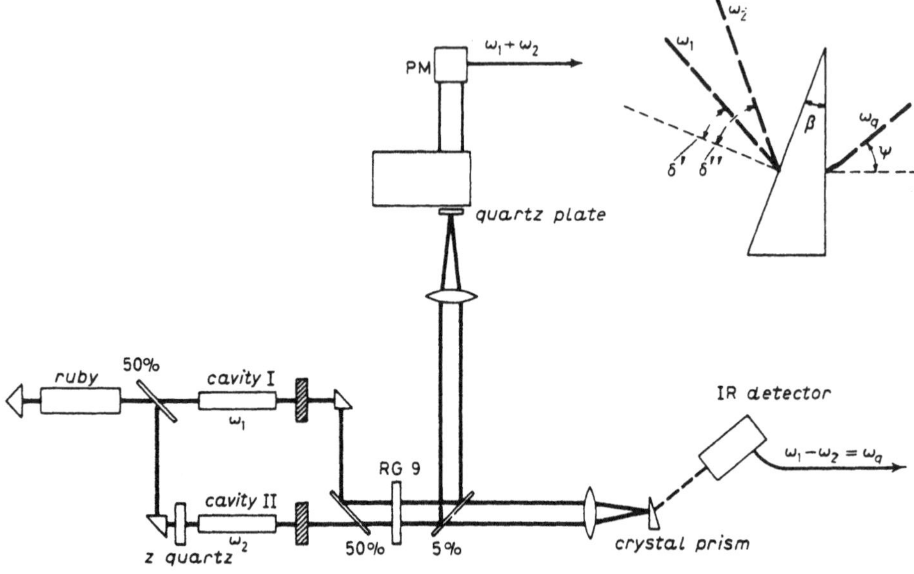

Fig. 1 Schematic diagram of experimental setup to measure polaritons by DFG from De Martini [11].

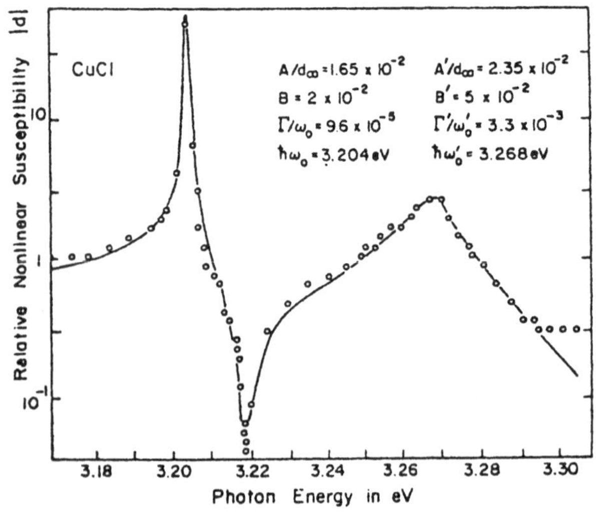

Fig. 2
Relative nonlinear susceptibility of CuCl from Haueisen and Mahr [12]. An anharmonic oscillator model with two resonant frequencies is used to fit the experimental data.

effects, which are as in the case of SHG expressible by the mismatch ΔK. A few years later Coffinet and De Martini [16] incorporated in their experimental set-up the necessary kinematic possibilities to achieve phasematching. In these experiments the authors studied the coherent excitation of the phonon polariton in GaP.

CARS as a $\chi^{(3)}$ process can be understood by considering energy and K-conservation as scetched in Fig 3. In the energy diagram we see that 2 photons ω_1 and a photon ω_2 $(<\omega_1)$ interact to generate a polarization at the antistokes frequency $\omega_{AS} = 2\omega_1 - \omega_2$. If $\omega_1 - \omega_2$ is tuned to a phonon resonance, $\chi^{(3)}$ shows a resonance if the right polarizations of the photon beams are selected, thus reproducing the Raman spectrum.

The main difference to *spontaneous* Raman effect lies in the fact, that CARS is a coherent effect with phasematching for only a small solid angle. As indicated by the wavevector diagram we expect the angle Θ to be rather small (1/2 to 1° in solids) since all frequencies ω_1, ω_2 and ω_{AS} are close together and in a region of normal dispersion far off any resonance.

Taking advantage of the possibility of spatial discrimination and the much higher efficiency of the coherent signal one can gain 9 orders of magnitude discrimination against background fluorescence by CARS as compared to spontaneous Raman scattering as discussed by Begley et al. [17].

Though CARS seems to have a lot of advantages, difficulties arise in detecting weak resonances because of a strong background from nonresonant contributions. This is a serious drawback of $\chi^{(3)}$ spectroscopy and follows from the additional degrees of freedom to induce a nonlinear polarization by the large spectrum of nonresonant

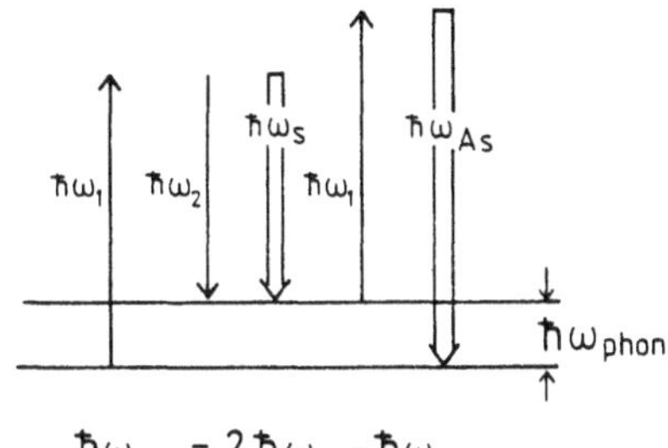

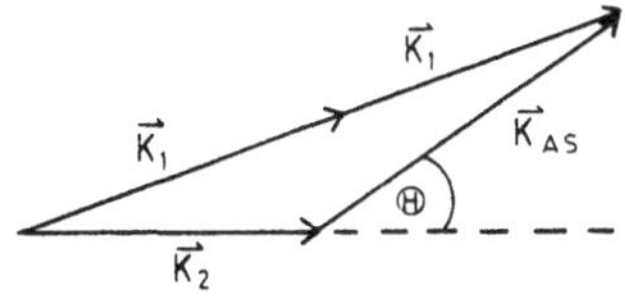

Fig. 3
Energy and $\vec{K}$-conservation diagram for CARS.

transitions. The total $\chi^{(3)}$ can be split up into a nonresonant $\chi^{(3)}_{NR}$ and a resonant $\chi^{(3)}_{R}$

$$\chi^{(3)}(\omega_{AS}) = \chi^{(3)}_{NR}(\omega_{AS}) + \chi^{(3)}_{R}(\chi_{AS}) \tag{6}$$

with

$$\chi_R = \frac{A}{\{\omega_0 - (\omega_1 - \omega_2) - i\Gamma\}}$$

Since the real part of $\chi^{(3)}_{R}$ changes sign, when $\omega_1 - \omega_2$ is tuned through the phonon resonance ω_0, the interference with the nonresonant contributions leads to a typical dispersion like lineshape in $|\chi^{(3)}(\omega_{AS})|^2$. As an example measurements of Levenson [18] on calcite are presented in Fig. 4. As discussed in detail by Shen [19], one

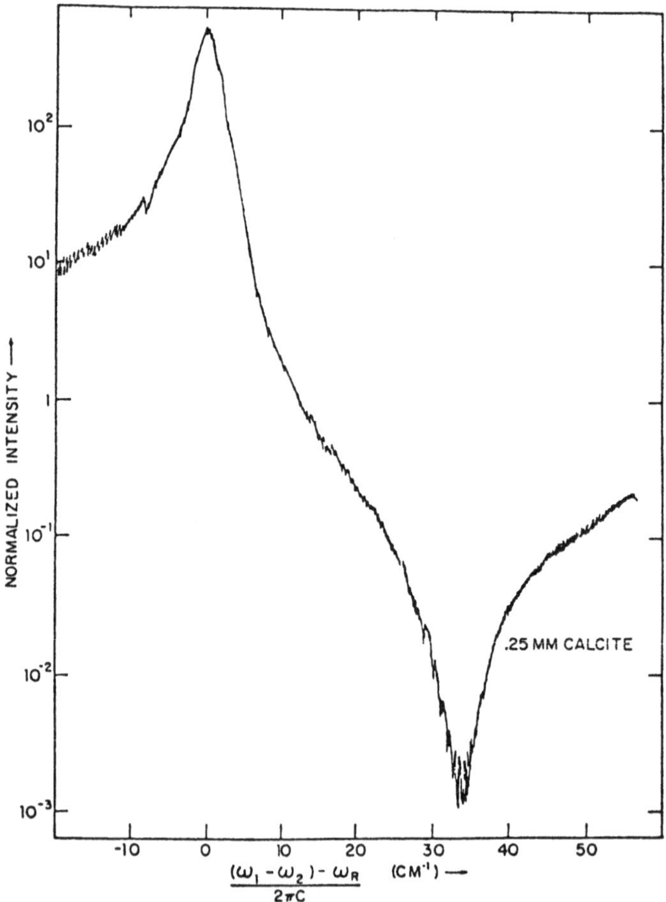

Fig. 4 Resonance structure of the 1088 cm^{-1} lattice mode of calcite as measured by CARS (after Levenson [18]).

gets information on $A/\chi^{(3)}_{NR}$ and Γ from the position of the maximum ($\Delta\omega_+$) and minimum ($\Delta\omega_-$). These experiments should yield accurate results, since only frequency measurements are involved. To deduce absolute values of $\chi^{(3)}_{NR}$ A and Γ have to be known from other data.

Though these interference effects can be used to derive valuable information on nonlinear susceptibilities they cause signal to noise problems, if one is looking for a rather weak resonant signal on a strong nonresonant background. Oudar et al. [20] were able to increase the sensitivity of CARS for weak Raman lines by using a special polarization arrangement to suppress the nonresonant background. The authors were able to measure Re $(\chi^{(3)}_R)$, Im $(\chi^{(3)}_R)$ and $|\chi_R|^2$ separately. Another interesting concept, which takes advantage of interference between *two* Raman resonances, is discussed by Lotem et al. [21]. For this method, however, one needs 3 different input frequencies, two of them being tunable.

The other extrem to do CARS type experiments is to use only one photon energy $\hbar\omega$ in a forward *degenerate* mixing experiment. This method of active nonlinear spectroscopy (ANLS) is very successfully applied by Maruani et al. [22] to study biexcitons in CuCl and CdS. Because of the fact that the energy of the biexciton is very close to twice the energy of the exciton ($E_{biex} = 2\,E_{ex} - E_B$, E_B (CuCl) = 36 meV) one can get an almost triple resonance in $\chi^{(3)}$ by tuning the laser to $E_{biex}/2$, which is then close to E_{ex}. The principle of the experiment and the relevant diagrams for energy- and wave-vector conservation are scetched in Fig. 5. Both pump

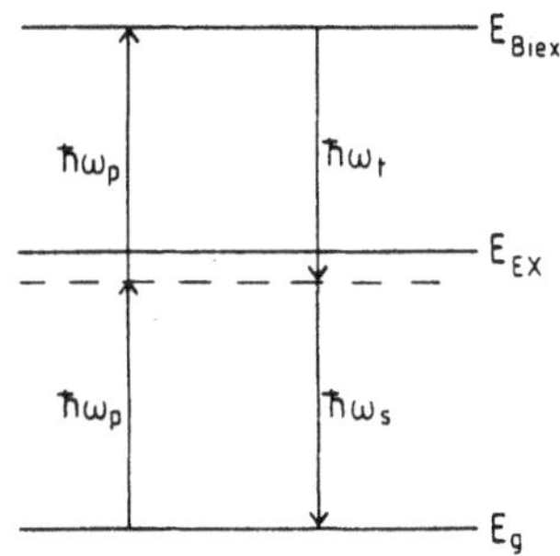

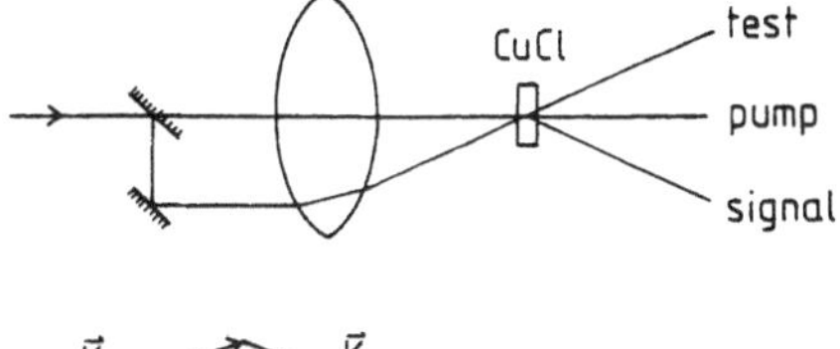

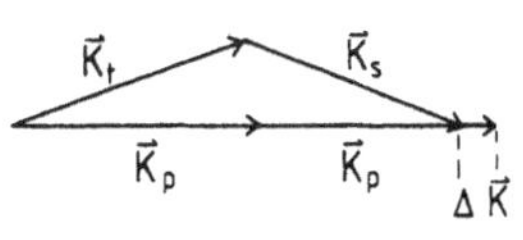

Fig. 5
Principle of the method of forward degenerate optical mixing (after Maruani et al. [22]).

photons ($\hbar\omega$) are close to the exciton resonance and together ($2\hbar\omega$) they are tuned through the biexciton resonance. Despite the strong dispersion in the energy region of $\hbar\omega$ a condition close to phase-matching can be achieved by choosing the angle Θ between the pump and test beam as small as possible. Using a double lock-in technique the authors manage to suppress the scattering of the intense pump beam into the signal channel sufficiently. As an example of their results we present a typical spectrum for CuCl in Fig. 6. The pronounced asymmetry with the much lower intensity on the high energy side of the biexciton resonance at 3.186 eV is explained by the authors assuming a Fano type interference.

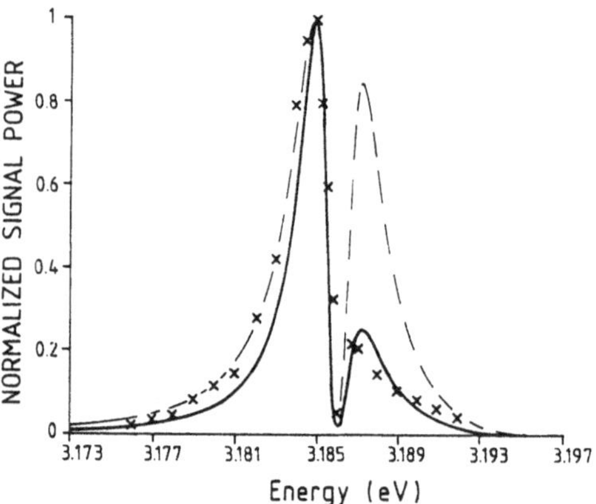

Fig. 6
Comparison of experimental results for CuCl with theoretical models. Dotted line: best fit for a 200 μm thick sample disregarding Fano type interference. Full line: coupling with continuum is included (after Maruani and Chemla [22]).

5 $\chi^{(2)}$ and $\chi^{(3)}$ interference

Having discussed $\chi^{(2)}$ and $\chi^{(3)}$ effects separately we will briefly consider cases where both terms contribute simultaneously. A necessary condition to observe such an effect is of course that the crystal lacks a center of inversion, otherwise $\chi^{(2)}$ would be zero. Even with this necessary condition being fulfilled one would naively expect the $\chi^{(3)}$ contributions to be much smaller than a $\chi^{(2)}$ term, which would make it difficult to observe interference effects. Taking advantage of resonance enhancement in $\chi^{(3)}$ by choosing a two-photon resonance ($\omega_1+\omega_2 = \omega_{ex}$) one can indeed observe such an effect. In Fig. 7 we present two-photon data on CuCl. The two-photon absorption which is proportional to the total imaginary part of $\chi^{(3)}_{TEP}$ (transverse exciton polariton), was measured by Fröhlich et al. [23]. By changing K vectors between the two photon beams the authors were able to measure part of the upper exciton polariton and the longitudinal exciton of the Z_3 resonance in CuCl. The principle of the method is shown in Fig. 8, where we have plotted for the special case of $\hbar\omega_1 = \hbar\omega_2$ the tuning curves for different angles between the beam directions. For $\Theta = 180°$ one can excite transitions at $K = 0$, where the transverse polariton (TP) and the longitudinal exciton (LE) are degenerate. In the actual TPA data of Fig. 7 $\hbar\omega_1 \sim 2\hbar\omega_2$, we thus get already a splitting between TP and LE for $\Theta = 180°$. For small angles ($\Theta \leqslant 30°$) between the K directions no two-photon absorption of the upper transverse exciton could be detected. This is in good agreement with theoretically considerations as was pointed out by Boggett and Loudon [24] and also discussed in detail by Kramer and Bloembergen [25]. These authors were even able to resolve in a *double* resonance configuration the interference between the longidutinal Z_3 exciton and the phonon polariton.

As discussed in detail by Fröhlich et al. [23] and Uihlein et al. [26] this interference effect follows quite naturally in the microscopic approach of two primary polaritons which fuse by anharmonic interaction to a third excitation. This might be again a polariton or a longitudinal exciton. The fusion rate can be measured as an attenuation of one of the primary polariton beams as is done in the standard TPA experiment. If the created particle is again a polariton, it can also be observed by sum frequency generation. Haueisen and Mahr [27] were able to measure resonances on the upper polariton to higher K-values (up to $\Theta = 0$, Fig. 7) by second harmonic generation.

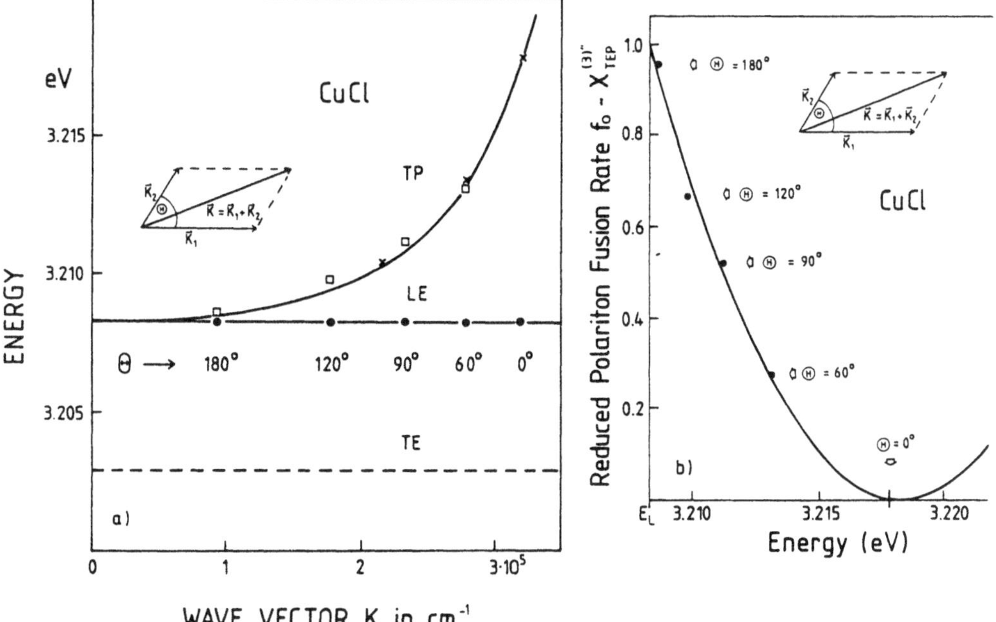

Fig. 7 TPA and SHG data of first exciton of CuCl.

a) polariton diagram: open squares and closed circles from TPA [23] and crosses from SHG [27].

b) Normalized fusion rate which is proportional to the imaginary part of $\chi^{(3)}_{TEP}$ as derived from TPA.

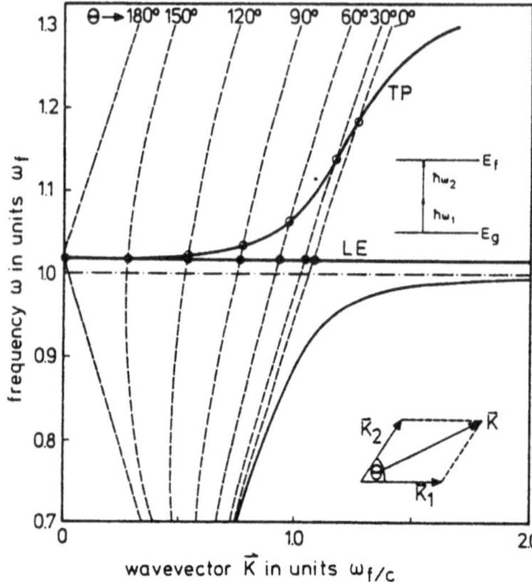

Fig. 8
Schematic diagram for K-space spectroscopy by TPA as calculated in a two oscillator model.

The pronounced frequency dependence of the fusion rate (Fig. 7) follows naturally in the microscopic model as an interference between the transition amplitudes to the photon and exciton part of the created polariton, respectively.

6 Two-photon absorption (TPA)

Having discussed up to now mainly experimental situations where nonlinear effects were detected by the *emission* of light at certain frequencies and propagation directions, we will now turn to nonlinear *absorption* processes. Assuming e. g. that the sum of two photon energies $\hbar\omega_1$ and $\hbar\omega_2$ is tuned into a resonance with the final level $\hbar\omega_f$ we derive from equ. 1 an imaginary $\chi^{(3)}$. If we consider in the summation over j only the resonant term, which we call f, we are left with a double summation over i and k. As pointed out by Bloembergen [28] this can be written in the following form:

$$\chi^{(3)}\,(\omega_f = \omega_1 + \omega_2) = \mathrm{i}\,B \left| \sum_n \left(\frac{x_{gn} \cdot x_{nf}}{\omega_{ng} - \omega_1} + \frac{x_{gn} \cdot x_{nf}}{\omega_{ng} - \omega_2} \right) \right|^2 \tag{7}$$

As discussed in detail by Mahr [29] the two-photon absorption constant $\alpha^{(2)}$ is proportional to this imaginary part of $\chi^{(3)}$. The same result is easily derived by second order perturbation theory as was done by Goeppert-Mayer [2] 50 years ago. The first TPA experiments with a tunable total photon energy $\hbar\omega_1 + \hbar\omega_2$ were done by Hopfield et al. [30]. The rather simple microscopic description of two-photon processes allows immediately conclusions on the magnitude of the effects and the final states involved. If parity is a good quantum number one derives from equ. (7), that the final state f must have the same parity as the ground state g. In this respect TPA is equivalent to Raman effect, where a vibrational or electronic level at the difference frequency is excited. Selection rules are the same for TPA and Raman. The additional degrees of freedom of having two photon beams, whose polarizations can be chosen independently from each other, allow to assign quantum numbers to the excited states studied. There is a standard formalism [31] to do these assignments. Using linear- and circular polarized light one can get a unique assignment even if different transitions overlap, as was demonstrated for the cubic group 0_h by Fröhlich et al. [32]. Direct forbidden transitions in solids are specially suited to be studied by TPA as discussed by Loudon [33]. Considerable improvement in the experimental techniques of TPA [34] made it possible to resolve the fine-structure of the even parity excitons in Cu_2O. As an example we present in Fig. 9 the OPA and TPA spectrum of Cu_2O. For details of the interpretation of the different exciton states we refer to the original papers by Fröhlich et al. [35] and Uihlein et al. [36]. Additional fine-structure can be studied by application of a magnetic field [35, 36]. In the following we will present some new results on the 1S exciton in a magnetic field up to 8.5 Tesla. From measurements in different polarizations one can assign Γ_5^+ to the two-photon allowed excitons. This is demonstrated for the n = 3 doublet in Fig. 10. The high quality crystal [37] allows an almost complete suppression of TPA in the (001)-parallel configuration. In a magnetic field the threefold level Γ_5^+ is expected to split into three components, which we can classify by quantum numbers ±1 and 0. The polarization selection rules have now to be worked out for

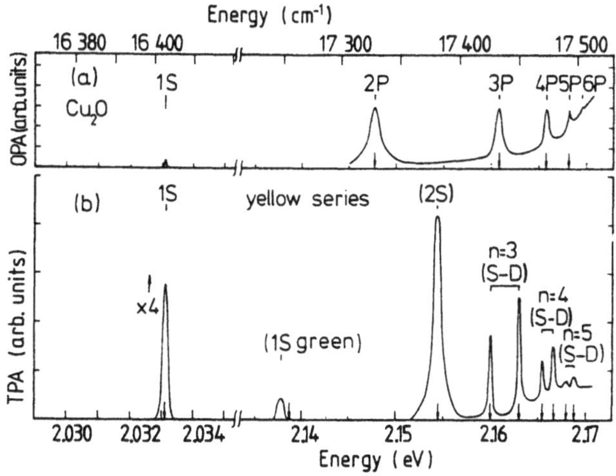

Fig. 9
One photon absorption (OPA) and two-photon absorption (TPA) of Cu_2O.

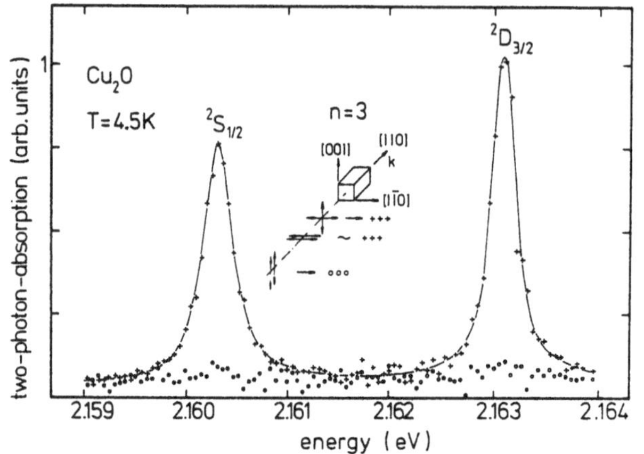

Fig. 10
TPA polarization dependence for n = 3 dublett in Cu_2O. Crosses: polarization of one laser parallel (001) and of the other laser parallel (1$\bar{1}$0) direction, the same results are gained with both lasers being polarized parallel (1$\bar{1}$0). Open circles: both polarizations parallel (001).

the different components of Γ_5^+. The results are given for a general field direction $\vec{b} = (b_1, b_2, b_3)$ and Faraday configuration in the following equations:

$$\mathrm{Os}\,(0, RR) = \mathrm{Os}\,(0, LL) = 1 - 4A + 9C \tag{8}$$

$$\mathrm{Os}\,(0, RL) = \mathrm{Os}\,(0, LR) = 9C$$

$$\mathrm{Os}\,(+1, RR) = \mathrm{Os}\,(-1, LL) = 9/2\,(A-C)$$

$$\left.\begin{array}{l}\mathrm{Os}\,(+1, LL) = \mathrm{Os}\,(-1, RR)\\ \mathrm{Os}\,(+1, RL) = \mathrm{Os}\,(-1, RL)\\ \mathrm{Os}\,(+1, LR) = \mathrm{Os}\,(-1, LR)\end{array}\right\} = 1/2\,(A-9C) \tag{9}$$

$$A = b_1^2 b_2^2 + b_1^2 b_3^2 + b_2^2 b_3^2 \qquad C = b_1^2 b_2^2 b_3^2$$

Os (M, PP′) gives the polarization dependent part of the two-photon oscillator strength of the component M (= 0, ±1) of Γ_5^+, where PP′ stand for the helicity of the two incomming beams with respect to the magnetic field $\vec{B}$ ($\vec{b} = \frac{\vec{B}}{|\vec{B}|}$).
As can be derived from these expressions, one can easily distinguish between all three components by just choosing the appropriate directions of the B-field and combinations of circular polarized light. Experimental results are shown in Fig. 11. Taking $\vec{b} = \frac{1}{\sqrt{2}}(110)$ we see from equ. (8) and (9) that e.g. Os (+1, RR): Os (−1, RR): Os (0, RR) = 9/8 : 1/8 : 0, which is well confirmed in Fig. 11. For the LL configuration the oscillator strengths for the M = +1 and M = −1 components are interchanged, the M = 0 component cannot be excited. In the $\vec{b} = (100)$ direction one can only excite the M = 0 component (Fig. 11). Linear polarized light cannot distinguish between the M = ±1 components. It is of not much use in the Faraday configuration, because the rather long crystals (d ~1 cm) used in TPA show considerable

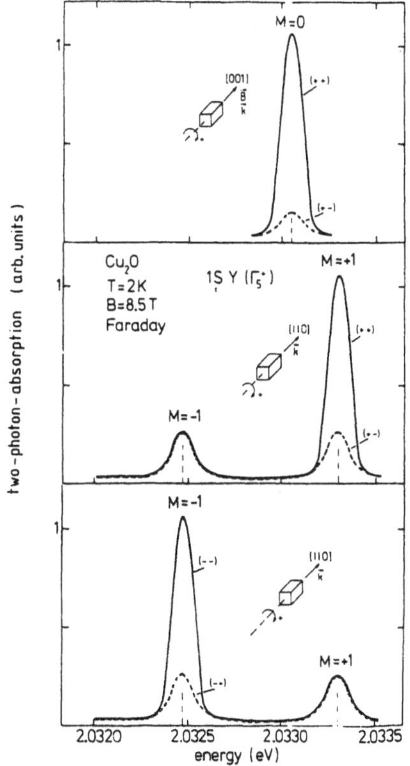

Fig. 11
TPA of the 1 S exciton (yellow series) of Cu_2O in a magnetic field of 8.5 T. All data are gained in Faraday configuration with polarizations and propagation vectors as indicated.

Faraday rotation for both beams $\hbar\omega_1$ and $\hbar\omega_2$, though they are far off resonance. Linear polarized light might be used in Voigt configuration. For completeness the polarization dependence for linear polarized light is given in equ. (10).

$$
\begin{aligned}
&\mathrm{Os}\,(0, \vec{\epsilon}_1\, \vec{\epsilon}_2) = (\vec{a} \cdot \vec{b})^2 \\
&\mathrm{Os}\,(\pm 1, \vec{\epsilon}_1\, \vec{\epsilon}_2) = 1/2\ \{a^2 - (a \cdot b)^2\}
\end{aligned}
\tag{10}
$$

$\vec{a} = (a_1, a_2, a_3)$ can be calculated from the polarization vectors $\vec{\epsilon}_1 = (l_1, m_1, n_1)$, $\vec{\epsilon}_2 = (l_2, m_2, n_2)$ by

$a_1 = m_1 n_2 + m_2 n_1$, $a_2 = n_1 l_2 + n_2 l_1$, $a_3 = l_1 m_2 + l_2 m_1$

The same polarization selection rules apply to the polariton in T_d-symmetry as discussed in detail by Fröhlich et al. [23]. By changing the direction $\vec{s} = \frac{\vec{K}}{|\vec{K}|}$ of the total K vector which corresponds here to the magnetic field direction $\vec{b}$ one can separately excite the longitudinal exciton (LE corresponding to the M = 0 component) and the transverse polariton (TP corresponding to the M = ±1 components). The predictions are very well confirmed by the experimental results as seen in Fig. 12. By application of a magnetic field one can split the two-fold transverse polariton as seen in Fig 13. In Voigt configuration on can even mix the LE and TP, which are degenerate at K = 0. It should be interesting to study this mixing as a function of the B-field and the initial ($\vec{B}$ = 0) LE-TP splitting for K ≠ 0.

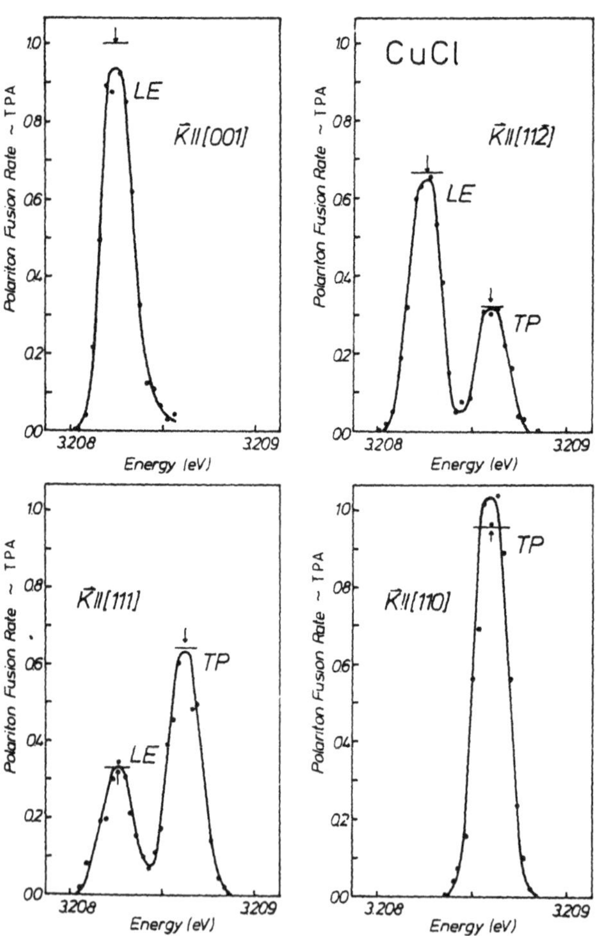

Fig. 12 TPA of longitudinal exciton (LE) and transverse polariton (TP) for different orientations of total K-vector (Θ = 180° s. Fig. 7). Polarization of both photon beams parallel (1 1̄ 0).

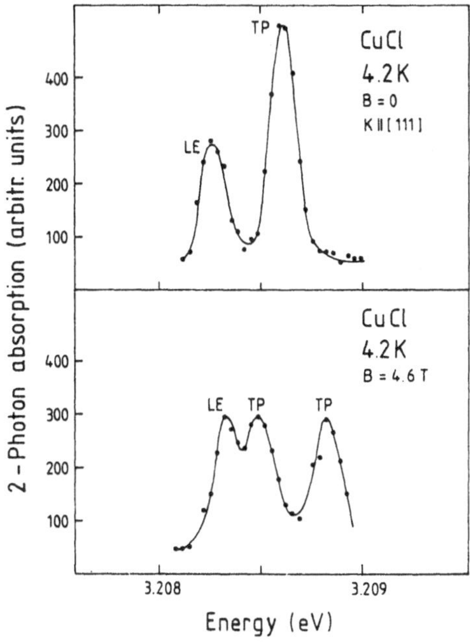

Fig. 13
TPA of the first exciton in CuCl in a magnetic field of 4.6 T. The measurements are taken in Faraday configuration with total K-vector parallel (111).

7 Conclusion

We have seen that many new spectroscopic methods can be brought up taking advantage of the additional degrees of freedom inherent in many-photon setups. Since spectroscopy is a tool to study resonances it is the task of the critical spectroscopist to select the appropriate spectroscopic technique. Many methods listed in section 2 will be only of limited interest because they can be replaced by simpler even linear ones, which yield the same information. Other nonlinear techniques will certainly be further improved. Systematic experimental investigations as recently published by Levenson and Eesley [38] are necessary to overcome the additional difficulties of the rather complex setups using more than one photon beam.

There are many other interesting application of nonlinear spectroscopy, which we have not mentioned. Two-photon Raman scattering, which is also known as hyper Raman scattering (HRS), is very succesfully applied to study dispersion curves of excitons [39]. Close to resonance conditions for the incoming photons and their sum yield strong signals. Nonlinear high-resolution spectroscopy of atoms and molecules has been very succesfully applied. For literature we refer to a review by Hänsch [40] and a more recent one by Ferguson et al. [41]. Another interesting field, which is reviewed by Brewer [42], is coherent optical spectroscopy. Transient effects, whose equivalents in spin resonance were studied many years ago, can now be observed by optical methods. Making use of high time resolution, which can be achieved in picosecond spectroscopy, very short relaxation processes can be successfully investigated. For literature we refer to articles by Kaiser and Laubereau [43].

References

[1] *P. A. Franken, A. E. Hill, C. W. Peters,* and *G. Weinreich,* Phys. Rev. Lett. **7**, 118 (1961).

[2] *M. Goeppert-Mayer,* Ann. Phys. **9**, 273 (1931).

[3] *D. S. Chemla,* Rep. Prog. Phys. **43**, 1191 (1980).

[4] *N. Bloembergen,* Nonlinear Optics (Benjamin, New York, 1965).

[5] *W. L. Smith, J. H. Bechtel,* and *N. Bloembergen,* Phys. Rev. B **12**, 706 (1975).

[6] Laser Induced Damage in Optical Materials: 1977 ed. *A. J. Glass* and *A. H. Guenther,* National Bureau of Standards Special Publication 509 (U. S. Government Printing Office Washington, 1978).

[7] *S. A. Akhmanov,* Nonlinear Spectroscopy, Enrico Fermi Course 64, p. 239, ed. *N. Bloembergen,* (North Holland, Amsterdam, New York, Oxford, 1977).

[8] *C. Flytzanis,* Quantum Electronics Vol. 1A, p. 9, ed. *H. Rabin* and *C. L. Tang* (New York: Academic Press 1975).

[9] *P. D. Maker, R. W. Terhune, M. Nisenoff,* and *C. M. Savage,* Phys. Rev. Lett. **8**, 21 (1962).

[10] *W. L. Faust* and *C. H. Henry,* Phys. Rev. Lett. **17**, 1265 (1966);
W. L. Faust, C. H. Henry, and *R. H. Eick,* Phys. Rev. **173** 781 (1968).

[11] *F. De Martini,* Phys. Rev. B **4**, 4556 (1971).

[12] *D. C. Haueisen* and *H. Mahr,* Phys. Rev. Lett. **26**, 838 (1971);
D. C. Haueisen and *H. Mahr,* Phys. Rev. B **8**, 734 (1973).

[13] *W. M. Tolles, J. W. Nibler, J. R. McDonald,* and *A. B. Harvey,* Appl. Spectr. **31**, 253 (1977).

[14] *M. D. Levenson* and *J. J. Song,* Coherent Nonlinear Optics, p. 293, ed. *M. S. Feld* and *V. S. Letokhov* (Springer, Berlin, Heidelberg, New York 1980).

[15] *P. D. Maker* and *R. W. Terhune,* Phys. Rev. **137**, A801 (1965).

[16] *J. P. Coffinet* and *F. De Martini,* Phys. Rev. Lett. **22**, 60 (1969).

[17] *R. F. Begley, A. B. Harvey,* and *R. L. Byer,* Appl. Phys. Lett. **25**, 387 (1974).

[18] *M. D. Levenson,* IEEE J. Quant. Electr. QE**10**, 110 (1974);
M. D. Levenson and *N. Bloembergen,* Phys. Rev. B **10**, 4447 (1974).

[19] *Y. R. Shen,* Reviews of Modern Physics **48**, 1 (1976).

[20] *J. L. Oudar, R. W. Smith,* and *Y.R. Shen,* Appl. Phys. Lett. **34**, 758 (1979); *J. L. Oudar* and *Y. R. Shen,* Phys. Rev. A **22**, 1141 (1980).

[21] *H. Lotem, R. T. Lynch, Jr.,* and *N. Bloembergen,* Phys. Rev. A **14**, 1748 (1976).

[22] *A. Maruani, J. L. Oudar, E. Batifol,* and *D. S. Chemla,* Phys. Rev. Lett. **41**, 1372 (1978);
A. Maruani and *D. S. Chemla,* Phys. Rev. B **23**, 841 (1981).
A. Maruani and *D. S. Chemla,* in Proceedings of the Fifteenth International Conference on the Physics of Semiconductors, Kyoto 1980, eds. *S. Tanaka* and *Y. Toyozawa* J. Phys. Soc. Jpn. **49**, 585 (1980) Supplement A.

[23] *D. Fröhlich, E. Mohler* and *Ch. Uihlein,* phys. stat. sol. (b)**55**, 175 (1973); *D. Fröhlich, E. Mohler,* and *P. Wiesner,* Phys. Rev. Lett. **26**, 554 (1971).

[24] *D. Bogett* and *R. Loudon*, Phys. Rev. Lett. **28**, 1051 (1972), J. Phys. C: Solid State Phys. **6**, 1763 (1973).

[25] *S. D. Kramer* and *N. Bloembergen*, Phys. Rev. B **14**, 4654 (1975).

[26] *Ch. Uihlein, D. Fröhlich* and *E. Mohler*, Polaritons, Proceed. of the first Taormina Research Conference on the Structure of Matter, p. 303, ed. *E. Burstein* and *F. de Martini* (Pergamon Press 1974).

[27] *D. C. Haueisen* and *H. Mahr*, Phys. Lett. **36A**, 433 (1971).

[28] *N. Bloembergen*, Nonlinear Spectroscopy, Enrico Fermi Course 64, p. 1, ed. *N. Bloembergen* (North Holland, Amsterdam, New York, Oxford 1977).

[29] *H. Mahr*, Quantum Electronics Vol. 1 A, p. 285, ed. *H. Rabin* and *C. L. Tang* (New York: Academic Press 1975).

[30] *J. J. Hopfield, J. Worlock*, and *K. Park*, Phys. Rev. Lett. **11**, 414 (1963); *J. J. Hopfield* and *J. Worlock*, Phys. Rev. **137**, A 1455 (1965).

[31] *M. Inoue* and *Y. Toyozawa*, J. Phys. Jpn. **20**, 363 (1965); *T. R. Bader* and *A. Gold*, Phys. Rev. **171**, 997 (1968); *M. M. Denisov* and *V. P. Makarov*, J. Phys. C **5**, 2651 (1972).

[32] *D. Fröhlich, B. Staginnus* and *S. Thurm*, phys. stat. sol. **40**, 287 (1970).

[33] *R. Loudon*, Proc. Phys. Soc. **80**, 952 (1962).

[34] *D. Fröhlich* and *M. Sondergeld*, J. Phys. E **10**, 761 (1977).

[35] *D. Fröhlich, R. Kenklies, Ch. Uihlein*, and *C. Schwab*, Phys. Rev. Lett. **43**, 1260 (1979).

[36] *Ch. Uihlein, D. Fröhlich* and *R. Kenklies*, Phys. Rev. B **23**, 2731 (1981).

[37] The high purity crystals were kindly supplied by *C. Schwab*, Laboratoire des Spectroscopie et d'Optique du Corps Solide, Université Louis Pasteur, 6700 Strasbourg, France.

[38] *M. D. Levenson* and *G. L. Eesley*, Appl. Phys. **19**, 1 (1979).

[39] *B. Hönerlage, A. Bivas*, and *Vu Duy Phach*, Phys. Rev. Lett. **41**, 49 (1978); *A. Bivas, Vu Duy Phach, B. Hönerlage, U. Rössler*, and *J. B. Grun*, Phys. Rev. B **20**, 3442 (1979); *T. Itoh* and *T. Suzuku*, J. Phys. Soc Jpn. **45**, 1939 (1978).

[40] *T. W. Hänsch*, Nonlinear Spectroscopy, Enrico Fermi Course 64, p. 17, ed. *N. Bloembergen* (North Holland, Amsterdam, New York, Oxford 1977).

[41] *A. I. Ferguson, J. E. M. Goldsmith, T. W. Hänsch*, and *E. W. Weber*, Laser Spectroscopy 4, p. 31, ed. *H. Walther* and *K. W. Rothe* (Springer Berlin, Heidelberg, New York 1979).

[42] *R. G. Brewer*, Nonlinear Spectroscopy, Enrico Fermi Course 64, p. 87, ed. *N. Bloemberggen* (North Holland, Amsterdam, New York, Oxford 1977).

[43] *W. Kaiser* and *A. Laubereau*, Nonlinear Spectroscopy, Enrico Fermi Course 64, p. 404, ed. *N. Bloembergen* (North Holland, Amsterdam, New York, Oxford 1977); *A. Laubereau* and *W. Kaiser*, Coherent Nonlinear Optics, p. 271, ed. *M. S. Feld* and *V. S. Letokhov* (Springer, Berlin, Heidelberg, New York 1980).

Magnetic Components for Microwaves and Optics

Hans J. Schmitt
Institut für Hochfrequenztechnik der RWTH Aachen, Aachen, Germany

Summary: Recent developments of new or improved devices underline the importance of more precise computer methods for device modelling and of materials architecture for better hardware realizations. This is illustrated by some applications which are based on particular advances of physics and technology: Management of saturation magnetization in YIG and substituted garnets for integrated selectivity, ferrites with high built-in anisotropy for mm-wave components and thin magnetic films with high Faraday-rotation for modulation in the optical part of the spectrum.

Introduction

The development of magnetic components for high frequency applications has reached a certain degree of maturity. Markets 1980 for components of > 1 GHz is around 30×10^6 \$ in USA alone, the world market about twice as large [1].

The major tasks of current research are:

– adaptation of components to modern semiconductor techniques
– increase of the frequency range
– realisation of novel system functions, especially for optical applications.

Progress is essentially coupled to two "arts": Refined CAD of components and associated circuitry and the development of materials technology.

1 Components CAD

1.1 Component Modelling

Conventional devices such as circulators and isolators constitute the major part of practical magnetic components. Recent activity in this field is mostly concerned with computer aided design techniques, aiming at automated design for a given set of operating conditions, frequencies, bandwidth, power, maximum loss etc. [2].

For most practical structures, computer managable descriptions of the passive circuitry are available. The main difficulty for magnetic device design rests in the description and modelling of the magnetic materials. A comparable modelling problem is encountered with the integration of active semiconductor elements. Semiconductors, however, can be considered as more or less concentrated elements and

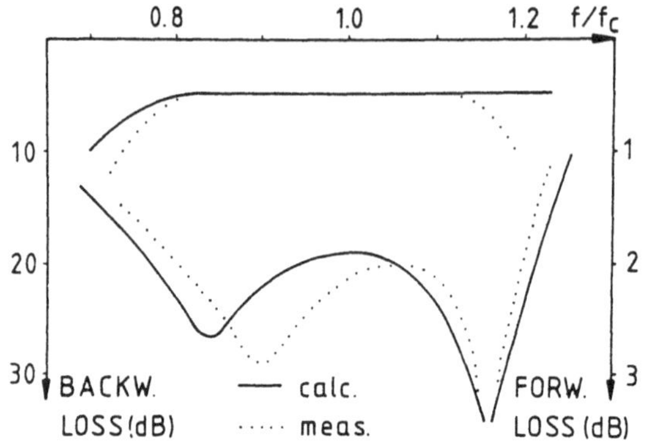

Fig. 1
Comparison of CAD-values and measured performance of a 3-port circulator [4].

normal statistical design procedures can be applied [3]. For magnetic materials, the problem is not so much one of stray in parameters, it is more related to basics: non-linearity, hysteresis, description of losses and often the plain complexity of mathematics for waves in gyrotropic media, along with scattering phenomena at interfaces to the reciprocal world.

The progress achieved is evidenced by the work on CAD of conventional junction circulators [4, 5]. Figure 1 shows representative values of computed and measured performance of a 3-port stripline circulator. The residual difference of up to a few % is mostly due to uncertain modelling of magnetic material, particularly with respect to losses. A similar variance still occurs with most other components such as isolators, phase shifters etc.

1.2 Magnetic Substrates

Complex mathematics, not easily adaptable to CAD is involved in the design of planar microwave circuits on magnetic substrates, e.g. for integrated phase shifters, couplers etc. The situation is mathematically involved because open guides like microstrip, slot- and coplanar lines already constitute a formidable problem even for isotropic dielectric substrates. The tremendous practical importance of this technology is witnessed by about 5000 publications listed in [6], for the period of 1948–1978, of which about 100 are concerned with magnetic substrates.

The most likely structure, the normally magnetized microstrip line has been solved e.g. by Borbush [7]. With a proper bias field the current distribution on the strip becomes asymmetrical (Fig. 2). This is used in a variety of so called edge guiding components e.g. for circulators and, by placing absorbing material at the edge, also for isolators.

Present research aims at a more complete description of fields and powerflow in other guides (slotlines, coupled lines) and on the development of managable mathematical expressions of their line parameters.

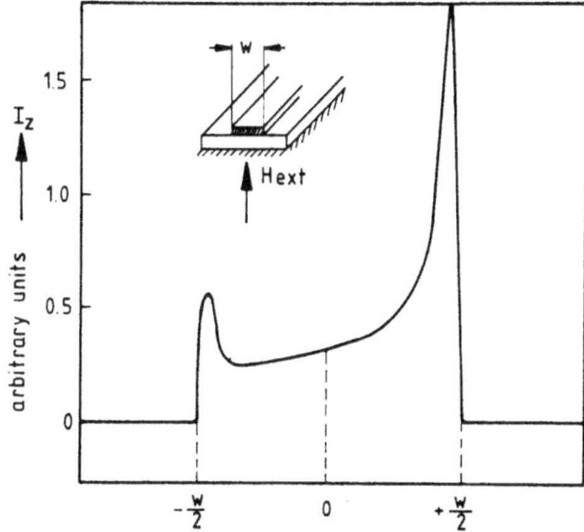

Fig. 2
Calculated distribution of axial current on microstrip for normal magnetization of ferrite substrate (Kapusta, unpublished).

2 YIG-Technology for Microwaves

A comprehensive book on "Magnetic Garnets" has recently been published by Winkler [8], including more than 2000 references to chemical, technological and applications aspects. Thus, only a few selected problem areas of particular relevance to microwaves will be illustrated in the following sections, the approach proceeding from the device point of view.

2.1 YIG-Filter

For the linewidth of pure YIG a certain practical limit [8] has been reached: The intrinsic linewidth is about 0,2 Oe (at 10 GHz), the smallest measured value is around 0,3 Oe, while values below 0,5 Oe can be maintained for practical applications as a matter of routine. Ferrimagnetic precessional resonance is used widely in single – and multi sphere fixed and tunable filters as well as in corresponding YIG-tuned transistor – and two terminal device (e.g. Gunn-) oscillators, respectively, for frequencies up to about 40 GHz [9]. Above this frequency magnetic fields required for resonance become excessive.

2.2 Substituted YIG

Until recently resonators for the lower microwave region (UHF and L-Band) posed a certain difficulty. For sphere resonators, the lowest useful frequency is

$$f_c \sim \gamma \frac{4\,\pi\,Ms}{3}$$

because of the need to saturate the magnetization, and to avoid additional losses in unordered magnetic systems [10]. Moreover, to prevent coincidence limiting and strong coupling to other magnetostatic modes one choses $f_c' = 3 f_c$ (e.g. ~ 5 GHz for pure YIG. Hence, for operation at low frequencies Ms has to be reduced by substitution of Fe by Ga, Al or similar nonmagnetic atoms. The problem arises, that in substituted YIG of the type $Y_3(Fe_{2-x}M_x)(Fe_{3-y}M_y)O_{12}$ both tetraeder- and octaeder sites are possible, with different magnetic properties. Because of size effects and electron-configuration Ga and Al prefer tetraedric positions; In, Sc etc. prefer octaedric sites [11]. In general there will be a certain mixture of both sites and hence a certain magnetization, which will depend on the conditions of crystal growing. The precise correlation with technology is not yet known in all details. The exact adjustment of particular values of 4πMs, however, is very important for practical filter devices in order to avoid spurious resonances and maintain required coupling

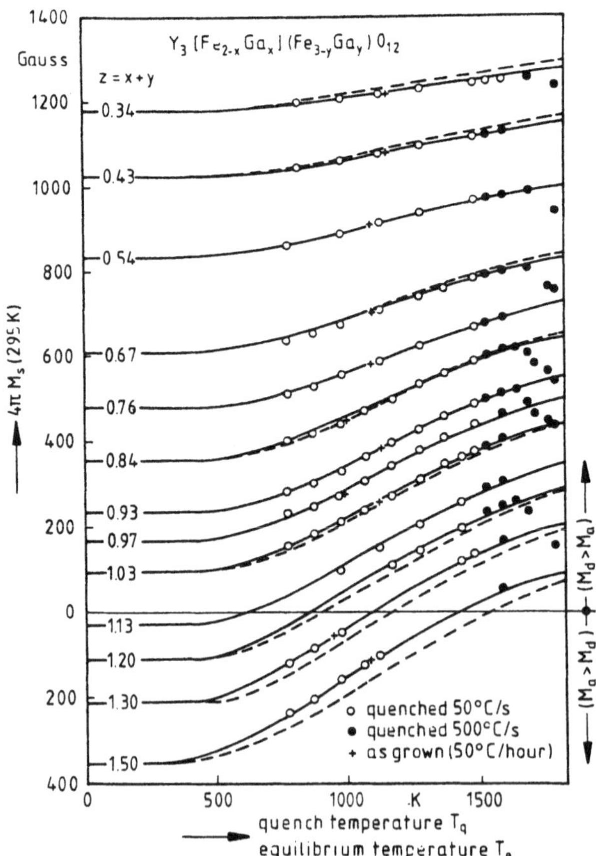

Fig. 3
Saturation magnetization of Ga-substituted YIG as a function of quenching temperature. Solid lines are calculated from molecular field theory.

values in multisphere filters for optimum in-band response. In several recent papers, Röschmann [11, 12] has shown, that the distribution to the different sites can be controlled by a subsequent thermal annealing process (~ 10 h, cooling rate > 50°/s). This allows a posteriori adjustment and, within wide limits, control of magnetization values. Figure 3 shows 4πMs for different Ga-substitutions as a function of annealing temperature. Similar values are obtained for Al. The kinetics of redistribution depends on the number of oxygen vacancies (Fig. 4): in N_2-atmosphere (large number of oxygen vacancies) accomodation times are 10^2-10^4 seconds at temperatures around 1000 °C [11].

The variation of 4πMs is large enough, e.g. $200 \leqslant 4\pi \mathrm{Ms} \leqslant 600$ Gauss to cover a wide range of applications in the frequency region around 1 GHz, i.e. for the practically important TACAN-filters starting from one single material composition. The elegant a posteriori control of 4πMs also offers the possibility to shift spurious resonances (the main resonance is more or less unaffected) out of the frequency range of interest. This is particularly important in order to achieve reasonably constant output – power in YIG-tuned oscillators. Also, the coupling of resonators can be modified via 4πMs change, which could simplify production of fixed-tuned filters with particular requirements of pass-band response. Such filters have been developed for 3 GHz in miniaturized form (< 0.5 cm^3) for applications as 'integrated selectivity' in electronically tuned TV-tuners [13]. These mini-filters can be readily incorporated into hybrid integrated planar circuits.

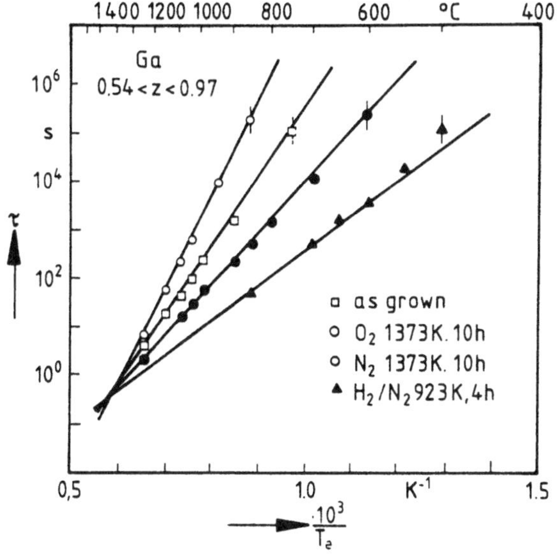

Fig. 4
Characteristic time for ionic redistribution as a function of temperature for different numbers of oxygen vacancies.

An alternative for applications at relatively low frequencies is offered by the $Ca_{2x}V_xBi_{3-2x}$ – garnets well known since more than 15 years [14]. With x ~ 1.3 the value of 4π Ms reaches several hundreds, while line-widths are just slightly larger than for YIG. In these materials the distribution of atoms is unique. The technology of production of such materials, however, has not yet reached the same routine as for YIG, mainly due to the higher temperature of the flux and the agressiveness of Bi-oxyde.

2.3 Planar Filters

Compared to the usual sphere-type filters, there are major advantages to be expected from planar magnetic resonators.

- The ferromagnetic resonance can be outside the spectrum of the other magnetostatic modes (Fig. 5).
- Planar resonators can be operated at almost arbitrarily low frequencies.
- The planar structure fits better – into the concepts of integrated rf-circuits technology.

LPE-technology now provides layers up to 30 μm with near perfect magnetic properties (YIG).

A disavantage, however, is the normally strong temperature dependence of the component. For the resonance of a disk, magnetized normal to the disk holds

$$f_R = \gamma [H_0 + H_A(T) - (1-3\,N_t)\,4\pi M_s(T)]$$

with N_t typically between 0.05–0.1. The temperature dependence of M_s enters nearly fully into the resonance condition. Recent investigations have shown, that within certain limits a compensation of the temperature effect may be possible via the temperature dependence of the anisotropy field H_A. For a given M_s, the influence of H_A can be fixed depending on the crystallographic cut, up to a maximum for magnetization in 111-direction.

The temperature dependence of f_R for a planar resonator made from a substituted garnet with x + y = 0.88 is shown in Fig. 6. With $H_{ext.}$ in 111-direction, compensation occurs near 35 °C. Again, thermal annealing provides a tool to shift this compensation region by a stimulated redistribution of substituted cations.

The incorporation of planar magnetic structures into integrated microwave circuits requires additional photolithographic steps [15], to provide the circuit connections from the normal dielectric substrate to the magnetic inserts, or special overlay technologies if magnetic disks are put on top of the substrate. Planar components could be advantageous, if large numbers of identical modules confined to a small volume or surface are required. Conceivably this could arise in connection with intelligent antennas etc. So far, however, there has not been a really convincing application need in the range of microwaves, in contradistriction to the case of optics, to be discussed later.

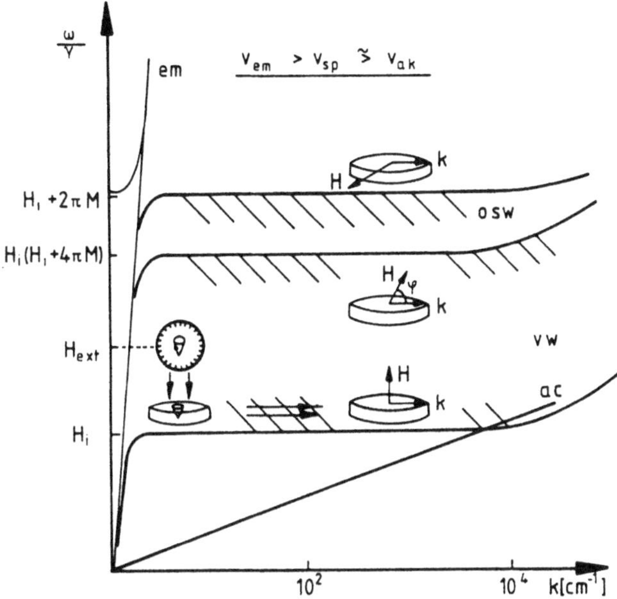

Fig. 5 Spin mode spectrum in disks: volume waves (vw) and surface-waves (sw) exist between electromagnetic waves (em) and acoustic waves (ac).

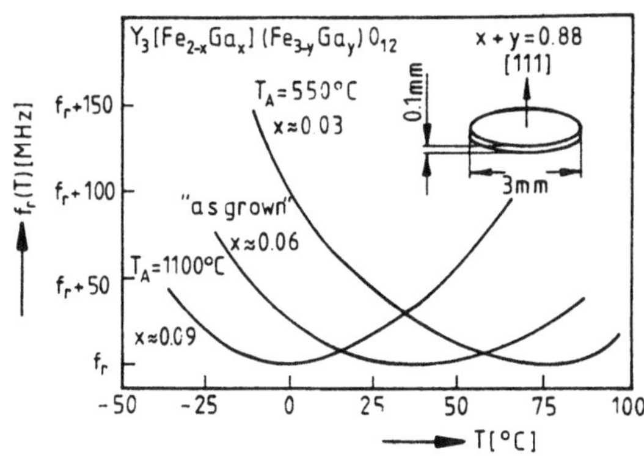

Fig. 6

Temperature compensation of the resonant frequency of Ga-substituted garnet disks. By means of controlled quenching temperatures T_A the compensation range can be shifted.

2.4 Delay Lines

Numerous investigations during the past years have been concerned with waves in YIG for delay and processing of high-frequency signals: Besides pure acoustic-waves (YIG at 3 GHz has a room temperature loss $\leqslant 0.1$ dB/μsec) also dispersive magneto-acoustic waves and magnetostatic volume – and surface waves have been investigated [16]. Several new results are given in [17].

In L-Band, operating at room temperature, YIG-delay lines offer delay times of several μsec with overall losses (including coupling) of a few dB. The important aspect is the large bandwith (time-bandwith product up to 10^5, e.g. for pulse compression [18], the possibility of parallel tapping (for correlators), and the wave dispersion, e.g. for real time frequency measurement of microwaves [19]. Present trends indicate, however, that this sort of signal processing will be replaced progressively by digital IC-techniques. Similarly, proposals for amplification of rf-signals by means of interaction between spin waves and drifting charge carriers have been superseded by progress in transistor technology.

2.5 Bubble-devices

Novel rf-phenomena have been observed in connection with magnetic bubbles, both in epitactic garnet films [20], and more recently also in hexagonal ferrites [21]. By means of microstrip – or microslot lines the rf-field (typically 1–3 GHz) may be concentrated in an area which is of the same order of magnitude as the bubble itself. The interaction of the rf-field with bubbles has been used for diagnostic purposes, Fig. 7 and for the generation of bubbles. Fig. 8 shows bubbles generated at a nucleation zone near the end of a slotline and 'thrown out' in a spiral fashion. A specu-

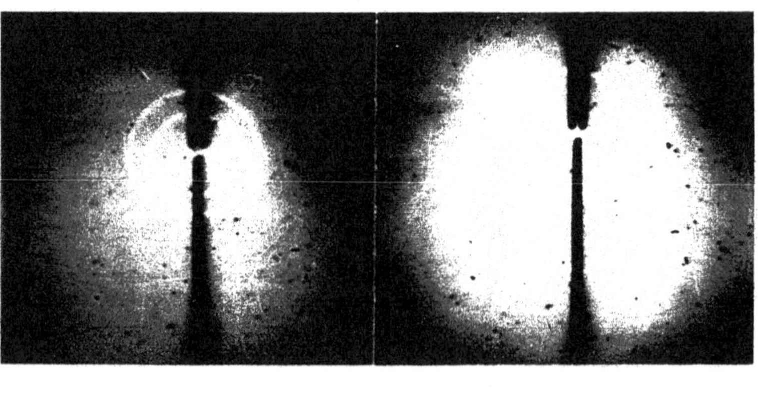

Fig. 7 Ring domains in garnet films excited by rf-power and bias field H_0

lation, that the reaction of bubbles on the rf-fields in turn could lead to novel microwave devices, in which properties may be controlled in a digital or analog form by shifting bubbles relative to the circuit did not materialize. The effects proved too weak.

At high magnetic fields, however, rf-excitation of spin precession in a low loss bubble layer can produce very wide opening of the precession cone. This in turn leads to a corresponding change of the optical Faraday rotation. Hence, thin epilayers may be used to visualize rf-field distribution in small geometries. Fig. 9 shows the intensity of transmitted light as a measure of rf-magnetic field distribution in the vicinity of a thinfilm-slotline [24].

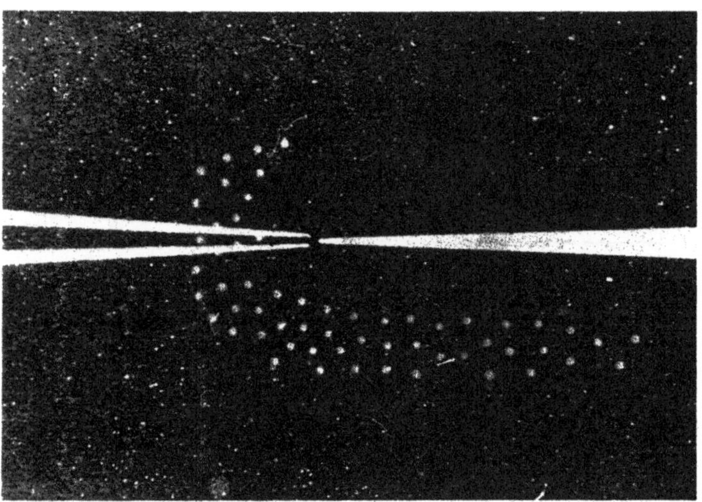

Fig. 8 Generation of magnetic bubbles by microwaves, fed to the tapered microstrip on garnet film.

Fig. 9 Visualization of rf-magnetic fields on slotline by use of Faraday rotation in garnet substrate. Slot width 20 μ.

3 Components for Millimeterwaves

For frequencies beyond 20–30 GHz ferrite components require bias fields of 10 000 Oe and more, which in turn require special and expensive magnets. On the other hand, the frequency range between 30 and 90 GHz gains in importance for short range communications, radiometry, spectroscopy and other measurement techniques, e.g. for semiconductors, where modern microstructuring possibilities lead to very high speed devices, to be characterized at correspondingly high frequencies.

Recently therefore, hexagonal ferrities have been taken up again [25, 26]. Because of their large anisotropy field, they require much smaller external fields and hence also promise simpler realisations of tunable components.

3.1 Hexagonal ferrites

$BaFe_{12}O_{19}$ has a linewidth of 16 Oe at 60 GHz, compared to an estimated value of the intrinsic linewidth of about 6 Oe (for comparison, YIG at 60 GHz has 2,2 Oe). This discrepancy indicates, that crystallographic or geometric imperfections resp. are present, which dissipate rf-energy via additional spinwave excitation.

Table 1 Hexagonal Ferrites

		$4\pi Ms$	H_A [Oe]	ΔH
YIG	$Y_3Fe_5O_{12}$	1800	66	0.35
M	$BaFe_{12}O_{19}$	4780	17 000	16
	$BaFe_{12-x}Al_xO_{19}$ (x = 1)	4000	20 000	20
	$Ba(Z_nT_i)_xFe_{12-2x}O_{19}$ x = 1	3500	11 000	40
	x = 2.5	1840	5 430	70

The anisotropy field of pure M-type-materials is around $H_A \sim 17\,000$ Oe; the magnetization $4\pi M_s \sim 4800$ Gauss.

By substitution of F_e by Z_n and T_i in $Ba(ZnTi)_xFe_{12-x}O_{19}$ the anisotropy H_A can be reduced, so that a smooth transition to the application range of YIG is obtained. Thus $H_A = 5400$ for x = 2.5. Substitution with Al increases the anisotropy field ($H_A = 22\,000$ for x = 2). Bulk crystals of Ba-Ferrites can be pulled at about 1200 °C from a high temperature melt by means of the 'top seeding' method, where BaO/B_2O_3 is used as a flux. Planar films can be produced by the LPE-method. As a substrate, various spinels can be used or alternatively again hexagonal crystals of different composition. The layer can be brought on by isothermal dipping in substituted solutions (0.5–3 μ/min). Sofar, the high frequency properties are not quite satisfactory and not yet comparable to bulk crystals especially concerning homogeneity. Loss values tend to be a factor of 3 higher than for bulk materials.

3.2 Filters and Oscillators

A tunable (60–90 GHz) single resonator bandpass-filter has been developed by Lemke et al. [27]. The single crystal Ba-ferrite sphere (H_A = 16 400 Oe) has a diameter of about 0.3 mm. The sphere is attached in an iris between two orthogonal waveguides (Fig. 10). It is placed into a small glas capillary for simple orientation of the easy axis in an external magnetic field. The polarisation of the rf-field is linear, the filter reciprocal. Bandwith is 250 MHz with an in-band transmission loss of 4–7 dB. Both waveguides are short-circuited immediately behind the sphere, in order to reduce the frequency dependence of coupling and to obtain a wide tuning range. The external magnetic field could be raised up to 15 000 Oe. Similar filters with 2 resonators have been reported more recently.

In somewhat lower frequency ranges 2- and 3-sphere filters have already been realized with substituted (Sc and Yb) ferrites. In the transition range to YIG (18–40 GHz) filters with Zn_2-Z-Ferrites have been investigated. Losses sofar are still somewhat higher than for pure YIG.

Tunable oscillators operating at 67–73 GHz have been realized by combination with a Gunn-device. The Gunn-element is placed into a hole in the dielectric of an image line made from high impedance Silicon, the Ba-ferrite sphere with $Q \sim 800$ (vs. $Q_0 \sim 10\,000$ for pure YIG) into another hole near the coupling end of the guide. Output power is about 10 mW, noise spectrum and smoothness of the outputlevel are distinctly better than for electronically tuned Impatt-oscillators.

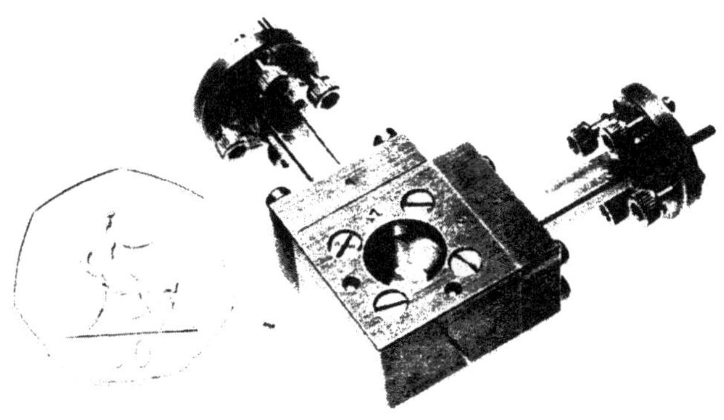

Fig. 10 Tunable single sphere ($BaFe_{12}O_{19}$, $\emptyset \sim 0.36$ mm) filter for 50–85 GHz; bandwidth is 250 MHz.

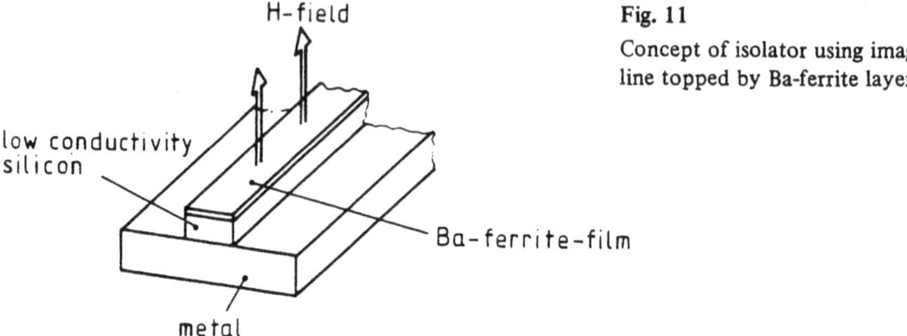

Fig. 11
Concept of isolator using image line topped by Ba-ferrite layer.

3.3 Planar Components

Ba-ferrites in planar form could be used for mm-wave isolators etc. Sofar no definitive results have been published. Difficulties with LPE-films arise in part due to the possible inhomogeneities of the materials (especially concerning Fe^{++} distribution) which cannot be fully corrected by annealing and may lead to increased losses. In general, the substrate of LPE-layers has sufficiently small losses in the millimeter-wave range, so that it may be used directly as a dielectric guide, for example in an image line (Fig. 11). With this technique integrated magnetic structures may be realized. In the very high GHz region these integrated structures may prove very important because of the small size of components and the very delicate mechanics involved.

3-port circulators using small disks of bulk material are under development in various Laboratories.

4 Magneto-Optic Applications

In contradistinction to typical microwave systems optical applications are very well adapted to parallel signal processing: A photography, for instance, consists of about 10^8 image points, which also may be considered as a corresponding number of parallel communication channels. Display, print or storage can be realized by an ensemble of many identical modulation-elements with lateral dimensions given by the wavelength of light or by the resolution capacity of the receiver respectively. Planar magneto-optic films are well suited for such optical control-systems, where structuring into many, small identical elements can be achieved by means of IC-related technologies. Basic physical principle is the differential Faraday rotation between opposite states of magnetization. One clear aim, therefore, was to enhance magneto-optic properties, i.e. Faraday rotation, by chemical changes. The most effective way proved to be a substitution of Bi into the crystal latice.

4.1 Thermomagnetic Switching

Several new developments are based on the thermomagnetic properties of thin garnet films. LPE-films ($Gd_3Fe_{5-y}Ga_yO_{12}$) on GGG-substrate show uniaxial anisotropy:

- The magnetization is always perpendicular to the film surface. Both magnetization directions are characterized by a different Faraday-rotation, so that the local direction of magnetization can be optically detected using polarized light.
- The homogeneous layer exhibits a statistical domain pattern, which can be easily influenced by ambient parameters, such as temperature, pressure a.s.o. For practical applications, therefore, the layer is structured into a pattern of small (1–100 μ width) single-domain cells. Within these islands, the orientation of magnetization is quite stable with respect to external influences (Fig. 12). A switch-over of the magnetization of individual islands can be achieved by use of the thermomagnetic effect near the compensation point. For this purpose, the cell is heated-up about 30–40 °C by means of a short (1–10 μsec) heatpulse, while simultaneously an external magnetic field is applied.

Fig. 12 Stable magnetization pattern in structured garnet film on GGG-substrate. Outer rim unstructured with statistical domains.

- At the compensation temperature the heatpulse reduces the an isotropy, this effects a momentary reduction of the stability. At the same time $4\pi M_s$ is increased, so that already relatively small fields of 100–200 Oe are sufficient to switch magnetization of the islands.

4.2 Optical Memories

For magneto-optic data storage heating of the cell is achieved by means of a focussed laser beam. Light is absorbed in the layer and converted into heat. The storage layer therefore, must have a certain absorption, but on the other hand for read-out the light losses must not be too large. Typical layer thicknesses are about 5 μ, cell sizes $10 \times 10\,\mu^2$ which can be switched with about 1 mW laser power in a matter of μsec.

Various systems have been developed, typically with 10^7–10^8 cells (bits) organized in groups on a slide-sized wafer. In order to adress individual data blocks, this wafer can be mechanically moved across the laser beam [28]. For higher speeds electronic laser beam deflection has also been suggested [29]. The data rate is between 100 kbit/sec and several Mbit/sec, depending on adressing principle and organisation of the memory. These system aim to substitute fixed-disk stores and 'floppy disks' and compete, in a way, with bubble memories.

Besides mono-crystalline garnets also amorphous layers of certain metal alloys (Gadolinium-Cobalt-Iron) show uniaxial anisotropy and magnetic compensation [30]. In these layers domains remain inherently stable, so that thin films need not be specially structured. Quite recently, these films have been improved to such a degree, that they may be considered for memory application. In ternary alloys Faraday rotation could be increased by Bi-substitution by a factor of 2 [31, 32] (Fig. 13), so that in layers of some 100 A, thickness with sufficient transparency a direct optical recording and read out has become possible. These materials are significantly cheapter and simpler than monocrystalline garnet wafers.

4.3 Magneto-Optic Display

By substitution of Bi into Gallium-Gadolinium-YIG the Faraday rotation can be dramatically increased (Fig. 14). Maximum values are about $10^5\,°$/cm for green light. Hence, with a layer thickness of 5 μ one reaches $\Theta_F \sim 50°$ corresponding to an efficiency of 10–15% for transmission between polarizers, i.e. about comparable to a conventional colour-slide.

An electronic display component has been suggested by Hill and coworkers [33]. In order to switch individual garnet-islands the heatpulse is generated electronically by means of a thin film resistance structure. Distance between individual elements of the light modulation matrix is about 80 μ. The small thickness of the layer allows heat-up in less than 5 μsec. By proper adressing procedures a whole row or column can be switched in this time period, thus in principle allowing to display images at

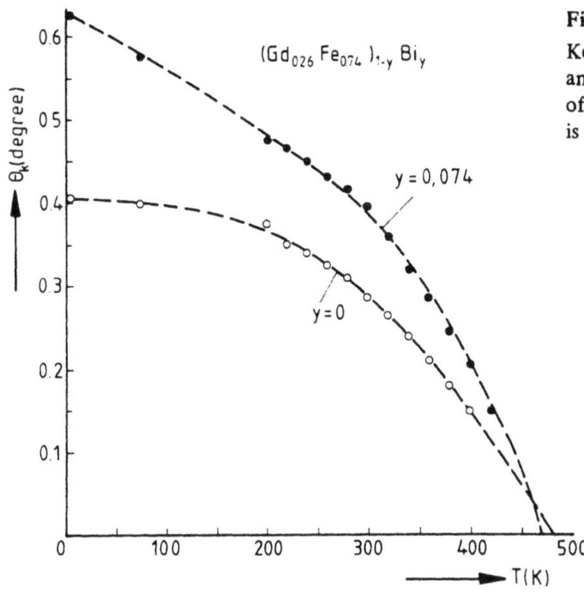

Fig. 13

Kerr-rotation in Bi-substituted amorphous GdFe as a function of temperature. Faraday rotation is also increased by Bi.

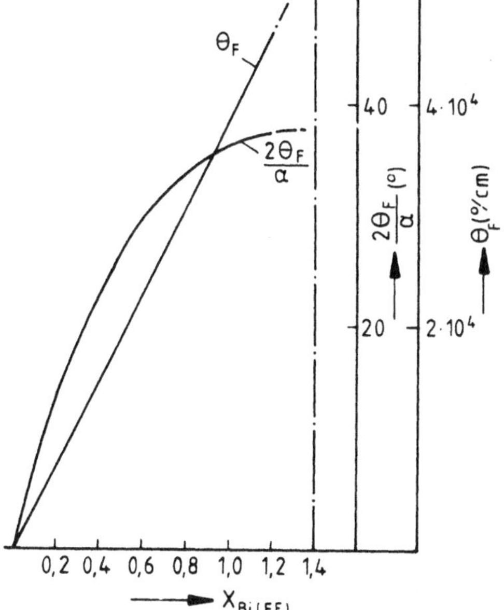

Fig. 14

Faraday-rotation and Figure of merit in Bi substituted GdFe-garnet layers, λ = 560 nm.

TV-speed. Incoherent light sources can be used for illumination (Fig. 15). With integrating receptors (e.g. for the eye about 20 msec) grey tone pictures can be displayed by choice of the duration of the different magnetization states by electronic coding. A display of colour pictures requires the superposition of 3 separate systems. A simpler colour coding may also be achieved by exploiting the dispersion of the Faraday rotation using only two subsequent switched modulation matrices.

A certain problem still arises in the blue region of the spectrum due to excessive absorption. Here iron borates maybe a better choice for the modulation wafer [34].

In these systems, the information remains magnetically stored, hence, a flickerfree display can be achieved even without external image repeaters or auxiliary stores. The size of the display is governed by limits of projection. With present technology crystals of 3″ can be pulled, so that also small direct view pocket displays could be conceived.

To adress individual image points or rows, resp., a magnetic field is switched on locally by "integrated" coils formed by thinfilm conductor strips. Adressing of the heatpulse proceeds via a separate x-y-conductor system. The "and" requirement of magnetic field and heatpulse practically excludes crosstalk, thus no complicated electronic compensation is necessary. In fact, only a small corner of a cell need be

Fig. 15
Iron-garnet light modulation matrix with 64–64 elements and electronically switched pattern.

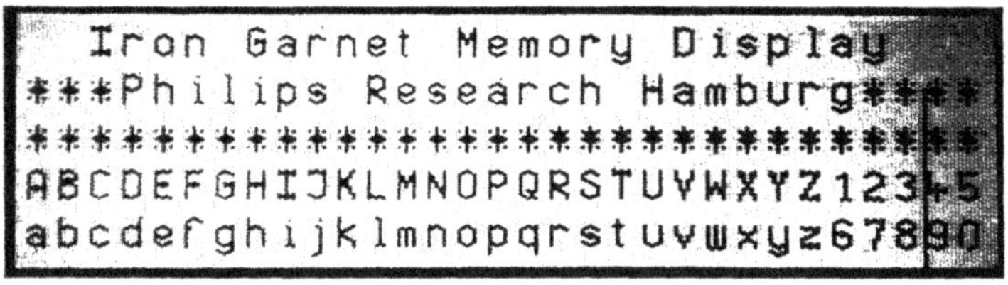

Iron Garnet Memory Display
Philips Research Hamburg*
****
ABCDEFGHIJKLMNOPQRSTUVWXYZ12345
abcdefghijklmnopqrstuvwxyz67890

Fig. 16 α-numeric output displayed by magneto-optic component

heated to serve as a nucleation zone for domain switching. Thermal power of a single switching operation is about 1 μWsec. Switching of images at TV-speed thus requires control power of about 3–4 Watt.

First applications are likely in the computer field, e.g. for display of a small number of lines α-numeric information, or with electronic typewriters (Fig. 16).

4.4 Magneto-Optic Printer

The application of similar components for hardcopy devices and intelligent copiers is already quite advanced. For this purpose row shaped modules of 256 cells each, spaced 80 μ apart, have been developed, combined with the control-ic and magnetic field coil on a small substrate. By combination of several modules also longer rows can be realized, e.g. up to A 4 with facsimile standard etc.

For printing, the recording material is moved mechanically transverse to the light-modulator row, for single copies e.g. photographic film, for office printers e.g. a photoconductor drum with electrostatic transposition to normal paper (Fig. 17). The recording of a picture taken with a TV-camera is shown in Fig. 18. Greytone is realized by time coding. Printing speed is essentially governed by the illumination lamp and the sensitivity of the recording material. Magnetically the speed limit is about 50 000 lines/sec.

Advantages of the optical modulation rows versus CRT's are the potentially higher resolution (2000 points and more) and the exact positioning of picture points, which allows an optimal coding of the image information.

4.5 Components for Optical Communication

Single magneto-optic cells have also been considered for external digital modulation of light-beams in optical communication systems. It turns out, however, that thermo-magnetic switching is too slow for most practical applications. Higher speeds up to 1 Mbit/sec can be obtained by reversible displacement of domains by means of magnetic fields generated in an evaporated conductor loop. Modulation depth can be adjusted by means of the polarizers [35].

LISA - Electrophotographic Printing

This page has been printed by electrophotography using a solid state linear light switching array (LISA) as ptical print head (printing light source).

LISA, developed by the Philips GmbH Forschungslaboratorium Hamburg (PFH) in cooperation with the Valvo Roehren- und Halbleiterwerke (RHW), produces a line of 1536 light dots arranged at a pitch of 125 um. This matches the standard Facsimile resolution of 8 dots per mm. LISA is electronically controlled by integrated circuits at a low voltage level. In order to print this text, 2400 lines of light dots were exposed within 1.7 s which corresponds to a recording speed of 1430 dot lines per s.

Fig. 17 Sample of magneto-optic printing. Number of lines refers to full page A4.

Fig. 18 Greytone reproduction with magneto-optic modulation rows, 256 elements, recording on Polaroid.

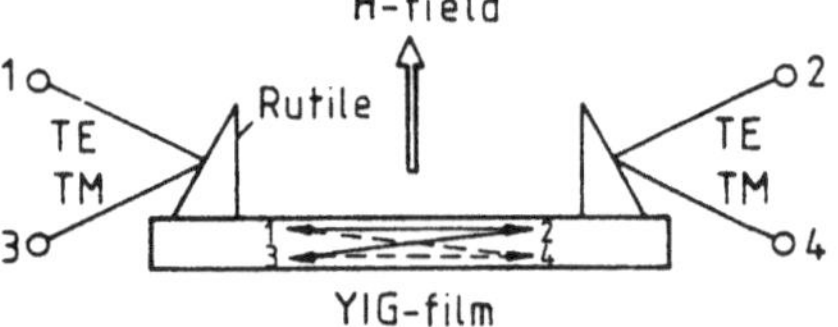

Fig. 19 Circulator for integrated optics based on unidirectional TE-TM-mode conversion [37] ——— TE path, ----- TM path.

The application of epitactic garnet layers as planar waveguides for integrated optics has been investigated some years ago. They appear suitable in the IR with losses of about 8 dB/cm at 1.15 μ and 3 dB/cm at 1.52 μ for Sc doped YIG [36].

Light is coupled into these waveguides as TE or TM-waves by small rutile prisms. Both waves are coupled in the gyrotropic material, so that mode conversion is observed to occur periodically with the length of the interaction region (Fig. 19). By selective coupling of modes non-reciprocal components can be realized, e.g. a 4-port circulator as suggested by Yamamoto [37].

The future importance of such components will depend on the further development of integrated optical systems.

Conclusion

Magnetic components for microwaves have reached a high degree of maturity. Computer aided design procedures and refined materials know how, especially in the area of single crystalline garnets have improved component performance e.g. for circulators, filters and oscillators.

With growing interest in the millimeter wave range hexagonal ferrites with high anisotropy gain in importance. These are becoming available with high quality both as bulk crystals and as epitactic layers, and first components have been developed.

Present interest centers also on optical applications of structured epitactical garnet layers for memories, printers and displays. By improvement of magneto-optic properties promising efficiencies have been obtained. The relatively simple technology, modular combinations in conjunction with electronic control and unlimited stability are the major advantages for future applications.

References

[1] US-Markets Forecast 1980, Electronics 134 (1980)

[2] *I. A. Cermak*, IEEE Trans. MTT-**22**, 155 (1974)

[3] *R. H. Jansen*, AEÜ **33**, 226 (1979)

[4] *G. Filipsson* and *B. Hansson*, Proc., 7. EuMc Copenhagen, 637–641 (1977)

[5] *Z. Uzdy*, IEEE Trans, MTT-**28**, 1134–1136 (1980)

[6] *I. Wolff*, 'Microstrip Bibliography', 1948–1978, Verlag H. Wolff, Aachen (1979)

[7] *J. Borbush*, AEÜ 73 (1977)

[8] *G. Winkler*, 'Magnetic Garnets', Vieweg Verlag (1981)

[9] *P. Röschmann*, NTG-Diskussionstagung Darmstadt, 29.–30. September (1977)

[10] *Röschmann, P.* Philips Techn. Rev. **32**, 322 (1971)

[11] *P. Röschmann*, J. PHys. Chem. Solids **41**, 1 (1980)

[12] *P. Röschmann, V. Tolksdorf* and *F. Welz*, IEEE Trans. MAG-**14**, 704 (1979)

[13] *P. Röschmann*, IEEE Trans. MTT-**21**, 52 (1973)

[14] *N. Ogasawara*, IEEE Trans. MAG-**9**, 1253 (1973)

[15] *R. D. Pringle, J. M. Owens, I. T. Simpson* and *I. F. Morton*, Proc., 4. EuMc Montreux, 590 (1974). Microwave Exhibitions and Publishers Ltd., Sevenoaks, U.K.

[16] *Z. M. Bardai, J. D. Adam, J. M. Collins* and *J. P. Parch*, AIP. Conf. Proc. **34**, 268 (1976)

[17] *J. A. Owens*, Microwave Systems News **9**, 44 (1979)

[18] *J. H. Collins* and *G. G. Neilson*, Electronics Letters **1**, 234 (1965)

[19] *H. J. Schmitt* and *J. K. Vogel*, German Patent App. 1766820 (1974)

[20] *M. Dötsch, H. J. Schmitt* and *J. Müller*, Appl. Phys. Letters **23**, 639 (1973)

[21] *F. Haberey, R. Leckebusch, M. Rosenberg* and *K. Stahl*, IEEE Trans. MAG-**16**, 681 (1980)

[22] *H. Dötsch* and *H. J. Schmitt*, Appl. Phys. Letters **24**, 442 (1974)

[23] *H. Dötsch*, J. Mag. Mat. **4**, 180 (1977)

[24] *H. Dötsch*, Mikrowellen Magazin **2**, 116 (1978)

[25] *G. Winkler* and *H. Dötsch*, Proc., 10. EuMc Brighton, 13 (1979) Microwave Exhibitions and Publishers Ltd. Sevenoaks, U.K.

[26] *F. Haberey, R. Leckebusch, M. Rosenberg* and *K. Sahl*, NTG-Fachbereichte **76**, 'Magnetische Werkstoffe und Bauelemente in der Nachrichten- und Datentechnik', VDE-Verlag (1980), 251

[27] *M. Lemke, W. Hoppe, W. Tolksdorf* and *F. Welz*, Microwaves, Optics and Acoustics **3**, 253 (1979)

[28] *I. Sander*, Proc. LASER 79, 4. Inter. Kongr. Opto-Elektronik, Munich (1979)

[29] *B. Hill, I. Sander* and *G. Much*, Optica Acta **24**, 495 (1977)

[30] *P. Chaudhari, J. J. Cuomo* and *R. J. Gambino*, IBA J. of Res. and Dev. **17**, 66 (1977)

[31] *P. Hansen* and *M. Urner-Wille*, J. Appl. Phis. **50**, 7471 (1979)

[32] *M. Urner-Wille, P. Hansen* and *K. Witter*, IEEE-Trans. MAG-**16**, 1188 (1980)

[33] *B. Hill* and *K. P. Schmidt*, Philips J. Research **33**, 211 (1978)

[34] *J. G. White, A. Miller* and *R. E. Nielsen*, Acta Cryst. **19**, 1060 (1965)

[35] *U. Killat* and *G. Rabe*, Proc., 4. Europ. Conf. Opt. Communication Genua, Sept. (1978)

[36] *P. K. Tien, D. P. Schinke* and *S. L. Bank*, J. Appl. Phys. **45**, 3059 (1974)

[37] *S. Yamamoto* and *T. Mukinoto*, J. Appl. Phys. **45**, 882–888 (1974)

Festkörperprobleme XXI (1981)

Applications of the Josephson-Technology

Wilhelm Jutzi
Institut für Elektrotechnische Grundlagen der Informatik, Universität Karlsruhe, Germany

Summary: Integrated digital circuits with Josephson junctions are described in detail, since they seem to have the largest economical impact. Only a brief survey is given on magnetometers, millimeter wave amplifiers, oscillators and mixers with Josephson junctions. The advantages of Josephson compared with semiconductor technology for high speed, low power, high packing density digital circuits are emphasized.

After a phenomenological description of basic electronic properties of single Josephson junctions digital circuits with interferometer devices are explained. As an example a memory cell for nondestructive read out is chosen where the binary information is represented by the absence or presence of a flux quantum. Different types of cell layouts are discussed. Limits of miniaturisation of interferometer type circuits with Josephson junctions are extrapolated and are compared with those of semiconductor circuits.

1 Introduction

Unusually large research and development programs are founded in the United States and Japan to investigate the application potential of the Josephson effect, predicted in 1962 [1].

The Josephson technology seems to outperform the semiconductor technology in the domain of integrated digital circuits for computers with ultra high three dimensional packing density and overall processing speed. In the domain of analog circuits Josephson junctions are already used for voltage standards or magnetometers.

In this paper I will try to give a survey of the applications of the Josephson technology with main emphasis on digital applications. In section 2 essential properties of single Josephson junctions are described. In section 3 interferometers with two Josephson junctions are discussed as the basic device for many circuit applications. In section 4 lead and niobium junction fabrication is briefly reviewed. Section 5 is devoted to magnetometers as a salient analog application. Section 6 concentrates on single flux quantum memory cells for destructive and nondestructive read out. Logic gates are treated in section 7. Finally in section 8 the miniaturisation properties of digital circuits with field effect transistors and Josephson junctions are compared.

2 Single Josephson junctions

A Josephson junction with a native tunnel oxide [2] is presently the most promising candidate for most applications. In Fig. 1 such a Josephson junction as an electronic switch is sketched. Two superconducting metallic layers M2 and M3 are isolated by a very thin oxide with a thickness between about $2 < t_{ox}/nm < 6$. The junction has a length l in the direction of the gate current I_G and a width w perpendicular to it. Isolated from the junction a superconducting ground plate M1 is arranged to reduce electromagnetic coupling between adjacent junctions and lines. On top of the junction an isolated and superconducting control line is used to generate a magnetic control field H_C inside the tunnel oxide.

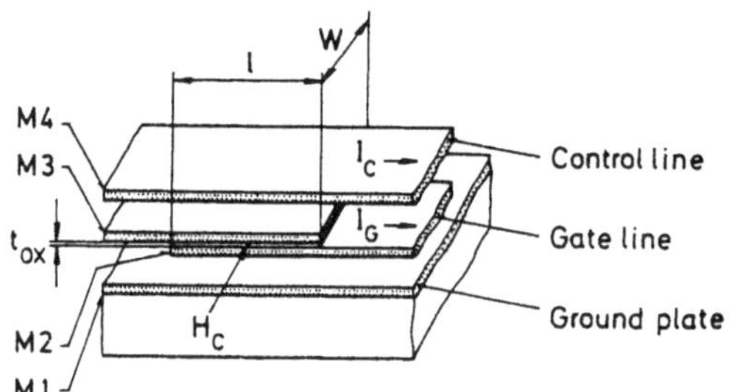

Fig. 1
Sketch of a Josephson junction as an electronic switch. There are four superconducting layers M 1 – M 4.

Josephson junctions with lead alloy or niobium electrodes are normally operated at roughly one half the transition temperature T_C near 4.2 K in liquid helium at atmospheric pressure to cut down cooling costs and temperature sensitivity of the maximum Josephson current. For lead alloy electrodes at 4.2 K the gate current I_G through the tunnel oxide versus the voltage U_j between the junction electrodes is sketched in Fig. 2.

A gate current below the maximum Josephson current I_0 can tunnel through the junction oxide without generating a dc voltage drop. This is called the zero voltage state of the junction. On trespassing the maximum Josephson current I_0 a voltage in the order of the energy gap voltage, $U_g = 2\,\Delta/e$, rises within a few picoseconds. For lead alloy electrodes the gap voltage is $U_g \simeq 2.5$ mV. The tunnel resistance R_{NN} defined with normally conducting electrodes is measured at large currents.

If the gate current is reduced sufficiently below the maximum Josephson current the junction switches back into the zero voltage state at a minimum Josephson current I_{min}. At a given current of a current source $I_0 > I_G > I_{min}$ there a two stable states of the Josephson junction: the zero voltage state and the voltage state. The hysteresis area can be reduced with a shunt resistance between the junction electrodes.

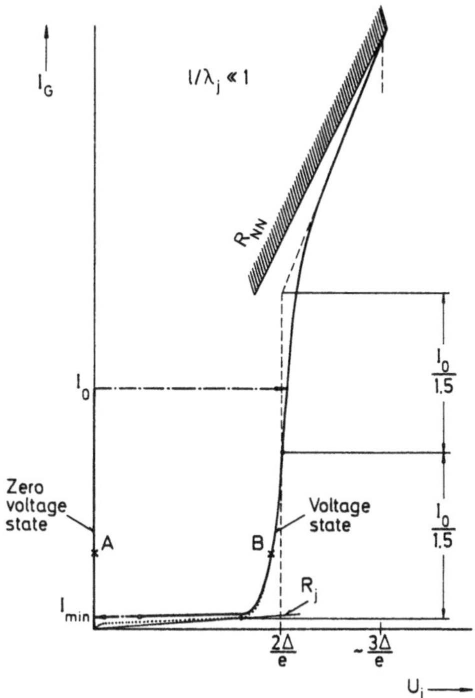

Fig. 2

Junction current I_G versus junction voltage U_j. The maximum Josephson current in zero magnetic field is I_0.

The maximum Josephson current increases exponentially with decreasing oxide thickness t_{ox}. The lower limit of the Josephson current is given by thermal fluctuations [2]:

$$I_R = 2\pi \frac{kT}{\Phi_0}.$$

k is the Boltzmann constant, T is the absolute temperature and $\Phi_0 = 2.07$ mV ps is one flux quantum. For on effective noise temperature T = 4.2 K the lower limit is $I_R \simeq 0.2\ \mu A$. Therefore a Josephson current can only be observed if the oxide thickness is sufficiently small and the junction area sufficiently large.

The maximum Josephson current can be changed by the magnetic control field H_C between the junction electrodes, as sketched in Fig. 1. The corresponding control flux is

$$\Phi_C = \mu_0 H_C l d$$

where the effective separation is

$$d = t_{ox} + \lambda_2 + \lambda_3.$$

λ_2 and λ_3 are the London penetration depths in the lower and upper junction electrodes, respectively. The maximum Josephson current of a junction with a sufficiently small lengh l and with a constant oxide thickness is zero for a control flux Φ_C equal to one flux quantum Φ_0. The maximum Josephson current versus Φ_C is described by $\sin x/x$ where $x = \pi \Phi_C/\Phi_0$, as shown in Fig. 3, if the magnetic self field of the currents through the junction can still be neglected. The self field must be taken into account if the junction length is larger than the Josephson penetration depth [2]:

$$\lambda_j = \sqrt{\frac{\Phi_0}{2\pi\mu_0 d j_{max}}} .$$

The Josephson penetration depth is $\lambda_j = 13\ \mu m$ for a maximum Josephson current density $j_{max} = 1\ KA/cm^2$ and a effective separation $d = 0.15\ \mu m$.

For a ratio $l/\lambda_j = 3.4$ the currents in a long Josephson junction and the normalized maximum Josephson current versus the normalized control flux or normalized control current are sketched in Fig. 4a and b. The maximum occurs at a negative control current where the self field and the external field partially compensate.

The main lobe with no flux quantum between the junction electrodes is called the $N=0$ mode. The side lobe is called $N=1$ mode if one flux quantum with a magnetic field opposing the control field is stored in the junction. The $N=0$ and $N=1$ modes overlap in a small area in the current plane. For a certain bias control current either the $N = 0$ or $N = 1$ mode can exist for $l/\lambda_j = 3.4$. The overlapping zone increases with increasing l/λ_j. Short junctions with $l/\lambda_j = 1$ have symmetric gate characteristics corresponding to the dashed line in Fig. 4b and vanishing small overlapping zones.

The dynamic behavior of very short junctions i.e. of point junctions being practically independent of a magnetic field is often represented by a mechanical pendulum model as sketched in Fig. 5. A point mass is connected via a bar with a wheel. A

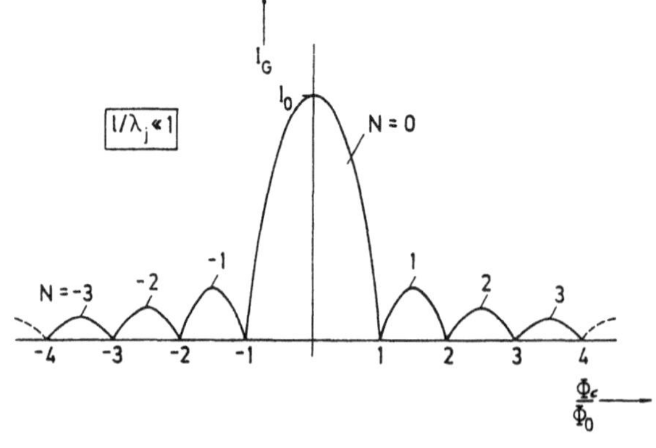

Fig. 3
Maximum Josephson current versus flux between the junction electrodes for negligible self field ($l/\lambda_j \ll 1$). The number of flux quanta in the junction is N.

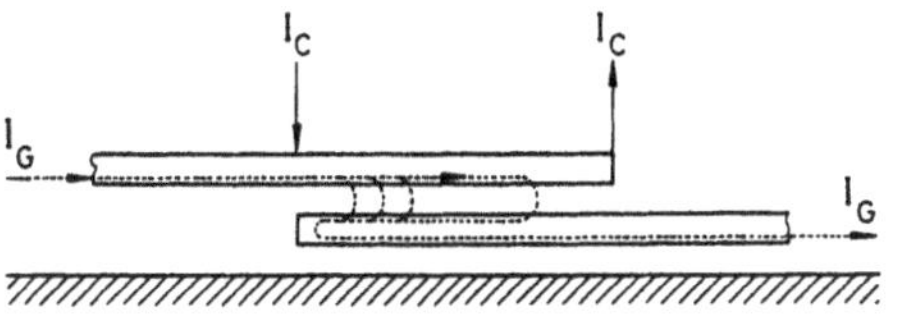

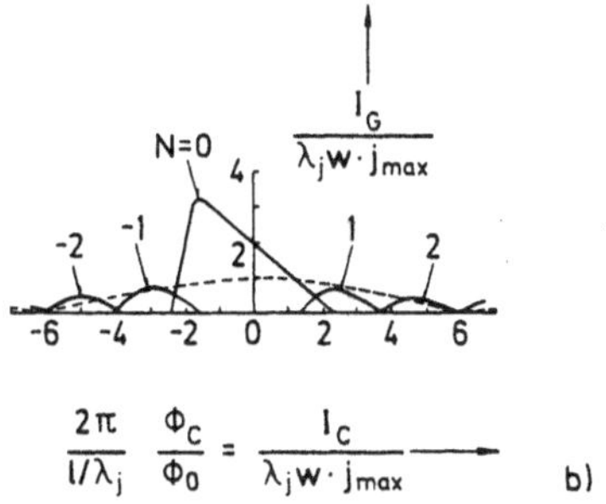

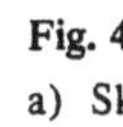

Fig. 4

a) Sketch of currents in a long Josephson junction.

b) Normalized maximum Josephson current versus normalized control current of a long Josephson junction with $l/\lambda_j = 3.4$ as a solid line. The dashed line corresponds to a short junction with $l/\lambda_j = 1$.

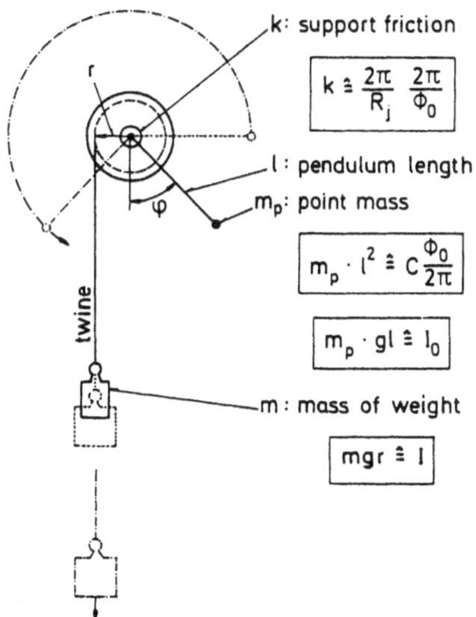

Fig. 5

Mechanical pendulum model of a point Josephson junction.

weight at a twine may rotate the wheel by a certain angle $\phi < \pi/2$ until an equilibrium state is reached. The stable states with $d\phi/dt = 0$ correspond to the zero voltage states of the junction. If the weight is large enough, so that $\phi > \pi/2$ the pendulum will rotate. The state $d\phi/dt \neq 0$ corresponds to the voltage state of the junction. The support friction damps the pendulum; it is a nonlinear function of the angular velocity $d\phi/dt$.

The electrical equivalent circuit of a point Josephson junction having the same differential equation as the mechanical pendulum model is shown in Fig. 6. It is an RC-circuit in parallel with a Josephson current source

$$I_j = I_0 \sin \phi$$

where ϕ is the phase of the order parameter. The phase is proportional to the time integral over the voltage U_j across the junction. A dc voltage U_j corresponds to a constant angular frequency $d\phi/dt = \omega = 2\pi f_j$ or a Josephson frequency

$$f_j = 483.6 \text{ GHz} \frac{U_j}{\text{mV}}.$$

The capacitance per area $C_F = C/wl$ is independent of voltage and in the order of $5\ \mu F/cm^2$ at $j_{max} = 1\ KA/cm^2$ for lead alloy junctions. The shunt resistance depends on voltage. It is often approximated by the resistance R of the I_G, U_j characteristic measured in a sufficiently large magnetic field to suppress any zero voltage current through the junction.

The damping of a point junction is described with the McCumber damping parameter [3]:

$$\beta = 2\pi \frac{R_j^2 C}{\Phi_0} I_0$$

where R_j is the linear approximation of the subgap shunt resistance (at $U_j = 2$ mV for lead alloy junctions). Large β – values indicate small junction damping corresponding to small values I_{min}/I_0 or large hysteresis area. The junction hysteresis practically vanishes for $\beta < 1$ [3].

The above described RCSJ-model (RC Shunted Junction) is an astonishing useful tool to visualized and to design analog and digital circuits with current densities $j_{max} < 10\ KA/cm^2$. Only at very high current densities a modification of the simple equivalent circuit in Fig. 6 might be needed [4, 5].

Most digital circuits presently operate with Josephson junctions of large hysteresis. The switching time from the zero voltage state to the voltage state, namely to the

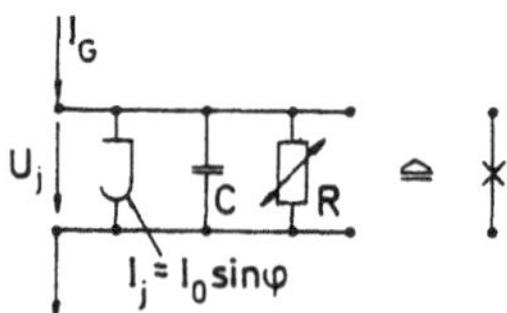

$$\varphi = \frac{2\pi}{\Phi_0} \int_0^t U_j \, dt$$

Fig. 6

Electrical equivalent circuit of a point Josephson junction.

gap voltage is indicative of the speed potential. A simple first order approximation of the 10 %–90 % rise time is found in neglecting the Josephson sin ϕ-term. The rise time is smaller than the R_jC-time constant as the voltage is limited at $2\,\Delta/e$ by the sharp increase of the quasiparticle current. For $V_m = R_jI_0 \gg 2\,\Delta/e$ the estimate is useful:

$$t_r \simeq 0.8\,\frac{2\,\Delta}{e}\,\frac{C_F}{j_{max}}.$$

The rise time decreases with increasing Josephson current density. With $2\,\Delta/e$ = 2.5 mV, $C_F = 5\ \mu F/cm^2$ and $j_{max} = 5\ KA/cm^2$ a rise time $t_r = 2$ ps is calculated. Rise times in the order of 30 ps have been measured at the resolution limit of semiconductor sampling oscilloscopes at 300 K [6]. A rise time oscillogram is shown in Fig. 7. Recently a resolution of 6 ps has been achieved using a sampling head with Josephson junctions integrated on the same chip as the device to be tested [7, 8].

A detailed investigation of the voltage rise reveals that after a current step exceeding the maximum Josephson current the voltage first rises very slowly and stays below 10 % of $2\,\Delta/e$ during the so called turn-on delay and then rises very fast between 10 % and 90 % according to the above estimate. The turn-on delay depending on current overshoot is calculated and can be estimate with the RCSJ-model, i.e. in taking into account the sin ϕ-term [9, 10].
The sin ϕ-term represents a strong nonlinearity between current and voltage. It is exploited for three main analog applications.

I. Oscillators.

A single voltage bias yields constant oscillations in the microwave or milimeter wave range. Josephson oscillators have been operated with energies only about 10 times the thermodynamical limit of $4 \cdot 10^{-23}$ J [11, 12]. The minimum

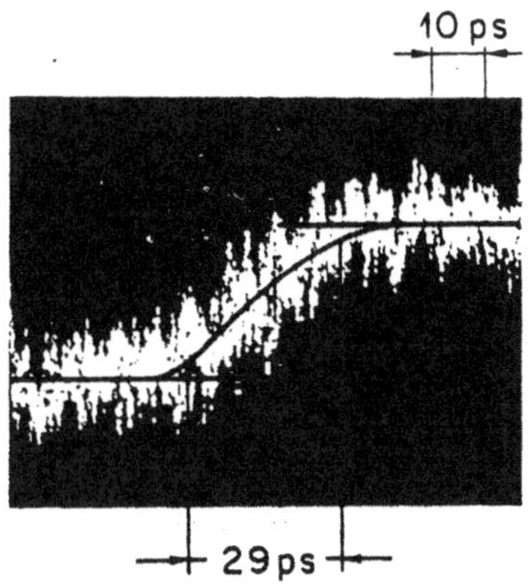

Fig. 7
Measured switching transient from zero to about 2.5 mV of a lead alloy Josephson junction with a Josephson current density $j_{max} \cong 5\ KA/cm^2$.

energy for semiconductor ring oscillators at 300 K is much larger and in the order of 10^{-15} J [13].

II. Voltage standard.

A dc and an ac voltage bias of a Josephson junction modulates the phase ϕ. The $\sin \phi$-term of the Josephson current source can be represented by a Bessel series, a phase modulation spectrum. A dc component of this spectrum appears as a current step in the IU-characteristic every time on integer number of the applied frequency is equal to the Josephson frequency corresponding to the applied dc voltage. In this way given frequencies can be precisely transformed into voltages. Since ultra high precision frequencies are available voltage standards with Josephson junctions are used [14, 15].

III. Mixers, amplifiers.

Two or more microwave or millimeter wave frequencies applied at strong nonlinearities of the junction allows for up and down mixing, detection and parametric amplification with low noise [16–20].

The experimental work in this field of analog applications mainly of the national bureaus of standards and of universities in many countries is very promising. However, the required number of voltage standards and even microwave and millimeter wave devices with Josephson junctions is limited compared with the expected huge number of circuits for digital applications in a Josephson computer.

3 Interferometers with two Josephson junctions

The sensitivity of the maximum Josephson current to an external magnetic field is improved in increasing the coupling area l · d of a single junction by a superconducting loop in parallel. A loop with one junction is called "ac squid" and with two junctions in parallel "dc squid", as sketched in Fig. 8.

The abbreviation "squid" stands for superconducting quantum interference device.

Both, dc and ac squids have been used for analog and digital applications [21–24]. Only dc squids are treated here as they seem to outperform ac squids in many respects.

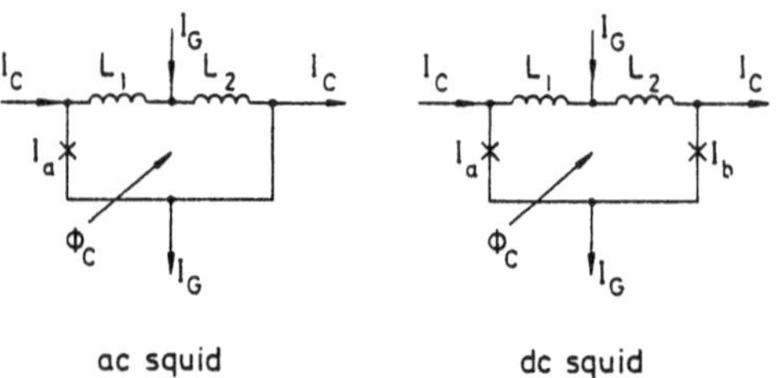

Fig. 8

Electrical equivalent circuit of an ac and dc superconducting quantum interference device (squid). Either a control current I_C or a control flux Φ_C is applied.

Dc squids are often called interferometers with two Josephson junctions. A structure with two unequally large Josephson junctions is sketched in Fig. 9. The junctions are connected via an inductive "bridge" on top of a superconducting ground plate. The superconducting control line is coupled magnetically with the interferometer inductance.

The magnetostatical properties of an interferometer with two point junctions are determined essentially by three design parameters:

1. The characteristic phase

   $$\lambda = 2\pi \frac{LI_b}{\Phi_0}$$

   where the $L = L_1 + L_2$ is the total interferometer inductance and I_b the maximum Josephson current of the smaller junction in zero magnetic field.
2. The ratio of the maximum Josephson currents of both junctions in zero magnetic field $a/b = aI_0/bI_0 = I_a/I_b$.
3. The tap ratio $p = L_1/L$ indicating the insertion point of the interferometer current I_G.

If these three parameters are kept constant the normalized maximum Josephson currents $i_G = I_G/I_0$ versus the normalized control current $i_C = I_C/I_0$ are identical for the same number N of flux quanta in the loop. An example is given in Fig. 10 for the parameters $\lambda = \pi$, $a/b = 1$ and $p = 0.5$. If no flux quantum exists in the loop, the normalized maximum Josephson current is defined by the boundary of the area marked by squares, around the origin of the current plane. This flux quantum state is called the N = 0 mode. If one flux quantum associated with a circulating current is stored (as sketched on the right of Fig. 10a) the boundary of the corresponding N = 1 mode is obtained in shifting that of the N = 0 mode by $2\pi/\lambda$ along the

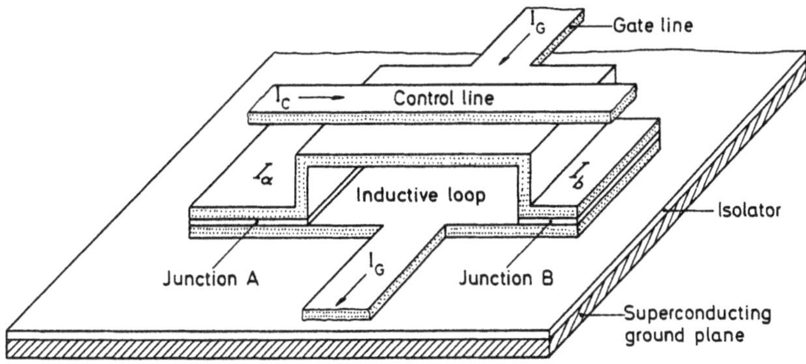

Fig. 9 Sketch of an interferometer with two Josephson junctions as an electronic switch. The maximum Josephson currents of junction A and B are I_a and I_b, respectively.

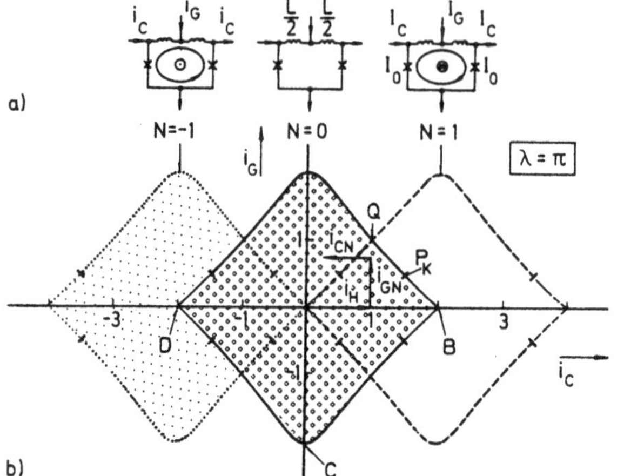

Fig. 10
a) Equivalent circuits of a two junction interferometer with a maximum Josephson current ratio $a/b = I_a/I_b = 1$ and the characteristic phase $\lambda = 2\pi L I_0/\Phi_0 = \pi$.
b) Normalized maximum zero voltage interferometer current $i_G = I_G/I_0$ versus the normalized control current $i_C = I_C/I_0$ for the flux quantum states N = 0 and N = ± 1.

positive i_C-axis. In other words the control current difference ΔI_C between two adjacent maxima times the interferometer inductance L is one flux quantum. All modes of an interferometer with two point junctions have the same shape in contrast to a single long Josephson junction where the maxima decrease with increasing N as demonstrated in Fig. 4. For $\lambda = \pi$ the envelopes of all modes vary between $1.05 < |i_G| < 2$; two adjacent modes overlap in a broad range.

The interferometer always enters the voltage state after trespassing the mode envelope. Below the envelope after leaving a mode boundary the interferometer switches into the voltage state for instance between Q and P_K in Fig. 10b and into the mode N = 1 between P_K and B on emitting a very short voltage pulse. The critical point P_K approaches Q if the interferometer damping is increased. The points P_K and Q practically coincide for a McCumber damping parameter $\beta = 1$. The critical point in Fig. 10 corresponds to $\beta = 136$.

4 Junction fabrication

Planar tunnel oxide Josephson junctions with pure lead electrodes can easily be fabricated by thermal oxidation. Extremely low subgap exess currents can be achieved. Unfortunately the are destroyed in most cases after the first cycle between 4.2 K and 300 K. This disadvantage is practically eliminated in using a fine-grained- In, Au, Pb alloy and a counter electrode with Pb-Bi. More than 600 cycles can be perfomed for a cumulative failure of 1 % [25, 26].

The reproducibility of the very thin tunnel oxide is achieved by a sputter etching process in a low-pressure rf oxygen discharge [27]. The oxidation rate drops with time until it is equal to the sputter rate; the corresponding oxide thickness saturates with time and depends only on the ratio of gas pressure and voltage at the cathode in the chamber. Niobium-niobium oxide junctions with lead counter electrodes are getting very attractive because the standard deviation of the maximum Josephson current on the chip is smaller and the cyclability is considerably larger than of lead alloy junctions [28, 29]. On the other hand, since the capacitance C_F of the niobium-oxide Nb_2O_5 is about 3 times larger than of the lead alloy oxide $PbO + In_2O_3$, niobium junctions for interferometers must have a smaller area and a larger current density for the same switching speed. Moreover niobium is more difficult to structure than lead. On the other hand promissing niobium devices are investigated such as edge junctions structured in a reactive plasma [28].

Many other materials and electrode configurations have been proposed for Josephson junctions. In most cases the fabrication perfection is still insufficient for integrated circuits.

5 Magnetometers

The basic features of an interferometer with two Josephson junctions as a very sensitive tool to measure small magnetic fields are briefly discussed here. As early as in 1964 experimental work has been started [21]. Very recently the sensitivity of dc squid magnetometers have been improved by several orders of magnitudes [30] in applying a thin film fabrication technology developed by IBM for integrated digital circuits with Josephson junctions [25].

The equivalent circuit of a dc squid magnetometer is shown in Fig. 11a. The shunt resistance of each junction R is designed for a McCumber damping parameter $\beta = 1$, i.e. to eliminate hysteresis of the I_G, U_I-characteristic. The corresponding magnetostatic characteristic of the dc squid for a characteristic phase $\lambda = \pi$ in Fig. 10 indicate a maximum modulation depth of the maximum Josephson current of about 50 % as sketched in Fig. 11b. For a bias current $I_B < 2\,I_0$ the interferometer voltage U_I is modulated as a function of control current or normalized flux as deduced in Fig. 11c. The dc squid magnetometer is a flux to voltage transducer.

The interferometer voltage U_I increases with increasing shunt resistance R at a given bias current I_B. Therefore the voltage change is also proportional to R:

$$\partial U_I \sim R\,\partial I_G \sim \frac{R}{L}\,\partial(I_G L) \sim \frac{R}{L}\,d\Phi.$$

The shunt resistance R generates white voltage noise:

$$\frac{\overline{U_I^2}}{\Delta f} = 4\,kT\,R$$

corresponding to the flux noise:

$$\frac{\overline{\Phi^2}}{\Delta f} \simeq 4\,kT\,\frac{L^2}{R}.$$

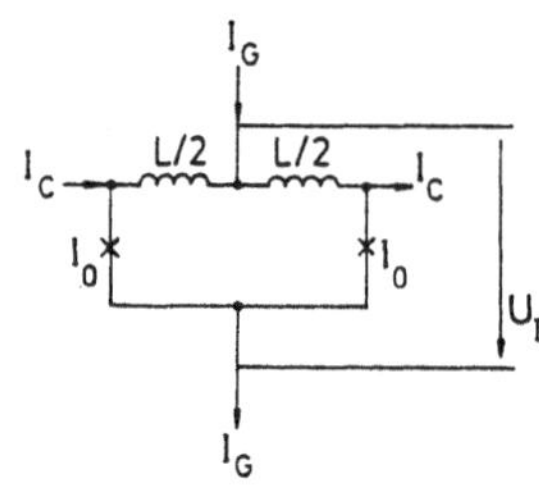

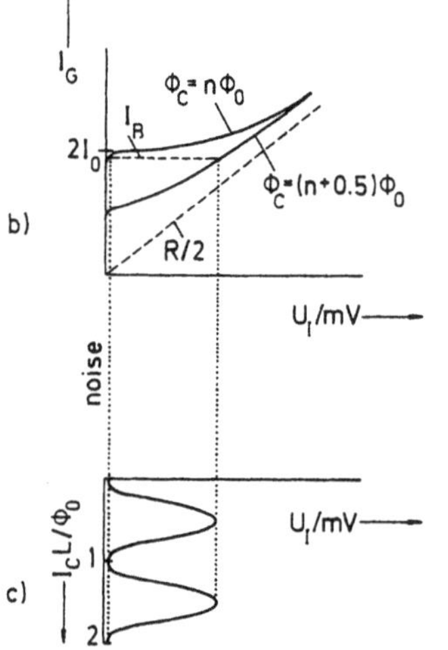

Fig. 11

a) Equivalent circuit of an dc squid magnetometer.

b) Current voltage of a magnetometer for a characteristic phase $\lambda = \pi$ and $\beta < 1$. The two solid lines correspond to the maximum and minimum currents as a function of control current or flux. Near the ordinate the zero voltage currents are slightly reduced by noice.

c) Interferometer voltage U_j versus normalized control flux LI_C/ϕ_0.

The energy sensitivity of magnetometers is defined [22]:

$$\epsilon = \frac{W}{\Delta f} = \frac{\overline{\Phi^2}}{2L}\,\frac{1}{\Delta f} \simeq \frac{1}{2}\,4\,kT\,\frac{L}{R}.$$

For $\lambda = \pi$ i.e. $L = \Phi_0/2\,I_0$ and with $\Phi_0 = h/2\,e$ the energy sensitivity is approximately

$$\epsilon \simeq \frac{1}{2}\,h\left[\frac{kT/e}{RI_0}\right]$$

where h is Planck's constant and e the electron charge. The expression in brackets goes to one if $RI_0 \approx kT/e = 360\,\mu V$ at 4.2 K. This condition has been approximately fulfilled in shunting lead alloy planar tunnel oxide Josephson junctions of a current density $j_{max} = 1\ KA/cm^2$ with integrated resistors as shown in Fig. 12. The loop inductance is $L = 11.5$ pH, the shunt resistance is $R = 2$ ohm and the maximum Josephson current is $I_0 = 100\,\mu A$. The achieved energy sensitivity $\epsilon = 5$ h is very close to the quantum limit [30].

Very recently niobium nanobridge dc squids have been fabricated. A photograph is reproduced in Fig. 13 [31]. The cross section of the bridge is $80 \times 30\ nm^2$; the

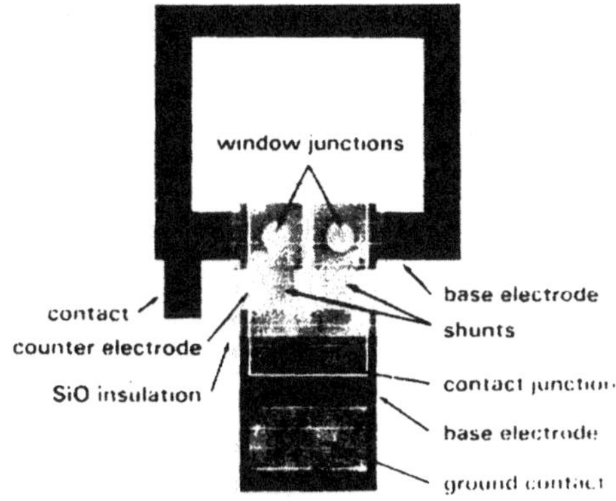

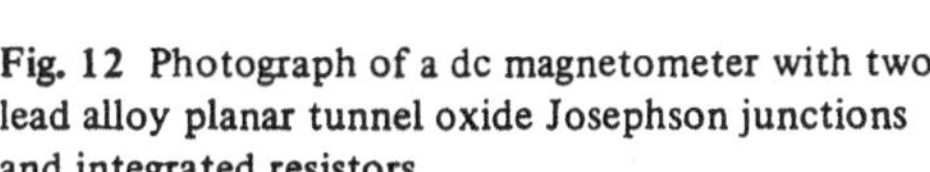

Fig. 12 Photograph of a dc magnetometer with two lead alloy planar tunnel oxide Josephson junctions and integrated resistors.

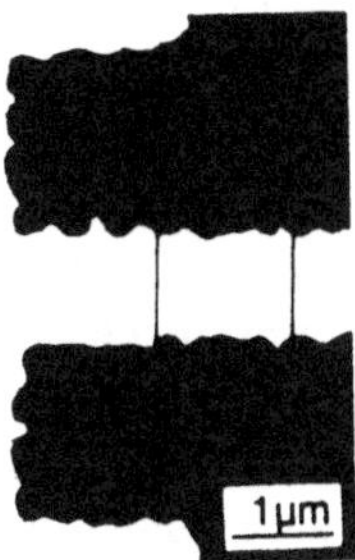

Fig. 13 Photograph of a dc magnetometer with two niobium nanobridges.

maximum zero voltage current density is $j_{0c} = 1.4\ \mathrm{MA/cm^2}$. The characteristic phase is $\lambda \simeq 2\pi$. The best energy sensitivity reached is $\epsilon = 3$ h very close to the claimed limit $\epsilon = \frac{1}{2}$ h [31].

6 Memory cells

One of the most challenging applications of the Josephson technology will be a high speed computer with about 10^8 instructions per second [32, 33]. Since a major fraction of the computer hardware cost is spend for memory units, memory cells and chips must be optimized to a high degree.

Memory cells have been investigated in detail where the binary information is represented e.g. by clockwise or counter clockwise circulating currents in a superconducting loop. The stored currents correspond to a large number of flux quanta [34, 35]. The magnetic field of the circulating currents is sensed nondestructively by a Josephson junction underneath the loop in a separate sense line. Two junctions in the loop are needed to write the information [34, 35]. In a recent nondestructive read out (NDRO) cell design the junctions have been replaced by two interferometers [36]. The write gate in the loop is a three junction interferometer [37] and the sense gate an asymmetric two junction interferometer [38]. Only a few flux quanta (<8) carry the information. A chip with a capacity of 4×1024 NDRO memory cells is expected to have an access time $T_{acc} = 0.5$ ns at a maximum power dissipation $P_{chip} = 6$ mW [36].

Besides the above high speed low capacity cache memory chip a main memory chip with high capacity and low power is needed, on the expense of speed if necessary. For this purpose interferometer memory cells are very promising where the binary information is represented by the absence or presence of a single flux quantum [39]. The information is stored in a two junction interferometer with a characteristic phase $\lambda = \pi$, a junction area ratio a/b = 1 and a tap ratio p = 0.5 and a damping parameter $\beta > 100$ according to Fig. 10. In the center of the overlapping zone of the N = 0 and N = 1 mode a dc bias control current $i_H = 1$ is applied.

A photograph of an implemented cell [39] is shown in Fig. 14. The interferometer inductance is realized with M2 and M3 layers separated by a SiO-layer of about 200 nm thickness. The minimum line width is 2.5 μm. The cell area is $A = 1500\ \mu m^2$ for a maximum Josephson current density $j_{max} = 2.2\ KA/cm^2$. This main memory cell area is smaller than the cache memory cell area but still large because three control lines are needed. One control line is used to bias the interferometer. The two other lines are required for random access of a cell array with large margins [39].

The quantum state of the cell is read out in applying the nominal currents i_{GN} and i_{CN} according to Fig. 10. If a flux quantum corresponding to the N = 1 mode has been stored the interferometer switches into the voltage state, as the boundary has been trespassed on top of points P_K in Fig. 10. If no flux quantum have been stored, the interferometer remains in the zero voltage state. Thus the "0" and "1" signals can easily be dicriminated. At the fall of the drive currents i_{CN} und i_{GN} the interferometer returns into the zero voltage state with about equal probability to trap one or no flux quantum. Therefore the information must be rewritten, i.e. the cell is read out destructively. To write the information the mode boundaries are crossed below the critical point P_K in Fig. 10 avoiding the voltage state.

Prototype chips with 2 X 1024 interferometer memory cells including address decoder and sense amplifier have been implemented and successfully tested.

Measurements indicate the feasibility of a 16 X 1024 bit memory chip with an access time of $T_{acc} = 15$ ns. For storage the power dissipation is zero. During access the chip power dissipation would be $P_{chip} = 40\ \mu W$. i.e. 10^5 times smaller than that of a semiconductor chip of comparable capacity and access time. A photograph of the prototype chip [39] is reproduced in Fig. 15.

Fig. 14
Photograph of a two junction interferometer memory cells for destructive read out. There are three control lines, one for bias and the two others for random access.

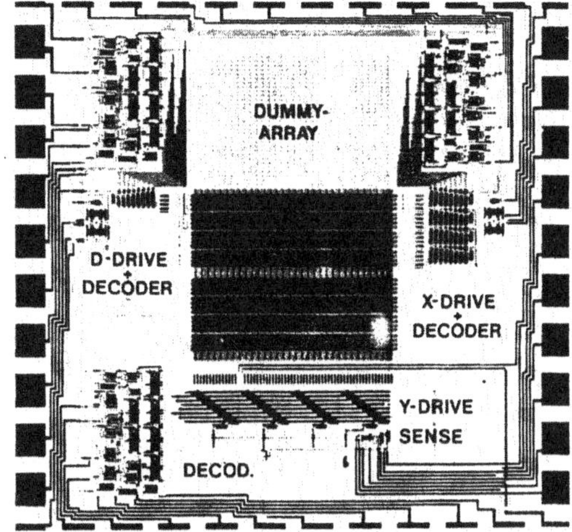

Fig. 15
Photograph of a prototype chip with 2 × 1024 interferometer cells and peripheral circuits.

Single flux quantum memory cells can also be designed for NDRO [40, 41] in doubling the characteristic phase to $\lambda = 2\pi$, and in using unequally large Josephson currents a/b = 2. For these parameters and a tap ratio p = 0.5 the magnetostatic gate characteristics of the N = 0 and N = − 1 mode are shown in Fig. 16b. The overlapping zone of the two modes is larger than for $\lambda = \pi$ in Fig. 10 and *includes* the origin. A bias current is no more needed to store nonvolatilly the binary modes N = 0 or − 1. The information of the 0-mode is read out in trespassing the mode boundary on top of P_K with the indicated nominal pulses i_{GN} and i_{CN}. The position of the critical point P_K in Fig. 16 corresponds to a damping parameter $\beta = 136$. If the N = 0 mode has been stored, the voltage state with about 2.5 mV is generated. The zero voltage state is regained without trapping a flux quantum in switching off first the control current and then the gate current. If the N = − 1 mode has been stored the zero voltage state is maintained. Thus nondestructive read out is achieved. The NDRO-cell area can be smaller than in [39] as the bias control line is eliminated. A high density version NDRO cell with only a single control line is discussed in [41, 42]. A two junction interferometer for $\lambda = 2\pi$, a/b = 2, p = 0.5 and a single control line has been implemented in The Institut für Elektrotechnische Grundlagen der Informatik, Karlsruhe. A photograph is shown in Fig. 17.

The control line has a width w = 10 μm, corresponding to the minimum line width. The interferometer area comprises 51 lithographic squares i.e. $A = 51 \cdot 100\,\mu m^2$. Cell arrays very close to the above described concept have been implemented in Japan. The predicted NDRO operation has been verified experimentally [43].

The storage and processing of binary information in shift registers for instance for self test of VLSI-chips [44] is often required. Single flux quantum shift registers

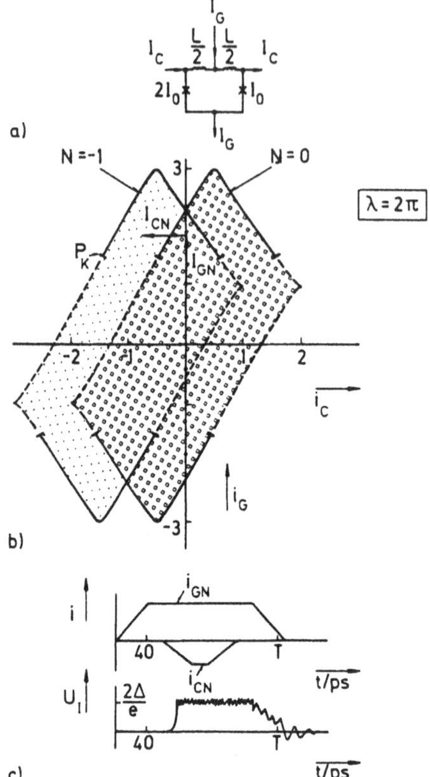

Fig. 16

a) Equivalent circuit of a interferometer memory cell with a current ratio a/b = 2 and a tap ratio p = 0.5.

b) Threshold curves of the normalized gate current versus normalized control current for $\lambda = 2\pi$ and $\beta = 136$.

c) Nominal drive currents for non-destructive read out and sense signal versus time.

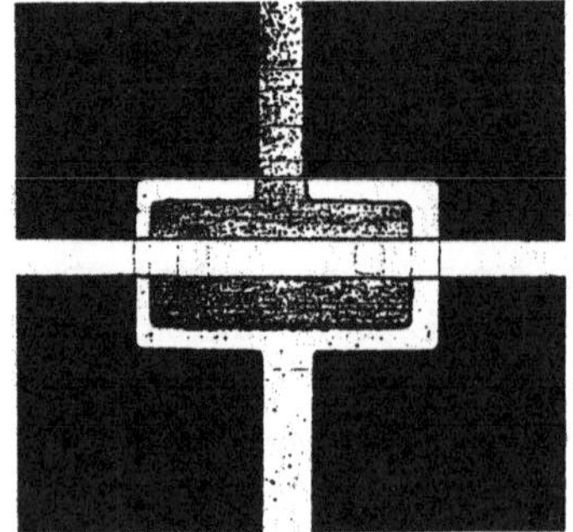

Fig. 17

Photograph of an implemented interferometer with two unequal window Josephson junctions for a/b = 2 and $\lambda = 2\pi$. The control line width is 10 μm.

have been proposed about 10 years ago [45]. Recently margins of a 5 junction interferometer as a shift register model as sketched in Fig. 18 have been simulated [46]. In each interferometer loop the characteristic phase $\lambda = 2\pi$ allows for a storage of a single flux quantum without bias current. Each loop is coupled with a local magnetic control field. The quantum state of the interferometer with 4 loops is described by a 4 digit quantum number. The binary number 0100 marks a flux quantum only in the second loop. All binary modes are represented in the normalized current plane of Fig. 19 if a control current is coupled only with the 2. loop. Inspection of the diagram reveals that one or no flux quantum can be shifted from the first *or* the third loop into the second loop without generating unwanted flux quanta. For a maximum Josephson current density $j_{max} = 10\ KA/cm^2$ a shift of a single flux quantum from one loop to the other in both directions is feasible within 4 ps and ± 25 % drive current margins [46].

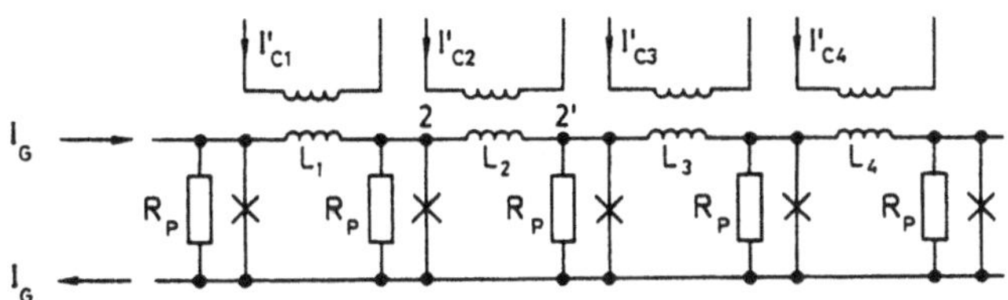

Fig. 18 Equivalent circuit of an interferometer shift register model with separate control lines at each loop.

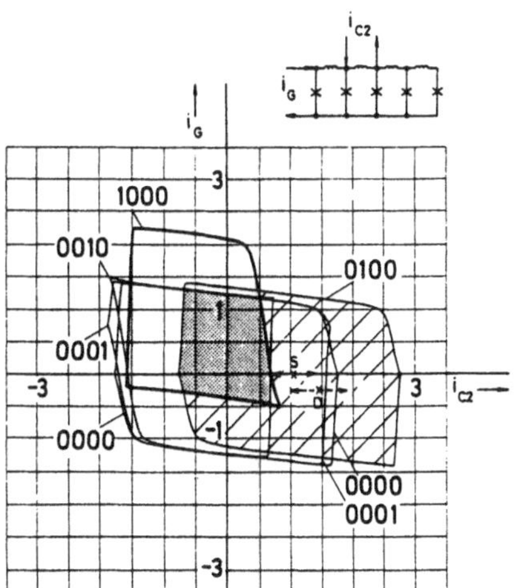

Fig. 19
Equivalent circuit and normalized threshold gate current i_G versus normalized control current at the second loop i_{C2} of a five junction interferometer with $\lambda = 2\pi$.

7 Logic circuits

Memory and logic circuits are needed for digital processing. Logic circuits must be provided on a large scale integration memory chip for instance to decode the binary address of each cell in the array.

Most of the present logic circuits with Josephson junctions operate in the latched mode, i.e. the logic output is stored independently of the input signal duration and is reset by a clock pulse [47, 48]. Only a few nonlatching concepts have been published e.g. in [49]. The input signals are either inductively [47, 48] or galvanically [48, 50, 51, 52, 53] coupled. Often devices with galvanic and inductive coupling are combined in a logic gate to achieve large operating tolerances and galvanic decoupling of input and output [53, 54]. If the input signal is fed to a grounded counter electrode of a junction a separate metallisation M4 for the control line is not needed, although input and output are galvanically decoupled [55].

Interferometers as switching elements outperform single Josephson junctions if operating tolerances must be increased on the expense of circuit area [37, 56]. The advantages of symmetric and asymmetric interferometers have been worked out [38, 41]. As an example only a three junction interferometer [37] is discussed here. The equivalent circuit and the magnetostatic characteristics are shown in Fig. 20. The characteristic phase is $\lambda = 2\pi/7$. Between the main binary modes 00 and 11, the mode 01 and 10 have small maxima of the maximum gate currents so that the interferometer switches independently of the circuit damping into the voltage state within a large range of gate control currents. Since a large control current overshoot can be tolerated, the interferometer is especially suited as input device. A scanning electron micrograph is reproduced in Fig. 21. Two separate control lines folded to a "hairpin" are seen on top of the device. They enter and leave on the left side of the

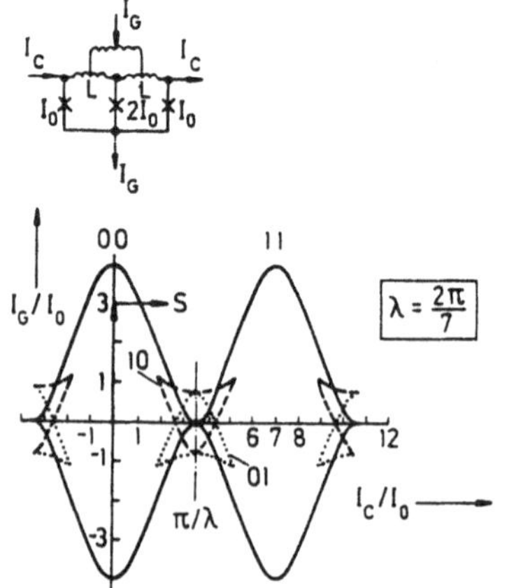

Fig. 20
Equivalent circuit and threshold curves of a non-uniform three junction interferometer with $\lambda = 2\pi/7$.

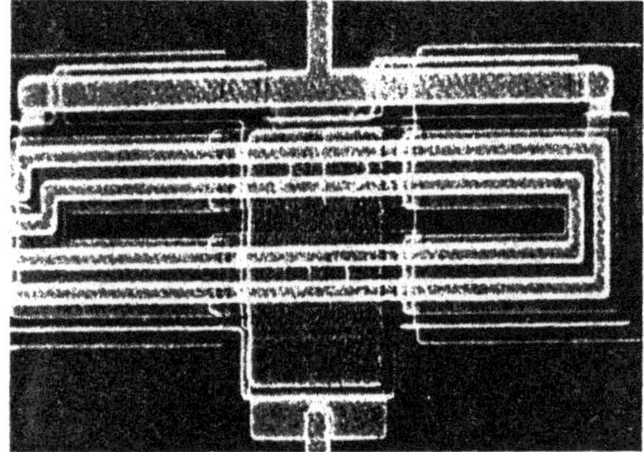

Fig. 21
Photograph of an inductively coupled three junction interferometer according to Fig. 20.

picture. Below the control line in the center of the picture, four circular window Josephson junctions are arranged. The lower two in parallel represent the center junction and the upper ones the outer junctions of the equivalent circuit in Fig. 20. The two interferometer inductances are implemented by folded strip lines to reduce the required gate area. The two strips at the insertion point of I_G on top of the picture are integrated resistors to cut down internal resonance amplitudes. For a minimum line width $w = 2.5\ \mu m$ and a current density $j_{max} \simeq 1\ KA/cm^2$ the gate area is $A_G = 2340\ \mu m^2 \simeq 375\ w^2$ [37].

Complete logic families, i.e. AND-, OR- and NEGATION-gates are available. The NEGATION-gates are timed inverters [57].

OR and AND gate circuits with the described three junction interferometer are sketched in Fig. 22. A "voltage source" of about $4\ \Delta/e$ provides a gate current bias via the decoupling resistance of 25 ohm. The OR-gate interferometer has two control lines A and B. If one control line is activated the interferometer switches to the voltage state and forces a current into the output line with a characteristic impedance $Z_0 = 6\ \Omega$. The output line is matched and can activate many other gates in series.

The two input AND-gate comprises two interferometers with one inductively coupled control line, each. The interferometer B switches into the voltage state only after interferometer A has switched and current transferred into interferometer B.

OR- and AND-Gates can drive many other gates in series. The so called serial fanout is large. Moreover, the operation of more than two output lines in parallel is possible [53, 54].

The average nominal gate delay and power dissipation of AND- and OR-gates are $t_G = 43$ ps and $P_G = 4.8\ \mu W$, respectively [53]. The power dissipation per gate area is $P' = 0.2\ W/cm^2$. The corresponding power delay time product is $W \simeq 200$ aJ. It is about a 1000 times smaller of present 2 fan-in, 2 fan-out semiconductor gates with gate delays below 100 ps [58, 59].

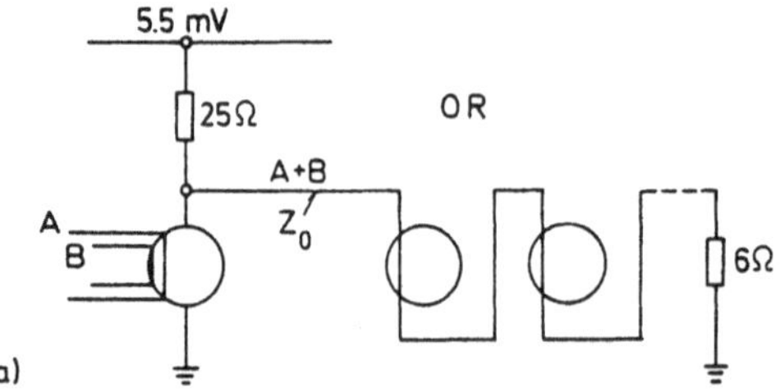

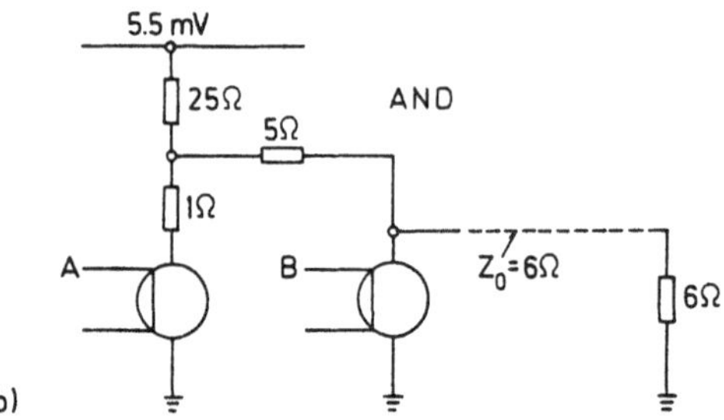

Fig. 22
Logic circuits with inductively coupled three junction interferometers and matched output lines: a) OR-gate and b) AND-gate.

Therefore the potential three dimensional packing density of digital circuits with Josephson junctions is about 1000 times larger than that with present semiconductors at 300 K because the heat extraction capability of forced air cooling at 300 K without cooling towers [60] is $Q_A \simeq 2\ W/cm^2$ and of liquid helium at 4.2 K is $Q = 0.8\ W/cm^2$.

Liquid cooling of semiconductor circuits near 300 K corresponds to an upper limit $Q_L = 20\ W/cm^2$ [61, 62] so that the two dimensional packing density can be increased as compared to forced air cooling by the ratio $Q_L/Q_A \approx 10$. However it seems rather difficult to improve the three dimensional packing density by the same ratio, since semiconductor chips cannot simply be immersed in the coolant liquid as superconductor chips in liquid helium and since towers, cooling tubes are space consuming.

Higher three dimensional packing density is required especially for large and fast computers in order to cut down the average distance between gates limiting the minimum effective delay per gate. Large and fast computers must have a small volume [32].

The main present potential advantage of logic circuits with Josephson junctions are summarized in table 23.

LOGIC GATES

PRESENT ADVANTAGES
OF THE

JOSEPHSON-TECHNOLOGY (4.2 K) AS AGAINST THE
SEMICONDUCTOR- TECHNOLOGY (300 K)

AT COMPARABLE LOGIC DELAYS BELOW 100 PS :

1. 1000 TIMES SMALLER POWER DISSIPATION OF HIGH FAN-IN / FAN-OUT LOGIC GATES, HENCE HIGHER THREE-DIMENSIONAL PACKING DENSITY AND HIGHER OVERALL PROCESSING SPEED.
2. MATCHED TRANSMISSION LINES.
3. SUPERCONDUCTING TRANSMISSION LINES OF SUBMICRON CROSS SECTIONS.
4. SUFFICIENTLY LARGE TOLERANCES BY EXPLOITING FLUX QUANTISATION.
5. MORE RELIABLE OPERATION OF LARGE COMPUTERS AT 4.2 K.

Fig. 23 Present advantages of the Josephson technology for logic gates.

8 Miniaturisation

Future lithography e.g. with electron and ion beam writing, x-ray exposure allow for an implementation of submicron minimum line widths at large aspect ratios [63]. Computer aided design and fabrication methods and tools pioneered for semiconductor circuits can be used often without change for superconductor circuits. Single devices with weak link cross sections 80 nm × 30 nm have already been operated [31].

Therefore the miniaturisation potential of the present 2.5 μm minimum line width Josephson and semiconductor technology is to be compared even though only very tentative extrapolations are possible.

The scaling properties of interferometers with Josephson junctions are investigated since they are the key of present technology. As a test vehicle a nonvolatile interferometer memory cell for NDRO [40] is chosen. The essential design data are $\lambda = 2\pi$, a/b = 2, p = 0.5 and $\beta > 80$.

A layout sketch for a minimum line width $l_1 = 1\ \mu m$ and for a maximum Josephson current density $j_{max} = 77\ KA/cm^2$ is shown in Fig. 24. The sketched window junction areas are rectangular, although implemented junction areas are rather oval. One control line is drawn. Another control line of slightly smaller width may be arranged on top of the lower one [64]. The metallisation thicknesses are about 0.2 μm. The maximum oxide thickness between the strips of the interferometer inductance is 0.2 μm. The effective separation including the London penetration depths is $d_B = 0.4\ \mu m$. The cell area is $A_{cell} = 51\ \mu m^2$. The number of lithographic squares $F' = 51$ is as large as that of the implemented interferometer in Fig. 17 and corresponds very closely to that of present highly sophisticated static volatile CMOS RAM flip flop cells [65] operated at 300 K. During read out of the interferometer the power dissipation is $P = 3.7\ mA \times 2.5\ mV = 9.4\ \mu W$. The power dissipation per cell area is $P' = 18\ W/cm^2$. Since one cell out of many thousands is activated at a time, the power of one cell can be extracted by liquid helium via the surrounding chip area (at least 23 A_{cell}). The energy of the cell is $W \simeq 0.5\ L\ I_b^2 = 1.3\ aJ = 3.23 \cdot 10^4\ W_g$ where $W_g = kT \cdot \ln 2$ is the thermodynamical limit to store binary information [62]. Therefore flux fluctuations are negligible.

The very important drive and fabrication tolerances cannot be discussed here, because they depend on future process sophistication. For this extrapolation the assumption is made that they do not change if the essential parameter λ, a/b, p and especially β can be kept constant. Moreover, the read out voltage 2.5 mV and the effective separation of the inductance $d_B = 0.4\ \mu m$ are kept constant if all lateral dimensions are scaled by K.

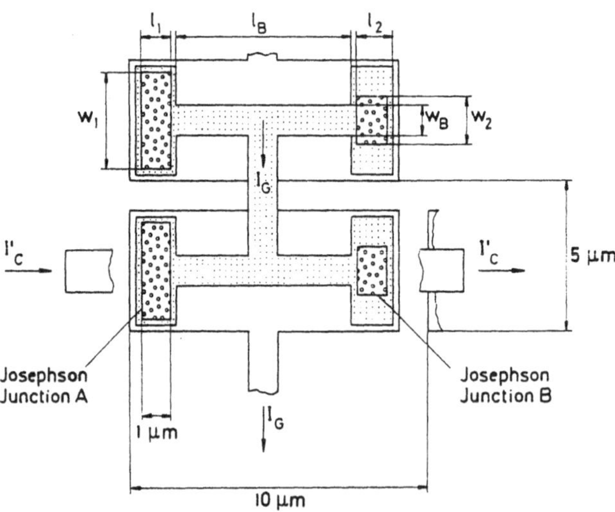

Fig. 24 Layout sketch of two interferometer memory cells for NDRO as a miniaturisation test vehicle.

With the above assumptions and the notation in Fig. 24 the scaling laws for interferometer circuits are:

$$A_{cell} = A_{cell\,1} \cdot K^2,$$

$$L \approx \mu_0 \frac{l_B}{2 + w_B K/d_B},$$

$$I_b = j_{max}\, w_2 l_2 \approx \frac{\Phi_0 \lambda}{2\pi\mu_0} \frac{2 + w_B K/d_B}{l_B K},$$

$$j_{max} \approx \frac{\Phi_0 \lambda}{2\pi\mu_0} \frac{2 + w_B K/d_B}{w_2 l_2 l_B K^3},$$

$$P' = \frac{3 I_b\, 2\Delta/e}{A_{cell}} \approx \frac{3\Phi_0 \lambda\, 2\Delta/e}{2\pi\mu_0 A_{cell\,1}} \frac{2 + w_B K/d_B}{l_B K^3},$$

$$W \approx \frac{1}{2} L I_b^2 = \frac{\Phi_0 \lambda}{4\pi} I_b \approx \frac{1}{2\mu_0} \left(\frac{\Phi_0 \lambda}{2\pi}\right)^2 \frac{2 + w_B K/d_b}{l_B K}.$$

For effective separations much smaller than the lateral dimensions $d_B \ll w_B$ the scaling laws simplify. In table 25 the scaling laws for semiconductor field effect transis-

		SEMICONDUCTOR (FET)	SUPERCONDUCTOR (INTERFEROMETER WITH JOSEPHSON JUNCTIONS)
GEOMETRIC DIMENSIONS	LATERAL	K	K
	VERTICAL	K	CONST, D_B = 0.4 µM ≪ W
LAYER CONDUCTIVITY		K^{-1}	
VOLTAGE		K	2Δ/E
CURRENT		K	CONST
JOSEPHSON CURRENT DENSITY			K^{-2}
LINE CURRENT DENSITY		K^{-1}	K^{-2}
POWER		K^2	CONST
POWER DENSITY		CONST	K^{-2}
LINE RESISTANCE		K^{-1}	0
CAPACITANCE		K	
INDUCTANCE			CONST
LINE DELAY		CONST	K
ENERGY		K^3	CONST

Fig. 25 Main scaling laws of semiconductor field effect transistors and interferometers with Josephson junctions.

tors and interferometers with Josephons junctions are summarized. The power dens ty of FET-circuits stays constant while the power density of interferometer circuits decreases with K^{-2} with increasing dimensions. The power delay time product or the energy of semiconductor devices scale with K^3. The interferometer energy unfortunately does not scale. On the other hand the line delay of open ended transmission lines on semiconductor chips is proportional to strip resistance and capacitance. This delay does not scale and will finally dominate all other delays [66].

GaAs-field effect transistor circuits at 4.2 K [67] with superconducting layers (Nb instand of Mo) could be operated with lossless transmission lines. Lossless lines, however, should be matched with their characteristic impedance to avoid multiple reflections. Unfortunately the maximum characteristic impedance of strip lines on the chip near 100 ohm is much smaller than the load resistance of miniaturised field effect transistors [59].

The delay on matched superconducting transmission lines scales with K. The power in matching loads scales with K, too, if the voltage and the vertical dimensions are kept constant.

Beside scaling laws absolute numbers of characteristic parameters of semiconductor and superconductor technology must be compared. Therefore the scaling behavior of the interferometer test vehicle in Fig. 24 is investigated as a function of w in Fig. 26. The required quadratic increase of cell area with F' = 51 lithographic squares

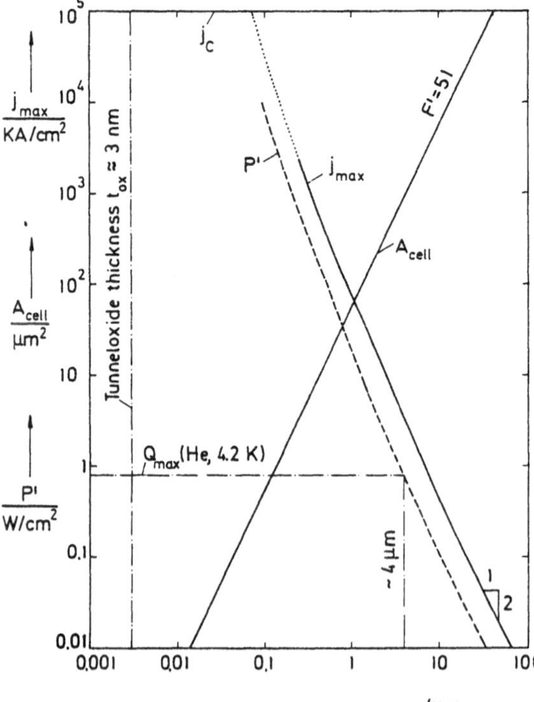

Fig. 26
Maximum Josephson current density j_{max}, cell area A_{cell} and power dissipation per cell area P' versus minimum line width w.

is plotted. The corresponding maximum Josephson current density is represented as a solid line in the measured range [68, 69]. The maximum measured value $j_{max} = 2\ MA/cm^2$ [69] corresponds to $w \simeq 0.24\ \mu m$ and a cell area $A_{cell} \simeq 3\ \mu m^2$. The maximum Josephson current density must stay below the critical current density of the electrodes j_c. For lead alloys it is $j_c \simeq 100\ MA/cm^2$. The power density increases strongly with miniaturisation. For $W = 4\ \mu m$ or $A_{cell} = 816\ \mu m^2$ the maximum heat extraction capability of liquid helium is already reached if all cells are activated simultaneously. Since only a few cells are activated at a time power dissipation does not seem to limit miniaturisation.

The active (not standby) power dissipation of a selected chip divided by the chip area of actual implementations or designs is plotted versus minimum line width in Fig. 27.

The numbers refer to the list of references. The encircled crosses represent, static memory chips, the crosses dynamic ones. The reference numbers [36, 39] mark chip designs of the Josephson technology. The main memory version with an access time $T_{acc} = 15$ ns [39] is operated more than 4 orders of magnitudes below the

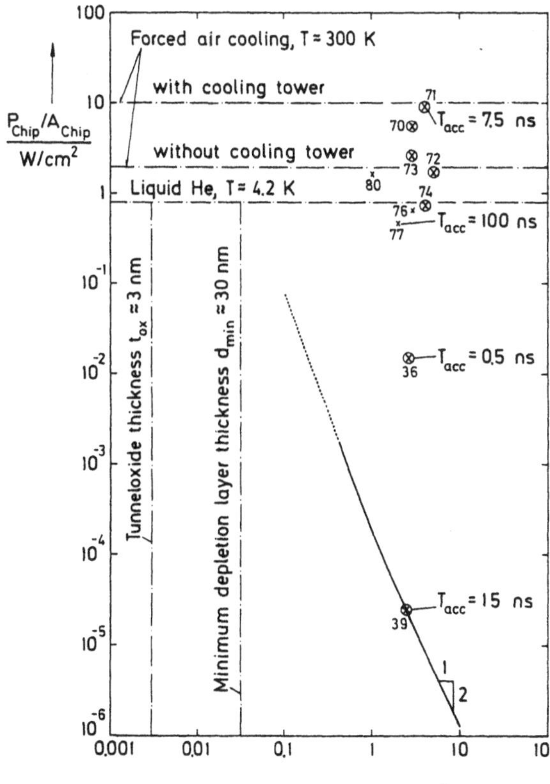

Fig. 27
Active chip power dissipation P_{chip} per chip area A_{chip} versus minimum line width w. Semiconductor FET memories: static [70, 71, 72, 73, 74] and dynamic [76, 77, 80]. Static Josephson memories prototypes [36, 39].

thermal limit. Its power dissipation is about 10^5 times smaller than semiconductor memory chips of comparable access time [70]. The chip power density for a smaller minimum line width than 2.5 μm is very crudely extrapolated with the assumption that it is proportional to the cell power density. In the whole range of foreseeable miniaturisation $0.24 < w/\mu m < 2.5$ the chip power density is sufficiently small.

The chip energy to access the information of one binary cell, the quality factor $W' = P_{chip} \cdot T_{acc}/C$, where C is the number of cells per chip, is plotted in Fig. 28 versus minimum line width. The encircled crosses represent memory chips with static cells and the crosses represent chips with dynamic cells. The quality factor roughly scales with energy according to table 25, i.e. K^3. Near the lower limit of the power delay time product or minimum energy of a single semiconductor device $W_{min} = 0.5$ fJ [61] strong deviations from the scaling law are expected. The minimum depletion layer thickness $d_{min} = 30$ nm for a doping level $10^{18}/cm^3$ in silicon [61], the minimum energy W_{min} and the quality factor $W' \sim K^3$ outline the potential miniaturisation range of the semiconductor technology.

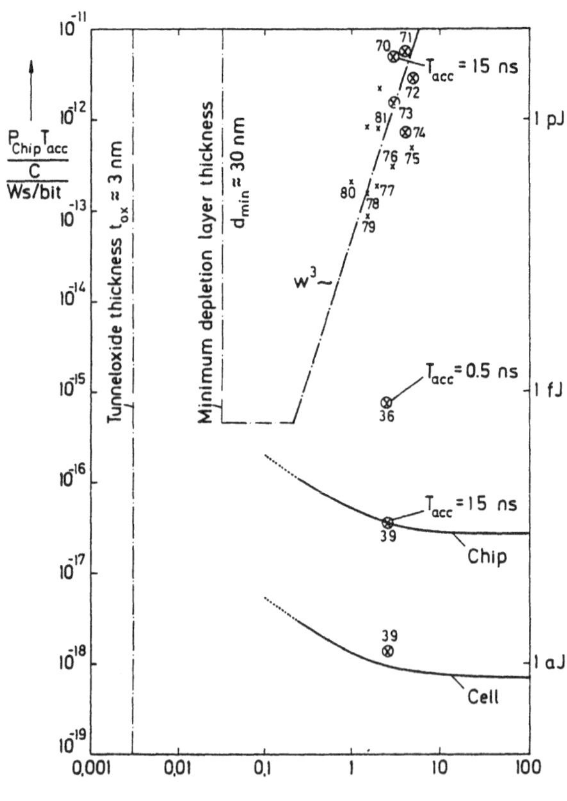

Fig. 28
Active chip power dissipation P_{chip} times acces time T_{acc} over the number of binary cells on the chip C versus minimum line width w. Semiconductor FET memories: static [70–74], dynamic [75–81]. Static Josephson memory prototypes [36, 39].

The energy of the test vehicle memory cell of Fig. 24 is plotted versus w as a solid line in Fig. 28. There is a slight decrease with increasing w. The reference number 39 marks the cell energy W of the 16 KBit prototype. The chip energy per cell W' is about 20 times larger than W owing to losses and delays in peripheral circuits on the chip. W versus w is extrapolated with the assumption of a proportionality between W and W' and plotted as a full line in Fig. 28. In the interesting range $w > 0.24\,\mu m$ the energy W' remains below the minimum energy of a single semiconductor device at 300 K; at comparable access times the quality factor of semiconductor circuits at 300 K and of Josephson circuits at 4.2 K differ by 5 orders of magnitudes for $w = 2.5\,\mu m$ and at least one order of magnitudes for $w = 0.24\,\mu m$.

The above scaling laws are modified if instead of interferometer circuits direct coupled logic circuits [51] are investigated where not inductances, but resistances an Josephson junctions must be miniaturized. If the resistor thicknesses and the currents scale with K and if the voltage is kept constant, the maximum Josephson current density and the power density scale with K^{-1} instead with K^{-2} for interferometer circuits. Therefore for a given speed the energy and power density gap between digital circuits with Josephson junctions and with semiconductor devices at 300 K may even be larger than extrapolated above for submicron dimensions.

9 Conclusion

The Josephson technology enters an increasing field of very promising applications. Digital applications will have an important commercial impact if large Josephson computers can be built with a reasonable cost performance advantage as against semiconductor computers. In any case, the Josephson technology is a very powerful vehicle to push the development of ultra high performance and very large scale integration circuits.

References

[1] *J. B. Josephson,* Phys. Lett. Vol. 1 (July 1962), p. 251.

[2] *L. Solymar,* Superconducting Tunnelling, Chapan and Hall Ltd, London 1972.

[3] *D. E. McCumber,* J. Appl. Phys., Vol. 39, No. 7 (Feb. 1968), p. 3113.

[4] *R. E. Harris,* Phys. Review B, Vol. 13, No. 9 (May 1976), p. 3818.

[5] *H. Lübbig,* SQUID 80, W. de Gruyter & Co., Berlin, New York, (1981), p. 1

[6] *W. Jutzi, Th. Mohr, M. Gasser, H. P. Gschwind,* Electronic Lett. 8 (1972), p. 589.

[7] *S. M. Faris,* Phys. Lett. 36 (1980), p. 1005.

[8] *D. B. Tuckermann,* Appl. Phys. Lett. 36 (12) (June 1980), p. 1008.

[9] *E. P. Harris, W. H. Chang,* IEEE an Magnetics, Vol. MAG-77, No. 1 (Jan. 1981), p. 603.

[10] *R. E. Jewett* and *T. van Duzer,* IEEE an Magnetics Vol. MAG-17, No. 1 (Jan. 1981), p. 599.

[11] *P. N. Langenberg, D. J. Scalapino, B. N. Taylor,* Proc. IEEE, Vol. 54, No. 4 (April 1966), p. 560.

[12] *C. V. Stancampiano*, IEEE Trans. on Electron Devices Vol. ED-27, No. 10 (Oct. 1980), p. 1934.

[13] *N. Yokoyana, T. Mimura, M. Fukuta, H. Ishikawa*, Digest of ISSCC 1981, p. 218.

[14] *P. Gutmann, V. Kose*, PTB-Mitt. 90 (1980), p. 1 (Jahresbericht der Physikalisch-Technischen Bundesanstalt 1980)

[15] *P. L. Kautz, G. Costabile*, IEEE Trans. on Magnetics, Vol. MAG-17, Nr. 1 (Jan. 1981), p. 780.

[16] *D. G. McDonald, R. L. Peterson, C. A. Hamilton, R. E. Harris* and *R. L. Kautz*, IEEE Trans. on Electronics Devices, Vol. ED-27, No. 10 (Oct. 1981), p. 1945.

[17] *R. L. Richards, T. M. Shen*, IEEE Trans. on Electronics Devices, Vol. ED-27, No. 10 (Oct. 1981), p. 1909.

[18] *Y. Taur*, IEEE Trans. on Electron Devices, Vol. ED-27, No. 10 (Oct. 1981), p. 1921.

[19] *M. T. Levinsen, N. F. Pedersen, O. H. Soerensen, B. Dueholm*, and *J. Mygind*, IEEE Trans. on Electron Devices, Vol.-ED-27, No. 10, (Oct. 1980), p. 1928.

[20] Russer AEÜ 23 (1969), p. 417–420.

[21] *R. C. Jaklevic, J. Lambe, A. H. Silver* and *J. E. Mercereau*, Phys. Rev. Lett., Vol. 12 (1964), p. 159.

[22] *J. Clarke*, IEEE Trans. on Electron Devices Vol. ED-27, No. 10 (Oct. 1980), p. 1896.

[23] *C. A. Hamilton, F. L. Lloyd, R. L. Kautz*, IEEE Trans. on Magnetics, Vol. MAG-17, No. 1 (Jan. 1981), p. 577–582.

[24] *W. Jutzi*, Cryogenics, (Feb. 1976), p. 81.

[25] *J. H. Greiner, S. Basavaiah* and *I. Ames*, J. Vac. Sci. Tech. 11 (1974), p. 81.

[26] *H. C. W. Huang, S. Basavaiah, C. J. Kircher, E. P. Harris, M. Murakami, S. P. Klepner, J. H. Greiner*, IEEE Trans. on Electron Devices, Vol. ED-27, No. 10 (Oct. 1980), p. 1979.

[27] *J. H. Greiner*, J. Appl. Phys. 42 (1971), p. 5151.

[28] *R. F. Broom, S. I. Raider, A. Oosenbrug, R. E. Drake, W. Walter*, IEEE Trans. on Electron Devices, Vol. 27, No. 10 (Oct. 1980), p. 1998.

[29] *J. C. Villegier, G. Matheron, P. Coeure*, IEEE Trans. on Magnetics, Vol. MAG-16, No. 5 (Sept. 1980), p. 1230.

[30] *M. B. Ketchen, R. F. Voss*, Appl. Phys. Lett., Vol. 35 (1979), p. 812.

[31] *R. F. Voss, R. B. Laibowitz* and *A. N. Broers*, Appl. Phys. Lett. 37/7 (Oct. 1980), p. 656.

[32] *W. Anacker*, IEEE Spectrum, (May 1979), p. 26.

[33] *W. Anacker*, IBM J. RES. DEVELOP, Vol. 24, No. 2 (March 1980), p. 107.

[34] *W. Anacker*, IEEE Trans. on Magnetics, Vol. MAG-5, No. 4, (Dez. 1969), p. 968.

[35] *R. F. Broom, W. Jutzi, Th. O. Mohr*, IEEE Trans. on Magnetics, Vol. MAG-11, No. 2 (March 1975), p. 755.

[36] *H. H. Zappe*, IEEE Trans. on Electron Devices, Vol. ED-27, No. 10 (Oct. 1980), p. 1970.

[37] *H. H. Zappe*, IEEE Trans. on Magnetics, MAG-13, (1977), p. 41.

[38] *H. Beha*, Electronic Lett. Vol. 13 (1977), p. 216.

[39] *P. Gueret, A. Moser, P. Wolf*, IBM RES. DEVELOP., Vol. 24, No. 2 (March 1980), p. 155.

[40] *H. Beha, W. Jutzi*, Offenlegungsschrift 2735133, (Feb. 1979).

[41] *H. Beha*, Digitale Speicherzellen mit Josephson-Kontakten, VDE-Verlag GmbH, Berlin, 1981.

[42] *H. Beha,* Intermag Conference 1981, Grenoble

[43] *M. Yamamoto* and *A. Ishida,* Proc. Intl. Conference on Solid-State Devices, Tokyo, Japan, (Aug. 1980), p. 109.

[44] *F. F. Tsui,* IBM J. RES. DEVELOP., Vol. 24, No. 2 (March 1980), p. 243.

[45] *T. A. Fulton, R. C. Dynes* and *P. W. Anderson,* Proc. IEEE, Vol. 61, No. 1 (Jan. 1973), p. 28.

[46] *H. J. Beha, W. Jutzi,* and *G. Mischke,* IEEE Trans. on Electronic Devices, Vol. ED-27. No. 10 (Oct. 1980), p. 1882.

[47] *D. J. Herrel,* IEEE J. of Solid State Circuits, Vol. SC-10, No. 5 (Oct. 1975), p. 360.

[48] *G. Gheewala,* IBM RES. DEVELOP., Vol. 24, No. 2 (March 1980), p. 130.

[49] *W. Bächtold,* IEEE Int. Solid State Circuit Conf. Dig. Tech. Papers. (Feb. 1975), p. 164.

[50] *T. A. Fulton, S. S. Pei, L. N. Dunkleberger,* Appl. Phys. Lett., Vol. 34, No. 10 (May 1979), p. 709.

[51] *T. Gheewala, A. Mukherjee,* IEDM Techn. Digest, Wash. DC, (Dec. 1979), p. 482.

[52] *S. Takada, S. Kosaka, H. Hayakawa,* J. Appl. Phys. Vol. 19 (1980), Suppl. 19-1, p. 607.

[53] *T. R. Gheewala,* IEEE Trans. on Electron Devices, Vol. ED-27, No. 10 (Oct. 1980), p. 1857.

[54] *H. Beha, W. Jörger, W. Jutzi,* IEEE Trans. on Magnetics, MAG-17, No. 1 (Jan. 1981), p. 587.

[55] *S. Hasuo, H. Suzuki, T. Yamaoka,* IEEE Trans. on Magnetics, Vol. Mag-17, No. 1 (Jan. 1981), p. 583.

[56] *H. Beha, W. Jäckel,* Digest of Intermag Conference, Grenoble 1981.

[57] *T. R. Gheewala,* IBM J. RES. DEVELOP., Vol. 24, No. 2 (March 1980), p. 130.

[58] *K. Lehovec, R. Zuleeg,* IEEE Trans. on Electron Devices, Vol. ED-27, No. 6 (June 1980), p. 1074.

[59] *K. Suyama, H. Kusakawa, M. Fukuta,* IEEE Trans. on Electron Devices, Vol. ED-27, No. 6 (June 1980), p. 1092.

[60] Siemens-System 7800, Bestell-Nr. D 16/6000-01 (Oct. 1978).

[61] *O. G. Folberth,* IEEE J. of Solid State Circuits. Vol. SC-16, No. 1 (Feb. 1981), p. 51.

[62] *R. W. Keyes,* Proc. IEEE, Vol. 69, No. 2 (Feb. 1981), p. 267.

[63] *Y. Tarui,* IEEE Trans. on Electron Devices, Vol. ED-27, No. 8 (Aug. 1980), p. 1321.

[64] *P. Wolf,* IBM Technical Disclosure Bulletin, Vol. 21, No. 4 (Sept. 1978).

[65] *O. Minato, T. Masuhara, T. Sasaki, H. Nakamura, Y. Sakai, T. Yasui, K. Uchibori,* IEEE Trans. on Electron Devices, Vol. ED-27, No. 8 (Aug. 1980), p. 1591.

[66] *J. D. Meindl, R. N. Ratnakumar, L. Gerzberg, K. C. Saraswat,* Digest of the ISSCC 1981, p. 36.

[67] *W. Jutzi,* Archiv Elektr. Übertragung, Vol. 25 (1971), p. 595.

[68] *S. Niemeyer, V. Kose,* Appl. Phys. Lett. 29 (1976), p. 380.

[69] *L. Jackel, E. Hu, R. Howard, L. Fetter, D. Tennat,* IEEE Trans. on Magnetics, Vol. MAG-17, No. 1 (Jan. 1981), p. 295.

[70] *T. Wada, O. Kudoh, Y. Nagahashi, S. Matsue,* IEEE J. Solid-State Circuits, Vol. SC-13, No. 5 (Oct. 1978), p. 635.

[71] *K. Kawarada, M. Suzuki, H. Mukai, K. Toyoda, Y. Kondo,* IEEE J. Solid-State Circuits, Vol. SC-13, No. 5 (Oct. 1978), p. 656.

[72] *A. Hotta, Y. Kato, K. Yamaguchi, N. Honma, M. Inadachi,* IEEE J. Solid-State Circuits, Vol. SC-13, No. 5 (Oct. 1978), p. 651.

[73] *H. Ishikawa, M. Yamamoto, H. Tokunaga, H. Toyokura, F. Yanagawa, K. Kiuchi, M. Kondo,* IEEE Trans. on Electron Devices, ED-27 (Aug. 1980), p. 1586.

[74] *O. Minato, T. Masuhara, T. Sasaki, H. Nakamura, Y. Sakai, T. Yasui, K. Uchibori,* IEEE Trans. on Electron Devices, ED-27 (Aug. 1980), p. 1591.

[75] *K. Natori, M. Ogura, H. Iwai, K. Maeguchi, S. Taguchi,* IEEE J. Solid-State Circuits, Vol. SC-14 (April 1979), p. 482.

[76] *H. Masuda, R. Hori, Y. Kamigaki, K. Itoh,* IEEE Trans. on Electron Devices, Vol. ED-27, No. 8 (Aug. 1980), p. 1607.

[77] *J. Y. Chan, J. J. Barnes, C. Y. Wang, J. M. Deblasi, M. R. Guidry,* IEEE J. Solid-State Circuits, SC-15, No. 5 (Oct. 1980), p. 839.

[78] *S. Matsue, H. Yamamoto, K. Kobayashi, T. Wade, M. Tameda, T. Okuda, Y. Inagaki,* IEEE J. Solid-State Circuits, Vol. SC-15 (Oct. 1980), p. 872.

[79] *T. Mano, K. Takeya, T. Watunabe, N. Ieda, K. Kiuchi, E. Arai, T. Ogawa, K. Hirata,* IEEE of Solid-State Circuits, Vol. SC-15, No. 5 (Oct. 1980), p. 865.

[80] *F. Yanagawa, K. Kiuchi, T. Hosoya, T. Tsuchiya, T. Amazawa, T. Mano,* Trans. IEEE on Electron Devices, Vol. D-27 (Aug. 1980), p. 1602.

[81] *M. Kondo, T. Mano, F. Yanagawa, H. Kikuchi, T. Amazawa, K. Kiuçhi, N. Ieda, H. Yoshimura,* IEEE J. Solid-State Circuits, Vol. SC-13, No. 5 (Oct. 1978), p. 611.

Contents of volumes published previously

Festkörperprobleme Band I (1960)
(Ed. F. Sauter)

Festkörperprobleme Band II (1962)
(Ed. F. Sauter)

Festkörperprobleme Band III (1963)

(Ed. F. Sauter)

Festkörperprobleme Band IV (1964)

(Ed. F. Sauter)

Festkörperprobleme Band V (1965)

(Ed. F. Sauter)

Festkörperprobleme Band VI (1966)

(Ed. O. Madelung)

Seite

Festkörperprobleme Band VII (1967)

(Ed. O. Madelung)

Seite

Festkörperprobleme Band VIII (1968)

(Ed. O. Madelung)

Festkörperprobleme Band IX (1969)

(Ed. O. Madelung)

Teil I. Referate des Fachausschusses „Halbleiter"
Part I. Plenary Lectures of the Professional Group "Semiconductor Physics"

Teil II. Hauptvorträge der Europäischen Tagung des IEEE
Part II. Invited Papers of the European Meeting of the IEEE

Festkörperprobleme Band X (1970)

(Ed. O. Madelung)

Seite

Festkörperprobleme Band XI (1971)

(Ed. O. Madelung)

Seite

Festkörperprobleme Band XII (1972)

(Ed. O. Madelung)

Seite

Festkörperprobleme Band XIII (1973)
(Ed. H. J. Queisser)

Festkörperprobleme Band XIV (1974)
(Ed. H. J. Queisser)

Festkörperprobleme Band XV (1975)

(Ed. H. J. Queisser)

Festkörperprobleme Band XVI (1976)

(Ed. J. Treusch)

Festkörperprobleme Band XVII (1977)
(Ed. J. Treusch)

Festkörperprobleme Band XVIII (1978)
(Ed. J. Treusch)

Festkörperprobleme Band XIX (1979)

(Ed. J. Treusch)

Festkörperprobleme Band XX (1980)

(Ed. J. Treusch)